ENERGY AND THE ENVIRONMENT

JOHN M. FOWLER
Visiting Professor of Physics
University of Maryland

McGraw-Hill Book Company
New York St. Louis San Francisco Auckland Düsseldorf Johannesburg
Kuala Lumpur London Mexico Montreal New Delhi Panama Paris
São Paulo Singapore Sydney Tokyo Toronto

ENERGY AND THE ENVIRONMENT

1 2 3 4 5 6 7 8 9 0 SGDO 7 9 8 7 6 5

Library of Congress Cataloging in Publication Data

Fowler, John M.
Energy and the environment.

1. Power resources. 2. Pollution. I. Title.
TJ153.F65 333.7 74-34001
ISBN 0-07-021720-3

This book was set in Press Roman by Publications Development Corporation.
The editor was A. Anthony Arthur;
the designer was Publications Development Corporation.
The cover was designed by J. E. O'Connor.
The printer was Segerdahl Corporation;
the binder, R.R. Donnelley & Sons Company.

ENERGY AND THE ENVIRONMENT

TO KAY

Contents

Foreword

This book is an effort to bridge the large and dangerous gap between science and public policy. Energy is absolutely essential to the operation of society, and since a large and complex social effort is required to produce and use it, it follows that these activities ought to be governed by public policy specifically developed for that purpose. In the last few years it has become painfully self-evident that there is no national, publically determined energy policy in the United States. We are now fitfully engaged in remedying this lack, and in the process it has become clear that there is often little rational connection between the policies that govern the production and use of energy and the scientific knowledge upon which they ought to be based.

On the one hand the government urges us to save energy; on the other hand it proposes to cut down the trackage of railroads in the Northeast–forcing a shift of freight from railroads to trucks, even though simple factual data tell us that the latter are by far the less efficient means of converting the energy contained in diesel fuel into the movement of goods. In the same way power companies, while purporting to cooperate in the energy conservation program, continue to urge their customers to adopt electric heat, even though–from simple physical laws and empirical data–we know that this is the least efficient way to convert the latent energy of a fuel into space heat.

In a rational social order the policies that govern the production and use of energy would be based on public knowledge of the relative benefits and costs (as measured, for example, in depleted resources and degraded environment). This would require, of course, that the people of the U.S. obtain the necessary scientific knowledge during their participation in our educational system or through less formal means. How little has been done, thus far, to achieve this goal is made plain by the fact that books, such as this one, that assemble the scientific background necessary for an understanding of the energy problem in a form that is accessible to the interested student and citizen, are still rare and badly needed.

This book is, then, a response on the part of a member of the scientific community to a social need for information. It reflects John Fowler's experience, not only as a practicing physicist and teacher of physics, but also as a distinguished participant in the work of the Scientists' Institute for Public Information, a group long devoted to the task of giving the public the scientific knowledge needed to understand, and to participate in the development of, the numerous issues of public policy that are now based on the data of science.

Barry Commoner
Washington University

Preface

Energy and the Environment was not written for specialists; I have been especially careful to make it accessible to the reader who finds science and its mathematical language alien territory. The primary goal has been to collect and interpret the background information which all of us need to deal with the complex energy/environment issues and choices we will face in the next two or three decades.

Although this book is suitable for casual reading or for supplementary reading in an environmental course, I have been determined, at the same time, to provide sufficient depth and data to make it an appropriate text for more concentrated study. I hope that *Energy and the Environment* will be read as a book, and that some readers will be drawn to it and through it by the importance of the twin crises it describes, rather than by academic assignment.

The energy story is developed at a rather leisurely pace, beginning with a vision of an energy-rich 1985 and then going back to the beginning of the universe to introduce and describe the various forms of energy. From this beginning the focus narrows to the production and consumption of energy in the service of man and then broadens to include the impact on the natural environment of man's fast-growing demand for energy.

The book contains not only the definitions and examples of the scientific concepts and laws which are needed to understand energy and its environmental interactions, but also historical resumes which highlight the changes in energy production and consumption, a detailed picture of the present situation in the United States (and a sketch of the world picture) and finally, projections into the future, not only of energy demands, but of the lines of development in science and technology which can give us the means of meeting those demands. Also included is a description of some of the political and economic mechanisms which are a necessary part of our energy-environment options.

The 15 chapters are given quantitative support by numerous graphs and tables. They are supplemented by five Appendices for the student who wishes to go more deeply into the scientific background of the energy crisis. Appendix 1 presents the simple mathematical ideas necessary to a full interpretation of the quantitative material. Appendix 2 provides more insight into the meaning of the terms energy, work, and power and relates them to force. Appendix 3 deals with heat and heat engines and with the basic concepts necessary to their understanding. Appendix 4 provides the basic vocabulary of electricity and describes, in simple terms, how electrical machines such as generators, motors and transformers work. The last Appendix provides not only the scientific basis of the energy conversions which take place in nuclear reactors (including the breeder reactors) but also provides insight into the controversy over the effects of the radioactivity from these devices and the other dangers they may pose for us.

The primary goal of this book is to provide a quantitative description of the twin crises of energy and the environment. There is, however, a secondary goal: to demonstrate the contributions that science can make to the understanding of the complex problems of society. I have tried to use the energy-environment theme to show the importance of quantitative reasoning, of estimation and modeling. It is not my goal to make more scientists of the readers. I hope, however, that students will be influenced by these examples to apply quantitative, logical analysis to the present and future developments in this area. More importantly, I hope that it will help readers distinguish between reasoning based on science and the offerings of propaganda and polemic in the flood of words yet to be written about these crises.

The task of planning and writing this book has been pleasant and stimulating. I have benefited by the advice, assistance, and encouragement of many people. Kay Mervine has served as unofficial editor, critic, and booster. Jack Farnsworth of McGraw-Hill encouraged the idea of a book that was both reader and textbook. Margaret Fowler and my children have put up with the more than usual amount of separation that this writing effort has entailed.

I must also thank Barry Commoner for writing the forewood and more importantly for the years of colleagueship in the efforts we have both made to use science in the battle to preserve a livable environment.

This book grew from a course of the same title which I taught as a Visiting Professor at the University of Maryland–College Park. I want to acknowledge the hospitality and encouragement I received from Howard Laster, Chairman of the Department of Physics at that university. Much of the final writing was done while serving as a Visiting Professor in the Department of Physics at California State College–Bakersfield. I was supported and encouraged by Dan Detwiler, Alan Greene and Leo Baggerty, and many other friends there.

Finally, I wish to thank Penny Vorhees who did such a splendid job of converting my handwriting into legible type.

John M. Fowler

ENERGY AND THE ENVIRONMENT

1

1985

In this energy-rich year, 1985, it is difficult to remember the great concern over the Energy Crisis of the early 1970s. Most of us have forgotten the first shortages in natural gas which caused some industrial and commercial difficulties in 1971 and warned us of the real crunch to come. The memory of the cold winter of 1972 and the colder ones of 1973 and 1974 has faded; the thermostats which many of us turned down to 66°F on those winter days have now moved back toward 70°F. Forgotten, too, is the discomfort during the "brownout" summers of 1974 and 1975; our hand is once again on the air conditioner control.

We drive to and from work in our "personal-size" cars; since the demise of their dinosaur-like ancestors, we no longer call them "compact." The ready availability of gas for their 35 mile per gallon engines (some are still nonelectric) has wiped out our memories of angrily waiting in long lines for gas in the winter and spring of 1974 as well as the threats of gas rationing which followed. We pay the present high prices for gasoline with no residue of the shock which followed the dramatic rise of prices in 1973 and 1974. And in the cities, the electrified mass transit systems have even taken away much of the smell of automobile exhaust, let the sun through again, and helped us forget.

The great legal battles have subsided; the long drawn-out skirmishes between the "nuclearphobes" and the "nuclearphiles" over the siting of nuclear power stations and the confrontation between environmentalists and the energy industry over the Alaskan pipeline are forgotten. The Alaskan pipeline shown in Fig. 1-1 was authorized by Congress in 1973 and now pours two million barrels of "North Slope" oil per day into West Coast bound tankers. Nuclear power reactors are a familiar part of our landscape as well as our oceanscape. The Floating Nuclear Plant shown in Fig. 1-2, which is located off the New Jersey Coast near Atlantic City, produces 2,000 megawatts (mw) of electrical power for Philadelphia and New Jersey.

The early 1970s were a time of transition; humankind seemed to have entered a kind of middle age. We faced the disturbing fact that the abundance of the earth was limited; not only were diamonds and copper becoming scarce, but even the fossil fuels[1] such as coal and petroleum were, in fact, limited and running out.

[1] Fossil fuels are so named because they are the remains of material which was once living. Through a process of compression and heat which has taken place over millions of years, these remains have been changed into complicated and valuable compounds of carbon and hydrogen.

FIGURE 1-1 An artist's concept of the Alaskan Pipeline. This almost 900-mile-long pipeline connects the North Slope oil fields around Prudhoe Bay to the port of Valdez. It has a diameter of 48 inches and is expected to carry 2 million barrels of oil a day by 1985. (Courtesy of Alyeska Pipeline Service Co.)

The 1970s also hammered home the knowledge that the process of converting the chemical energy of the fossil fuels for our use threatened our environment. By burning these fuels, which then made up 96 percent of our energy supplies, in generating plants and automobile engines, we were loading our air with pollutants

FIGURE 1-2 The floating nuclear plant ordered by the Public Service and Gas Company of New Jersey to be placed in the Atlantic 2.8 miles offshore and 12 miles northeast of Atlantic City, N.J. It will have two nuclear generating units each with a capacity of 1,150 megawatts. (Courtesy of Office of Information Services, USAEC.)

and damaging our health, property, and crops.

But the 1970s are gone, and with them many of those unsettling concerns. The 250 million Americans now happily consume almost twice as much energy as those 205 million worried Americans of 1970. We still burn coal; in fact, its use has increased from about 500 million tons in 1970 to nearly 900 million tons this year, but the noxious sulphur oxides released into the air over our cities are not much in evidence.

Coal is now a multipurpose fuel. Much of it is burned at electric power plants located right by the mine, or mine mouth plants, as they are called, and the fumes only extend to the near neighbors, New Mexican Indians or Dakota ranchers, whose newfound prosperity may have numbed their olfactory senses. Sulphur-free gas and liquid fuels made from coal are beginning to flow in appreciable quantity through the nation's pipelines and to successfully compete with high-priced natural gas and oil. Even the long *unit trains* carrying fuel to the metropolitan power stations have a load that looks like coal but is actually the new solvent refined coal which is sulphur-free and high in energy content. It is one of the many products developed by the multipurpose coal factories, or Coalplexes.

We still burn oil and natural gas; 8 billion barrels of oil were burned in 1985—almost twice our 1970 consumption—and more than 3 trillion cubic feet of natural gas, half again our 1970 consumption. There have been changes in the sources and mode of delivery. The switch to foreign sources of petroleum begun in the late 1960s has continued; some 40 percent of these fuels came in from overseas this year and 15 billion United States dollars flowed in the opposite direction. This energy is shipped in by super tanker, arriving at one of the deep water ports like the one shown in Fig. 1-3. Not only do the huge oil tankers from the Mideast arrive there, but so do some of the new refrigerated tankers carrying liquified natural gas (LNG) from the Soviet Union or Algeria. Much of this LNG is delivered to the big Texas offshore port.

FIGURE 1-3 The Six-Mile Superport planned for the mouth of the Delaware Bay will allow supertankers and other deep draft ships to unload in deep water, 6 miles from the coast. Oil will be transported to shore through an underwater pipeline parallel to the connecting causeway. (Courtesy of Hudson Engineers, Inc. of Philadelphia who developed this preliminary study with its own funds.)

FIGURE 1-4 (a) The high-voltage transmission system as of 1970 consists of 300,000 miles of line.(Courtesy of the Office of Information Services USAEC)

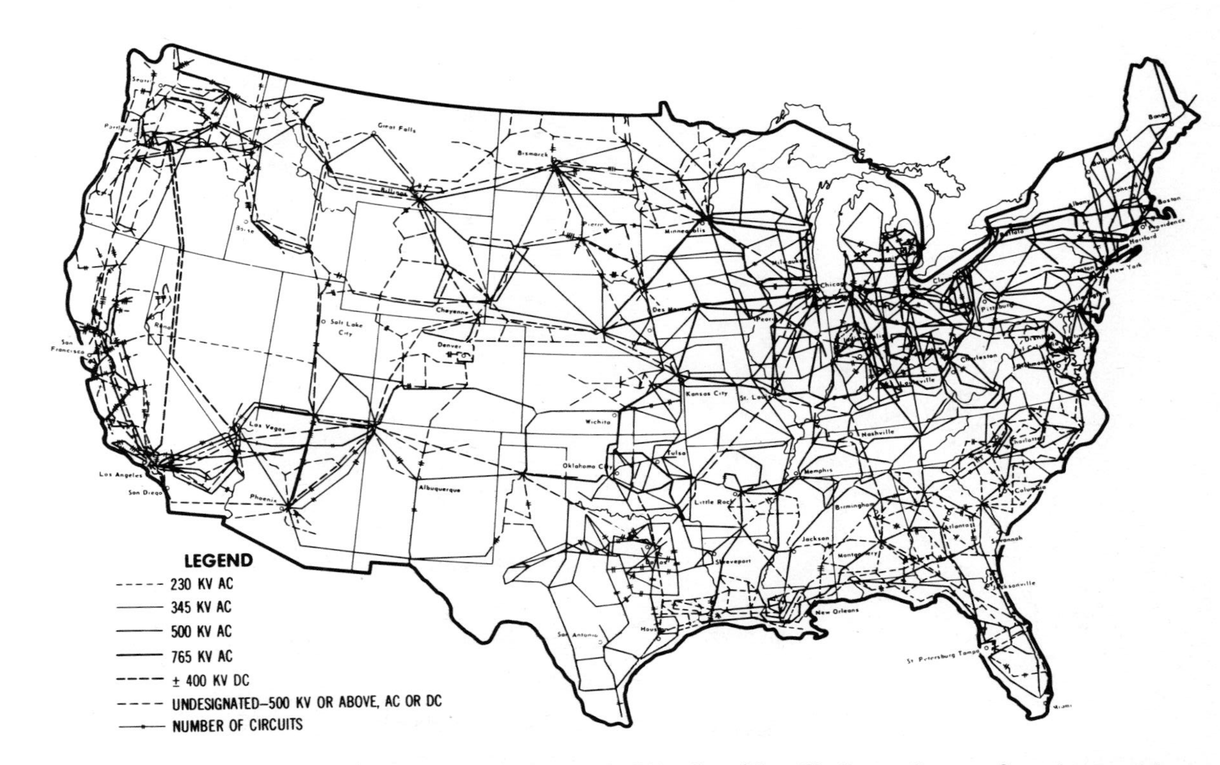

FIGURE 1-4 (b) The projected growth by 1990 will increase this to 500,000 miles of line. The lines radiate out from the generating centers in the coal plains of Montana and Wyoming, and in the Four Corners area, showing the growth in importance of "mine-mouth" generation. (Courtesy of the Office of Information Services, USAEC.)

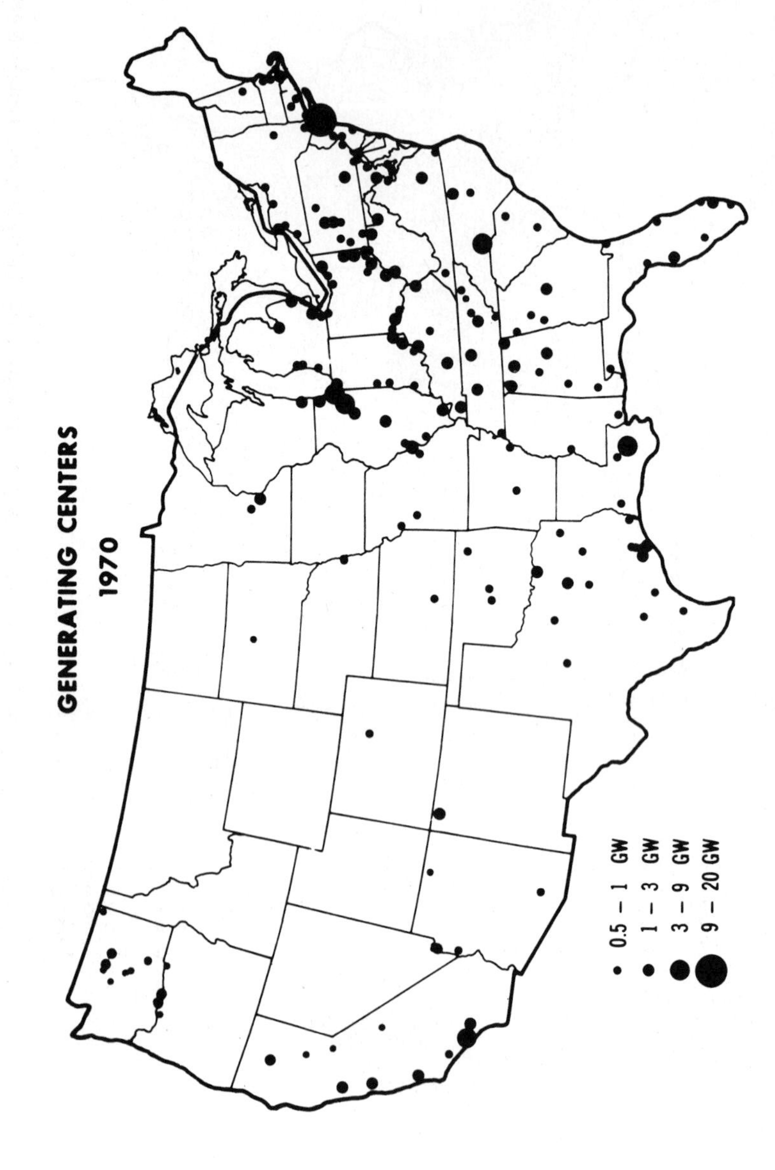

FIGURE 1-5 (a) The location of generating centers of electric power as of 1970. [A gigawatt (gw) is 1000 megawatts of or 10^9 watts.] (Courtesy of National Power Survey, Federal Power Commission, 1970.)

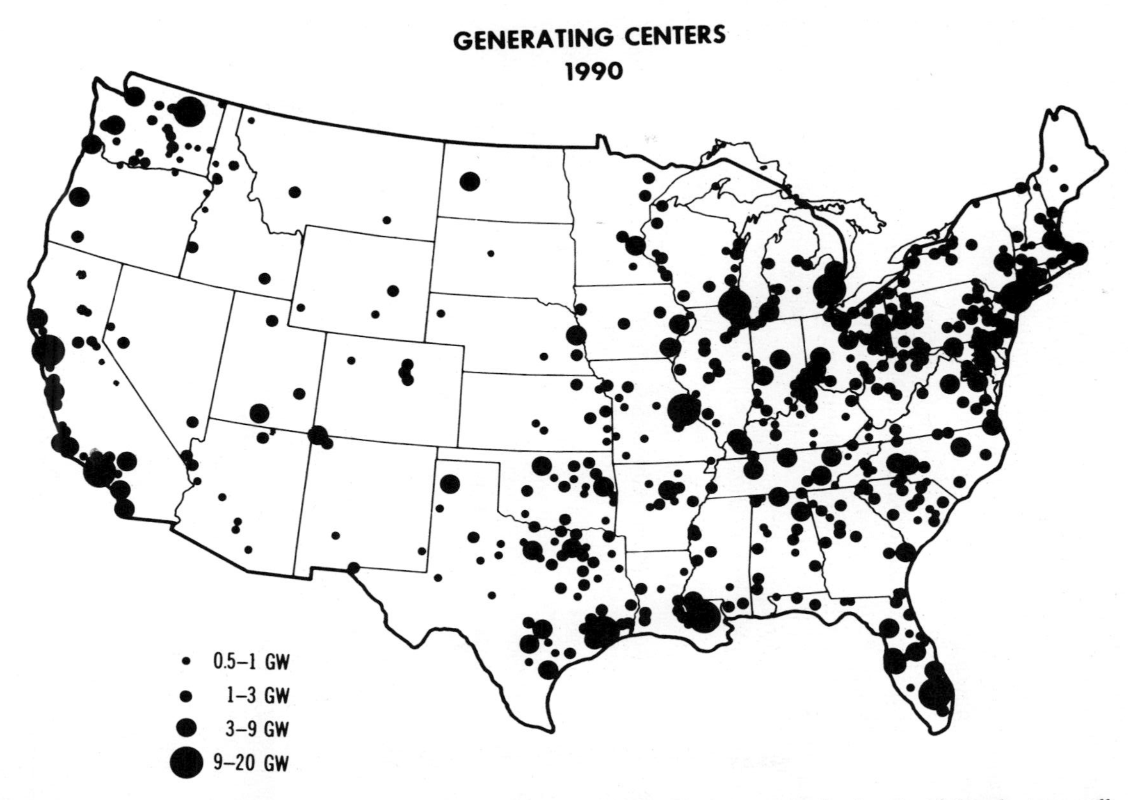

FIGURE 1-5 (b) The projected location of generating centers as of 1990. The increase in the density of the plants as well as in generating capacity is shown. The clustering of plants around cooling water is also distinctive. (Courtesy of National Power Survey, Federal Power Commission, 1970.)

Electricity, the all-American energy, has continued to lead all other forms in its rate of growth. The increase in electric energy consumption from a total of 1.6 trillion kilowatt-hours (kw-hr) in 1970 to almost 4 trillion kw-hr in 1985 has been made possible by a comparable growth in the generating and transmission facilities. The comparison with the 1970s is striking; Figure 1-4 compares the transmission systems then and now and Fig. 1-5, the number of generating plants.

In large part, this increase in electric power has been provided for by the construction of the many fossil fuel burning plants we see around our country; only in the Northwest has the building of new hydroelectric plants been of importance. The strange pumped-storage plants, with their elevated reservoirs alternately filled and emptied, now appear in many places. They are not a source of electric energy, however, but a place to temporarily store it for future use.

The biggest change, of course, has come in the area of nuclear power. In 1970 some 19 of these plants accounted for 2 percent of the total electrical generating capacity. Presently, almost 300 nuclear plants account for about 40 percent of that capacity. Many of them are now grouped together in "nuclear parks" much the same as the one shown in Fig. 1-6. Nuclear reactors have come into their own elsewhere, particularly in Western Europe where some 600 billion kw-hr of electricity are generated from nuclear fuels as against only 90 billion kw-hr in 1970.

The nuclear future is also bright. The first three demonstration breeder reactors are now in service and are not only each producing hundreds of megawatts of electric power, but are producing more plutonium fuel than they use, some 40 tons per year. The nuclear industry and its governmental supporters are toasting their success and reminding us that we now can, if we need to, use these breeders to produce electric power from the uranium containing "black shale" that makes up Tennessee's Appalachian Mountains.

Perhaps even more encouraging news, certainly to those who cherish the Appalachian mountains as they are, comes from Princeton and Livermore, the first two American laboratories to have produced a controlled thermonuclear reaction. At both these laboratories, pilot, fusion-powered generators are being tested, so far without any insurmountable engineering problem. Although they run on scarce

FIGURE 1-6 A "nuclear park" planned for Bonsal, N.C. The four nuclear generating units, which will be on line by the early 1980s will produce almost 4000 megawatts of electric power. (Courtesy of National Power Survey, Federal Power Commission, 1970.)

lithium at this stage of their development, the popular press is already pointing to the ocean's unlimited deuterium as the future fuel of this nuclear horn of plenty.

The "bullish" energy market is not an American monopoly. Although we still get a large share—more than a quarter of the total energy consumed—worldwide consumption is increasing faster than in the United States and is now almost double the 1970 total. The worldwide use of coal has doubled since 1970 and now stands at near 4 billion tons per year. Our balance of trade is helped by the exportation of some of our domestic coal. Worldwide oil production has almost doubled, going from around 15 billion barrels per year in 1970 to a present 27 billion barrels. Electric power has also achieved worldwide popularity; the total of about 5 trillion kw-hr consumed in 1970 has risen to 15 trillion kw-hr this year.

There are some comparisons with the 1970s that are better not made. Most of us have luckily forgotten how cheap energy was then. We would rather not compare our nearly 1 dollar per gallon gasoline with the 30 cents per gallon gasoline that the independent gasoline dealers (remember them?) were offering as late as 1972. It is hard to believe that natural gas, for which the residential consumer paid about 1 dollar per 1,000 cubic feet in 1970 is now selling for 3 dollars per 1,000 cubic feet, or that 90 cents per gallon heating oil sold for 20 cents per gallon as late as 1972. We no longer hear the boasts about the ever-decreasing costs of electricity; from an average rate of 1.5 cents per kw-hr in 1970, it is now passing 5 cents per kw-hr. The small user, the residential consumer, pays close to 7 cents for the energy which cost him 2 cents 15 years ago. The electric bill is not a welcomed communication.

There are still those who give us hope that the bill can be reduced by tapping the free power provided by the sun and the hot interior of the earth. Out on the Arizona desert the "new sun worshippers," as the press has come to call the solar energy enthusiasts, have completed their 1,200 acre "energy farm" and its row upon row of tiny "greenhouses" are now capturing their small share of the sun's energy. The concentrated heat from the sun is transported by long pipes of heated liquid sodium to central conversion plants where the superheated steam turns a turbine and produces electricity. Some of this electricity in turn is used to electrolyze water into hydrogen and oxygen. Already, this pilot farm is providing both electricity and hydrogen fuel to energy-hungry Tucson and Phoenix and the country is intrigued by Tucson's experimental hydrogen-powered "minibuses."

The sun is also being put to work on a smaller scale all across the South and Southwest. Some 30,000 houses now show the distinctive black, glass-enclosed, solar collectors on their roofs. In these houses, more than 90 percent of the energy for heating, hot water, and even air conditioning is drawn from this inexhaustible source.

Geothermal power, tapping the earth's heat, has also come into its own. The California Geysers Valley and similar areas elsewhere in the United States are now producing 19,000 mw of electric power, about 2 percent of the total United States generating capacity. Even more impressive is the new installation in the Imperial Valley where hot water is pumped from the ground, flash-distilled, and the resulting few million acre-feet of fresh water is pumped into the Colorado River to meet the conditions of our old treaty agreement with Mexico.

There have been impressive geothermal resource discoveries in other parts of the world; the use of inexpensive electricity along the Rift Valley in Africa, for instance, has caused quick industrial expansion in many of the countries surrounding it. The United Nations Building in New York City also serves to demonstrate the practicability of this energy source; it is now warmed by water circulated 20,000 feet beneath it to tap the earth's heat.

America's song is the hum of the electric generator; the discordant cries of the "prophets of doom" are muted. Nevertheless, the careful listener can still hear them. The informed observer looks out on the land of plenty and sees beneath the layered technology, the inevitable strains that Commoner warned of in *The Closing Circle.*[2]

The signals of strain are somewhat changed. It is no longer the billowing clouds of sulphurous smoke from power plants that warn us (although many power plants still billow smoke). It is, rather, the inevitable but invisible heat that rises into the atmosphere, and is then carried away in our streams or dumped in the ocean that worries the present crop of "ecology freaks." They do not care for the present condition of the warm, green Ohio, the once beautiful Susquehanna, or that big "cooling pond" which used to be the Chesapeake Bay. It worries them that the climate over the great American cities is now dominated by the "heat island" effect of trapped sunlight and waste heat.

The environment has suffered elsewhere. The strip mining of the Black Mesa which made some Indians wealthy but drove the Hopi's and Navajoes from that ancestral home, is being repeated on a larger scale in the North Central Region: Wyoming, Montana, and the Dakotas. Those once unplowed plains have been ripped by huge coal shovels and now lie, by the thousands of acres, in near ruin under the smokey skies. The ranches have been replaced by a complex of generating plants which produce as much electricity as did all of Japan in 1970. Along with the growing number of coal gasification plants and the huge Coalplexes, they use most of the water of the entire Yellowstone Basin.

Oil shale, to which this country turned for that precious liquid in an effort to gain some independence from the Mideastern power plays, now provides nearly 1 million barrels of oil a day. Those who remember the primitive beauty of the Green River region of Colorado are offended, nonetheless, by the 1.5 million tons of crushed, cooked rock discarded each day and the huge surface mines eating into the once lovely mountain sides.

The flow of oil from other sources has had its own effect. Ecologists are still dismayed at the damage done to the delicate Alaskan tundra by the brief break in the Alaskan pipeline, but the tourists attracted to the Alaskan interior by the road paralleling that pipeline threaten more widespread and permanent damage. Fishermen are finding reduced catches in the Gulf of Alaska, where the recent tanker collision was the most newsworthy of a continuing series of oil spills. Their complaints are echoed from the continent's eastern shore; the tidal marshes of Delaware and Maryland are being slowly poisoned by oil spills at the supertanker port (see Fig. 1-3) and by the increasing number of offshore wells along the Atlantic Coast.

[2]Barry Commoner, *The Closing Circle,* (New York: Knopf, 1973).

The more than doubling of electric power production has raised power transmission line acreage from 4 million acres in 1970 to more than 8 million acres (see Fig. 1-4), creating aesthetic as well as land-use problems.

The benefits of abundant energy are themselves not evenly spread and enjoyed. The doubling or tripling in price has increased the gap between the high and low income groups in our country. A different sort of economic threat comes from the Middle East where the tens of billions of dollars flowing into that region each year from the oil-hungry countries dominate the world money market.

Spurred by abundant energy, the plastic and synthetic fabric industries continue to expand and dump more and more unnatural and undegradable solid waste onto the earth. Still worse, in the eyes of some pessimists, is that the increase in jobs and in food production brought about by abundant energy has caused the world birthrate to turn upward again.

Even the nuclear horn of plenty has spewed out trouble along with largess. The dire predictions of the doom sayers have not come true; most of the reactors have been good clean neighbors. The one serious accident which fouled thousands of acres of upstate New York and killed a few citizens is now almost forgotten (except by the insurance companies). However, in the salt mines of Kansas there are now 1,000 tons of radioactive waste piled up, an increase of almost a hundredfold over the stored radioactivity with which the early 1970s had to contend. The one radioactive water contamination which caused a scare in some small Kansas towns has been corrected and we are assured it won't happen again. Still, there are those who wonder how we, as a society, will be able to guarantee the monitoring necessary to keep the reactors safe for the next several hundred years. To add to this, we have now begun to store plutonium-containing waste, and its potential danger may last for thousands of years. These long-term worries no longer concern the popular journalists. What does concern them, however, is the fate of that truckload of almost pure, bomb-grade plutonium which was hijacked on the way from the reprocessing plant. The sudden aggressiveness of a certain small Third World nation may be a clue to its eventual destination.

The preceding paragraphs are fiction. They sketch a credible future, but only one of a range of alternate futures. We can guarantee neither the occurrence of the successes nor the failures we have pictured. We have no guarantee at this time that fusion will be successfully controlled by 1985, or that coal conversion techniques will be commercially feasible by then. Nor can we predict with any certainty that the Alaskan pipeline will suffer a break or that a serious reactor accident will occur.

The true picture of 1985 is lost in the mists of the future. The decisions which will shape that future, however, are being made today. They are decisions which will not wait; they must be made now. There is an inevitable delay between the decision and the result when we are dealing with energy. If we want more electricity in 1985 we must begin to build the generating plants by 1978. If we are to have a liquid fuel from coal, we must increase the support of research and development and build the pilot plants immediately. Conversely, many of our prophecies are self-fulfilling. If we project that our electric power needs will double by 1985, then the utility companies will begin immediately the 6 to 10 year task of building the facilities to generate and distribute the additional power. Given the existence of these facilities in 1985, power consumption *will* double; the power will be sold, the generators will not stand unused.

It is my hope that the facts and basic principles, the approach and the philosophy of this book will help provide some guidelines to that future.

In the chapters which follow we will deal with energy as a central theme. It is a rich theme; it will begin in the bare abstract realm of physical science, but will quickly branch from there to geology and engineering as we look beyond definitions and physical principles to the sources of energy and the techniques of conversion. From there we will track the theme into the complex structure of society as we examine the uses of energy, the effects of the release of all this energy, and of some of its poisonous byproducts. We will examine, in the murky depths of our crystal ball, some of the directions that may lead toward a brighter 1985. New technology which could get us more and cleaner energy from existing fuels as well as the potential of new, large sources in the sun and sea well be described. Some suggestions and examples of much needed change in the ethic of energy usage, of ways to increase the effectiveness with which we utilize our energy resources and to reduce the present appalling waste, will be explored.

The 1970s have taught us that we have not really made ourselves masters of the environment. We have become aware, some of us for the first time, of the interlocked ecological cycles which sustain life and of the damage that is done by interferring with or overloading those cycles. By choosing energy for study we are looking at only one dimension of this multidimensional ecological web. That is a dangerous simplification; in nature, the whole is always greater than the sum of its parts. However, if we must choose one dimension out of the many, energy is the right one, for it is at the base of it all; it fuels the engine of life.

2

Energy

> And the earth was without form, and void; and darkness was upon the face of the deep. And God said "Let there be light": and there was light.[1]

Energy, in the form of light, is the first gift to the earth recorded in the Bible. In Greek mythology, Prometheus, the Titan descendent of Uranus (Heaven) and Ge (Earth), stole fire from the gods to preserve mankind and was condemned by Zeus to suffer eternal torture. Whether as light or fire, the histories of civilization begin with energy.

The beginning which astronomers now reconstruct, and from which the myths of the future may be woven, is a violent one. With the great modern optical and radio telescopes it is possible to look far backwards in time. The light fades, of course, long before we can see the beginning, but what can be seen looks like the aftermath of an explosion; and, in the faint radiation background that fills all of space, we may even be hearing echoes of that cosmic bang.

To define energy and describe some of the important roles it plays in the universe, we will go back to the beginning, look for energy there, and try to follow it as it changes form and role down through time to the present.

IN THE BEGINNING THERE WAS ENERGY

There is growing evidence from the astronomers that the universe at time zero was a tightly packed mass of some basic elementary particle—perhaps neutrons. This mass was compressed to an astounding density, 10^{15} (1,000,000,000,000,000) times as dense as water. In this "Cosmic Egg," as Asimov[2] has named it, was concentrated all the matter and all of the energy of the universe. Since Einstein has shown us that matter can be converted to energy, and since energy can never be created nor destroyed, that "Cosmic Egg" had in it all the energy ingredients for the life and death of the universe.

That energy can neither be created nor destroyed was a fact hard won from nature over several centuries of head scratching and experimentation. It now forms one of the foundation stones of our understanding of the physical universe and is

[1] *Genesis* 1:2-3.

[2] A very readable account of the origins of the universe can be found in Isaac Asimov, *The Universe,* (New York: Walker, 1966).

called the *First Law of Thermodynamics.*[3] The First Law of Thermodynamics, also known as the *Law of Conservation of Energy,* states that:

> In a closed system, energy can be neither created nor destroyed.

The "Cosmic Egg" was such a *closed system*; there was no interaction between it and other systems which may have existed (though we are unaware of them) and, as far as we know, the universe which grew from it is still a closed system.

While the amount of energy in the universe remains constant, it can and does change form. In the "Cosmic Egg" energy was present in at least three forms: in the mass of the neutrons themselves, in their motion as they continuously collided under the immense pressure of the forces which drew them together, and in the radiation that must have filled what little space remained between the compacted neutrons.

The "Cosmic Egg" was remarkably short lived; it would, in fact, have exploded in the time it has now taken you to read about it. Yet from the first, its energy took on forms which demonstrated the two major categories: *kinetic energy* and *potential energy.* Kinetic energy is energy in transit, the energy of motion. In the "Cosmic Egg," there were at least two forms of kinetic energy: the radiation energy carried by gamma rays, moving with the speed of light and being absorbed and reemitted by the neutrons, and the random chaotic motions of the colliding neutrons. The latter, random motion, we call *heat energy.* The temperature of that primeval mass surely measured several million degrees.

Potential energy is stored energy, energy waiting to be used. It can be stored in many ways, as we shall see, following the primeval energy of the "Cosmic Egg" down through the eons to the present. In the beginning, however, it was stored in only one way, in the mass of the neutrons themselves.

The Big Bang

Packed with so much energy, the "Cosmic Egg" would have been very unstable, and the explosion which blew it apart was surely the biggest bang there ever was. As the matter of this egg went flying in all directions out into empty space, several changes took place. Its temperature dropped rapidly, and, as the space between the neutrons increased, the chaotic motions of collision became the organized outward motion of all the neutrons. Many of the neutrons, now freed of the enormous pressures, changed into protons and electrons and formed hydrogen, the first atom. Energy was changing its form; random heat energy was converted to kinetic energy of motion of the outward rushing atoms. More striking was the conversion of some matter into energy; as each neutron changed into a proton and an electron, energy was released. When an electron and a proton became bound into a hydrogen atom, more energy was released. The most important energy conversion that was taking place, however, was the continuous conversion of the kinetic energy of the explosion into gravitational potential energy. The mass spread out into space against the attractive force of gravity (see Appendix 2). It was as if invisible springs connecting all the bits of matter were being slowly stretched. Even now, most of the energy

[3] *Thermodynamics,* or the dynamics of heat, is that part of physics that deals with the energy conversions between heat energy and other forms.

of the universe still resides in the gravitation field of all that dispersed mass, in those "stretched springs" so to speak. When the universe begins to collapse, as this particular theory says it will, all the matter of the universe will come crashing back on itself to form the mass and energy of the new "Cosmic Egg" for the next universe.

Stars and Galaxies

All this happened 10 to 20 billion years ago: the explosion, the first stages of the expansion of the shattered egg, the formation of protons and electrons, all probably happened quickly, within a few hours. A short time later (on the scale of astronomy), perhaps tens of millions of years, the still-expanding hydrogen gas cloud filled large regions of space. Soon, chance fluctuations in the density of this gas began to cause local condensations to take place. As the density of such a condensation grew, it drew together most of the gas in its immediate neighborhood. A neighborhood here measures millions of *light years* across. (A light year, 6×10^{12} miles, is the distance light will travel in a year.) A separate and immense entity was soon formed in space. A galaxy was born.

Such a *protogalaxy* received its share of the original energy of the universe. Kinetic energy was carried in by the atoms pulled into it by the increasing gravitational attraction and, as we shall see, there was more energy stored in the atom's nuclei waiting to be released.

The blob of gas that was the protogalaxy was itself in motion. It was shrinking in size and becoming more spherical under the inward pulling forces of gravity. More importantly, for its final shape, it was beginning to spin, set rotating by the countless impacts of atoms as a merry-go-round can be spun by riders jumping on it.

Thus, over a few more tens of millions of years the great cloud of hydrogen took the distinctive galactic shape shown in Fig. 2-1, flattening out to a disk with a bulbous center and trailing great spiral arms behind. Then a new phenomenom appeared. Within the thickened gas of the galaxy new condensations began to form from spheres of gas that contracted on themselves. As the atoms were pulled into the contracting sphere, the kinetic energy of their "falling" motion was converted

FIGURE 2-1 This bright galaxy in Ursa Major with its conspicuously dense central area and spiral arms is similar to the galaxy which contains our solar system. (Courtesy of McGraw-Hill.)

through the increasing number of collisions, to random motion, or heat energy. The temperature of the gas began to rise; much of this heat energy was radiated away, but in the center of the sphere, the temperature approached 20,000,000°F. At this temperature, with the brightness of millions of "hydrogen bombs," the *thermonuclear reactions* turned on, the hydrogen at the center began to burn, and a star was born.

Energy from the Nucleus

The light of the stars which grace our night skies signals the second most important form of energy in the universe, nuclear energy. To see how it comes about we will reel our film backwards and see what happened in that contracting sphere of hydrogen which became a star.

As more and more gravitation potential energy was converted to heat energy, the atoms collided more and more strongly; in fact, the atoms were soon stripped of their electrons and the collisions were then between bare protons. Protons, electrically charged and alike, repel each other strongly; but as the temperatures mounted and the force of the collisions grew, the protons approached each other more and more closely during collision, and a point was reached, at temperatures of around 20,000,000°F, where the protons got so close together during a collision (within about 10^{-12} centimeters (cm) of each other) that the short-range nuclear forces could take over. These nuclear forces, strongly attractive, then pulled the two protons together to fuse them into a new nucleus. When this happened, when *fusion* took place, not only was the original kinetic energy of the colliding protons released, but some of the mass of the particles themselves was converted to energy. Such a reaction between nuclear particles is called a *thermonuclear reaction,* since it is triggered by heat.

The sequence of nuclear reactions which provides the enormous energy of a star, such as our sun, is a multistep process, the net accomplishment of which is to convert four hydrogen nuclei (protons) into a helium nucleus. The helium nucleus is slightly less massive by 0.73 percent than the total of the four protons. This extra mass has been converted into energy according to the rules of the famous Einstein formula, $E = mc^2$. First it is converted into the organized kinetic energy of motion and then into the random motion of heat energy.

It is these reactions which account for the enormous energy of our star, the sun. They turned on when gravitational contraction produced sufficient heat. This new source of energy so increased the pressure in the sun's interior that gravitational contraction was halted and the sun stayed at its present size.

Our star was probably not formed of the primordial hydrogen from the "big bang." That occurred 10 to 15 billion years ago, while the sun is only about 5 billion years old. It was probably formed out of hydrogen gas enriched with elemental debris from earlier stars which had exploded at the end of several cycles of more complicated nuclear reactions, in which not only helium, but all the rest of the elements, had been formed. From this richer stuff around the early sun, the various planets, including our home planet, number three from the sun, were probably formed.

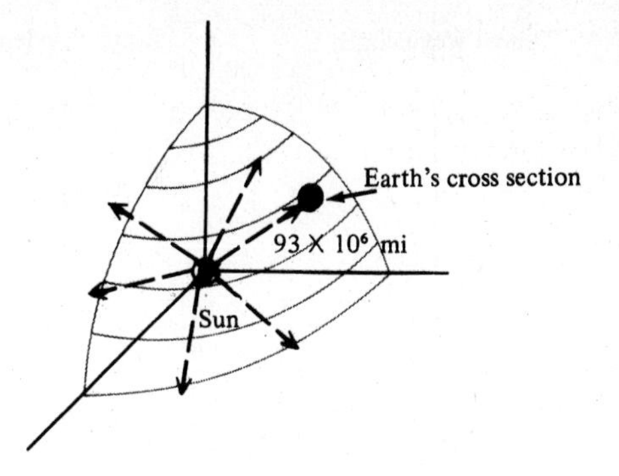

FIGURE 2-2 The sun radiates energy uniformly in all directions. The earth's share is that fraction of the total given by the ratio of its cross-sectional area to the surface area of a sphere with a radius of 93 million miles.

We have skipped over many fascinating details of the formation of the solar system in order not to lose sight of energy. The interested reader can fill in the details from any of several popular books on astronomy.[4] Let us now focus on the energy relations between the earth and the star which gives it life.

The sun is emitting energy at an enormous rate: 4.6 million tons of mass per second are converted into energy. This, however, is an insignificant rate of loss for this massive star; it will take 20 billion years to use up just 20 percent of the sun's hydrogen. This energy is radiated into space in all directions. At the distance of the earth, 93 million miles away, it has spread out over the surface of a sphere of this radius. How much of it strikes the earth? The answer is contained in Fig. 2-2; the earth's fractional share is determined by the ratio of the earth's cross-sectional area to the total surface area of that huge sphere. Putting this relation in numerical terms, we find that only 4 parts in 10 billion of the power emitted by the sun strikes the earth.[5] If we look back at the 4.6 million tons or 9.2 billion pounds of mass which the sun burns up each second, our share is about the energy equivalent of 4 pounds of the sun's mass per second.

[4]Fred Hoyle, *Frontiers of Astronomy,* (New York: New American Library, 1957). Chapter 6 gives an account of this theory and Chap. 12 discusses the formation of the elements in exploding stars.

[5]The spherical surface has an area of $4\pi R^2 = 4\pi \times (93 \times 10^6 \text{ mi})^2$; the earth's cross-section is $\pi r^2 = \pi \times (4{,}000 \text{ mi})^2$; therefore, the ratio is approximately

$$R = \frac{\pi \times (4{,}000 \text{ mi})^2}{4\pi \times (93 \times 10^6 \text{mi})^2}$$

$$= \frac{4 \times 10^6}{(93)^2 \times 10^{12}}$$

which, if we consider $93 \triangleq 100$, becomes

$$R \triangleq \frac{4}{(100)^2 \times 10^6}$$

$$\triangleq \frac{4}{10^{10}}$$

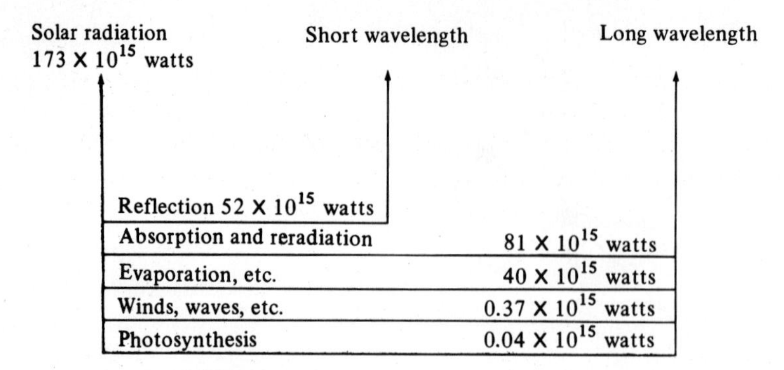

FIGURE 2-3 Major flow paths of solar energy, 75 percent is reflected, or absorbed and reradiated. Of the remainder, about 2 percent drives the wind and waves, while only one tenth of 1 percent is used for life in the process of photosynthesis.

This is not to complain; much energy passes us by, but a still enormous amount reaches us. If we express energy per second in terms of electrical power,[6] the earth receives energy at the top of its atmosphere at the rate of 173 trillion kw. This is enough energy in about 21 minutes to satisfy the whole greedy human race for a year (at the 1970 level of consumption).

Although the earth receives only four parts in 10 billion of the total energy radiated from the sun, our usable share is even less. Figure 2-3 shows the relative amounts of the sun's energy which are reflected by the atmosphere, absorbed, and reradiated in the atmosphere or in the oceans. From this, we see that only one tenth of 1 percent of the original amount is used in that great engine of life, photosynthesis.

ENERGY FOR LIFE

Let's again take up the trail of energy through the eons. We left it 5 billion years ago with the formation of the sun. We will skip over the 4.5 billion years during which the earth was formed, mountains were raised and washed away by the great seas, and life began. We will take up the trail, again, 300 million years ago, in the middle part of the Late Paleozoic Era. Plants have emerged from the sea and the tropical marshlands are covered by ferns, horsetails, and mosses grown to enormous size. Man is still far in the future, his genes are beginning their evolution in some of the creatures of these jungles. (See Fig 2-4 for artist's conception.)

A small fraction of the sun's energy which beat down on these strange jungles was used by the huge fern trees and other vegetation in their life processes. They converted the radiant kinetic energy of sunlight into *chemical potential energy,* the third great form of potential energy.

The energy-converting process of photosynthesis can be summarized in chemical shorthand as

[6]Note that power is the time rate at which energy is delivered or consumed. The watt is a unit of power, the rate of flow of energy. Many electrical appliances are rated by the flow of energy they require, for example, 100-watt light bulbs. The kilowatt is 1,000 watts. These terms, their measurement and units, and the relations between units are discussed further in Chap. 3 and in Appendix 2.

$$CO_2 + H_2O + \text{energy} \longrightarrow C_x(H_2O)_y + O_2$$

which, in words, states that in a plant, carbon dioxide (CO_2) and water (H_2O) are combined, with the addition of energy from sunlight, to form the carbohydrate group $C_x(H_2O)_y$ and oxygen (O_2). $C_x(H_2O)_y$ is a general formula for the carbohydrate group which is an important part of the molecule of sugars and starches. Ordinary sugar, for instance, is $C_{12}(H_2O)_{11}$. This is an oversimplification of a stepwise process which proceeds more correctly as

$$CO_2 + H_2O + \text{energy} \longrightarrow \text{intermediate products}$$

$$\text{intermediate products} + \text{energy} \longrightarrow C_x(H_2O)_y + O_2$$

The radiant energy from the sun, kinetic energy (energy on the move), is caught by plants and used to break up the molecules of CO_2 and H_2O and rearrange their atoms. To form carbohydrates from CO_2 and H_2O, forces must be operating so that molecules are pulled apart and atoms moved around. Since the forces which hold atoms and molecules together are electrical forces, the chemical potential energy which we have just introduced is a form of *electrical potential energy.*

It is certainly potential energy. The carbohydrate formed with the help of the sun's energy is either food or fuel, depending on the use we make of it. Eating or burning turns the photosynthesis reactions around; carbohydrates in the plant

FIGURE 2-4 A picture of the Pennsylvanian Diorama at the Smithsonian Museum of Natural History showing the "Coal swamps" of 300 million years ago, where decay and compression produced coal. (Courtesy of Frances M. Hueber, Curator, Division of Paleobotany, Museum of Natural History, Smithsonian Institution, Washington, D.C.)

sugars or starches combine with oxygen to re-form CO_2 and H_2O, releasing energy in the process.

$$C_x(H_2O)_y + O_2 \longrightarrow CO_2 + H_2O + \text{energy}$$

The fern and moss jungle of 300 million years ago grew in a rich environment. Much of the world at that time had an almost tropical climate, wet and warm. In addition, the atmosphere was very rich in CO_2, released from the depths of the earth by many active volcanoes and hot springs.

Let us now focus down on one spot on Paleozoic earth, a swampy river delta in southern Illinois only a few feet above the level of the ocean which covered Missouri to the west and parts of Kentucky to the south. This swampy delta was formed of sediment washed down from the mountain ranges to the east and north. It was covered by a gloomy jungle where, for thousands of years, fern trees and huge mosses grew, died, and fell into the shallow water from which they emerged. Some of them decayed. Bacteria consumed their carbohydrates and recombined the carbon C with O_2 to form CO_2 which was returned to the atmosphere to feed future plants in the familiar cycle of life and death. Some of the plants, however, were covered with water, and the oxygen-requiring bacteria could not work on them. Over tens of thousands of years, huge amounts of this undecayed plant material built up into the spongy mass we call *peat.* (A modern peat bog, 6 feet deep, can be found in regions of the Dismal Swamp of Virginia and North Carolina.)

In peat, the *anaerobic* bacteria (those which do not require oxygen), plus the pressure and accompanying heat caused by the overbearing sediment, drove off some of the water, oxygen, nitrogen, and miscellaneous plant products so that the percentage of energy-rich carbon was increased. Peat, when burned as fuel, as it is in parts of the world, releases about 6,000 Btu per pound.[7]

In our Illinois river delta the thick layer of peat was finally buried under tons of sea-born sediment and further changes took place. The pressure increased greatly and, along with it, the temperature. The peat was greatly compacted, perhaps by as much as a factor of 16 (in other words, a thickness of 16 feet became compressed to a thickness of one foot). More and more of the water, nitrogen, oxygen, and other materials were driven out so that more and more pure carbon was released from its earlier molecular combination. The percentage of carbon increased from an original 50 percent in the living material to 75 or 80 percent. The peat was now compacted into a hard, black mineral; the southern Illinois coal basin was formed.

What had happened to the energy? The radiant energy from the sun was converted to chemical potential energy in the plants. Through the compaction and reactions under pressure and heat, a great concentration of this energy had taken place. Thus, in the high-grade coals, there are 11,000 to 12,000 Btu of energy per pound compared with, for instance, the 5,000 or so Btu per pound of the original wood.

[7]The Btu, or British thermal unit, and other energy units such as the Calorie and the kw-hr are defined and compared in Appendix 2 and in Chap. 3. For comparison, 1 pound of gasoline has an energy value of about 21,000 Btu. (There are about 6 pounds in a gallon of gasoline.)

Man Enters the Scene

The Illinois coal deposits, with their tremendous energy locked in the carbon atoms, lay buried for millions of years. As the earliest settlers came to Illinois, they found scattered about black outcroppings of this coal thrust to the surface by mammoth foldings of the earth's crust. Those early settlers knew what it was; coal had been used in Europe as early as the twelfth and thirteenth centuries. But they were not very interested in it as fuel, for all about them was the abundance of wood—wood which needed to be cleared from the land.[8]

Not until the settlers had chopped their way across the land and wood no longer lay at their doorstep did they turn to coal. As the country became more and more industrialized, coal consumption grew, from perhaps 8 million tons in 1850 to 56 million tons in 1875 and 270 million tons in 1900. It reached a peak of 633 million tons in 1945 but production has since decreased, amounting to about 450 million tons in 1970.

Let us go back to our saga of energy and follow a ton of coal mined from that Illinois deposit in the 1970s. Had it been pulled from the ground thirty years earlier, odds are it would either have been made into coke[9] for the smelting of iron ore or would have ended up in the firebox of a steam-driven locomotive. In 1970, however, its likely destination was an electric power plant, perhaps one on the outskirts of Chicago; the one whose plume of smoke you might have noticed while flying over that city.

In that ton of coal were some 24 million Btu of energy. Dumped into the huge furnace of the power plant the coal burned efficiently; most of the heat energy released (85 to 90 percent) went into the boiler which turned water into steam. The other 10 to 15 percent of the released heat, along with the sulphur impurities in the coal, went up the stack. (This wasted heat which warmed the air above the power plant, is of interest in itself. We will look at it in greater depth in our discussion of thermal pollution.) Inside the boiler, the super-heated steam with a temperature of 1000°F, was sent against the giant turbines to turn them, thus turning the electric generators which they drove to produce electricity.

Our story is almost over. We have followed this infinitesimal bit of the energy from that "Cosmic Egg" down through the ages from sun to plants to coal and now to electricity. We must keep careful track of what is left. Ten percent of the 24 million Btu (that is, 2.4 million Btu) were lost as heat to the atmosphere. We lost a lot more in the turbine; the steam went in at 1000°F, but it left still hot; it

[8]Schurr and Netschert, *Energy in the American Economy 1850-1975,* (Baltimore: Johns Hopkins Press, 1960). An example of the reluctance of the early settlers to use coal comes from an account recorded in 1840:

> David Handy purchased some lands near the Blount County line. On this land there was coal in the bed of the Warrior River. He built two flat-boats in the fall and loaded them with coal and floated them to Mobile. In Mobile nobody would buy the coal. He had to give it away and send a negro along with every bucketful to show the people how to light and burn it.

[9]Coke is a coal product made by heating coal in the absence of oxygen. The volatile gases are driven off and the carbon concentration is greatly increased. Charcoal is made from wood by a similar process.

was considerably above the 212°F boiling point of water when it was exhausted from the turbine. Some 60 percent of the heat energy was not used to turn the turbine but instead was carried to a nearby river by the power plant's cooling system. Thirteen million Btu were lost thay way.

The generator converted the kinetic energy of the turbine into that most important form of kinetic energy, electricity. This was done very efficiently, so that practically all of the remaining 8.6 million Btu went out into the high voltage transmission lines as electrical energy to be distributed throughout Chicago.

Who can say where it ended? Electricity has many varied uses. We can estimate with some assurance that 10 percent of it (about 1 million Btu of energy) was lost in the transmission and distributing lines; it was converted to heat and warmed the surrounding atmosphere. The remaining 7.7 million Btu was used throughout the city, in air conditioners, cooking stoves, water heaters, motors, industrial plants, and the like. In the end, all the kinetic energy of this electricity ended as heat energy, directly converted in ranges and heaters or converted by friction in motors.

Let's look back now with the help of Table 2-1. We began with 24 million Btu of chemical potential energy in our ton of coal, lost about 2.5 million Btu up the stack as heat when we burned it, lost 13 million Btu as heat to the river due to the inefficiency of the turbine, approximately another million Btu went as heat into the atmosphere from losses in the transmission lines, and finally the remaining nearly 8 million Btu did its job and ended up again as heat in the air above Chicago.

We need but one more night to finish the story, a clear night with no clouds between Chicago and deep space. The heated air and the warm water of the river now finally give back to the universe the loan made so long ago. The heat is radiated away from the earth and travels with the speed of light out into those seemingly empty infinite depths; but space is not empty. Here and there in those depths are atoms and molecules of gas and dust, the raw material of the stars, thinner now in the aging galaxy. These widely scattered particles absorb the heat energy traveling out from earth and increase their motion by the amount of that energy. The universe becomes a bit warmer.

A Warming Trend

In the course of following energy through time, we saw it converted from one form to another, sometimes kinetic energy, sometimes potential energy. We began

Table 2-1 Where the Coal's Heat Went

Distribution	*Amount (M Btu)**	
Original total heat	24	
Lost with stack gases (10%)		2.4
Lost to cooling water (60% of remainder)		13.0
Electrical output	8.6	
Lost in transmission (10%)		0.9
Total losses		16.3
Useful work	7.7	

*M Btu = million Btu

with radiant energy and mass energy in the "Cosmic Egg," a condition so special that there is not much we really know about it. After the explosion, the major conversion must have been to gravitational potential energy as the mass went rushing out into space against the attractive force of gravity. In this conversion, much of the energy was left in the random motion of the dispersing atoms; heat energy remained.

The second energy conversion we considered took place in a cloud of gas and dust pulled together by these gravitational forces. In this case, particles moving in toward the cloud's center converted gravitational potential energy into kinetic energy. Packed together as they were, they collided incessantly and this orderly motion was converted into the random motion of heat energy.

When the nascent star's interior became hot enough, the thermonuclear reaction began. This is a conversion of nuclear potential energy to radiant energy and again, heat. We then followed some radiant energy to the earth, where much of it was converted to heat energy by absorption into the atmosphere and onto the earth's surface, while a small fraction was converted to chemical potential energy in photosynthesis. Finally, we examined the conversion of the chemical potential energy of coal to heat energy, then into the mechanical energy of the turbine and eventually into electrical energy. In the end it all became heat.

Energy cannot be created or destroyed; that was the statement of the First Law of Thermodynamics. None of the energy we have followed was destroyed, or created; it was only converted from one form to another. But at each conversion there was a *heat tax*[10] imposed; some of the energy went from a "useful" form (either kinetic or potential) into heat energy, in which its usefulness has been decreased. The conversion to heat energy is a one-way street. While it is possible to convert any form of energy to heat with 100 percent efficiency, to reverse the process is difficult. As we saw in the case of the steam turbine, in the conversion to mechanical energy of the heat released by burning coal, we lost 60 percent of it.

The one-way nature of energy conversion is also a law of nature. For our study of the "energy crisis," more important than the First Law we have just reviewed, is the *Second Law of Thermodynamics,* which states that:

> No device can be constructed which, operating in a cycle (like an engine), accomplishes *only* the extraction of heat energy from a reservoir and its complete conversion to mechanical energy (work).

There are many other ways to state this important law which will be examined in Chap. 4. Its consequence, however, should already be clear. Heat energy cannot be completely converted to mechanical energy. In any conversion some of it is irrevocably lost; it remains in the form of heat and cannot be reclaimed for useful purposes.

On a universal scale, this need not worry us. It is a true projection that all that primeval energy of the "Cosmic Egg" will finally wind up as heat and that the universe shows a warming trend. But we are tens of billions of years away from that end. At the scale of our operations on earth, with our limited supplies of fos-

[10]This term is used by H. Odum in *Environment, Power and Society,* (New York: Wiley, 1971).

sil fuels, the consequences are more immediate. We will examine some of them in the chapters which follow.

SUMMARY

In this chapter we have made the distinction between kinetic energy, energy of motion or energy in transit, and potential energy, energy which has been stored in some way. The translational energy of a moving object or the radiant energy of light are examples of kinetic energy. The three primary forms of potential energy are related (necessarily, as we shall see) to the three primary forces in the universe: gravitational, electric, and nuclear. Most important to us on earth is that form of electric potential energy we call chemical energy, the potential energy of foods and fuels. We also introduced heat energy, a form of kinetic energy in which the motion is random; for example, the motion of the molecules of a hot gas colliding and rebounding chaotically.

The two most important "limiting laws" of energy were also introduced, limiting laws because they describe what can and cannot happen in interactions involving energy. These laws of thermodynamics can be stated in the following manner:

First: Energy can neither be created nor destroyed.

Second: No device can be constructed which operating in a cycle (like an engine) accomplishes *only* the extraction of heat energy from a reservoir and its complete conversion to mechanical energy (work).

From man's point of view, the First Law says: "You can't get something for nothing" and the Second Law says "You can't break even." The Second Law describes the heat tax imposed on all of our energy conversions. We have, therefore, focused our attention in the study of the "energy crisis" not on the using up of energy—it will never "run out"—but on the irrevocable degradation which accompanies its conversion from one form to another as we use energy to fuel our industrial society.

3

Energy on Earth

> Spaceship Earth was so extraordinarily well-invented and designed that to our knowledge humans have been on board it for two million years not even knowing that they were on board a ship . . . so superby designed as to be able to keep life regenerating on board despite the phenomenon, entropy, by which all local physical systems lose energy.[1]

In the previous chapter we followed energy across vast reaches of space and time. In this chapter we will narrow our focus to earth and reconnoiter energy on this small planet, "Spaceship Earth."

This analogy, while it does carry important implications of stewardship and cooperation, does fail in some respects. While we must survive on earth-grown food, breathe earth's atmosphere and drink its water, earth, unlike a spaceship, is not a closed system. Life on earth could not be sustained very long without the constant flow of energy from the sun. As was shown in Fig. 2-2, this energy drives most of the earth's engines (winds and ocean currents, for example) and provides the substance of life as it goes through the many steps in its ultimate conversion to heat energy.

ENERGY SOURCES

Earth is not entirely without energy resources of its own. There is a generous quantity of heat energy stored in its molten interior. There is also both kinetic and potential energy stored in the earth-moon-sun system.

In addition, a relatively small amount of solar energy has been stored in the carbon atoms of the fossil fuels (coal, petroleum, and natural gas). A potentially much larger energy reserve is locked into the nuclei of the very light and the very heavy elements. We are just beginning to learn how to release this energy by fusing light nuclei together (in the fusion reaction) or splitting the heavy nuclei (in the fission reaction). These mechanisms will be discussed at greater length in later chapters.

The flow of energy to and from earth seems to be in balance. Despite all the time that has passed since the sun and earth were formed, there is no evidence

[1] R. Buckminster Fuller, "Operating Manual for Spaceship Earth," (New York: Simon & Schuster, 1969).

that the sun's energy is being stored, or that earth is heating up; nor is there evidence that earth is cooling down.

An overall picture of the energy balance of earth is given in Fig. 3-1. Several comparisons can be made immediately: the almost complete dominance of the sun's energy input is obvious. It is also obvious that the major energy flows don't involve man or life itself directly. Of the 173 trillion kw of power striking the top of earth's atmosphere, 35 percent is reflected immediately back out into space; another 40 percent is absorbed by the molecules of the atmosphere, changed in wavelength, and then reradiated away as infrared radiation; and the remaining 25 percent is mostly taken up by the large-scale movements on the planet's surface: wind and ocean currents, and phenomena such as the evaporation of water, which temporarily store energy.

Although these large-scale phenomena affect life, which would exist in much different fashion, if at all, without them, life relies directly on only 2 or 3 hundredths of a percent of this incoming energy.

In the preceding chapter we tried to establish a concept of energy as something that is sometimes static (locked in an atom or molecule, a nucleus, or even a gravitational field) and sometimes active (carried by moving particles or by light waves). To be put to use it must be in its active form; in other words, it must be kinetic energy. Plants absorb the kinetic energy of sunlight falling on them and convert it by photosynthesis to chemical potential energy. An automobile is set into motion only when the static, chemical potential energy of gasoline is converted, by

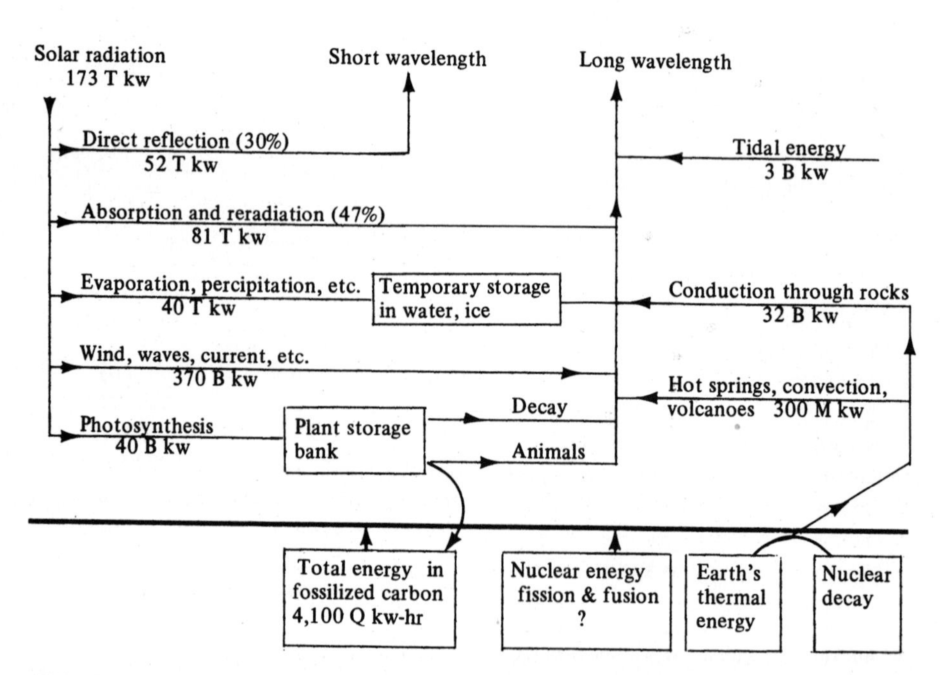

FIGURE 3-1 The energy sources for the earth. The continuous sources such as tides, tidal currents, and the geothermal sources are small in comparison to solar energy.

burning the gasoline in the engine, to kinetic (heat) energy and then to mechanical kinetic energy.

When we speak of energy sources, we are usually speaking of potential energy. It is difficult to store kinetic energy for any long period of time. A planet moving in its orbit about the sun and spinning on its own axis does, however, "store" mechanical kinetic energy, in a sense. Heat energy is stored in the molten interior of the earth; using the same principle, it is stored for shorter periods in an insulated hot water heater. There are always steady losses of this heat energy to the surroundings; there are no perfect insulators.

On a universal scale, there are only three primary forms of potential energy: gravitational, electrical (chemical), and nuclear potential energy. Most of the energy of the universe is still in the form of gravitation energy. Furthermore, the continuing expansion of the universe tells us that the conversion of kinetic energy to gravitational potential energy, which began with the "big bang," is still going on. Some of that energy was won back in the condensation of galaxies and stars, but most of it waits there to fuel the collapse of the universe which some theorists say will not occur for at least tens of billions of years from now.

The light in the universe or, more properly, the *radiant energy,* comes from the conversion of nuclear potential energy in the stars. Its character, or color, determined by the amount of energy contained in each radiated *photon,* is influenced by chemical factors such as the kinds of gases which make up the outer, radiating surface of the stars. Other than that small effect, however, chemical energy does not play a large role in the universe. It is, however, most important on earth.

ENERGY FOR MAN

As Fig. 3-1 shows, there are five primary forms of energy used by man on earth: solar energy, geothermal energy (the energy of the heated core of the earth), tidal energy, chemical energy, and nuclear energy. The first three of these are kinetic forms while the last two are stored potential energy. Of these, solar energy alone can be considered a continuous source, the life expectancy of the sun being what it is. The others must be thought of as depletable sources. Let us consider each in turn.

Solar Energy

We have already examined this form of energy in some detail. Its ultimate source is the thermonuclear reaction in the sun in which hydrogen is fused to form helium. It arrives at earth's atmosphere as radiant energy, in a sense, as pure energy since it needs no matter to carry it through space.

Radiant energy is not easily described. It is carried through space in concentrated form by photons, discrete chunks of energy which have an identity only while they are moving (there is no such thing as a photon at rest). The energy of a photon is either absorbed completely or the photons are reflected without energy loss when they strike something. Photons are characterized by two measures;

their *wavelength,* which determines how they interact with matter, and their *frequency,* which determines how much energy each one carries. (The physics of photons and other wave motions will not be discussed in depth in this text. The interested reader should turn to any elementary physics text.)

The wavelength of a periodic disturbance such as a water wave or a radio wave, is the distance between repeating points, for example, the distance between the peaks or valleys of a water wave. Wavelengths of radiation differ greatly. A photon of ultraviolet radiation has a wavelength of about 10^{-5} cm while a photon of infrared radiation may be 100 times larger. The wavelength of visible light falls in between these two. Since the wavelength of a photon is related to the size of the object which can absorb it, the photons of visible ultraviolet and infrared radiation can be absorbed by atoms and molecules, whereas it takes a long aerial to absorb photons such as those in radio waves whose wavelengths may be hundreds of meters.

The frequency of a wave measures the speed of its vibration which, in turn, determines its energy. High-frequency photons carry a relatively large amount of energy. Thus, γ-ray and X-ray photons vibrating at frequencies of 10^{18} vibrations per second, carry enough energy to do physical damage to the body while radio waves of 10^{6} vibrations per second do no damage. Light waves at 10^{14} vibrations per second carry enough energy to cause chemical changes in molecules.

Figure 3-2 shows the wavelengths of the important components of incoming solar radiation. We see that the peak of the radiation at earth's surface is in the visible wavelength, that is, most of the incoming radiation is visible light. It is surely no coincidence that life forms developed which responded to those wavelengths.

Radiant energy which strikes a chemical substance can interact in several ways: (1) it can bounce off, that is, be reflected; (2) it can be absorbed and set the molecules or atoms into vibratory motion themselves and then be reradiated, either unchanged in wavelength or at longer wavelengths; (3) it can be absorbed and then dissipated within the substance as heat; (4) finally, it can be absorbed and produce a chemical change.

Examples of each of these mechanisms are shown in Figs. 2-2 and 3-1. Much of the incoming energy is directly reflected (interaction 1). The energy absorbed in

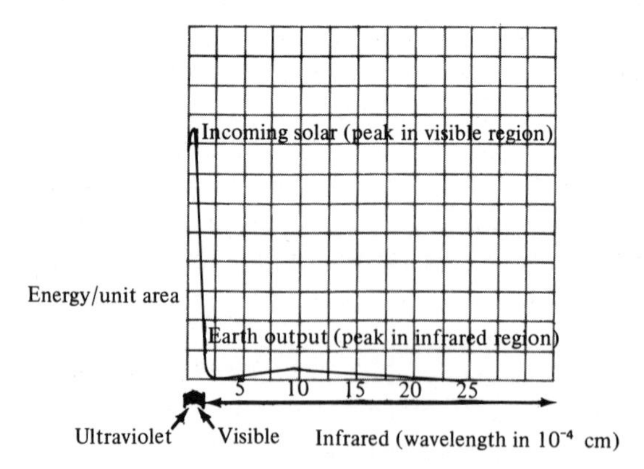

FIGURE 3-2 Energy from the sun comes in at short wavelengths (peaked in the visible) and after fueling the earth's heat engine is degraded to the longer wavelengths of infrared radiation.

the atmosphere is either reradiated (interaction 2) or dissipated as heat. Heating (interaction 3) is particularly important at the surface of the earth. Heat absorbed by land provides the major heat input into the lower atmosphere. It is this converted solar energy which provides the kinetic energy for the great trade winds and all the other winds of the "atmospheric engine." Windmills are thus driven, indirectly, by solar energy. The combination of unequal heating, the earth's gravity, and the spin about its axis are responsible for the atmospheric and oceanic motion.[2]

The selective absorption of radiant energy in the atmosphere (interaction 2) has important consequences. Life is made possible by the absorption of most of the ultraviolent radiation at the top of the atmosphere. If this energetic radiation were not stopped by the thin layer of ozone there, no living thing could exist in the sunlight.

The atmosphere does not absorb very much radiation at the visible wavelengths; it is transparent to them. When this radiant energy is absorbed by plants or soil at the earth's surface, however, it is reradiated at longer wavelengths, in the infrared region. Figure 3-2 also displays the wavelength at which energy is radiated away from earth, illustrating the strong shift of this energy to longer wavelengths. At these longer wavelengths, the atmosphere is no longer transparent; instead, the carbon dioxide and water vapor molecules absorb these wavelengths, thus acting as a kind of barrier against them. This is fortunate, for without this atmospheric barrier, the earth, at night, would radiate most of the energy gathered during the day back out into space so that nights on the earth would be very cold indeed. Instead, the absorbed energy is reradiated so that about half of it comes back to earth, maintaining, in this way, a much milder temperature. This *greenhouse effect* (named for its similarity to the way glass windows in a greenhouse trap the heat radiation) will be discussed in Chap. 9 where we ponder the climatic effects of increasing the carbon dioxide in the atmosphere by the burning of fossil fuels.

There is another effect of solar energy of great importance to man. Heat absorbed by the ocean, or any other body of water, can give individual water molecules enough energy to escape the liquid. This is the process we call *evaporation.* This water vapor is then lifted by rising heated air and is carried by the winds until it falls as rain. By this lifting of great masses of water away from the surface of the earth, some of the solar energy is turned to gravitational potential energy. A small amount of this gravitational potential energy is converted to mechanical energy when water runs through a turbine on its way back to the sea. Thus, hydropower is a second indirect consequence of solar energy. Because of its importance we treat it as a primary form in the discussion which follows (see Fig. 3-3).

We have already described the process of photosynthesis in Chap. 2. This process is an important example of interaction 4. There are other effects of lesser importance. The ultraviolet absorbing molecule, ozone, is formed when radiation splits an oxygen molecule O_2 (two oxygen atoms) and the free atom recombines to make ozone (O_3). In Chap. 7, when we investigate the effect of sunlight on

[2]For more information about these mechanisms, see A. H. Oort, "The Energy Cycle of the Earth," *Scientific American,* **223:** 54, September 1970.

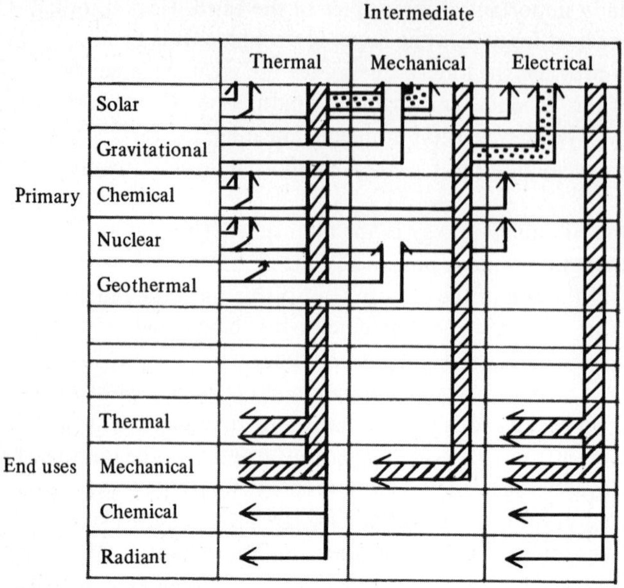

FIGURE 3-3 Pathways of energy conversion in our industrial society. The dominant conversions are to thermal and mechanical energy (heat and work). The conversion pathway to electric energy is a long one: thermal to mechanical to electrical. Electricity, a true "intermediate" form can in turn be converted to any of the end uses.

some of the chemical pollutants which make up smog, we shall mention some photochemical reactions whose effects are not so fortuitous.

Chemical Potential Energy

Whenever energy can be gained by rearranging the atoms in a molecule or by combining free atoms, chemical potential energy is available. The most important example of that is the reverse of the photosynthesis reaction discussed in the previous chapter. When the carbon, hydrogen, and oxygen molecules in carbohydrates and sugars are *oxidized,* either by burning or by the complicated process of digestion, CO_2 and H_2O are formed and energy released:

$$C_x(H_2O)_y + X\,O_2 \longrightarrow (CO_2)_x + (H_2O)_y + \text{energy}$$

There are other important burning processes: free carbon in coal, for instance, combines with oxygen to form carbon dioxide and release energy; free hydrogen combines with oxygen to form water.

$$C + O_2 \longrightarrow CO_2 + \text{energy}$$

$$H + O_2 \longrightarrow H_2O + \text{energy}$$

Thus, both carbon and hydrogen are fuels.

There are other chemical sources of energy. If strips of zinc (Zn) and copper (Cu) metal are put in a mixture of sulphuric acid (H_2SO_4) and H_2O, the Zn, because it can be more tightly bound to the SO_4 than can the H_2, replaces it, forming $ZnSO_4$, and releases energy. This energy can be used to drive electrons through a wire connected between the Zn and Cu strips. This is the conversion of chemical potential energy to electrical energy which takes place in a battery (we discuss this process further in Appendix 4).

There are many other sources of chemical energy of present and future use. At present, however, their importance is dwarfed by the fossil fuels. It is on this non-renewable source that our present industrial society depends for virtually all (96 percent in 1970) of its energy.

Nuclear Energy

Energy can also be stored in the nuclei of atoms and released by rearrangements of the nuclear constituents, protons and neutrons. We presented one example in the previous chapter, the thermonuclear reactions in the sun. In these fusion reactions, light nuclei (those made up of one or a few protons and neutrons) combine to make heavier ones, and energy is released. In the reactions which take place in the sun, the net result is that four protons fuse to make a helium nucleus (two protons and two neutrons). As we have said, the helium nucleus has less mass than the combined mass of four protons and this mass difference is converted to energy. (In fact, mass is converted to energy in chemical reactions also, but the energy released per particle is so small that this mass loss is virtually unmeasurable. In nuclear reactions the mass loss per particle is a million times larger.

The same kind of energy release can be obtained at the other end of the table of the elements, the heavy end. If a heavy nucleus, such as uranium, can be made to split into two medium mass nuclei, energy is again released. The two medium mass nuclei have less mass then the original uranium so that in this reaction the missing mass is also converted to energy.

There are many possible energy-releasing nuclear reactions involving both light and heavy elements. Fusion reactions involving helium, carbon, and even iron take place at the higher temperatures in the more massive stars and are responsible for the energy release of these hotter stars. This mechanism is believed responsible, by some astronomers, for the formation of the heavier elements.

To be of use to man, however, the fission and fusion reactions must be subject to his control. We have devised ways to control the splitting of uranium and are diligently searching for a way to produce a controllable thermonuclear fusion reaction. We will postpone a discussion of these mechanisms, however, to later chapters.

Tidal Energy

There is energy stored as a complicated mix of both potential and kinetic energy in the earth-moon-sun system. The potential energy is gravitational, stored in the separation of the three bodies; the kinetic energy is stored in the rotation of the bodies about each other and about their own axes. The earth converts some of this primary energy to kinetic energy through the mechanism of *ocean tides.* Tides are caused by the gravitational attraction of the moon (and, to a smaller extent, the sun) on the earth's waters: being flexible they can move toward the moon on the near side and away from it on the far side. The converted energy is dissipated as heat energy through friction, warming the oceans and their shores, although imperceptibly.

Man has for some time been intrigued by the possibility of converting *tidal energy* to his use, by, for instance, trapping water with a dam at high tide and letting it do work in running back to the seas. Such projects are feasible only in places where geologic peculiarities cause very high tides. The first major tidal electric plant began operation in 1966 in the la Rance estuary on the Brittany coast of France. The average tidal range here (up to down) is about 26 feet. The power plant consists of 24 units of 10,000 kw capacity each; the estimated yearly output is about 500 million kw-hr.

There has been longtime interest in a tidal energy project using Passamaguoddy Bay on the United States-Canadian boundary. So far, the studies conducted have not provided convincing evidence of commerical feasibility. As we shall see in Chap. 13, we do not expect projects such as these to contribute very much to our total energy supply.

Geothermal Energy

The interior of the earth is molten. Some of this heat reaches the earth's surface, conducted through the rocks of the earth's crust and, occasionally, brought directly to the surface by hot springs or volcanoes. This is stored heat energy, with the earth's crust providing the insulated container. Geologists are not all in agreement as to how it reached its initial molten state; some say that earth was formed molten and then cooled off, others favor a cool-born earth, heated, at least in part, by the rain of gravity-attracted debris. There does seem to be agreement, however, that its heat is maintained against losses through its surface by the decay of radioactive materials.

There are small amounts of radioactive materials such as uranium, radium, and thorium in the earth's crust and, it is believed, in the earth's interior as well. These radioactive atoms spontaneously radiate some of their mass and energy. (Some of these reactions are described in Appendix 5.) As these radiations are stopped in the earth's material, their kinetic energy is converted to heat. Although the concentration of radioactive material in the earth is quite small, there is enough heat energy produced in this way to balance that lost through the surface. Thus, geothermal energy can be thought of as "natural" nuclear energy.

There are many regions of the earth in which molten material is near the surface. If this occurs in porous and water-filled rock, then steam is generated. If, in addition, there is a hard, impermeable rock layer above this, the steam will build up high pressure. By drilling a *steam well* it is possible to get the energy to turn a turbine and generate electricity. There are commercial installations of this type in Italy, New Zealand, and California. The thermal energy of the earth is available in many other forms, as hot water for instance; it can also be obtained by drilling into the earth itself, which increases in temperature 48°C for every mile below the surface. We will examine many of the possible applications of geothermal energy in the chapter on future sources of energy.

THE CONVERSION OF ENERGY

As we have emphasized and illustrated already, energy is stored as potential energy, but to be used, it must be converted to kinetic energy. In Fig. 3-3, the major pathways of energy conversion from the primary forms we have just discussed are shown. We have distinguished between intermediate forms and end uses in this figure. The intermediate forms are all kinetic forms. The end uses may be either kinetic or potential. For instance, when the chemical potential energy of gasoline is converted to thermal energy by burning it in an engine, the thermal energy is an intermediate form and must itself be converted to mechanical energy before it is used. In an oil furnace, however, the thermal energy is the end use. Similarly, a water mill converts gravitational potential energy to mechanical energy, an end use. In a hydroelectric plant, however, the mechanical energy is an intermediate step and further conversion to electricity takes place. Electricity is, of course, an intermediate form itself. Why it is such a popular intermediate form is immediately obvious from Fig. 3-3: it can be converted easily to any of the desired end uses.

Figure 3-3 displays only those conversion pathways which play a major role in the present industrial society. Thus, the conversions of solar energy in the atmosphere, oceans, and biosphere, which we have discussed earlier, are not shown. Instead, we have added as a primary source, the gravitational potential energy which the sun stores by lifting water from the lakes and oceans. Because of its relative insignificance commercially, we do not show tidal energy as a primary source. Its conversion pathway is, of course, from mechanical to electrical.

Work and Power

Energy is neither created nor destroyed; it is only converted from one form to another. Figure 3-3 shows the paths of conversion, but we have said nothing, so far, of the mechanisms of conversion. To discuss mechanisms we must introduce the concept of *work.*

Work is a measure of the amount of energy which is converted from one form to another. It is a term usually associated with man's use of energy, with a conversion for some useful purpose. Work, therefore, represents a change in energy, or energy in transit from one form to another. It is similar to heat energy in this respect, as we shall see in Chap. 4. Since the total energy of a system cannot change, it is clear that work is done by or on a *part* of a complete system. When, for instance, a man draws an arrow back in a bow, he does work on the bow, storing *elastic potential energy.* (Elastic potential energy is actually another form of electric potential; it is stored in the changed orientations of molecular combinations within the deformed material.) He provides this energy ultimately at the expense of some chemical energy from molecules within his muscles. In other words, chemical energy decreases and elastic potential energy increases. When the arrow is released, the bow does work on it, giving up as mechanical kinetic energy, *most* of that stored elastic energy. (That "most" is an important qualification which we shall get to shortly.) Thus, when the potential energy of a part of a closed system increases, work has been done on it by the rest of the system; when the potential energy of a part of the system decreases, it has done work on the rest of the system.

There are many examples of work: a weightlifter does work when he increases the potential energy of a barbell by lifting it; an automobile engine does work by moving the car down the road; a warehouse laborer does work in moving the boxes in his warehouse to a new location. Similarly, work is done on a battery when it is charged and that battery can then do work on the automobile engine; work is done by pumping water up into a tank and that water can do work when it flows back down.

In his stimulating book, *Environment, Power and Society*,[3] Howard Odum distinguishes two kinds of work, *processing work* and *storage work*. In processing work, potential energy is completely converted to kinetic energy as matter is moved about and rearranged; in storage work, potential energy is converted to kinetic and then reconverted to potential energy. The automobile, the warehouse laborer, and the battery starting an engine, in the previous examples, are all doing processing work, while the lifted barbell, the charging battery, and the pumped water are examples of storage work. The reader might try to classify other forms of work as one or the other or a combination of these two types.

At the end of Chap. 2 we summarized the evidence for the Second Law of Thermodynamics which rules out the possibility of building a device to convert heat completely into work. Stated another way, this law assures us that all energy conversions in nature are irreversible. In simple language this means that you cannot convert energy from one form to another and then convert it completely back to the original form. If water runs down from the tank, for instance, converting gravitational potential energy to kinetic energy, it cannot then be made to run all the way back up into the tank by reconverting that kinetic energy. The reason for this impossibility is clear; some of the potential energy which the water had at the top of the hill was converted to heat (by friction within the pipes, for example) and this heat cannot then be reconverted completely to work. In any conversion there is, as Odum puts it, a heat tax (see Chap. 2); some of the potential energy will be converted to a quantity of heat energy which cannot then be recovered for useful work. We will examine this most important consequence in more detail in Chap. 4.

The Role of Force

The concept of work is intimately associated with that of a *force*. Intuitively, when we think of work we think of pushing or pulling something, of exerting a force on it. Our intuition is correct: if work is done *on* a part of a system, a force is exerted *on* that part; if work is done *by* a part of a system, then a force is exerted *by* this working part on something else. To pick a few of our previous examples: the archer exerts a force *on* the bow, bending it and increasing its potential energy; the warehouse laborer must exert force when he does work *on* the boxes by moving them around. In the same spirit the bow exerts a force on the arrow when the archer releases the bowstring and does work on the arrow.

[3] H. Odum, *Environment, Power and Society*, (New York: Wiley, 1971).

Energy cannot be converted without the existence of forces. When processing work is being done, that force is exerted on the moving matter. When storage work is being done, the force must be exerted to overcome a resisting force. The storage of potential energy must be accomplished against a force. This is a most important point. It is illustrated by the lifting of the barbell against the downward pull of gravity or the bending of a bow against the resistive forces within the bow.

We must anticipate here a possible point of confusion. As Appendix 2 explains in more detail, forces always come in pairs, the famous "action-reaction" forces which Newton described:

For every action there is an equal and opposite reaction.

But these paired forces *do not operate on the same object*. You push *on a wall* and the wall pushes back *on you*; you pull *on a doorknob* and the doorknob pulls back *on you*. The resistive forces against which we must work to store energy are different. Take the example of the weightlifter; he lifts the barbell, exerting a force on it. He works, however, against the pull of the earth's gravity and that gravitational force is also applied on the barbell; it is *not* a reaction to his action, and it is not, as we shall discuss later "equal and opposite" to the force the weightlifter exerted.

The application of a force is necessary for work to be done, for the energy of part of a system to be either increased or decreased. This is not enough, however. The force must act through a distance; there must be movement. Furthermore, the force must act along (or opposite to) the direction of motion. Again, we refer to examples: the weightlifter applies a force in the same direction (up) in which the barbell is lifted; the warehouse laborer can do work only by pushing or pulling the moving box in the direction he wants it to move.

We can state this same conclusion in a negative way: forces which act at right angles to the direction of motion do no work. The earth, after all, exerts a steady gravitational pull on a circling space station, but does no work on it (see Fig. 3-4). The energy of the space station does not change: it neither speeds up nor slows down. Since gravitational force is at right angles to the direction of motion, it cannot change that motion.

The amount of work done depends, therefore, on both the force applied and the distance moved. There is the additional qualification that the force must be along the line of motion. We have examined the extreme cases in which the force was along the line of motion or at right angles to it. In the former case, the work is given by the product; force times distance, in the latter, the work is zero.

Most of the instances of interest to us in this book will be of these extreme cases. The reader who is interested in the complete treatment of the relation between force, work, and distance, including the intermediate case of a force applied

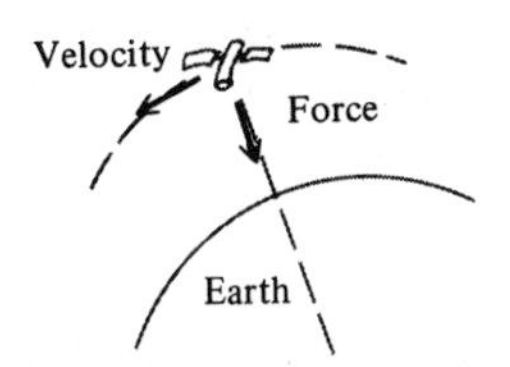

FIGURE 3-4 The force of gravity which holds the satellite in its circular path around the earth is at right angles to the path of motion and therefore does no work.

at some angle to the line of motion (other than 0° or 90°), is referred to Appendix 2. For our purposes, therefore, work is defined as the product of the applied force times the distance along which this force is applied, that is:

$$W = F \times D$$

and these two are required to be parallel. There is, as we shall see later, an important force which always acts perpendicularly to the direction of motion, and, thus, does no work: the magnetic force which will play an important role in the discussions of electricity in Chap. 6.

Power

Since work involves both a force and a distance, it is only a small step to inquire about the time it takes for the motion to occur. This inquiry leads us to the concept of *power*. Power is a measure of the flow of energy; of the rate at which energy is converted from one form to another, or equivalently, a measure of the rate at which work is done. We write this symbolically as:

$$P = W/t$$

where W is the amount of work done in the time, t.

Conversely, work is equal to power times time, that is:

$$W = P \times t$$

It is often not enough just to get work done; the quickness with which the task is performed is also important. "Time is money" and so we are willing not only to pay for the work, but also to pay for the speed in accomplishing the task.

Perhaps the most common example of the recognized importance of power is in the automobile. A car of a given mass, moving at a certain speed, has a definite amount of kinetic energy. (The quantitative formula relating kinetic energy to mass and speed is given in Appendix 2.) Thus, it takes a definite amount of work to provide this kinetic energy. But it makes a difference, at least to some people, whether it takes 10 or 20 seconds to reach that speed. To reach 60 miles per hour in 10 seconds takes *twice* the power from an engine as it does to reach it in 20 seconds. Thus, the horsepower rating of a car is a measure of the rate at which it can produce kinetic energy. It follows, of course, since the energy comes from the burning of gasoline, that the gasoline is consumed at twice the rate (gallons per second, for instance) during the quick acceleration as during the slower acceleration.

Generally speaking, whenever work is done at a continuous rate, it makes sense to talk about power. We speak, for instance, of the power the earth receives from the sun, 173×10^{15} watts (see Fig. 3-1). We rate electric heaters and motors by power. We pay for the power capabilities of devices when we buy them (a 1000-watt heater costs more than a 500-watt heater, a 1 horsepower motor more than a 0.5 horsepower one). We pay for the energy, however, as we use it.

Units

We have defined the basic terms energy, work and power to be used in our discussion of the production and consumption of energy and its interaction with the environment. For that description to be quantitative, however, we must have units. We must be able to say how much energy is consumed or wasted, or how much electric power generating capacity will be needed in 1980, for example.

There are many different units for energy and work. Since work represents a change in energy, it is expressed in the same units as energy. In Appendix 2 all of the commonly used units of energy and power and the numerical relationships, or *conversion factors* between them are defined (see Table A2-3). For most of the data presented in this book, it will be convenient to use a unit of heat energy since 96 percent of our energy is derived from the burning of some form of fuel. Again for convenience, we will use the engineering unit, Btu, since most of the published data are expressed in this way. This unit has already been referred to in Chap. 2, and, in a sense, defined by equating the heat energy in a pound of coal to approximately 12,000 Btu. A more accurate definition is: one Btu is the heat energy which raises the temperature of one pound of water, one degree Fahrenheit. (For the energy equivalent of other fuels and for some representative large and small amounts of energy see Appendix 2, Table A2-4).

It follows, of course, from the relationship between work (energy) and power, that power units will be energy units divided by time units. Thus, Btu per day is a unit of power. For most readers, the power unit which comes most easily to mind is horsepower, which is approximately the steady power output of a strong horse. Most of our data, however, will be presented in the electrical unit of power, the watt, or in its multiples, the kilowatt (1,000 watts) or the megawatt (1 million watts). When discussing electrical energy we will also use the hybrid unit kilowatt-hour (kw-hr), power times time. These terms are defined in Appendix 2 and examined again in Chap. 6. In Appendix 2 we have also gathered together some representative data on power outputs of various power-producing systems and of various animals, including men and horses. We will, however, redefine these units as we need them in the chapters which follow.

One Last Review

We have come a long way. In Chap. 2 we watched energy change form in the kaleidoscope of time; in the first part of this chapter we examined the primary forms of energy on earth and the important paths of conversion of those primary forms to intermediate forms and end uses. We have discussed work and power. By now you may have noticed that although we have introduced most of the important forms of energy, nowhere have we given a short, pithy "definition" of energy; nowhere is there a sentence stating that "Energy is the . . . " This is not an oversight. Energy cannot be precisely defined in this way. It is probably best defined by the First Law of Thermodynamics: energy is a quantity, having the dimensions of a force times a distance, which is conserved in all interactions within a closed system. If that does not satisfactorily answer the question "What is energy?" then the reader will have to be content with the answer: "It is heat, work, gravitational potential energy, kinetic energy of motion, and so on" and by describing each of these forms individually. The only other true and terse statement that can be made about energy is that it can be converted entirely to heat.

We can now say a bit more about our assertion that there are three primary forms of potential energy in the universe: gravitational, electrical, and nuclear. The primacy of those three forms follows from the fact that there are only three kinds of forces in the universe: the gravitational force, which acts between masses; the electrical force, which acts between electrical charges; and the nuclear force, which acts between neutrons, protons, and other nuclear particles. (The nuclear force comes in at least two different forms and some physicists prefer to make a distinction between them. We do not find this necessary here.) All the other forces, those of springs and rubber bands, surface tension, the forces of exploding bombs, of earthquakes and winds, and the force which splits the atom, are but particular manifestations of these three. Thus, since a force is necessary to the storage of potential energy, it can only be stored in these three forms.

The conversion processes diagrammed in Fig. 3-3 proceed through these forces. The fast-moving water molecules of hot steam exert force against the turbine blades causing it to move. The turbine through its axle exerts a force on the rotor of the electric generator forcing it to turn in its magnetic field. The magnetic field exerts a force on the electrons moving with the wires of the armature and causes them to flow along the wire generating electric current. We will examine each of these conversions more closely in later chapters.

SUMMARY

In this chapter we have described the earth's five major energy sources: solar, gravitational, chemical, nuclear, and geothermal. We have also shown the paths of conversion from these primary sources to intermediate forms and end uses.

We introduced work as a measure of the increase or decrease in the energy of part of a system and related it to force by the expression:

$$W = F \times D$$

remembering that the force F and the distance D must be parallel for this expression to hold. Work is, therefore, done *on* part of the system when a parallel force is exerted on it over a distance and done *by* part of a system when it exerts a force on another part of the system.

Two kinds of work were defined: (1) processing work, the work of moving or rearranging matter where work is converted to heat energy by the forces, such as friction, which resist motion; and (2) storage work, in which potential energy is converted to kinetic energy and then back to potential energy. In such a conversion some of the original potential energy is converted to heat.

In addition to work, power was defined as the flow of energy or equivalently as the time rate at which work is performed:

$$P = W/t$$

where the power P is the ratio of the amount of work W done in the time t.

The concept of energy that emerges is one of a quantity which takes many forms, and which can be converted, with some heat loss, from one form to another.

Among these forms are the mechanical work which turns our engines and moves us and our commerce, and heat, which warms us and is a vital input to manufacturing. Through all these changes in form, however, the total amount of energy in the universe remains the same.

4

The Efficiency of Energy Conversion

In Klamath Falls, Oregon, hot water from a nearby geothermal source provides wintertime heat to 500 homes. On some southwestern ranches, windmills still pump water into drinking tanks for cattle. Working watermills with their huge stones grinding slowly to turn corn into cornmeal or wheat into flour can still be found in some rural areas, although usually preserved as historic relics.

From an energy point of view, these disparate examples have something in common: they are a few of the rare instances in which energy is used without conversion in the form in which it is found. As we have emphasized in the previous chapter (see Fig. 3-3), the primary forms of energy on earth (solar, chemical, nuclear, gravitational, and geothermal) are almost always converted to the intermediate forms (thermal, mechanical, or electrical) and then to end uses, most importantly work and heat.

The chief source of our energy at this midpoint of the 1970s is the fossil fuels. This energy is stored in the carbon atoms of coal, and the more complicated combinations of hydrogen, carbon, and oxygen in oil and natural gas. It is released when the carbon and hydrogen atoms are oxidized (burned) and is converted to a new and different form: thermal energy. Since 96 percent of the energy we use goes through this conversion to thermal energy, it is important to our understanding to investigate rather carefully what happens to energy when it passes through this "thermal bottleneck."

The conversion of energy to and from the thermal form has an importance beyond its connection to the fossil fuels: it is the conversion of mechanical energy to thermal energy that provides for the heat tax which we have mentioned earlier. Thus, when energy in any form is used to do work, either processing work or storage work (see Chap. 3), some of that energy immediately, and eventually all of it, ends up in thermal form.

In this chapter we will be concerned with the efficiency of the various energy conversion processes, the limits on efficiency when there are such limits, and the nature of the losses which occur in all conversion processes. In particular we will give much attention to the heat engines, such as the gasoline engine and the steam power plant, which are involved in almost half of the energy conversions important to our economy.

EFFICIENCY OF ENERGY CONVERSION

An *efficient* person gets a lot done with little effort; the definition of an efficient machine is the same. The *efficiency* of a machine or, more generally, of any process in which some energy or work is put in and some combination of work and energy comes out, is the ratio of the desired output (work or energy) to the input, that is:

$$\text{efficiency} = \frac{(\text{energy or work out})}{(\text{energy or work in})} \times 100\% \qquad (4\text{-}1)$$

This ratio is always less than one, since it is not possible to create energy. By multiplying the ratio by 100 percent the efficiency is conventionally expressed as a percentage.

We can draw an example from the automobile engine. If we burn a gallon of gas containing 130,000 Btu of chemical energy in an engine sitting on a test block and measure the mechanical energy which we obtain from that engine, we find that the mechanical energy output is only about 32,500 Btu. If we put these figures into Eq. 4-1, we find that:

$$\begin{aligned} \text{eff} &= \frac{32{,}500}{130{,}000} \times 100\% \\ &= 0.25 \times 100\% \\ &= 25\% \end{aligned}$$

The automobile engine does process work, moving matter. For process work, the efficiency is obtained from the ratio of the work output to the energy input. For storage work, the efficiency is given by the ratio of the stored energy to the input energy. Finally, in that third classification of conversions, where the desired output is heat delivered to a specific use such as the heating of a home or hot water or molten steel, the efficiency is the ratio of the delivered heat energy to the input energy.

While the definitions are similar, the limits on the efficiencies of these three types of conversions are very different and deserve separate treatment. Let us look at storage work first, for the necessity of less than 100 percent efficiency is clearest here.

A form of storage work of increasing importance is the so-called pumped storage hydroelectric plant. In these plants, which are discussed in more detail in Chap. 6, water is pumped up into a reservoir during periods of electric power surplus and then allowed to flow back down through the turbines and generate electric power when it is needed. It is found in this operation that 3 kw-hr of electric energy needed to pump water, which will in turn produce 2 kw-hr of energy on its return through the turbine. The efficiency is, therefore

$$\begin{aligned} \text{eff} &= \frac{2 \text{ kw-hr (output)}}{3 \text{ kw-hr (input)}} \times 100\% \\ &= 67\% \end{aligned}$$

The failure to reach 100 percent efficiency in storage operations is not a failure of man's ingenuity and engineering skill, but a failure decreed by nature. It can be understood from two different points of view. Let us examine a simple version of pumped storage, the lifting of a bucket of water to some distance above the earth's surface. In this operation we do work by forcing the bucket to move against the gravitational attraction of the earth. The stored energy, which we can get back as work by pouring the water on a waterwheel, is the product of the gravitational force and the height.

The work done in lifting the water, however, is the product of the *lifting force* times the distance. The lifting force must be greater than the gravitational force; if it is not, the two forces will be balanced and the bucket will not move. The lifting force must get the bucket moving, keep it moving against air resistance, and then stop it when it reaches the proper height. The starting and stopping work is *not* stored by the lifting process; it is wasted, dissipated as heat in the air and in the arm and body muscles of the person doing the lifting. The amount of energy wasted will get smaller and smaller as the lifting is done more and more slowly, but the efficiency can never be 100 percent; the lifting force will always have to be greater than the gravitational force if the bucket is to move.

The conversion to and from electrical energy is another large category of energy conversions. We will discuss it in detail in Chap. 6 and in Appendix 4. What we will find will be no surprise; in both the generator, which converts mechanical energy to electrical energy, and the motor, which converts electrical energy to mechanical energy, the conversion is accomplished against an opposing force and, therefore, as was true in the example of a lifted bucket, the working force has to be a bit greater than the opposing force so that some energy is necessarily lost.

It is always true that energy storage must be accomplished in opposition to a resisting force. The force which causes the energy to be stored must be greater than this resisting force and therefore, the work done (the product of the storing force times the distance) will be greater than the energy stored (the product of the resisting force times the same distance).

The second way to look at failure to reach 100 percent efficiency is in terms of losses. It is not possible to convert or even transfer energy without losing some of it; losing it, that is, from the process for which it was intended. We pointed out the losses in the muscles while lifting the bucket of water. There are others: the bucket must push the air aside, that takes energy; the water sloshes back and forth in the bucket, that takes energy, and none of this energy is stored for future use.

In the example of battery charging, both points of view are illustrated. Unless the motive force, or the voltage of the charger, is greater than that of the battery, electric current cannot be forced to flow through the battery in a direction opposite to that of its normal flow. Thus, the charging work must be greater than the output energy. It is the losses that account for this difference; the flow of the electric current through any conductor is resisted and energy is lost in the process and not stored in the battery (see Chap. 6 and Appendix 4).

In the second category, process work, we have a multitude of examples. Most of our mechanical energy is expended in this way, in transportation, in the nearly infinite variety of tasks motors do for industry, from moving conveyer belts to stamping out car bodies. Mechanical energy, by definition, involves motion, or in other words, motors, wheels, pulleys turning, pistons pumping up and down, loads rolling, sliding, lifting, and falling. Whenever there is motion, however, there is resistance to motion; something bumps, rubs, or rolls against something else. The only apparent exceptions are motions in the near vacuum of outer space where the planets and galaxies move in almost complete freedom from resistance, and in the inner space of atoms where electrons move with similar freedom. We get no work from these motions, however, the therefore, need not consider them.

We generally classify this resistance to motion as a force, the *force of friction.* This force of friction must be overcome if motion is to persist. We take great pains to reduce this force by polishing and oiling surfaces that touch, providing bearings, and streamlining machines that move in air or water so that the fluid passes easily over them. We can never get rid of friction completely, however. Thus, to move things, a force must be supplied over a distance to overcome the friction, and this work (force times distance) is lost from profitable use. Again, we cannot expect 100 percent efficiency.

In the case of electrical energy converters there are, in addition to the frictional losses involved in causing the motor and generator to turn around, several losses which are connected with the flow of electric current. The most important of these, the *resistive loss* in the wire which we just mentioned, is crudely analogous to friction. The electrons which make up the current "bump into" the electrons of atoms in the current-carrying material, and lose energy.

The third kind of efficiency, *heating efficiency*, is quite different. There is no law of nature which decrees that we cannot convert chemical energy or any other form of energy completely (100 percent) to heat energy. We do not accomplish it in the burning of fuels, for practical reasons having to do with the physical difficulty of burning bulk materials. We can get 100 percent conversion efficiency, for instance, with electricity; a heating wire in water will give all its electrical energy to water in the form of heat and can be 100 percent efficient. Nonetheless, when we describe furnaces and water heaters, for example, we will not find them rated as 100 percent efficient, for we are interested in the heat energy delivered for a specific use and, therefore, the heat that goes up the smokestack or leaks through insulation is lost from the purpose intended, thus decreasing the efficiency.

Some Typical Conversion Efficiencies

Now that you are prepared to expect less than 100 percent efficiencies, you can look at Fig. 4-1 which shows representative efficiencies of a host of energy-converting devices. One sees that the "to and from" conversions of electricity stand at the top of the list and, along with the devices that convert the chemical energy of fuel to heat, dominate the entire upper third of the chart of efficiency. Those important devices which convert heat energy to mechanical energy, such as the steam turbine, diesel engine, and aircraft gas turbine, fall below 50 percent efficiency. The automobile engine, which, by itself accounts for one quarter of our energy use, comes in at 25 percent efficiency. Three-fourths of the energy you buy and put into your gasoline tank does not help you get down the road. This inefficiency contributes greatly to the automobile's relatively unimpressive performance in terms of energy per passenger mile and also to some of the environmental problems we will look at later. (This will be explored at greater length in Chap. 5.)

We should also point out at least two other important efficiencies. While the

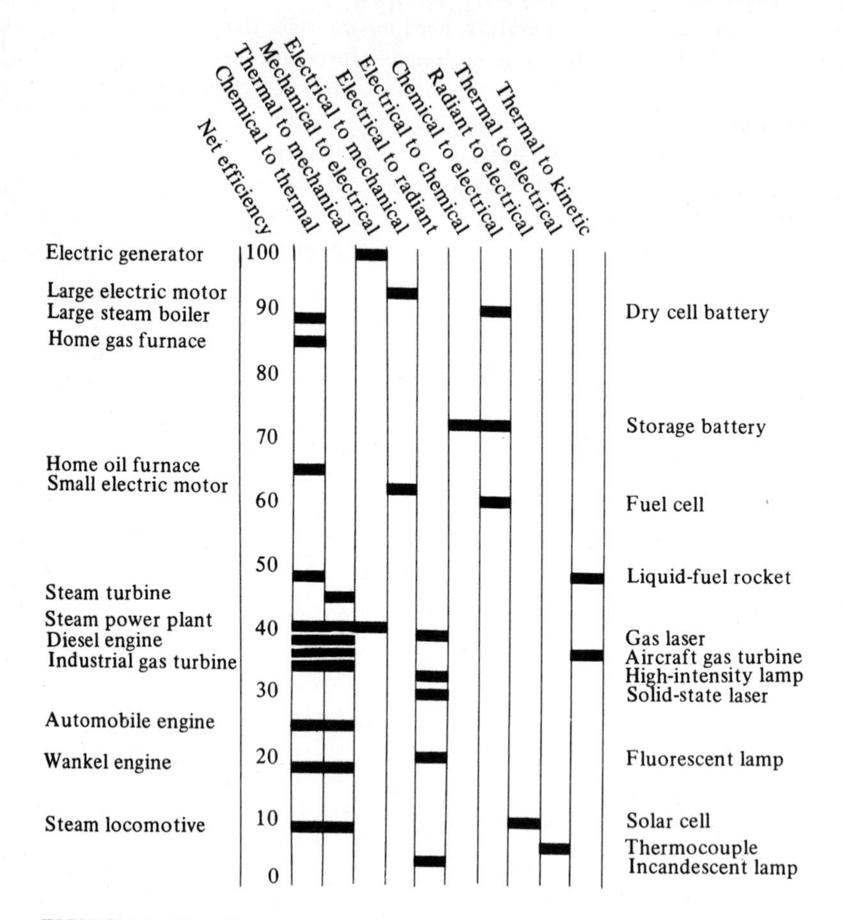

FIGURE 4-1 The efficiencies of various energy convertors. The high efficiencies of chemical to thermal and electrical to mechanical (and vice versa) stand out; the "heat engines" all fall below 50 percent. (Source: "The Conversion of Energy," Claude M. Summers, Copyright © 1971 by Scientific American Inc. All rights reserved.)

efficiency of "to and from" conversion of electricity is high, the steam power plant, which generates 77 percent of that electricity, operates, at best, at 40 percent with a national average of about 33 percent. Lighting, which accounts for perhaps 20 percent of the total electrical energy used, depends on fluorescent lamps (20 percent efficient) and incandescent lamps (5 percent efficient). Since fluorescent lamps now provide about 70 percent of the nation's light (the rest divided between the more efficient high intensity lamps and the inefficient incandescent lamp) the overall lighting efficiency is about 13 percent.

The case of lighting reminds us of a deeper point which applies in the other cases as well. To obtain a fair evaluation of the efficiency with which a certain task is performed, we should really go back to the original source of energy and compare the amount of energy resources which we consume in order to get a desired out-

Table 4-1 Energy System Efficiency of Electric Lighting (from coal-fired generation)

Step	*Efficiency of step (percent)*	*Cumulative efficiency (percent)*
Production of coal	96	96
Transportation of coal	97	93
Generation of electricity	33	31
Transmission of electricity	85	26
Lighting, incandescent (fluorescent)	5 (20)	1.3 (5.2)

put. What is more useful to us in trying to understand the entire energy picture, is the *system efficiency* of the often multistep conversion process involved. To illustrate: in the case of lighting, coal must be dug from the ground (using some energy), transported, burned in a boiler, converted to steam, the thermal energy of the steam converted to mechanical energy (usually in a steam turbine), the mechanical energy converted to electricity, the electrical energy transported to the home and then converted to light. These steps, with their corresponding efficiencies are shown in Table 4-1. Table 4-2 provides comparison data on system efficiencies for the heating of water by electricity and by natural gas.

We see progressive losses at each step of the way in all these systems. In the heating of water, it becomes particularily clear that a comparison of just the final heating efficiencies gives a much distorted picture of the energy costs of the two systems. We will look at some of these system efficiencies again as we consider ways to save energy in Chap. 14.

There are two final comments to make before we tackle the important class of energy conversions to and from the thermal form. The first is: "It's been worse." The second is: "It's not going to get much better."

Figure 4-2 shows the trend of some important energy conversions from 1850 to the present. The conversions are to mechanical work (the motor and transportation are the major contributors here), to electricity, to process heat (that is, heat used in industrial processes such as that used in the cracking of petroleum to make gasoline) and to comfort heat, (the heating of homes, offices, and schools, for example).

Table 4-2 Energy System Efficiency Water Heating

Step	*Efficiency of step (percent)*	*Cumulative efficiency (percent)*
ELECTRIC (coal-fired)		
Production of coal	96	96
Transportation of coal	97	93
Generation of electricity	33	31
Transmission of electricity	85	26
Heating efficiency	92	24
GAS		
Production of natural gas	96	96
Transportation of natural gas	97	93
Heating efficiency	64	60

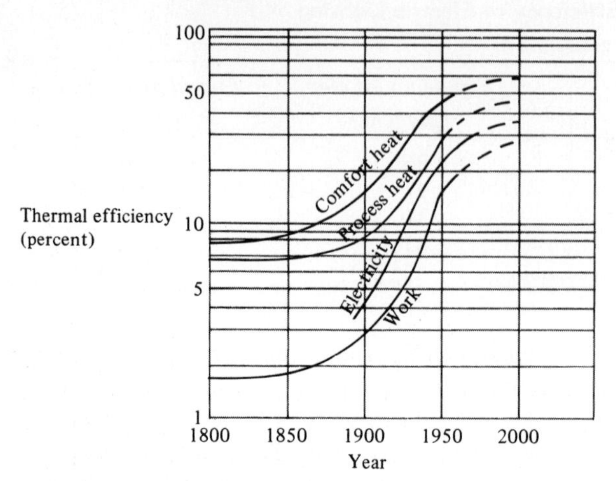

FIGURE 4-2 The efficiency of these important classes of energy conversions rose sharply from 1850 to 1950, but improvements will now be harder to obtain. (Source: "Energy and Power," Chauncey Starr, copyright ©1971 by Scientific American, Inc. All Rights Reserved.)

We see that there has been dramatic improvement in the efficiency of conversion over that time period; mechanical work efficiency improved by a factor of ten, from approximately 2 percent to approximately 20 percent; electric power production improved in efficiency from about 4 percent in 1890 to near 33 percent in 1970 (an eight-fold increase), and the two chemical-to-heat conversion efficiencies improved by similar amounts. These improvements are due to a variety of factors: improvements in furnaces (the fireplace being only about 10 percent efficient), improvements in generators, change in rail transportation from steam engines (10 percent efficient) to diesels (35 percent efficient), and so on.

The other side of the picture is also presented. We cannot expect similar great improvements in the future. There are limits to efficiencies; 100 percent is an absolute limit and for those devices such as the heat engines, which convert thermal energy to mechanical energy, there are even more stringent limits. Efficiencies will continue to improve; there is strong economic motivation for this, but there is now less room for improvement. The areas of largest potential gain are in cutting down the multiple steps in the electrical generating system and in improving on the heat engines. We will come back to this point later.

The trend of efficiency for one of these—the automobile gasoline engine—is worth more attention. Fig. 4-3 shows average miles per gallon data on automobiles since 1920. Miles per gallon is not a measure of the *thermal efficiency* of an automobile, the efficiency with which the heat energy is converted to mechanical energy. That efficiency has improved a little (from 22 to 25 percent) since 1920. The number of miles traveled per gallon depends upon the weight of the car and upon the number of "life support systems," such as air conditioners, which are hung on it. In terms of the energy cost of the moving of a car along a road, however, miles per gallon is a true measure of efficiency. The trends of the data of Fig. 4-3 deserve our attention. During the Depression, the average miles per gallon rose with the introduction of leaded gas and high compression engines. Since 1940 it has dropped steadily as cars have increased in weight and horsepower and as automatic

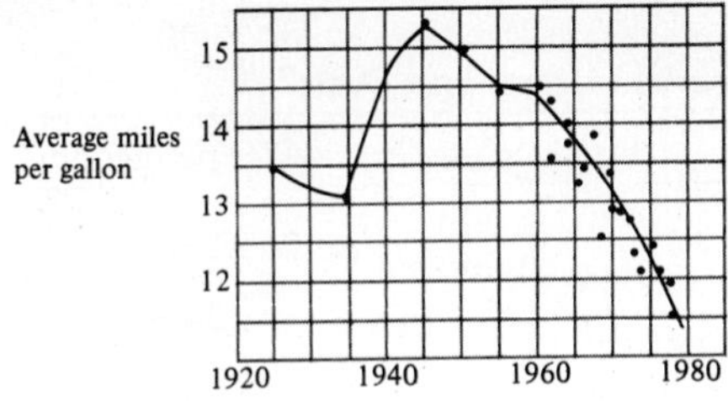

FIGURE 4-3 Improvements in gasoline and engines caused an increase in the mi/gal used from 1930 to 1940. Since then, increase in weight and energy draining auxiliary systems have caused a steady decrease. [Source: U.S. Environmental Protection Agency.]

transmission, air conditioners, and the like have been added.

We are being informed by a powerful advertising campaign that the average miles per gallon of 1973 cars and those which follow will drop even further due to the emission control equipment. The following succinct response to these complaints was made by then Environmental Protection Agency (EPA) Administrator, William Ruckelshaus:

> *Q.* What about Detroit's argument that emission controls cut gasoline mileage and thus exacerbate the fuel crisis?
>
> *A.* The 7 percent average fuel penalty on 1973 automobiles due to emission devices is a significant penalty. But if they make that argument and don't mention air conditioners, which have a fuel penalty between 9 percent and 20 percent, or automatic transmissions, which have a 6 percent penalty, or the weight of an automobile that goes from 2,000 pounds to 5,000 pounds and has a fuel penalty of up to 150 percent, then they're not leveling with people. If we're worried about conserving fuel, then we ought to start regulating some of these things. I think the last one you get rid of is the emission control device. That's the one with the most social benefit.[1]

Table 4-3 compares the rated miles per gallon under simulated urban driving conditions of representative 1974 cars. There is no evidence that the downward trend is not continuing.

The system efficiency of automobile transport shows the wastefulness of this important component of our economy even more clearly. For this we must take into account the extraction of petroleum from the ground, its refining, transport, and the thermal and rolling efficiency of the car. Rolling efficiency is the efficiency with which the engine's power is transmitted to the wheels. These efficiencies are summarized in Table 4-4.

The pressure of rising gasoline prices may, at some future date, reverse the trend of Fig. 4-3 and bring smaller, more efficient cars into popularity. The growth of American Motors sales in 1974 early seems a foretaste of this. There is not, at this time, evidence of any determination on the part of the automobile industry to improve the thermal efficiency of the automobile engine itself. It is interesting to note that the rotary engine, or *Wankel engine* which is coming into use, is actually less efficient than the present piston-driven engine (see Fig. 4-1). We will comment on that engine later. To gain some further understanding of the efficiency of auto-

[1]William Ruckelshaus, "Environment: The Administration's Environmental Czar Sizes Up the Next Two Years," *Business Week,* February 24, 1973.

Table 4-3 1974 Gas Mileage from EPA Suburban-Urban Test Cycle*

Model	*Manufacturer*	*Weight class (lb)*	*Miles per gallon*
Civic	Honda	2,000	29.1
Corolla-1 sedan	Toyota	2,000	24.8
Datsun B210	Nissan	2,250	24.9
Super Beatle	Volkswagen	2,250	20.9
Fiat 128-sedan	Fiat	2,250	17.4
Volkswagen	Volkswagen	2,500	23.7
Opel 1900	Opel	2,500	18.2
Vega Hatchback	Chevrolet	2,750	24.6
Dodge Coltwagon	Mitsubishi	2,750	22.8
Pinto	Ford	2,750	22.8
Mazda RX3 Wagon	Toyo Kogyo	2,750	10.8
Mustang	Ford	3,000	20.1
Datsun 2607	Nissan	3,000	16.1
Volvo 144	Volvo	3,000	16.1
Gremlin	American Motors	3,000	15.9
Maverick	Ford	3,000	15.0
Plymouth Compact	Chrysler-Plymouth	3,500	16.7
Nova Hatchback	Chevrolet	3,500	15.2
Javelin	American Motors	3,500	13.2
Torino	Ford	4,000	14.0
Pontiac Ventura	GM Pontiac	4,000	9.9
GTO	GM Pontiac	4,000	8.9
Matador	American Motors	4,500	10.0
Cutlass S	GM Oldsmobile	4,500	9.5
Plymouth Intermed.	Chrysler Plymouth	4,500	9.2
Cougar	Ford	4,500	9.5
Ford	Ford	5,000	10.7
LeSabre	GM-Buick	5,000	10.4
Plymouth	Chrysler-Plymouth	5,000	10.4
Silver Shadow	Rolls Royce	5,000	9.3
Chrysler	Chrysler-Plymouth	5,000	9.1
Bonneville	GM-Pontiac	5,000	7.8
El Dorado	Cadillac	5,500	10.4
Buick Estate Wagon	GM-Buick	5,500	9.6
Ford Station Wagon	Ford	5,500	9.5
Chrysler Wagon	Chrysler-Plymouth	5,500	8.9
Lincoln	Ford	5,500	7.9
Toronado	GM-Oldsmobile	5,500	7.8

*Source: *1974 Gas Mileage Guide for Car Buyers* (Washington, D.C.: U. S. Environmental Protection Agency, February 1974).

mobile engines and to support our pessimism about improvement, we will now look in detail at the important group of energy converting devices to which they belong, the heat engines.

THE ONE-WAY STREET

There is an efficiency missing from the very complete data of Fig. 4-1: the conversion of mechanical energy to thermal energy. It has little economic importance.

Table 4-4 Energy System Efficiency Automobile

Step	*Efficiency of step (percent)*	*Cumulative efficiency (percent)*
Production of crude oil	96*	96
Refining of gasoline	87*	84
Transportation of gasoline	97*	81
Thermal to mechanical-engine	25†-30‡	20-24
Mechanical efficiency-transmission (includes auxiliary systems)	50-60‡	10-15
Rolling efficiency	60‡	6-9

*E. Ayres and C. Scarlott, *Energy Sources, The Wealth of the World,* (New York: McGraw-Hill, 1952).

†Claude Summers, "The Conversion of Energy," *Scientific American,* **224**: 3: 153, 1971.

‡M. E. Campbell, "The Energy Outlook for Transportation in the United States," *American Highways,* **52**: 3: 25, 1973.

We rub our hands together to warm them and our ancestors, we are told, rubbed sticks together to start a fire; but there are no commercially important devices designed to accomplish this conversion. Mechanical energy is too costly. The efficiency of this conversion, however, can be 100 percent. When we try to turn this conversion around and convert heat energy to mechanical work, however, we get into trouble, for in that case we are trying to go in the wrong direction, on "nature's one-way street." It is for this reason that the efficiencies of the engines that depend on this conversion, the automobile engine, the steam power plant, or the diesel, are low on the chart of Fig. 4-1. The "one-wayness" of this conversion, perhaps the most important constraint on our use of energy, is tied to the special nature of heat energy. The story of how the mystery of heat was unraveled, of the intuition and experimentation that led to the proof that heat was a form of energy and interconvertible with other forms at a fixed rate of exchange, is a fascinating one. It is summarized in Appendix 3, along with the story of how the understanding of heat in these terms led to a statement of the First Law of Thermodynamics. We also have relegated to that appendix, to preserve a clearer story line here, the more precise physical description of temperature, specific heats, and so on, which fill out the broader picture of heat.

We will find it useful in this chapter to make a distinction between heat energy and internal energy. We will use the term "heat energy" to describe the energy in motion between a hot and cold body. Heat energy is thus energy in transit, analogous to the mechanical kinetic energy of motion. Internal energy is stored energy. This energy may show itself in an increase in the unorganized random motion of the molecules, but it may also be a form of potential energy; for instance, it may depend on the separation of the molecules.

If we raise the temperature of a pan of water by putting it in contact with a hot object (a stove top, for instance), we will say therefore, that heat energy is transferred from the stove top to the water and that the internal energy of the water is increased. What makes the distinction important is the fact that the increased internal energy of the pan of water could have been obtained in ways other than by a transfer of heat energy, by shaking it vigorously, for example, and it can be decreased in ways other than by the leaking of heat energy to a cold body; it could go to increase the potential energy of evaporated molecules, for instance. The distinction is important in understanding what takes place in heat engines as well as for an appreciation of the conversions of heat energy to other forms. In later chapters the distinction will not be so important and we will use the terms heat energy or thermal energy more loosely to describe the energy stored in the chaotic motion of molecules, whatever the source.

There is a great deal of interesting physics to be learned from a study of heat energy and internal energy, of the various mechanisms of heat transfer, the changes of state, melting, freezing, evaporation, and the like. For the understanding of the heat engine, however, which is our primary goal, the most important substance is the gas. Fortunately, it is also the simplest state of matter to understand. The most important properties, for our purposes, can be deduced from a model that treats the gas as being made up of identical noninteracting (no forces between them) marbles that collide elastically with each other and with the container walls. With this model we can explain how a gas can exert pressure and how a heated gas can convert some of its internal energy into mechanical work.

The internal energy of a gas that behaves in the simple way of this *ideal gas,* as the gas of noninteracting molecules is called, depends only on one quantity, the temperature. In Appendix 3 we show that, for such a gas, the temperature T is proportional to the average kinetic energy of the colliding molecules. For an ideal gas, all the internal energy will be in the simple kinetic form. In defining temperature in this way, we must use the so-called absolute temperature scale with $T = 0$ at *absolute zero.* The meaning and necessity of this distinction is also discussed in Appendix 3.

In addition to temperature there are two other quantities which are needed for a complete description of the state of a gas: *pressure* and *volume.* The pressure is the force per unit area, in pounds per square inch, for instance. This pressure comes from the molecules striking the walls of the enclosing vessel, or a piston in a cylinder. The volume must also be specified, for without a container to define its volume, a quantity of gas becomes a dispersed number of atoms or molecules.

These three quantities are related by the important equation which we write as:

$$PV/T = \text{Constant} \tag{4-2}$$

This expression, which is rigorously true only for the ideal gas we have described (but will serve us well at the level of precision of the descriptions of this chapter), shows how changes in these three quantities affect each other. If the temperature is held constant, for example, an increase in volume must be accompanied by a decrease in pressure, and vice versa. If the volume is held constant and the temperature increased, then P increases (see also Appendix 3).

Heat into Work

With the relationship shown in Eq. 4-2, we can now demonstrate how to accomplish that very useful trick of converting the random, unorganized motion of a

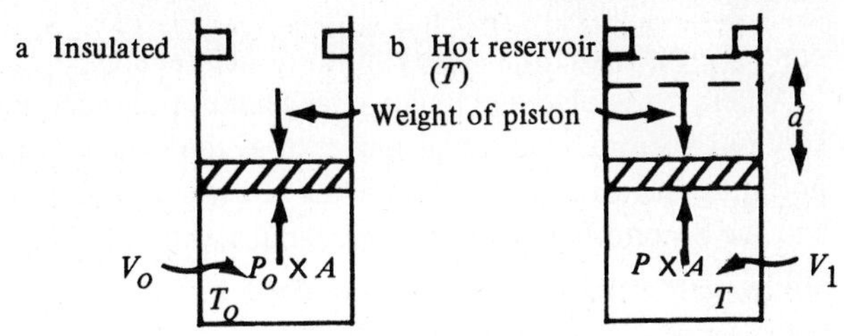

FIGURE 4-4 *(a)* The piston is supported at rest by the pressure of the gas. *(b)* The slight increase in temperature causes the piston to move up(still at the same pressure) and to do work.

heated gas into the useful and organized motion of an engine.

The traditional device for accomplishing this is the piston-cylinder arrangement shown in Fig. 4-4. Under the conditions represented in Fig. 4-4*a*, the piston is in equilibrium. The downward force of the piston (in this case, its weight) is just balanced by the upward force, which is the pressure P_0 (of the gas at the temperature T_0 and volume V_0) multiplied by the area of the piston, since force is the product of pressure and area.

Now the system is put in thermal contact with a *hot reservoir* at some constant temperature T, which is only slightly larger than T_0. The hot reservoir might be, in a practical example, the firebox of a steam plant or the core assembly of a nuclear reactor, but any larger constant temperature source of heat energy will do.

Since the temperature T of the reservoir is greater than the temperature T_0 of the gas, some heat energy will flow into the gas; the average kinetic energy of the molecules and, thus, the temperature of the gas, will increase slightly. From Eq. 4-2 we see that if T increases, either P or V (or both) must increase. The pressure of the gas cannot, however, increase because the instant it begins to rise (as the molecules strike the wall and the piston at higher velocities), the piston will begin to move upward, keeping the pressure the same. What will change, therefore, is the volume, which will increase as shown in Fig. 4-4*b*. As heat energy flows in from the reservoir, the piston will slowly move up until it reaches the top of the cylinder and the gas will now occupy a new volume V. The pressure and temperature will still be essentially the same as P_0 and T_0. Since a force F (the pressure P_0 of the gas on the piston times the area A of the piston) has operated over a distance D, work has been done.

Where did this work come from? It should be clear that its source was the heat energy Q_{in} put into the gas by the hot reservoir; in fact, since the internal energy of the gas, which is measured by its temperature, has not changed (the temperature was constant at T) the Law of Conservation of Energy requires that:

$$W_{out} = Q_{in}$$

The gas has, therefore, acted as a converting agent, absorbing the heat energy and transmitting it through random motion of the molecules to the piston where it produces an orderly motion. As we shall see later, it "goes against nature" to convert random motion to orderly motion. Therefore, this process is not a very efficient one. However, it works; and it forms the basis of heat engines. It is also possible, as shown in Appendix 3, to obtain mechanical work from the internal energy of a gas without a steady input of heat energy.

Whether the piston-cylinder arrangement produces work by converting heat energy put in from an external source or at the expense of internal energy, the important point to be recognized is the apparent conflict between these processes and the Second Law of Thermodynamics which, by our statement in Chap. 2, requires that:

> No device can be constructed, which operating in a cycle (like an engine) accomplishes *only* the extraction of heat energy from a reservoir and its complete conversion to mechanical energy (work).

The example we described did convert heat energy completely to work, but the process was not a cycle. To make it cyclic would require bringing the gas back to its original temperature, volume, and pressure. It is the necessity of returning the *working fluid,* in this case, the gas, to its original state that causes us to throw away energy and reduce the efficiency of the operation.

The process shown in Fig. 4-4 cannot produce net work in a cyclic manner. As is easily seen, if the process is reversed and the gas compressed back to its original state, the same amount of work will be required and there will be no net production. Actual cyclic processes are more complicated. One familiar example is the four-stroke cycle of the internal combustion engine shown in Fig. 4-5. The four strokes are as indicated: *intake,* in which the valve is open and the air-gasoline mixture admitted; *compression,* during which the piston, driven by the flywheel, compresses the gas to its maximum value and the mixture is ignited by a spark; *power,* during which the hot gas drives the piston down; and *exhaust,* during which the returning piston drives the burned gases out of the now open exhaust valve.

The Efficiency of Heat Engines

It was Sadi Carnot, a brilliant young French physicist in the early nineteenth century, who first recognized the importance of the cyclic operation of engines and abstracted the key components. He recognized that the properties of the working

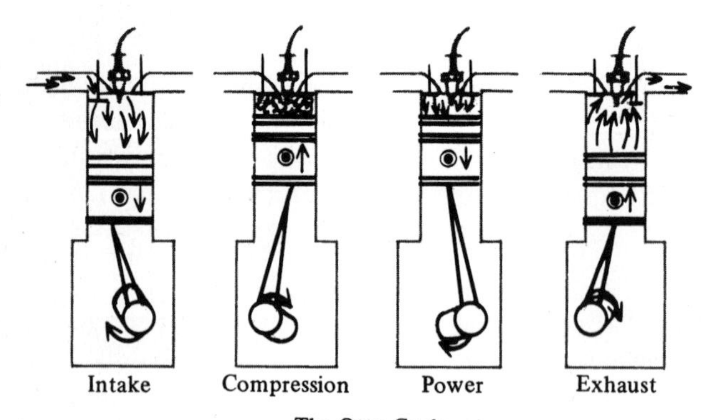

FIGURE 4-5 The four strokes of the familiar internal combustion (gasoline) engine. Only one of the four is a power stroke.

fluid were not important since it is always returned to its original state. The important steps are: (1) the absorption of heat energy Q_1, from a reservoir at high temperature T_1 (the boiler of the steam engine, for instance); (2) the production of mechanical work W; and (3) the rejection of a smaller amount of heat energy Q_2 to a reservoir at the lower temperature T_2 (the exhaust, or in a steam engine the condenser, in which the steam is cooled and liquified). This process is shown schematically in Fig. 4-6*a*.

Carnot stated the fundamental principle of a heat engine:

> The thermal agency by which mechanical effect may be obtained, is the transference of heat from one body to another at a lower temperature.

He further realized that the most efficient cycle would be a *reversible* one, one which could be run backwards as in Fig. 4-6*b*. In this reverse cycling, work W is put in, and heat Q_2 removed from the cold reservoir and transferred along with the work as Q_1 to the hot reservoir. (The discerning reader may recognize this as the mechanism of a refrigerator. We will discuss that application in the next section.) This definition of *reversibility* requires that there be no losses of work or heat energy, that the work obtained from Q_1 with the rejection of Q_2 be just enough to transfer Q_2 back to the hot reservoir. It should be clear from what we have said that reversible cycles do not occur in nature: there are always losses. By studying this theoretical cycle, however, Carnot was able to set limits on the efficiency of real cycles.

In Appendix 3 we discuss the concept of reversibility and irreversibility and the theoretical cycle that Carnot devised in more detail. We shall content ourselves here with stating his conclusions.

The efficiency of any heat engine is given by:

$$\text{eff} = \frac{W_{\text{out}}}{Q_{\text{in}}} \times 100\%$$

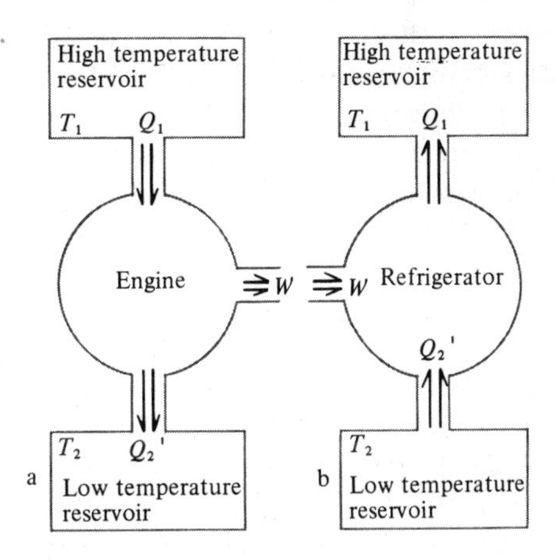

FIGURE 4-6 The reversible cycle studied by Carnot acts as an engine in *a* taking in heat energy Q_1, producing work W, and exhausting heat energy Q_2. In *b* it acts as a refrigerator or "heat pump" using work W, to transfer Q_2, from the low temperature to the high temperature reservoir. In both cases, $Q_1 = Q_2 + W$.

Since we are now considering the engine shown in Fig. 4-6*a* as a perfect engine (no friction or leakage losses) the work out must be, by the Law of Conservation of Energy, just:

$$W_{out} = Q_{in} - Q_{out}$$

The efficiency is, therefore[2]:

$$\text{eff} = \frac{Q_1 - Q_2}{Q_1} \times 100\%$$

or, dividing through by Q_1:

$$\text{eff} = 1 - \frac{Q_2}{Q_1} \times 100\% \qquad (4\text{-}3a)$$

The final deductive step which makes this efficiency equation of practical use is the following: since this is an ideal engine (its properties do not depend on the working fluid, for instance, steam or gas) the rate Q_2/Q_1 can only depend on the ratio of the temperature T_2/T_1; there is no other property available. Thus, we can set $Q_2/Q_1 = T_2/T_1$ and write the efficiency as:

$$\text{eff} = 1 - \frac{T_2}{T_1} \times 100\% \qquad (4\text{-}3b)$$

This is a most important result. Let us first examine it, along with Eq. 4-3*a* which preceded it, as a consequence of the Second Law. As we have been stating this Law, it denies the possibility of constructing a machine capable of drawing heat energy out of a reservoir and converting it completely to work; it labels as impossible, for instance, a steamboat which converts the heat of the ocean to motive power or a power plant that uses heat from the air to run its generators. The First Law denies the possibility of a machine that creates its own energy, "perpetual motion of the first kind." The possibilities denied under the Second Law, such as the steamboat running on ocean heat, are "perpetual motion of the second kind." Equations 4-3 make the reasons for this denial clear. Any heat engine must reject some heat Q_2 to a lower temperature reservoir or, equivalently, a heat engine must operate between a temperature difference T_1-T_2.

We can also see reinforcement of the concept of absolute zero in Eq. 4-3*b*. It is clear that T_2 cannot equal zero, for then the efficiency would be 100 percent, in violation of the Second Law. Thus, T_2 is unreachable. Similarly, there is no possibility of a temperature less than $T_2 = 0$, for then the efficiency would be greater than 100 percent. (This warns us, by the way, that, to use Eq. 4-3*b*, we must express temperatures on the absolute temperature scales described in Appendix 3.) We see that the other condition for 100 percent efficiency $T_1 = \infty$, is also unreachable since infinite temperatures are impossible.

Equation 4-3*b*, the efficiency equation, has practical consequences in that it guides us in the design of heat engines. To obtain maximum efficiency we must

[2] To simplify the notation here and in what follows we will use the subscript 1 to indicate input values and 2 to indicate output values, hence $Q_{in} = Q_1$, $Q_{out} = Q_2$.

put heat energy in at the highest possible temperature T_1 and exhaust it at the lowest temperature T_2 that we can. If we now pick up one other fact, we can see the full power of this efficiency relationship. The cycle that Carnot used and that we have assumed here is reversible. In Appendix 3 we prove that no other heat engine operating between temperatures T_1 and T_2 can have a higher efficiency than this reversible one. Therefore, Eq. 4-3*b* gives us an *upper limit* on the efficiency of any real engine operating between those two temperatures.

Let us consider two examples. The first is a steam engine which receives steam at $T_1 = 1000°F$ and exhausts it at $T_2 = 212°F$, the boiling point of water. These are realistic temperatures for actual steam power plants. To compute its limiting efficiency we must first convert the degrees F to a scale for which $T = 0$ is at absolute zero. This scale is the *Rankine* scale, explained in Appendix 3, with zero at -460°F. Thus, we must add 460° to T_1 and T_2 and then substitute them in Eq. 4-3*b*.

$$\text{eff} = 1 - \frac{672}{1460} \times 100\%$$

$$= 1 - 0.46 \times 100\%$$

$$= 54\%$$

Thus, the highest possible efficiency of any heat engine operating between these temperatures is 54 percent. Actual engines reach 40 percent, as was shown in Fig. 4-1. Since the temperature 1000°F and the resulting pressures of about 3,000 pounds per square inch are at the limit of capability of our present boiler and pipe systems (some experimental systems run at 1500°F) not much improvement in steam generator efficiencies can be expected. At 1500°F the limiting efficiency would be 65 percent.

Turbines. So far, we have only considered reciprocating (piston-driven) steam engines. They were the first heat engines to be invented and lend themselves to comparison with the "perfect" Carnot engine. However, almost all of the steam engines used in electric power generation are steam turbines. The conversion of heat energy to mechanical energy is even easier to understand here. A simplified version of a steam turbine is shown in Fig. 4-7. Steam is heated to high pressures and then directed at high velocity against a bladed wheel as shown. The transfer mechanism is the same as in the piston arrangement: the molecules strike the

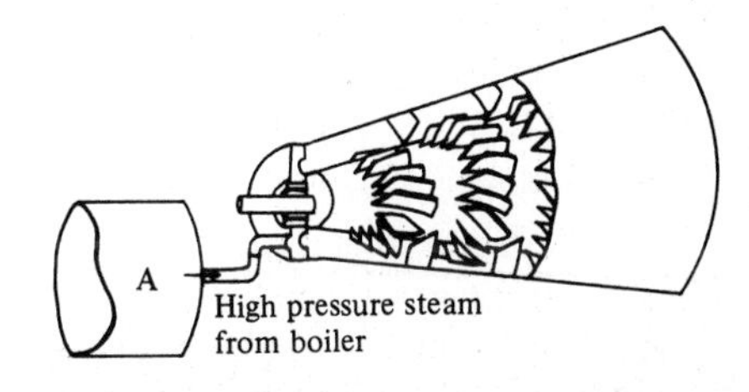

FIGURE 4-7 A steam turbine. High-pressure steam strikes the tilted vanes of the rotors and causes them to turn. The vanes along the sides deflect the steam to the next rotors. The volume of the turbine must increase from the input to the exhaust because as the steam gives up energy, its pressure decreases and its volume increases.

blades and bounce off of them, exerting a force which causes the blades to turn. In bouncing off the moving blades they give up energy, thus the temperature of the steam is lowered and the turbine operates as a heat engine. Equation 4-3*b* is, therefore, still operative; the temperatures we have just used for the efficiency calculation which gave the upper limit of 54 percent are, in fact, typical of steam turbines rather than reciprocating steam engines.

Since the steam turbine is such a simple concept and so like the waterwheel, one might wonder why it was not invented first. In a sense it was. The steam toy, Hero's aeolipile, built in Greece around AD 75 was a variation on the turbine principle. What delayed the practical turbine was the high velocities of the steam and, therefore, the high speeds of rotation of the blades which required very sophisticated construction—good bearings, blanced blades, and strong materials. Present-day turbines are *staged,* that is, have a series of rotating blades through which the steam flows, losing some energy and decreasing in pressure at each stage.

A different kind of turbine, the gas turbine, is also used in electric power generation and to power jet aircraft. Efforts are under way to develop gas turbine engines for trucks, trains, and buses. Its operating principle is similar, except that air, compressed to high pressures by a compressor which is turned by the turbine itself, is fed into a combustion chamber and heated by the continual burning of fuel such as oil. This high pressure combustion gas is directed against the turbine. The gas turbine is also a heat engine. Its efficiencies, however, are lower (about 20 percent) because the combustion gases are exhausted at high temperatures. (In Chap. 12 we will discuss some proposals to get more work from these hot gases.)

Another Limiting Efficiency. A final example of interest is the automobile engine. In this engine the input temperature T_1, the combustion temperature of the gasoline-air mixture is about 4000°F. T_2 is the temperature of the exhaust. The lowest value it can have is 70°F. Adding 460° to T_1 and T_2, the limiting efficiency is, therefore:

$$\begin{aligned} \text{eff} &= 1 - \frac{530}{4460} \times 100\% \\ &= 1 - 0.12 \times 100\% \\ &= 88\% \end{aligned}$$

In actual engines the temperature is very high only momentarily and then drops rapidly. There is much leakage of heat and the exhaust temperature is much higher than the 70° we chose here, perhaps as high as 1000-1500°F. With incomplete burning of fuel, hot exhaust, and leakage all contributing, the actual thermal efficiency of automobile engines is, as shown in Fig. 4-1, about 25 percent.

It is also interesting to note in Fig. 4-1, that the Wankel rotary engine, which has many apparent advantages over the piston engine (more power per pound, less emission of pollutants, and goes hmmmm instead of boing! boing!) does not have an efficiency advantage. This seems, at first, a contradiction. The pollution from a piston engine is, in part, due to incomplete burning of the air-gasoline mixture; it does not get maximum energy from its fuel.

How can the Wankel engine have a lower efficiency and yet emit less pollutants?

The answer comes in two parts. First, the lower efficiency is explained by Eq. 4-3*b*: the fuel mixture is burned more slowly and at a lower temperature. Thus, T_2/T_1 is a larger number and the efficiency is smaller. But the engine is also smaller so that more room is left for emission control devices. The lower fuel combustion temperature also reduces certain types of pollutants. Even with a lower thermal efficiency, the smaller rotary engine will give more miles per gallon and save energy since a lighter car can be built around it.

As a final note on the interpretation of heat engines and their efficiency, we point out the following. Gases at high temperatures have extractable energy in them; the higher T_1, the larger the amount of work that can be extracted. Energy is, therefore, wasted whenever a gas is allowed to cool down without doing work. In that sense, a home furnace whose interior heat is about 2000°F is an extremely wasteful device. This high temperature heat energy is allowed to degrade to about 70°F without doing work. The same sort of waste occurs in electrical home heating; the resistance heater may be at 1000°F in order to keep the room comfortable. A more efficient way to get heat into a room was first suggested in the nineteenth century. It works on the principle of the refrigerator and will be discussed in the next section.

Refrigerators and Heat Pumps: Heat Engines in Reverse

As early as 1852, Lord Kelvin, one of the great pioneers in thermodynamics, realized that a heat engine, if worked backwards, would "pump" heat from the cold reservoir to the hot reservoir. This is the principle of the refrigerator, but perhaps more importantly, the principle of the *heat pump* which can heat a home in the winter and cool it in the summer. In principle, a refrigerator works by the process shown in Fig. 4-6*b*. In practice, the process is somewhat more complicated. We start with a compressed gas at the temperature T_1 of the outside room. This gas is caused to expand (and therefore to cool) and is put in contact with the interior of the refrigerator. The cool gas absorbs an amount of heat energy Q_2 and expands further. We then insulate it and compress it, bringing it back to the original temperature T_1 and then compress it further, driving heat energy Q_1 into the warm outside. To accomplish this we put in work W and transmit heat energy $Q_1 = Q_2 + W$ to the outside.

The cycle shown in Fig. 4-6*b* is a perfect, reversible cycle. But even in this perfect cycle an input of work is always necessary; in fact, an alternative but equivalent statement of the Second Law is:

> It is impossible to construct a device that, operating in a cycle, will produce no other effect than the transfer of heat energy from a cooler to a hotter body.

Actual refrigerators follow the general pattern of the reversible cycle noted here, but with inevitable losses that make their cycles irreversible. A gas such as Freon becomes a liquid when compressed, and is cooled to room temperature. It is then circulated to the interior, absorbs heat energy, becomes a gas, expands and absorbs more heat energy, and is then pumped outside, where it is compressed and the excess heat is discharged to the room. The input work, provided by the refrigerator motor, is used to circulate the gas, compress it, and run the fan to cool it. There are, of course, frictional losses and heat leakage to make it irreversible.

The important application of this reverse cycle, and one that Kelvin recognized, was that such a device used input work to pump heat from one place to another. If you wanted to cool your house, you could circulate the cool, expanded gas indoors, pick up heat, and discharge it outdoors. If you wanted to heat your house, you could turn it around, circulate the cold, expanded gas outdoors, pick up heat, and pump the heat indoors—you refrigerate the outdoors, so to speak.

From the point of view of energy conservation, this is a much more economical process than burning fuel or using electricity to heat the house. The kilowatt of electrical energy converts to 3412 Btu of high temperature heat energy in an electric heater; but that same kilowatt would be able to move three or four times as many Btu from the outdoors to the indoors if it were used to power such a heat pump.

It may seem that such performance must violate some law of nature, but it does not. Again, we can turn to a gravitational analogy for help. The situation is analogous to a man with a garden above a stream and a reservoir above the garden as in Fig. 4-8. The reservoir is the equivalent to the high temperature heat. He can irrigate his garden by either letting the water run down from the reservoir directly or by letting this water, with its kinetic energy, turn a waterwheel which lifts water from the river. If the height of the reservoir above the garden is five times the distance of the garden above the stream, then a bucket of falling water can lift five buckets of water from the stream and use that, along with the reservoir water, to irrigate. The heat pump works in the same fashion lifting heat through the relatively small temperature difference between indoors and outdoors. Heat pumps have had some popularity in the temperate regions of this country. Their energy economy may make them more popular as the burning of fuels to release heat becomes more expensive.

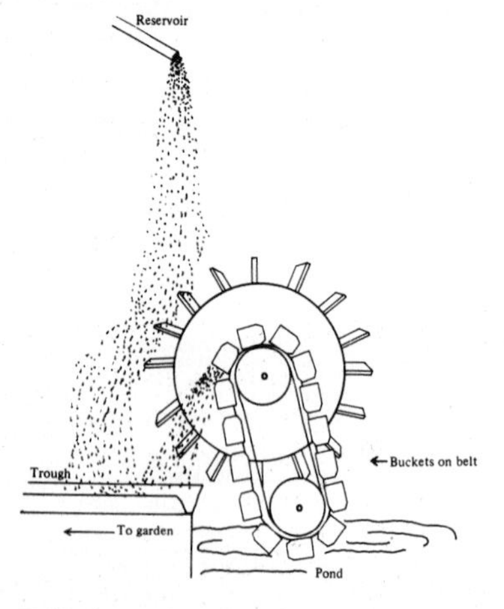

FIGURE 4-8 If the water from the reservoir is used to turn the waterwheel and lift water from the nearby lake, much more water can be provided for the garden than by using the reservoir water directly.

THERMODYNAMIC ECONOMY

Much of the focus of popular discussion of the Energy Crisis has been on increasing energy production: finding new oil fields, increasing the output of coal mining, and even developing new sources, perhaps with the aid of nuclear reactors now and nuclear fusion soon. The message hidden in this chapter, which we now wish to make explicit, is different. We hope, by the examination we have made of the two Limit Laws of Thermodynamics, to focus attention on the efficiency with which energy is used.

The first of these Limit Laws, the Law of Conservation of Energy, suggests this focus. We know that the problem cannot be the "running out" of energy, since energy cannot be destroyed. The problem must be in its utilization. The other Limit Law, the Second Law of Thermodynamics, is much more specific; in fact, through it we can define a quantity that measures what is actually being lost, the potential of the energy for productive conversion, and suggest some of the rules for a *thermodynamic economy*.

The quantity that gives us this deeper insight is called *entropy*. In Appendix 3 we have developed this concept rather thoroughly. We will summarize and simplify those remarks here. Entropy, as it is defined, is a property of an interacting system, which could mean a sample of a gas, an engine, an object falling to earth, and the like. It is similar to potential energy in that its absolute value must be arbitrarily set; it is the *change in entropy* during an interaction which carries important information.

There are several ways in which entropy can be characterized. Perhaps most important is its connection with the degradation of energy from a form in which it is available to do work to a form from which it cannot be transformed into work. As shown in Appendix 3, the change in entropy for an irreversible process is positive; in other words, entropy increases. In an irreversible process—and all spontaneous "natural" processes fall into this category—some energy is degraded into low temperature heat energy and, by the restrictions set by the Second Law, is lost for further conversion to work.

We have seen in earlier sections that it is impossible to convert heat energy into work with 100 percent efficiency. We can also state the Second Law in terms of entropy:

> In any conversion or transfer of energy within a closed system, the entropy of the system increases.

The concept of entropy, however, is capable of giving us more than an elegant restatement of the heat tax. It is more broadly interpretable in terms of a progression from order to disorder. The concept of order would apply, for example, to a crystal in which the atoms or molecules form a more or less regular structure, or to a gas in which a larger percentage of the atoms or molecules are moving in the same direction. The prototype, disordered, system would be a hot gas in which the atoms or molecules are moving chaotically in all directions.

It was the physicist Boltzman who pointed out that disordered states were more probable than ordered ones, and that the spontaneous direction of natural processes was from order to disorder. The state of a system requires its complete description at some point in time. The state of an ideal gas, for instance, is specified by giving the values of its pressure, volume, and temperature. He also showed that the entropy of a state of a system was mathematically related to its probability of existing. Thus, the spontaneous or natural direction for entropy change is to increase. A particularly suitable example of such a change is provided by the melting of ice. We can consider ice cubes dropped into warm water as a closed system: no additional energy is put in or taken out. The ice, with its molecules arranged in a crystalline structure, has a considerably higher degree of order than liquid water in which the molecules, while still bound together, have much larger randomly directed motions. The ice will melt when it comes in contact with the warm water; this is the *spontaneous direction.* It will cool the water in the process; the water must provide heat energy to disrupt the bonds that hold the ice crystal together. This, in turn, increases the order of the water, for the water molecules, as they become cooler, have less heat energy and therefore move less wildly about. The molecules in the ice, however, have gone from the high order of the solid to the disorder of the liquid and it is this change that dominates.

To see how this works in an energy-conversion example let us look at a simple, but hypothetical, conversion process. Suppose we have a certain amount of energy E in the form of the kinetic energy of the stream of marbles in shown Fig. 4-9*a*. They all have the same mass and velocity. Therefore, the total energy is just the sum of $\frac{1}{2}mv^2$ for each one. If we cover the piston with some material so that the marbles stick on impact, it is possible to cause them to give up all their kinetic energy to the piston in a highly efficient conversion.

Compare this to the example in Fig. 4-9*b* which is, again, a collection of moving marbles enclosed in a box. In this case, the total energy E is the same but now the marbles are moving randomly in all directions. We can still extract some of this energy from them; we can for instance, put a hole O in the end of the box and select those few molecules which are moving in the right direction, allow them to hit the piston, and give their energy to it as before. It is obvious, however, that we can only convert a small percentage of the total energy in this way.

Given energy due to organized motion (low-entropy energy) such as that possessed by the stream of marbles in Fig. 4-9*a*, we can convert it, with only small

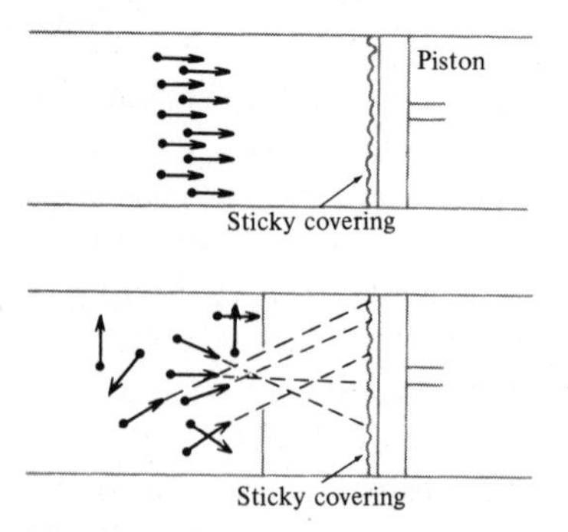

FIGURE 4-9 The "ordered" motion of the marbles in *a* can all be converted into energy at the piston while only a part of the disordered motion of *b* can.

Table 4-5 Energy Ranking Based on Entropy

Form of energy	*Entropy per unit energy*
Gravitation	0
Energy of rotation	0
Energy of orbital motion	0
Nuclear reactions	10^{-6}
Internal heat of stars	10^{-3}
Sunlight	1
Chemical reactions	1-10
Terrestrial waste heat	10-100
Cosmic microwave radiation	10^{4}

losses, to other organized forms such as the motional energy of the piston. The stream of marbles, however, is a highly improbable state; it could not have formed naturally, implying that it had to be forced. The disordered, high-entropy state, however, is a natural and probable form; the organized state will go toward this state with time as the marbles collide with each other and with the walls. The one-way nature of the conversion pathway from order to disorder is clear.

We can thus rank energy on an entropy scale from low entropy, and, therefore, highly available energy, to the high entropy and unavailable energy of random motion. Such a ranking for the more important energy forms in the universe is shown in Table 4-5. In this ranking done by Freeman Dyson[3], the highly ordered forms of gravitational potential energy and kinetic energy of massive objects stand first with zero entropy. The list below them depends on the inverse of their characteristic temperatures; nuclear reactions in which the energy releases (and temperatures) are very large, have low entropy, since this energy can be converted to other forms with high efficiency. T_{in} in the efficiency equation is very large. In terms of order, the very strong nuclear forces are not as affected by uncertainties in position due to random thermal motions as are the weaker electrical forces in chemical reactions. Thus, Table 4-5 provides a map of the one-way street of energy conversion; the spontaneous direction is from top to bottom.

Energy conversion can proceed in the opposite direction, producing a decrease in entropy. In the example of pumped storage hydroelectric power mentioned earlier we begin with the chemical energy of coal, burn it, convert the high entropy heat energy through a chain of conversions (mechanical to electrical and back to mechanical) to zero entropy gravitational potential energy. But we have paid a price, for at each step of the way, some energy became low-temperature heat energy and was lost for further conversion. The entropy of the pumped water was decreased, but if we consider the whole system (coal plus generating plant plus water) the entropy increased.

There are other interesting examples of decreased entropy; a charged storage battery, frozen ice cubes, even the process of life and growth. In each of these examples, order has been won from disorder and entropy has decreased. If the total system is considered, however, the total effect has been an increase in disorder. To charge a battery we must provide energy above and beyond that neces-

[3]Freeman Dyson, "Energy in the Universe," *Scientific American,* **224**: 50, September 1971.

sary to reform the chemical combinations in the battery plates. This energy goes to low-temperature heat energy as losses in the current-carrying wires, for instance. In freezing ice we increase the order and thus, decrease the entropy of the water in the ice cube trays by removing heat from it. This heat, however, has to flow into a substance (the expanded Freon gas) which is at a lower temperature. Thus, it raises the entropy and the disorder of this gas. Furthermore, we put electrical energy into the refrigerator through the motor and this energy is degraded to heat. In the life process, highly ordered structures are built from the much simpler structures of various chemicals, but to accomplish this, life takes in relatively low-entropy energy—sunlight and chemical energy—and gives off high-entropy heat energy. We see, therefore, that the only scarcities in the universe are of low entropy, of available energy and of order. This suggests that we begin to pay attention to thermodynamic economy, that we utilize the potential of energy and order as efficiently as possible.

We don't do badly with efficiency. The efficiency of our major conversion process—the chemical energy of fossil fuel to electrical energy—is not very much below its thermodynamic limit. There is some room for improvement as we have seen in the efficiency of the engines we use in transportation.

What about the potential of order? This is also stored on earth, for example, in any deposit of minerals in which the concentration is higher than the natural concentration. Iron ore, with 50 percent pure iron, is an example since the natural concentration of iron, as a percentage of all matter on earth, less than 1 percent. We buy order at the expense of energy when we refine this iron and make it into an automobile, of which many parts are essentially pure iron.

A practice of thermodynamic economy would cause us to look carefully at two aspects of this process. What do we do with this precious order? Do we preserve it by recycling the iron, or at least by storing all the iron in one place, for example, a junkyard? Obviously, there is room for improvement here.

We must also look at the efficiency with which we create order. Just as we derived an upper limit for the efficiency of energy conversion by looking at the "ideal" reversible process, we can derive "ideal" upper limits for the energy necessary to lower the entropy of iron ore to the pure ordered state of automobile parts. An example of such an application in the manufacture of automobiles was provided by R. S. Berry.[4] He finds that the ideal limit of energy needed to produce this order is only about 10 kw-hr while the actual expenditure of energy is 4000 kw-hr to 6000 kw-hr. In contrast to the efficiency of energy conversion which is within a factor of two of the theoretical limit, our efficiency of order creation is several hundred times less than the ideal. It appears, therefore, that great savings of energy are possible in the future if we focus research attention on metallurgy and find better means to purify metals and shape them into the objects we desire.

In the long run the entropy of the universe must increase; it is the "arrow of time" which marks the direction of past to future. What we must do, in our practice of thermodynamic economy, is to make sure that with each increase in entropy

[4] R. Stephen Berry, "Recycling, Thermodynamics, and Environmental Thrift," *Bulletin of the Atomic Scientists*, May 1972.

that our utilization of the potential of energy or order causes, is for a necessary end, and is accomplished with as high an efficiency as possible.

SUMMARY

Energy must be converted from one form to another before end use; from a potential to a kinetic form or from one kinetic form to another. At present, the most important conversion route is chemical potential (fossil fuels) to thermal energy and then to mechanical kinetic energy, the kinetic being an end use (in transportation, for example) or the intermediate step before electrical power generation.

The efficiency of an energy conversion is defined by:

$$\text{eff} = \frac{\text{energy (or work) output}}{\text{energy (or work) input}} \times 100\%$$

No efficiency either for storage or processing work can be as large as 100 percent; there are inevitable losses of input energy from the intended use. In particular, we considered storage work examples and pointed out the necessity of work being done against an opposing force if kinetic energy is to be converted to potential. Since the force provided must be larger than this opposing force, while the energy stored is the product of the opposing force times the distance, the failure to reach 100 percent efficiency becomes obvious.

We showed, in Fig. 4-1, some representative conversion efficiencies, and, in further discussion, pointed out the importance of using the system efficiency in analyzing the total energy picture; of following an energy consuming process from its beginning and looking at the efficiency of each step in order to arrive at an overall efficiency for the process.

Next, we turned to that most important class of conversion devices, the heat engine. We showed how the random molecular motion of a hot gas could be harnessed to produce work and obtained an expression for the limiting efficiency of the thermal to mechanical conversion. The important descriptive equation is:

$$\text{eff} = 1 - \frac{T_2}{T_1} \times 100\%$$

This equation gives the limit on the efficiency of heat engines operating between input temperatures T_1 and exhaust temperatures T_2.

Finally, we expanded on the limiting statement of the Second Law by introducing the concept of entropy as alternatively a measure of the degree of unavailability of energy, or a measure of progression from order to disorder. Entropy increases whenever an irreversible process occurs, and since all natural processes in closed systems are irreversible, one can restate the Second Law:

> In any conversion or transfer of energy within a closed system, the entropy of the system increases.

From a broader application of this principle, which includes the potential of order (as contained, for instance, in concentrated iron ore or refined aluminum) as well as energy, we developed, as a guideline for a thermodynamic economy, a motto of "zero entropy growth" to go with the other "zero growth" motto.

5

Energy in the Service of Man

Energy on the grand scale as described in Chap. 2 is impressive; the exploding "Cosmic Egg," the rush of matter to fill an empty universe, the blinding brilliance of burning stars, make for a dramatic story. Even after turning from the universe to our star, the sun, we are still awed by the abundance of energy: 14 million trillion Btu per day supplied to an earth which uses only 3,200 trillion Btu per day to provide for its thin layer of life. At this scale we are still describing a balanced cycle: the sun's energy falls on and through the atmosphere, fueling the great heat engines which move the wind and water. A small part, about 0.2 percent of the sun's input, goes through a conversion in photosynthesis. All of this energy, degraded in thermodynamic potential by the various energy conversions which have taken place, is eventually reradiated back out into the universe.

Man alone has departed from that cycle. He draws on the energy stored in the fossil fuels and is beginning now to tap other nonsolar sources of energy. Compared with the enormous daily delivery of solar energy, his demands are still small, but they are growing. That demand for energy, in fact, now threatens exhaustion of the fossil fuels, while the increasingly rapid conversion of their stored energy to heat has begun to make an unmistakable mark on the functioning of this planet (a story we tell in Chaps. 8 and 9).

It is now time to increase the magnification of our inspection and focus on this human species whose use of energy is so unique. In Chap. 4 we described the various energy conversions and the limits on their efficiencies. In this chapter we will look at the end uses, at the purposes for which this energy is converted to work and heat. We will also look at growth and change in the pattern of consumption and in the fuels which provide the energy.

ENERGY AND EVOLUTION

Man is evolution's most successful, as well as its newest, product. The measure of the success of a species in this roulette of life is numerical growth, and man's numbers have steadily grown. In Fig. 5-1 we show the shape of the human growth curve. The origin is, of course, lost; it stretches back perhaps as far as 3 million years to man's predecessor, whose "Skull 1470" (and this was probably a woman's skull) was found recently near Lake Rudolf in Kenya, East Africa. Man grew slowly over those hundreds of thousands of years; he numbered perhaps a few hundred thousand at the beginning of the Paleolithic Era a million years ago, and a few million by the time he was drawing pictures on the walls of the caves in France's Dordogne Valley, 25,000 years ago. As Fig. 5-1 shows, his numbers were nearing 100 million about 5,000 years ago as the great civilizations began to grow up around their birthplaces—the Nile Valley, the fertile delta of the Tigris and Euphrates, and similar fertile valleys in China and India. Then, in the last 1,000 years, we see the curve sweeping upward into the billions.

The energy consumption of man, which is also displayed in Fig. 5-1, kept pace with and fueled this numerical growth. Primitive man lived at balance with nature. He was part of the great cycle of life, eating plants and the plant eaters, claiming only his small share of the energy released by photosynthesis, and returning it, day by day, as heat and waste. His per capita consumption was the 2000-3000 Calories per day, which keeps the human body alive and active.

The discovery and use of fire, which predated the Dordogne cave artists by perhaps 70,000 years, gave a modest boost to the curve of consumption; the 50 or so pounds of wood he burned on chilly days and cooked with (at probably 1 percent efficiency) added perhaps 5000 more Calories, but did not alter the overall balance. Nor was the balance or the direct reliance on photosynthesis changed when

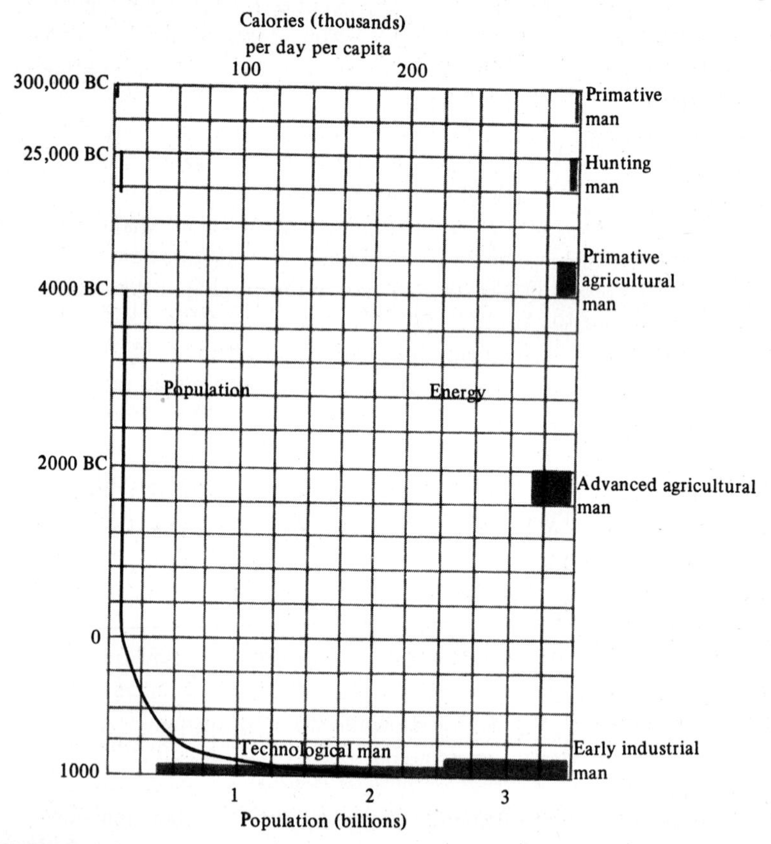

FIGURE 5-1 Growth of the human population has paralleled the growth in consumption of energy and both have spurted dramatically in this century. The bar graphs indicate the average per capita consumption of energy during the indicated period of history.

man domesticated the burro in the highlands of Iran (or some such animal somewhere else) around 5000 BC. That part of the animal's energy which he used for transportation and agriculture probably added another 4000 Calories, but this energy also was part of the sun-earth, input-output cycle discussed in Chap. 3.

Even at this low level of energy consumption (a little less than is used in an average home today in daily electrical energy alone) mankind produced impressive results. The massed energy of man and animals built Egypt's Pyramids and the awesomely scaled temples and palaces of Assyria. Sometime around 1000 BC, however, man made his next great energy breakthrough and began to harness the kinetic energy of wind, first to drive his ships and later to turn windmills. The Romans, at about the time of Christ, developed the waterwheel. The sailing ships speeded up exploration of the planet man was beginning to master; the machines that water turned began to amplify the productivity of a single man, thus leaving others free to take on the inward exploration of self.

About 1,000 years ago, the human population began to mass itself for the rocket trip to come (see Fig. 5-1). It was at about this time that man began to use, inefficiently and sparingly at first, the treasure that the sun had left locked up in the fossil fuels. Coal was burned in Europe in the thirteenth century and probably as early as the twelfth century. Even in this country, it appears that the Hopi Indians mined coal as early as AD 1000 and used it to heat their homes and to fire pottery. By the year 1700 the city of London consumed half a million tons annually. When the steam engine began to turn the wheels of the Industrial Revolution in the late nineteenth century, coal provided much of the 75,000 Calories per day per capita consumption of that period.

The final brief portion of Fig. 5-1 brings us to today. Both the population and the energy consumption have greatly increased in the last century. The energy consumption we show is that of the United States citizen of about 1970, "technological man," as we call him. He consumes, on this "affluent island," as Ralph Lapp colorfully puts it,[1] almost a quarter of a million Calories per person per day.

GROWTH AND CHANGE

Figure 5-1 shows us, in perspective, the slow growth in humankind's consumption of energy over the hundreds of thousands of years of our evolution as a species. What is of more concern to us in this book, however, is the record of the last century when as we see, the growth has been explosive (characteristic of that exponential growth discussed in Appendix 1).

Figures 5-2 and 5-3 present the energy consumption records for the world and the United States; the first figure gives total consumption and the second shows per capita energy consumption, or total energy consumption divided by the population. (Note that these are per capita per year data rather than the per capita per day data of Fig. 5.1.) These two figures introduce quantitative and qualitative features which will be typical of many of the presentations throughout the remainder of this book.

[1] Ralph Lapp, *The Logarithmic Century*, (Englewood Cliffs, New Jersey, Prentice-Hall, 1973).

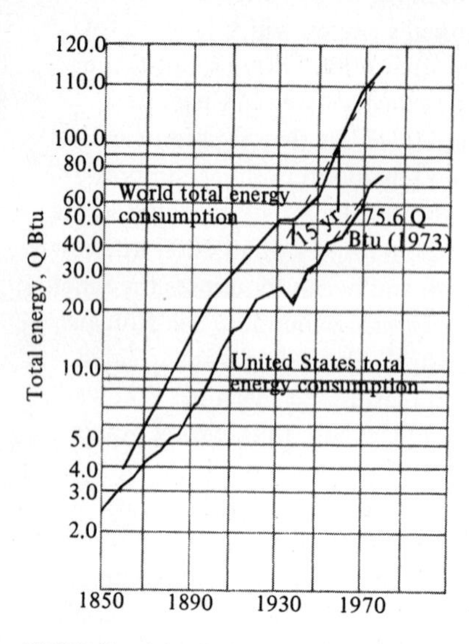

FIGURE 5-2 U.S. and world total energy consumption. At present, the U.S., with 6 percent of the world population, uses 36 percent of the total energy. World consumption is growing somewhat more rapidly than U.S. (Courtesy of Bureau of Mines.)

There are two quantitative features; the enormous size of the numbers (billions and trillions and quadrillions), and the switch from the Calorie, which we have used so far, to a new unit, the Btu. We used the Calorie in our introductory sections because it is a familiar unit, the same one dieters use to measure the energy content of food. From the comparisons shown in Fig. 5-1, however, we are now aware that food energy is a small part of the total, a few thousand Calories out of 250,000 Calories per day. Since our interest from now on will be focused on nonfood energy production and consumption, we will turn to the engineering unit of energy, the Btu, in which the energy of coal, oil, and other fuels is measured. It is rather easy to make this conversion; we have provided a list of the energy conversion factors in Table A2-3. From this we see that 1 Btu = 4 Calories. The United States, as we see in Fig. 5-2, consumed 68.8×10^{15} Btu of energy in 1970. That is a huge number. As we discuss in Appendix 1 we will abbreviate 10^{15} as Q (for quadrillion). We will also abbreviate one trillion (10^{12}) as T, one billion (10^{9}) as B and one million (10^{6}) as M. In this terminology, the total United States consumption in 1973 was 75.6 Q Btu and the per capita consumption that same year was 358 M Btu (or 0.358 T Btu).

In addition to these quantitative features, there is an important qualitative feature indicated in Figs. 5-2 and 5-3 which will be repeated often: the vertical scale is not a linear scale, rising equal amounts over equal distances, but a scale for which equal distances represent tenfold increases. This *semilogarithmic* scale is described in detail in Appendix 1. There are important reasons for using this scale; it allows the presentation of data which cover a large numerical range; the 1970 figure for total world energy consumption, 190 Q Btu, is almost 100 times larger than the starting point of the curve in 1850. It would have been difficult to

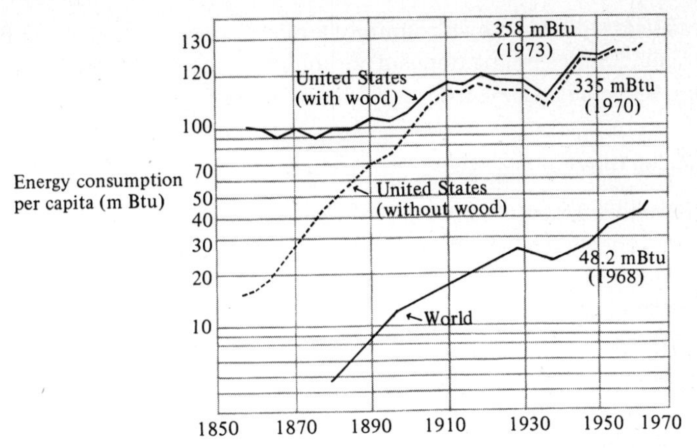

FIGURE 5-3 U.S. per capita energy consumption of energy during the period since 1850 has been greater than the world average and is presently almost six times greater. (Courtesy of Bureau of Mines.)

show that kind of growth in any detail on a linear plot.

The other important feature of the *semilog plot,* as this type of presentation is called, is that exponential growth curves (such as the population curve shown in Fig. 5-1) become a straight line in a semilog plot. Thus, in Fig. 5-2, the straight portions in the period 1860-1900 of the world total consumption curve represents pure exponential growth.

It is also useful to point out here one other important piece of information that can be easily read from these semilog plots. To the extent that they are approximately exponential curves (straight lines), they exhibit a constant *doubling time.* The doubling time is, as the name implies, the amount of time which it takes for the growing quantity to double. We can show this with the total world energy data shown in Fig. 5-2. We have approximated the growth from 1940 to 1970 by a dotted straight line. In 1938 the consumption was about 50 Q Btu. By using the dotted line approximation we can see that twice that, 100 Q Btu, was reached in 1953. Thus, the doubling time during this period of growth was 15 years. The reader can use the dotted approximation to the United States growth curve over the same period to verify a 19-year doubling time for total United States energy consumption.

The doubling of energy consumption in such a short time is staggering growth. That the world total consumption doubled in 15 years means that, in the past 15 years, as much energy has been consumed as was consumed in man's entire previous history; and it will double again the next 15 years.

The final comparison which these curves afford is of the great gap that exists between the per capita consumption on this affluent island and that in the rest of the world. The United States, with 6 percent of the world's population, uses about 36 percent of its energy. The per capita consumption in the United States is more than six times the world average (see Fig. 5-3). We also see that the world is catching up (its curve rises more steeply). We will look in some detail, in the last part of this chapter, at both United States and world patterns of use and at some of the imbalances within our own country. Our next concern is to look in finer detail at the United States production and consumption of energy, in order to determine where that energy came from and where it went.

Changing Input

We have thus far emphasized growth, but we must also look at change. In our study of energy we find these two interlocked; a change in fuel (the successful drilling for oil, for instance) leads to growth; then growth (the insatiable United States demand for oil, for instance) creates the pressure for more change and the pattern repeats itself.

There has been change at both the input and the output end of the energy flow diagram: in the relative importances of the various sources of energy available to man, and in the pattern of the consumption to which that energy flows, consider the input end first.

We will, as promised earlier, concentrate on the nonfood energy. The neglect of this tasteful form of energy is justified by the numbers. Of the nearly one quarter of a million Calories per capita per day which we ascribe to technological man, only 10,000 Calories come in the form of food, perhaps 2,200 Calories as plant matter directly eaten, 6,300 Calories as plant matter fed to meat animals which produce the 900 Calories we get from meat, and another 1,500 Calories wasted in one way or another. Where do the other 240,000 Calories come from? We will look at these fuel sources in the historical order of their importance, beginning with wood.

Wood. As this young country passed the middle of the nineteenth century, its major source of energy was the same source that had served stone age man—wood. Even though in Great Britain coal was already being used for most of its energy, wood, in unlimited supply from the forests of this new republic, accounted for 90 percent of its energy needs. It was used lavishly in home heating; it is estimated that 17.5 cords of wood per year were used in the years 1850 to 1860 heat a typical American home. (A cord is a stack of wood 4 feet by 4 feet by 8 feet.) The heat energy equivalence of this impressive woodpile is 2.5 times the amount of heat energy used to warm a typical home today; improvements in efficiency and in insulation account for the reduction. The fireplace was less that 10 percent efficient. Stoves, which are about four times as efficient, were not popular because of the necessity to chop the wood into small pieces and because a stove did not give off the cheery light that blessed a fire. In total, in 1850 it is estimated that 100 million cords of wood were burned, three quarters of it (75 million cords) in open fireplaces.

But wood was not just fireplace fodder; it also fed the growing industrial fires of this country. By 1850 river steamships and railroads had begun to be major factors in transportation; they burned 7 or 8 million cords of wood a year in the 1860s. The iron and steel industry also used wood; half the energy for iron smelting came from charcoal during this period, accounting for 1.5 million cords of wood per year. The charcoal used in iron smelting totaled some 700,000 to 750,000 tons. It is interesting to note that although charcoal is no longer used in

Table 5-1 Fuel Wood Consumption in 1879*

Use	*Cords (millions)*		*T Btu*		*Percent*	
Domestic		140.5		2,810		95.5
Industrial		6.7†		134		4.5
Railroads	1.97		39.5		1.3	
Steamboats	0.79		15.8		0.5	
Mineral operations	0.63		12.6		0.4	
Manufactures	1.86		37.2		1.3	
Charcoal	1.50		30.0		1.0	
Total		147.2		2,944		100

*Source: Schurr and Netschert, *op. cit.*, p. 53, Table 9.

†Includes wood used to manufacture charcoal for iron smelting.

industry, an estimated 650,000 tons were used by outdoor chefs in their barbeque grills in 1970.

The amounts and percentages of wood consumption before the turn of the century are provided in Table 5-1. As much as half of this wood came from virgin land being cleared for homesteading, about 5 million acres per year in the 1850s. The rest was hauled from the forests. The record of this fuel in the period 1850 to 1950 is shown in Fig. 5-4. The use of wood peaked around 1870 and fell off rapidly during the twentieth century. It is interesting to note, however, that even in the 1950s wood provided about 10 percent of the total energy to the residential sector and 70 percent of the rural heat.[2]

Coal. As we have remarked in Chap. 2, this country of immigrants had to relearn the use of coal. For that reason, the coal industry grew slowly during the early nineteenth century. By 1850, however, the demonstrated superiority of coal (on an en-

[2]Schurr and Netschert, *Energy in the American Economy 1850-1975*, (Baltimore: Johns Hopkins Press, 1960), pp. 56-57.

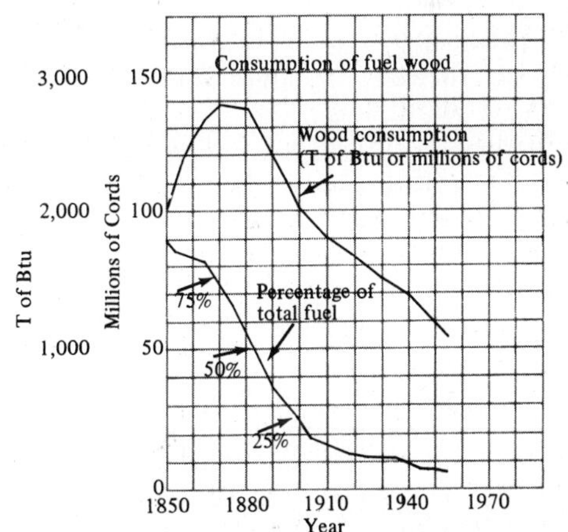

FIGURE 5-4 The consumption of wood as fuel in the U.S. The top curve shows consumption in millions of cords; the bottom curve shows wood use as a percentage of total energy. Although wood accounted for only 3 percent of the total energy in 1950, the total quantity used was almost half that used in 1850. (Source: Schurr and Netschert, *op. cit.*, p. 491, Table 1.)

ergy per volume basis) and the growing scarcity of wood (in the East, in particular) caused the beginning of the first of the great replacement cycles which have characterized the fuel mix during this past century. The growth and decline of the various fuels as percentages of the total fuel mix is shown in Fig. 5-5.

Anthracite, coal of almost 100 percent carbon content, was the most popular at first, as it is cleaner and less smokey. By 1870, however, the more plentiful and cheaper bituminous coal, a "younger" fossil, of somewhat lower carbon content but comparable heat value, had taken over. It is interesting to note that until 1850 more than half of this came from the great "Pittsburgh Seam," a long, deep seam that lay exposed along the rivers above that city. It was the access to this source that established the industrial importance of Pittsburgh.

The time scale of the replacement of wood by coal and of the replacement cycles that followed this one are shown in Fig. 5-5. The rapidity of the first turnover was dramatic; wood had fallen from 90 percent of the energy input in 1850 to 20 percent by the end of the century, while coal, over the same 50-year period, grew from 10 percent to 70 percent. One of the reasons for the swiftness of the replacement was geographic; there were rich seams of coal in the East, in Pennsylvania and the Appalachian Mountains region. In addition, it was in the East where high population density and industrial concentration created an energy demand that the vanishing forests could not satisfy.

Changes in industry and transportation also fed the demand for coal. The Bessemer process of steelmaking, while it could use charcoal, was more efficiently fed by coal-derived coke. The practice of moving the iron forge after the retreating edge of the forest was finally abandoned; and the great steel plants that were to be the backbone of the steel industry in the United States were situated at such spots as Pittsburgh, where navigable rivers flowed through coal country. The steam locomotive was also demanding more and more coal as the steel rails, now produced more cheaply by the coke-consuming open-hearth furnaces, reached out to the Western frontiers.

The story of coal can be read from Fig. 5-6. Production continued to grow through the early twentieth century and, as the coal beds of the East Coast began

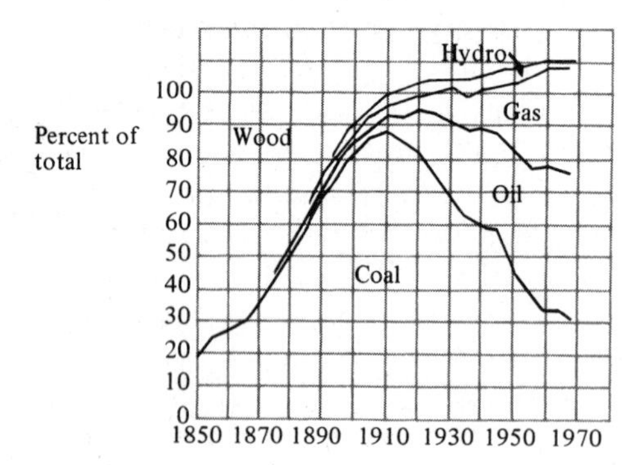

FIGURE 5-5 There have been two major replacement cycles in the past century, wood by coal (1850-1910) and coal by petroleum products (1910-1970). Shortages of the latter may force a third replacement cycle of fossil fuels by nuclear energy and a return to importance of coal.

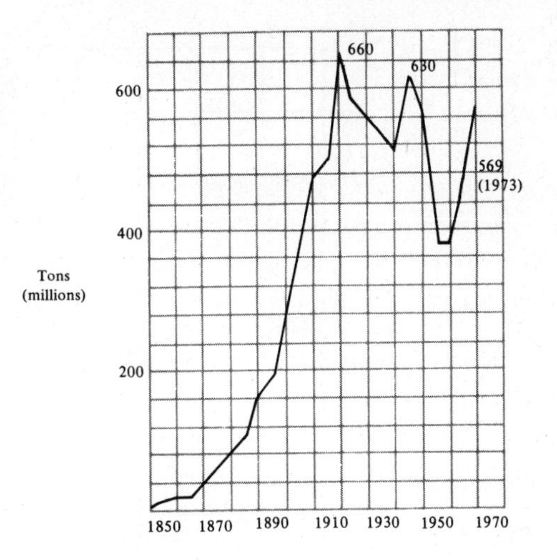

FIGURE 5-6 U.S. coal consumption. The decline in coal production after the 1920 peak was halted by renewed demand during World War II. There is evidence of a second revival due to present demand for coal by electric utilities. (Courtesy of Bureau of Mines.)

to produce more grudgingly, they were replaced by those in Ohio, Illinois, West Virginia, and other Midwestern states. Our coal wealth was such that not only did it energize our own industry, heat our homes, and power our trains, but it enabled us to become a major exporter as well. During the early peak in coal production, 1918, when 678 million tons were taken from the ground, we exported 27 million tons overseas; during the next peak, in wartime 1945 when 688 million tons of coal were mined, exports amounted to 94 million tons.

Who used this coal? The tonnage burned by the important consumers is compared in Fig. 5-7 where we can see the patterns of change that have dominated the coal industry. The decrease in coal used by railroads was brought about by external competition with gasoline-driven cars and trucks which reduced railroad passenger miles by 10 percent in the period 1925-1950 (after it had more than doubled in the preceeding 25 years) and by internal competition which saw the coal-burning steam engine replaced by oil-burning diesel engines at the end of World War II.

Fortunately for the mine operators, the country had already become hooked on electricity by then, and the rapidly growing demand for coal by the electric utilities replaced much of the lost demand from the railroads. The continued growth in electric energy consumption will, it appears, cause coal production to continue to rise toward a new peak even though another large consumer, the coke-burning steel companies (see Fig. 5-7) are converting to electric arc furnaces.

The importance of electrical utilities' use of coal is expected to continue to grow even into the nuclear era. Figure 5-8 shows the growth in electric utility consumption of the fossil fuels projected to 1990. The dominance of coal is apparent. In 1960 utility burning of coal accounted for 47 percent of the total domestic use; in 1970 its share had risen to 62 percent. The oil embargo of 1973 and 1974 caused several oil-burning plants to convert or reconvert to coal and further increased its dominance.

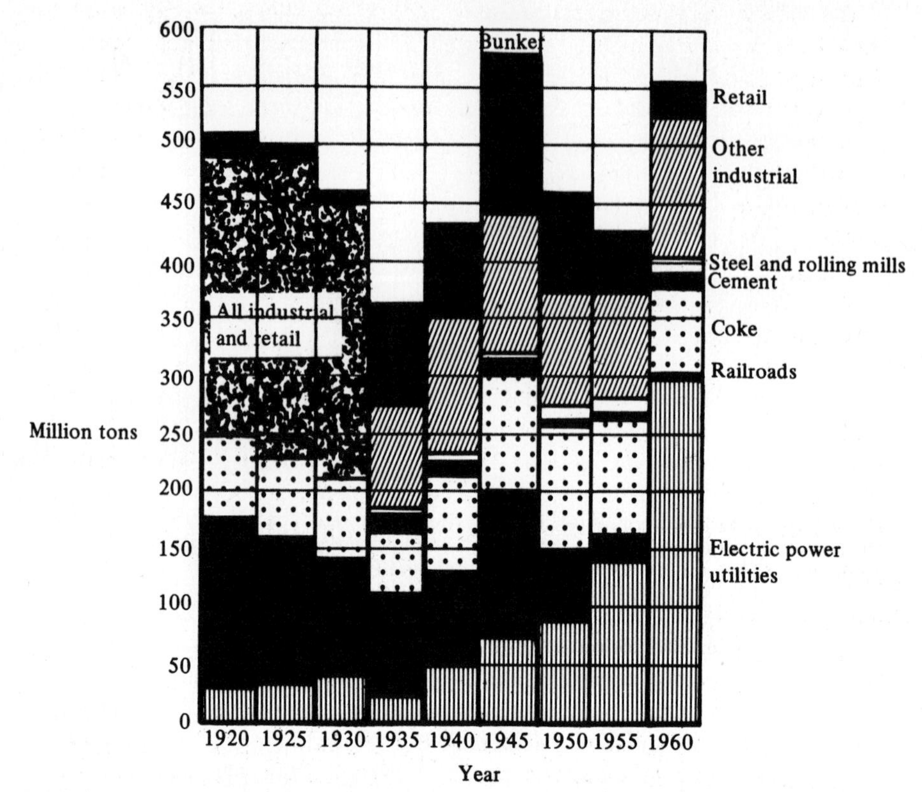

FIGURE 5-7 Coal consumption by the consumer class. Coal demand has survived the decline in demand for retail coal and the switchover from railroads to diesel. The most rapidly growing category is now electric power utilities. (Courtesy of Bureau of Mines.)

The emergence of this firm market has had an important influence on coal production techniques. It has made it economically feasible to go to large-scale strip mining and to mechanization of underground mining. It has also put a preminum on large mines, those with annual outputs of 500,000 tons or more, since the economies of mechanization and of transportation (special-purpose trains, for example) are only attained if a long period of high productivity for recovery of investment is anticipated.

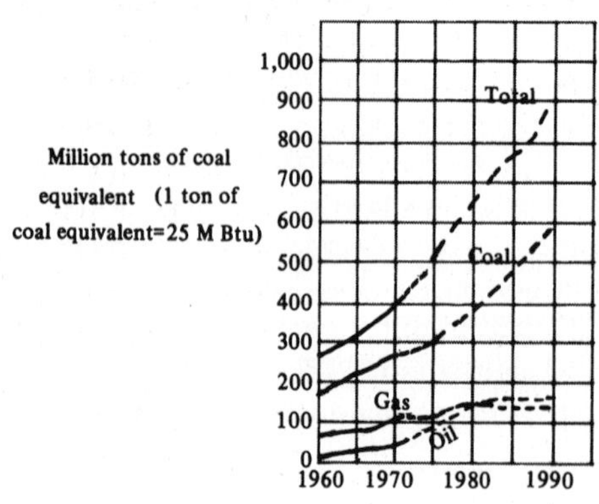

FIGURE 5-8 Project fossil fuel requirements for electric utility generation in equivalent tons of coal. (Courtesy of Federal Power Commission.)

The industry has also turned to *mine mouth* and *midpoint* generating stations. These latter are located midway between the mines and the load centers. Such plants need abundant coal and accessible water. The potential sites are limited; they exist only along the Ohio River where it cuts through the Appalachian coal basin, the Ohio and Mississippi Rivers where they traverse the southern part of the Illinois basin, and along the Arkansas River's cut through the Western interior basin. Potential sites are scarce in the West but two midpoint sites, the Mohave and the Navajo, which already have 3,760 mw generating capacity, are part of the controversial "Four Corners" development which we shall look at more closely in Chap. 9. We will also describe there the plans for developing the enormous coal reserves of the north central plains.

There is more to be said about coal: where it is and how much remains. Later on in the text we will deal with those questions as well as with the environmental effects of its mining and burning, and with the promise of new ways of using it, as liquid and gaseous fuel, for instance. For the time being, however, we will leave coal and look at the petroleum products where expanding use after the turn of the century led to the second great replacement cylce of Fig. 5-5.

Oil. Petroleum products are fossil fuels produced by decomposition of organic materials, but they are much younger fossils than coal. Like coal, they are also found in sedimentary basins (more frequently in marine basins than freshwater basins). The exact mechanism of their formation is still a subject of research; it is complicated by their migration through the rock. Their probable fossil youth, however, is substantiated by the discovery of droplets of oil in association with more or less unaltered organic material in some rocks and by, for instance, the known immediate production of methane (the major component of natural gas) from decaying organic material.

Petroleum compounds, when first formed, are large complicated hydrocarbon molecules much like those in the living cells from which they are formed. As they are buried and subjected to high temperatures and pressures, these molecules are *cracked* into lighter and more mobile ones ranging up through the light fuel oils to kerosine, gasoline, and the natural gases. The refining of crude oil reproduces this cracking process in a controlled way in order to obtain such preferred fuels as gasoline.

Oil and gas migrate, as we said, toward the surface. Some of it arrives there and escapes, creating the oil pools and gas flares which produced the first evidence of this fuel. Much of it is trapped by some kind of domed, impermeable formation and thus forms the pools sought by the exploration geologists. Finding these pools, however, is still almost a matter of luck; there seems to be no way of explaining, for instance, why rocks in Iran produce abundant oil while rocks of the same type and age in Australia are barren.

Like coal, petroleum's fuel potential has been recognized for a long time. Before 1000 BC the Chinese were drilling for gas and oil and using it for heating and lighting. Pitch from surface asphalt deposits was used in Babylonia and Egypt by 3000 BC, and the Babylonian Empire (2500-540 BC) created a minor fossil-fuel age using oil, asphalt, and some coal. As now, oil was used a great deal in those days by the military; the "Greek Fire," which was a major weapon of the Greeks, the Romans, and even Hannibal, was made from asphalt, melted and set aflame.

By 1000 AD the Burmese were drilling commercial oil wells; an oil well dug at Modena, Italy in 1640 flowed for 200 years and produced kerosene for street lighting in Genoa and Parma. Interestingly enough, some of the early oil was produced by distillation of oil shales, a process which is now being looked to as a resource of the future.

The first important United States well was drilled in August 1859 at Titusville, Pennsylvania, using, it is said, techniques not much better than those used by the Chinese in 1000 BC. Oil was struck at the Titusville well at a shallow 69.5 feet, and the United States oil age began. In 1860 half a million barrels were produced, and by 1870 production had risen to 4.2 million barrels and was already accounting for 1 percent of the total energy.

The explosive expansion of the oil industry was triggered, perhaps, by the success of that first drilling; but there was much more to it than this. Changing times demanded oil. In this country, and even more determinedly in Europe, mankind was struggling to augment sunlight and lengthen the day. Candles of tallow, and candles and lamps of spermaceti and sperm oil taken from whales were popular for indoor lighting. Coal gas (gas liberated by heating coal to make coke) was used both indoors and outdoors. It was immediately recognized that oil could be used for this purpose; in fact, kerosene from surface oil pools was already being used. By a simple and inexpensive refining process, 70 percent or so of crude oil could be converted to an illuminating oil such as kerosene. A year after the drilling of the first well, there were 15 refinery plants in that area and, by 1863, Pittsburgh itself had 60 such plants. It did not take long for the necessary facilities for storage and shipment to be built and for an export market to develop. By 1900 the United States exported one third of its domestic production. The record of production and consumption of oil in this country is shown in Fig. 5-9.

The interaction of growth and change is particularly obvious in the growth of oil consumption. The pressure of the enormous supply of this cheap and easily obtainable fuel led to the development of a variety of uses: first, kerosene for lamps, then, the heavier oils were burned for space heating and, of course, the important lubricating oils (not an energy use but then, as now, an important end use). The development, first in Germany in 1876, of the internal combustion engine changed the priorities of refineries. Gasoline became the desired fuel and the availability of this cheap and plentiful fuel contributed to the rapid acceptance of the automobile.

The changing priorities are reflected in refinery statistics. In 1900 almost 60 percent of the crude oil was refined to kerosine with 13 percent and 14 percent going to gasoline and fuel oil, respectively, and 9 percent into lubricating oils. By 1930 gasoline and fuel oil each took about 40 percent of the crude oil input, and kerosine and lubricating oils accounted for about 4 percent apiece. The present distribution is gasoline 53 percent, fuel oil 31 percent and kerosene 0.5 percent. The fact that gasoline and fuel oil (in spite of the resurgence of kerosene as jet fuel) still take up almost 90 percent of the market creates an inflexibility that we have become aware of in the 1970s. If fuel oil is short in the winter, then refineries produce more of it; but the increase can come only at the expense of gasoline and the shortage is shifted to the automobile driver.

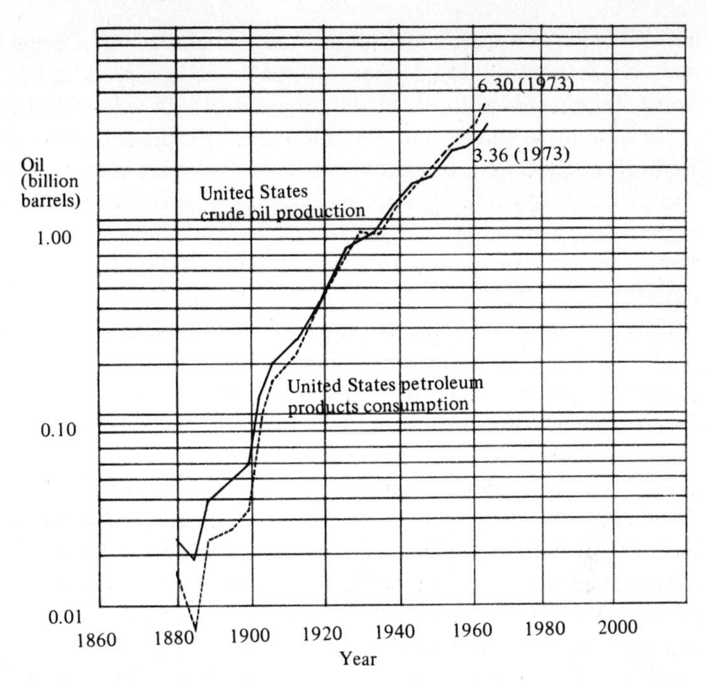

FIGURE 5-9 Production and consumption of oil. The U.S., which exported two thirds of its production in 1885, now imports more than a third of it. (Courtesy of Bureau of Mines.)

The switch from coal to petroleum products, the second of the fuel cycles, saw coal go from 76 percent of the total in 1910 to less than 30 percent by 1955, while petroleum increased its share over the same period from 10 percent to 65 percent. The rapid acceptance of these easily handled liquid or gaseous fuels has caused production to double about every eight years. But, as we see in Fig. 5-9 United States production of oil has not kept pace with our demand and we have, to our dismay, become more and more dependent on imports. We will examine this fact and its effects in detail in Chap. 11.

Natural Gas

Natural gas is the new popular fuel, booming from about 5 percent of the market in 1925 to 30 percent by 1970. It is, of course, not really a new fuel; it is a normal component of petroleum and has come through the same formation process. The Chinese, the Egyptians, and other ancient civilizations have made sporadic and minor use of it. Natural gas consists primarily of methane, a very simple hydrocarbon, and smaller amounts of other similar, but heavier and more complex, hydrocarbons. Some of these liquify at atmospheric pressure and these liquid materials, known as *natural gas liquids* (NGL), are becoming important fuels themselves. One of them is a natural gasoline which is usually added to refined gasoline; the fuels sold as bottled gas, such as propane and butane, are also NGLs.

The early history of natural gas is one of conspicuous waste. Although it had some application from the beginning as an industrial fuel in the Midwestern oil fields, the difficulty of storage and shipment resulted in much of it being burned at the well. It is estimated that 90 percent of the natural gas produced by the rich Oklahoma, Texas, and California fields in the early 1900s was burned or otherwise lost.

In addition to storage and transportation difficulties, the fluctuating demand for heat, which is natural gas' conventional conversion route, adds complication. It was not until a pipeline network connected the rich Southwestern and Western fields with the populations of the Midwest and East, and large underground storage systems (some of them in natural caverns) were built, that this fuel began to compete with the others.

The biggest boost came from two federally financed pipelines: the "Big Inch," a 24-inch diameter pipeline 1,250 miles long and the "Little Big Inch," 20 inches in diameter, 1,475 miles long which were built during World War II to connect the Southwestern oil fields to the Eastern coast. In the 20-year period from 1935 to 1955, total pipeline mileage increased from 167,400 miles to 448,770 miles, about a third of which are long-distance transmission lines. The estimated extent of pipelines in 1971 was 934,700 miles, 256,000 of which were transmission lines and 611,300 distribution lines. These distribution pipelines are calling themselves to our attention more and more through leaks and explosions, a new environmental hazard.

Natural gas has become a premium fuel; it provides (at least it did until the removal of federal price controls) cheap Btu; it is clean and requires a very simple furnace. It has become very popular in both residential and industrial heating and has made inroads into the electric utility market. Figure 5-10 shows the record of consumption since 1885.

In the residential market, natural gas has not only made inroads into the markets of coal and oil, but is challenging electricity. It has the advantage of efficiency, needing only a single conversion of 60 to 70 percent efficiency, rather than the several conversions that reduce the efficiency of electrical heating below 30 percent.

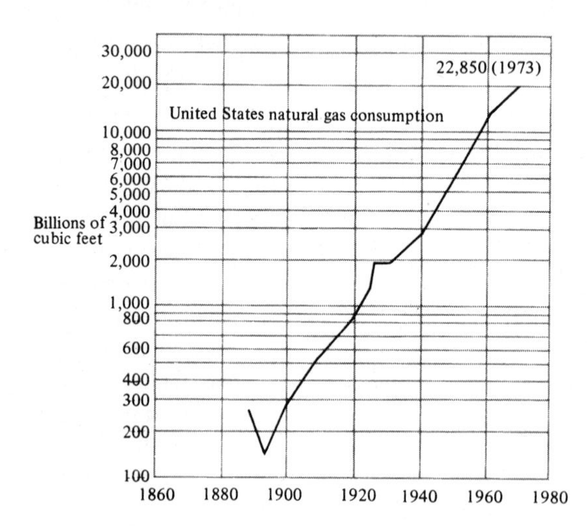

FIGURE 5-10 The consumption of natural gas has grown rapidly since 1940, doubling about every nine years. Shortages may be slowing this rapid increase in consumption. (Courtesy of Bureau of Mines.)

Gas also has an advantage in the economy of shipment which we will look at in the next chapter. We will also examine there the appliance-by-appliance competition between these two premium forms of energy. Later, we will look at the fuel-cell, which powered by gas, meets electricity head-on on its own home ground.

The future of natural gas is, however, uncertain; domestic supplies are limited and natural gas shortages in 1971 were the first indicators of the "energy crisis." We are now beginning to look toward foreign supplies, for instance, shipment of gas as liquids in refrigerated tankers, but the cost of such gas may open it to successful challenge by coal-derived gas or by a new fuel like hydrogen. It does not seem likely, however, that we will give up the convenience of gaseous fuels.

Other Energy Sources

In the middle of the nineteenth century, when Europe had already begun to rely heavily on the nonrenewable energy of the fossil fuels, the young United States industry was built almost entirely on renewable energy supplies: wood, wind, and water. In 1850 the sailing ship was still of major importance in ocean travel, and waterwheels, most of them along New England streams, were the most important "prime movers" for industry. In Table 5-2 we show the record of the 20-year period during which the coal-fired steamship and the coal- or wood-fired steam engine replaced these indirect, but low-powered, "solar engines."

To make the record complete, in fact, we should have added another solar engine, the working animal—chiefly the horse—to Table 5-2. Its contribution was important: its total horsepower was greater than the others in 1850 and 1860, and still equal in 1870. By 1900, however, farm animals accounted for only 23 percent of the total work performed, and by 1920 only 5 percent. This declining percentage did not reflect a decline in the amount of work from animals, which stayed roughly constant from 1880 to 1920, but rather the soaring growth in steam and later electric power sources.

It is tempting to bemoan the conversion we see reflected in the statistics from the late 1800s; from a nation which still lived in balance with nature, obtaining most of our Btu from wood and our horsepower from water, wind, and horse. It is good to look again, however, at Fig. 5-3; in 1850 the per capita energy consumption was 101.3 M Btu per year and wood, used very inefficiently and only for space

Table 5-2 Estimated Mechanical Work Output, 1850-1870*

	1850		*1860*		*1870*	
Energy source	*Billion hp-hr*	*Percent*	*Billion hp-hr*	*Percent*	*Billion hp-hr*	*Percent*
Wind	1.4	64	2.1	58	1.1	33
Water	0.9		1.3		1.7	
Coal	0.7	36	1.8	42	4.9	67
Wood	0.6		0.7		0.8	
Total	3.6	100	5.9	100	8.5	100

*Source: Schurr and Netschert, *op. cit.*, p. 511.

heating, contributed 91.9 M Btu in that total. In 1968, on the other hand, the per-capita energy consumption in this country was 330 MBtu per year and only 56 million of those Btu were used in the much more extensive space heating that now exists. As environmentally satisfying as that balance appears, we could not have built to the present standard of living relying on horses (and using a quarter of our agricultural acreage to feed them) and drawing only on the solar sources–wind, water, and wood.

PATTERNS OF CONSUMPTION

In the last decades of the nineteenth century industry and transportation together used only about 30 percent of the nation's total energy; the great bulk of the energy–most of it coming from wood–went into home heating. In 1968 industry and transportation used two thirds of the vastly larger total energy supply, and residential and commercial space heating altogether only used 18 percent. These comparisons signal the fact that, in the pattern of energy consumption, there has been change as radical and important as we have just seen in our study of energy sources. The nature of these changes and the portents we see in them for the future are the subject of this section. We will be particularly interested in looking carefully at the manner of energy consumption in the major sectors to see if we can find some uses to be considered as targets for the saving of energy, the topic of Chap. 14.

The sharing of total energy among the four major categories (residential, commercial, transportation, and industrial) in 1968[3] is indicated by the pie diagram shown in Fig. 5-11. Also displayed in that figure are the major end uses within the categories. As we see in that display, heat energy is overall the most important end use, a fortunate result since 96 percent of our energy comes from the burning of fuels.

[3]Report to the Office of Science and Technology, "Patterns of Energy Consumption in the United States," (Washington, D.C.: Stanford Research Institute, USGPO, January 1972). This represents latest complete data.

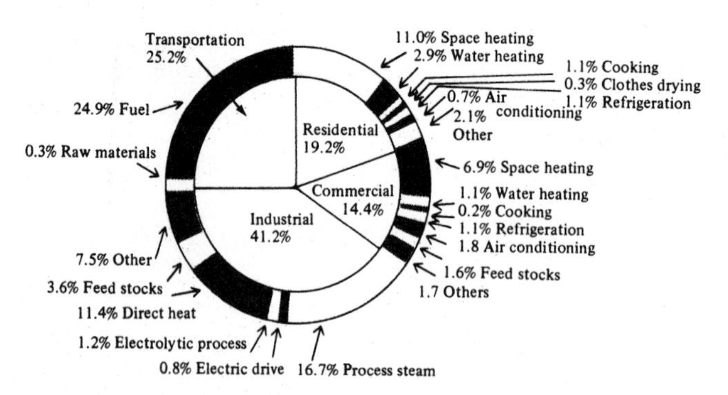

FIGURE 5-11 The industrial sector consumes almost half the total energy while propulsion energy (gasoline, etc.) for transportation alone amounts to a quarter of the total U.S. consumption. Heat energy is the most important end use. (Source: *Patterns of Energy Consumption, op. cit.*)

Table 5-3 Major End Uses of Energy Consumption*

End use	*Percent of total*
Transportation (fuel; excludes lubes and greases)	24.9
Space heating (residential, commercial)	17.9
Process steam (industrial)	16.7
Direct heat (industrial)	11.5
Electric drive (industrial)	7.9
Feedstocks, raw materials (commercial, industrial, transportation)	5.5
Water heating (residential, commercial)	4.0
Air conditioning (residential, commercial)	2.5
Refrigeration (residential, commercial)	2.2
Lighting (residential, commercial)	1.5
Cooking (residential, commercial)	1.3
Electrolytic processes (industrial)	1.2
Total	97.1

*Source: Report to the Office of Science and Technology, "Patterns of Energy Consumption in the United States," (Washington, D.C.: Stanford Research Institute, USGPO, January 1972), p. 7.

Another way of looking at energy consumption is provided by Table 5-3, which gives the percentages of the top 12 end uses. Of the 100 or so ways in which energy is used, these 12 accounted for 97 percent of the total in 1968 and the top 4 alone accounted for 71 percent.

We also see that heat energy accounts for 51.4 percent (adding in the small percentages for water heating and cooking). This half of our energy is spent efficiently: 60 to 70 percent of energy in this form is used in the task for which it is intended. The quarter of the total energy used in transportation is, in contrast, used very inefficiently, at 5 to 10 percent, as we have seen in Chap. 4.

We summarize the gross input-output statistics by the energy flow diagram shown in Fig. 5-12. These data show an overall efficiency of United States energy conversion of about 36 percent. We wasted, largely in the generation of electricity and in transportation, 40.4 Q Btu out of the total of 63.2 Q Btu consumed that year (1971). The total Btu equivalent of energy-producing material consumed was 67.8 Q Btu when we add in the coal and petroleum used as raw material in the chemical industry and as lubricants.

Taken together, Figs. 5-11 and 5-12 and the data of Table 5-3 present a solid outline of the gross pattern of energy consumption. Let us now look at the details.

Residential Consumption

In 1968, the 19.2 percent of the total energy consumed in the residential sector amounted to 11.6 Q Btu, the equivalent of 600 million cords of wood (140 million cords total were burned in 1880), or 444 million tons of coal, or 2 billion barrels of oil. This enormous amount of energy was spent among the end uses shown in Table 5-4, which compares the percentage share of these end uses in 1960 and 1968. In terms of actual fuels used, Table 5-5 provides the physical

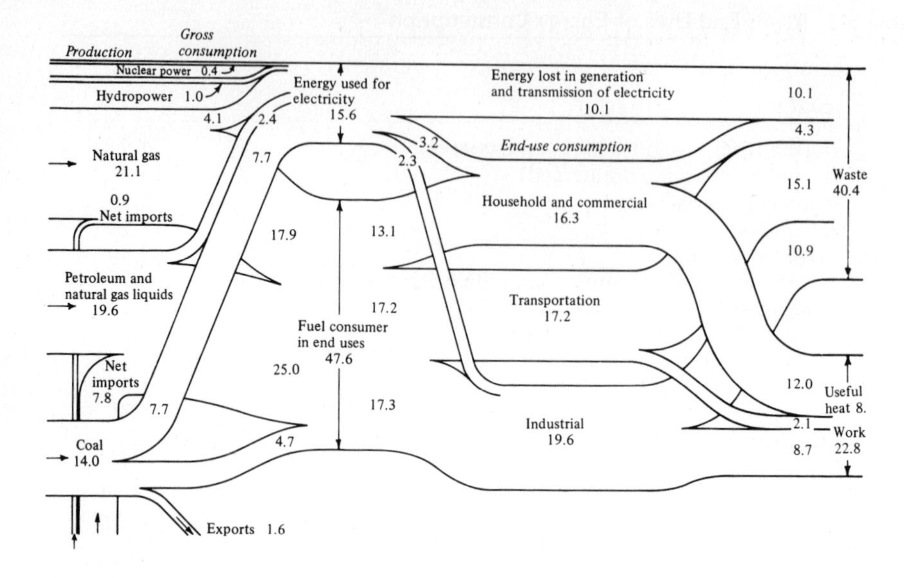

FIGURE 5-12 The flow of energy in the U.S. The overall efficiency of the conversion of potential energy to useful work in the U.S. is estimated at 30 percent. Consideration of losses in mining, refining, and transporting this energy would probably lower the efficiency even. (Courtesy Earl Cook, Texas A & M, 1973.)

amounts (tons, barrels, kw-hr) and the percentages of the major fuels consumed in this sector in 1968. We see that natural gas dominates the picture, followed by fuel oil and electricity. Note that to produce the 1.39 T Btu of electrical energy, it was necessary to waste, as lost heat energy, 2.43 T Btu.

Returning to Table 5-4, let us look at the major end uses. Space heating is the largest; it is, in fact, the only one in which energy in a form other than electrical plays an important role. Fuel oil, natural gas and electricity are all used in home heating with the latter two gaining control of the market. Fuel oil provided 53

Table 5-4 Comparison of End Uses*

End use	Percent of total 1960	1968
Space heating	60.8	57.5
Water heating	14.5	14.9
Cooking	7.0	5.5
Refrigeration	4.6	6.0
Air conditioning	1.7	3.7
Television	2.0	3.0
Clothes drying	1.2	1.7
Food freezing	1.0	1.9
Other	7.2	5.8
Total	100.0	100.0

*Source: Report to the Office of Science and Technology, "Patterns of Energy Consumption in the United States," (Washington, D.C.: Stanford Research Institute, USGPO, January 1972), p. 33.

Table 5-5 Fuel Use In Residential Sector (1968 data)*

Fuel	*Physical amount*	*Btu*	*Percent of total Btu*
Dry natural gas	4.5 billion ft^3	4,606	39.7
Fuel oil	3.4 million barrels	2,060	17.7
Natural gas liquids	172 million barrels	686	5.9
Kerosene	78 million barrels	446	3.8
Electricity	408 billion kw-hr	1,390	11.9
subtotal		9,188	
Electrical conversion losses		2,428	20.9
Total		11,616	100

*Source: Report to the Office of Science and Technology, "Patterns of Energy Consumption in the United States," (Washington, D.C.: Stanford Research Institute, USGPO, January 1972).

percent of the home heating energy in 1960 with natural gas providing 45 percent. Electricity accounted for less than 2 percent. By 1968, however, fuel oil provided only 45 percent, while natural gas had risen to 49 percent and electricity, the fastest growing of the three, had gained 7 percent of the market. We will look at electric home heating again in Chap. 6.

Of the other residential end uses, only in water heating, cooking, and clothes drying does any form of energy other than electrical energy play a role. We will defer, until the next chapter, the discussion of the competition between electricity and gas and will also look at growth in the numbers of appliances, as well as the rising amounts of electrical energy used by them. Thus, outside of space heating, the present story of residential energy consumption is a story of electricity which will be told later.

Overall, energy use in the residential sector grew at an average rate of 4.8 percent in the period 1960-1968, compared with the total energy growth rate of 4.3 percent. Such a growth rate predicts a doubling every 16 years. As to the efficiencies in this sector, there is a mixture. Almost two thirds of the energy used in space heating, for instance, is being used at high efficiency (50 to 60 percent) while the electrical energy which, because of the losses in generation and the inefficiency of air conditioners and the like, is used at less than 30 percent efficiency.

The Commercial Sector

The commercial sector is a "catch-all"; it includes everything which is not residential, industrial, or transportation. We will, therefore, not be able to say very much of a general nature about it. In 1968 the office buildings, small businesses, hotels and restaurants, recreational services, governmental institutions, schools and colleges, etc., which constitute this category used 8,766 T Btu, 14.5 percent of the United States total. Consumption grew at a yearly rate of 5.4 percent, the highest of any of the four categories.

The percentage shares of the various fuels is shown in Table 5-6 for 1960 and 1968; the breakdown of the end uses for those two years is in Table 5-7. Again,

Table 5-6 Fuel Use in the Commercial Sector*

Fuel	Percent of total 1960	1968
Coal	17.8	6.5
Heavy oil	13.7	14.1
Light oil	12.2	13.4
Asphalt and road oil	12.8	11.2
Natural gas	18.4	21.0
Electricity	25.1	33.8
Total	100.0	100.0

*Source: Report to the Office of Science and Technology, "Patterns of Energy Consumption in the United States," (Washington, D.C.: Stanford Research Institute, USGPO, January 1972), p. 63.

we see that control of the interior environment through space heating and air conditioning are the major items. The most rapid growth was in the "other" category, which consists for the most part of such electric energy consumers as lighting, motors, and elevators.

The fuel mix for space heating over the 1960-1968 period changed in the commercial sector also. Coal is losing the race, dropping from a third to about a tenth of the total, while both fuel oil and natural gas have gained conversions. Fuel oil use increased from 48 percent in 1960 to 58 percent in 1968 while natural gas use rose from 19 percent to 29 percent. Electric heating does not play as important a role here as in residential space heating, but its use is growing even in this sector.

As we have said, the commercial sector is so diverse that we will not attempt a detailed end-use analysis. One major component worthy of some further comment, however, is the commercial building: offices, schools, theaters, and so on. These units are huge energy consumers, of electrical energy in particular. The World Trade Center in New York City, for instance, uses as much electrical energy as the city of Schenectady. These buildings also use large amounts of energy for lighting and air conditioning. We will look at these end uses with an eye to energy conservation in Chap. 14.

Table 5-7 Comparison of End Uses*

End use	Percent of total 1960	1968
Space heating	54.2	47.7
Asphalt and road oils	12.8	11.2
Water heating	9.5	7.5
Air conditioning	10.0	12.7
Refrigeration	9.3	7.6
Cooking	1.7	1.6
Other	2.5	11.7
Total	100.0	100.0

*Source: Report to the Office of Science and Technology, "Patterns of Energy Consumption in the United States," (Washington, D.C.: Stanford Research Institute, USGPO, January 1972), p. 66.

Energy is also used in building construction, in the production of the concrete, steel, aluminum, and other materials used to build them. We can find, in present-day construction practices, evidence of energy extravagance, of the substitution of energy-intensive materials for energy-economical ones. In the construction of the Sears Roebuck building in Chicago a few years ago, for instance, 4 million pounds of aluminum were used as its outer "skin." Some 150 B Btu of energy were used to produce this aluminum (which would be 335 B Btu if the losses in electricity production, which makes up three fourths of the total energy, are taken into account). If, instead, stainless steel had been used, 5.75 million pounds of which would have accomplished the same purpose, only 75 B Btu would have been needed. We will discuss some other examples of the substitution of energy-intensive for energy-economic materials in the following section.

One final point to be made here is that the commercial sector, despite its diverse components, makes very specific energy demands. It is using more and more of the "modern" (and expensive) forms of energy such as natural gas for heating and electricity for lighting, air conditioning, motors, and so on. Therefore, the efficiency of energy conversion in this sector is similar to that in the residential sector.

The Industrial Sector

Industry's share of energy grew rapidly from the late nineteenth century until the middle of the twentieth century, remained at a more or less constant 40 percent up until 1970, and is expected to decline slightly over the next decades. The actual amount of energy used in industry has, of course, steadily increased; it was 18.3 Q Btu in 1960, 25.0 Q Btu in 1968, and 29.7 Q Btu in 1973.

In 1968 the six industries shown in Table 5-8 accounted for two thirds of the 25 Q Btu consumed. A breakdown by energy sources is shown in the same table

Table 5-8 Fuel Use in the Industrial Sector*

Industry group	*Coal*	*Natural gas*	*Petroleum products*	*Electricity*	*Total energy*
Primary metal industries	2,838	863	306	1,291	5,298
Chemicals and allied products	666	1,219	1,426	1,626	4,937
Petroleum refining and related industries	†	1,012	1,589	225	2,826
Food and kindred products	263	593	134	338	1,328
Paper and allied products	467	341	211	280	1,299
Stone, clay, glass, and concrete products	406	449	87	280	1,222
Subtotal	4,640	4,477	3,753	4,040	16,910
All other industries	976	4,781	721	1,572	8,050
Total	5,616	9,258	4,474	5,612	24,960

*Source: Report to the Office of Science and Technology, "Patterns of Energy Consumption in the United States," (Washington, D.C.: Stanford Research Institute, USGPO, January 1972), p. 83.

†Included in all other industries.

Table 5-9 Changes in Industrial Energy Sources*

Energy source	Percent of total 1960	1968
Coal	26.7	22.6
Natural gas	34.3	37.3
Petroleum products	19.3	18.0
Electricity	19.7	22.6

*Source: Report to the Office of Science and Technology, "Patterns of Energy Consumption in the United States," (Washington, D.C.: Stanford Research Institute, USGPO, January 1972).

and the changes in the relative importance of the various energy sources between 1960 and 1968 is provided by Table 5-9. The growth in the use of natural gas and electricity at the expense of the other fuels is apparent.

While the total amount of energy consumed in industry has been increasing, the efficiency with which the energy has been used to produce, for instance, a ton of aluminum, steel, or paper has also been improving. This improvement appears in the data for the individual industries and in the long-term ratio of *value added* to total energy consumed. We will take a brief look at both of these indicators. Value added is the value of the goods shipped by industry minus the sum of the costs of materials and energy used to produce them. The ratio of value added to total energy measures a kind of efficiency; perhaps one should call it an economic efficiency or energy productivity. Figure 5-13 shows that this energy productivity increased sharply between 1917 and 1954, then slowly until 1963, and now seems to have leveled out. There are, in fact, indications that since 1966 the ratio has been dropping; we will examine the causes and consequences of this in Chap. 11 when we try to make some projections of future energy needs. The evidence, however, seems to point to past improvements in the efficiencies of the various industrial processes but suggests that we are now nearing the limiting values of those efficiencies. The records of the various industries show these improvements specifically.

Primary Metals. As Table 5-8 shows, the primary metals industry and the chemical industry are the largest users in the industrial sector. Within the primary metals

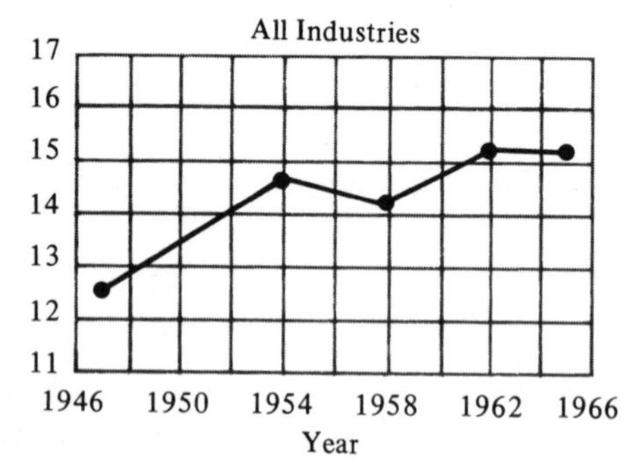

FIGURE 5-13 The ratio of value added (in dollars) to resource energy used is a measure of energy productivity. It increased fairly sharply after World War II but has remained almost constant since then. [Source: Commoner, Boksenbaum, and Corr, *Energy and Human Welfare*, Vol. III, (New York: Macmillan, 1974).]

Table 5-10 Fuel Use in Iron and Steel Industry*

Energy source	*Percent of total*	
	1960	*1968*
Coal	72.5	67.2
Fuel oil	8.5	5.6
Natural gas	12.3	17.3
Electric	6.7	10.1

*Source: Report to the Office of Science and Technology, "Patterns of Energy Consumption in the United States," (Washington, D.C.: Stanford Research Institute, USGPO, January 1972).

industry the production of iron and steel products accounted for 64 percent of the total in 1968. Energy is used by blast furnaces (most of them still use coke) to smelt iron ore and make steel, and to reheat pig iron or steel for further treatment.

The relative percentages of fuel used in the iron and steel industry in 1968 are compared with 1960 data in Table 5-10. Again, we see the picture of growth in use of electricity and natural gas at the expense of fuel oil and coal, although coal continues to provide the major fraction of the energy. The growth of electric energy consumption in the primary metals industry, as open-hearth furnaces are being replaced by electric arc furnaces, is contributing to the accelerating demand for more generating capacity. This demand is also fed by the increasing popularity of aluminum and the growing need for zinc, copper, and lead. We will examine the ways in which electric power enters into the production of these energy-intensive metals in the next chapter.

Table 5-11 summarizes the energy cost of representative metals. The energy cost of steel making, 26 M Btu per ton in 1968, provides an example of the improving efficiency we mentioned earlier; the energy needed to make a ton of steel was 30 M Btu in 1960. Table 5-11 also illustrates the potential energy savings in the recycling of metals. We will return to that point later.

The Chemical Industry. The chemical industry is a large, diverse category; some 215 of the 500 largest industrial corporations are active in some aspect of the manufacture and sale of chemicals. Within the chemical industry the use of energy varies

Table 5-11 Energy Costs of Metal Production*

Metal	*Energy cost (M Btu)*
Steel	26
recycled	3.6-5.5†
Aluminum	75‡
recylced	4.7
Lead	31
Copper	40
Zinc	46-52†

*Source: Report to the Office of Science and Technology, "Patterns of Energy Consumption in the United States," (Washington, D.C.: Stanford Research Institute, USGPO, January 1972).

†The large figure is for a large percentage of electrical energy and includes the losses.

‡If the coal is burned to make electricity, which accounts for 46.7 M Btu, is considered, this becomes 167 M Btu.

greatly. Rather than attempt to look at the energy cost of all the different processes, we will limit ourselves to plastics manufacturing. The replacement of paper and wood by various types of plastic is typical of the kind of technological change that has pushed us into environmental crises. Since plastics are not biodegradable, they contribute to solid waste problems and, since they are more energy-intensive than paper and wood, their manufacture draws down our energy supplies.

The raw materials for the manufacture of plastics are themselves fuel materials: ethylene and propylene, for instance, produced from natural gas. These raw materials account for about 75 percent of the fuel energy charged to the chemical industry sector. The four major plastic materials are: polyethylene, the oldest and still most popular form; polypropylene, used in "hot fill bottles" such as those for syrups; polyvinylchloride, a newer, clear plastic used also for bottles and other display items, and polystyrene, used in molded meat trays and hot drink cups. In Table 5-12 we summarize the energy cost of each of these.

It is clear from Table 5-12 that plastics, while they may be inexpensive, are not energy economic; polyethylene and polypropylene, for instance, are almost twice as energy-intensive as aluminum (see Table 5-11). The manufacturing of plastics also requires considerably more energy than the 36 M Btu per ton required for the production of paper. Plastics have an additional disadvantage in that, at least with present technology, they cannot be recycled. On the other hand, as the alert reader may suspect, much of the energy used to create their complicated molecular forms is stored as chemical potential energy and can be released by burning. Polyethylene and polystyrene have fuel values of 19,950 Btu per pound and 15,800 Btu per pound, almost double the 10,000 to 13,000 Btu per pound of coal. Problems with their incineration in conventional furnaces, and with the toxic nature of some of their combustion gases, have so far denied us this method of recovering their stored energy.

Petroleum. The petroleum industry is somewhat of an anomaly in that it produces a fuel, but uses energy to do it. Energy is required to refine the crude oil to the desired products, an average of 715,000 Btu per barrel of crude oil. Since crude oil has an energy value of 5.8 M Btu per barrel, 12 percent of the energy, mostly in the form of natural gas and oil burned for heat, is used in the refining process itself.

Paper and Allied Products. Paper production uses significant amounts of energy also, on the average, 23 M Btu per ton in 1967. In addition, production and en-

Table 5-12 Energy Costs of Plastics Manufacture

	*Energy**	
Type	*per ton (M Btu)*	*per pound (thousand Btu)*
Polyethylene	137.5	68.8
Polypropylene	144.5	72.2
Polyvinyl chloride	102.6	51.3
Polystyrene	45.2	22.6

*An additional 25%-30% more energy is expended in the final production of cups, trays, and the like.

ergy consumption have been increasing in this area of industry: its share of industry's total consumption was 4.1 percent in 1960 and 5.4 percent in 1968. At the same time, however, the efficiency of production was improved; it took 27 M Btu per ton to produce paper in 1950. This improvement could have been larger if the industry had not decreased the input of recycled waste paper from 28 percent in 1958, for instance, to 21 percent in 1967. Since pulp made from recycled waste uses less than one fourth the energy used by pulp made from virgin wood, considerable savings are possible if the recycling ethic can be made commercially attractive.

Stone, Clay, Glass, and Concrete. The various industries in this category use 9.4 percent of the industrial total. The cement and concrete industries use about 60 percent of the total with brickmaking (14 percent) and glass (16 percent) sharing the rest.

The amount of energy consumed per ton of finished material also covers a considerable range. Cement requires about 13.1 M Btu per ton, about half the amount required by steel. The requirements of other products in this category are summarized in Table 5-13. A return to brick buildings would, it appears, save energy.

The Food Industry. This industry's share of the total energy was 5.3 percent in 1968 and has been declining since. In this industry, however, the actual production, measured by value of products, increased in the period 1960-1968. This was accomplished in spite of a decrease in number of employees. The food industry seems to be at the stage of dramatic improvement in efficiency that many other industries passed through in 1947-1958. In that period of improvement, 1958-1966, the proportion of disposable income spent for food declined from 24 percent in 1958 to 21 percent in 1966. Unfortunately, in the 1970s, this period of lowered prices is now ancient history.

Food production is now very energy intensive. Instead of photosynthesis, the energy we now get from food comes largely from the fossil fuels. In primitive "slash and burn" agriculture the ratio of the return in food energy to human energy input can be as high as 15 or even 20 to 1.[4] By contrast, using the data for

[4] R. A. Rappaport, "The Flow of Energy in an Agricultural Society," *Scientific American*, **224:** 3: 116, 1971.

Table 5-13 Energy Requirements of Cement, Glass, and Clay Products*

Product	*Total energy (M Btu/ton)*
Cement	13.1
Glass	
Plate and bottles	14-18
Technical glassware	54
Handmade glassware	80
Porcelain	80
China	40
Tile	4
Bricks	1-6

*Source: Report to the Office of Science and Technology, "Patterns of Energy Consumption in the United States," (Washington, D.C.: Stanford Research Institute, USGPO, January 1972).

Table 5-14 The Total Energy and Employment Required to Deliver a Pound of Protein to the Consumer through Various Food Products in 1963*

Food product	*Producer's price,† 1963 (dollars/lb)*	*Total production energy to protein ratio (Kcal/lb)*	*Production energy to energy content ratio‡*
Meat products	0.50	32,600	6.3
Cheese, natural and processed	0.30	18,800	2.6
Fluid milk	0.12	51,200	6.1
Fresh or frozen, packaged fish	0.35	17,700	6.5

*Source: "Options for Energy Conservation," Bruce.

†Statistical abstract of the U.S. Department of Commerce, 1971; approximate.

‡Marketing energy not included, add about 10 to 15 percent; total caloric energy used.

the year 1963, Herendeen[5] calculated that, on the average, 1 Calorie of energy in the food put on the American table required 4.5 Calories to produce, when the energy cost of fertilizer, farm equipment, transportation, packaging, etc. is taken into account. All evidence suggests that this ratio is higher in the 1970s. Since protein in the diet is connected with energy, a comparision of protein sources is useful here. Table 5-14 compares, at 1963 levels, the price per pound, energy per pound, and their ratio for some popular United States protein sources. The energy-economic choice is obviously cheese.

Another interesting observation from this same University of Illinois study was reported by Hannon,[6] who computed the energy cost of the "fast-food stands." Taking McDonalds food chain for the example, he reports that, if the energy consumed in producing, cooking, and especially, wrapping the foods (in energy-intensive plastics) is taken into account, then this one chain used the energy equivalent of 12.7 M tons of coal, which is 33.2 T Btu, or 5 percent of the total energy used in residential cooking.

[5] Robert A. Herendeen, "An Energy Input-Output Matrix for the United States, 1963: Users Guide," Center for Advanced Computation, University of Illinois-Urbana, CAC Document No. 69, March 1973.

[6] Bruce Ingersoll, "Are Burgers Draining Our Nation's Energy," Chicago *Sun-Times,* October 30, 1972, from an interview with Bruce Hannon, Center for Advanced Computation, University of Illinois-Urbana.

Table 5-15 Industrial Energy Consumption by End Use, 1968*

End use	*Energy (T Btu)*	*Percentage of total*
Process Steam	10,132	40.6
Direct Heat	6,929	27.8
Electric Motors	4,794	19.2
Electrolytic Processes	705	2.8
Feedstock	2,202	8.8
Other	198	0.8
Total	24,960	

*Source: Report to the Office of Science and Technology, "Patterns of Energy Consumption in the United States," (Washington, D.C.: Stanford Research Institute, USGPO, January 1972).

We have looked at the share of the total industrial energy used by the various industry groups and looked in detail at specific industries. It is instructive, if one is interested in the overall efficiency of energy use, to look at the amounts of energy that go into various end uses irrespective of the industrial group. These data are displayed in Table 5-15. We see from this table that 68.4 percent of the total energy is used in producing heat for which the conversion efficiency is high; the next largest amount, 19.2 percent, is used in the electric motor. As we have pointed out, however, the increasing consumption of electricity portends a lowering of overall efficiency if the generation losses are taken into account.

Transportation

In 1968 transportation used 25.2 percent of the total energy, a slight percentage decrease from the 25.6 percent used in 1960, but an increase in total energy from 11.0 Q Btu to 15.2 Q Btu. The comparable figures are 24.8 percent and 17.6 Q Btu in 1973. In contrast to the other sectors, transportation is a one-fuel sector; only kerosene for jets and diesel oil for trucks and trains make significant showings alongside gasoline (see Table 5-16). It is also a one-unit system, for as we shall see, the automobile dominates all statistics.

The 25.2 percent share of total energy attributed to the transportation sector in Fig. 5-12 is somewhat misleading since only the energy of the fuels are taken into account. (Energy materials converted to lubricants and similar products amount to about 1 percent of the total for transportation.) A better perspective on the importance of this sector is provided by the breakdown of Table 5-17. In this table we include not only the energy used for direct propulsion but also for refining, manufacturing, and associated processes that are directly related to transportation. One sees that transportation and transportation-related activities consume closer to 40 percent of the net United States energy, that is, the total energy consumed minus that amount lost in the generation of electricity.

We can gain better perspective on the drain that transportation makes on petroleum products by translating the 15.2 Q Btu of energy in direct fuel use, into bar-

Table 5-16 Comparison of Energy Sources Used in Transportation*

	Percent of total	
Source	*1960*	*1968*
Coal	1	nil
Natural gas	3	4
Liquefied petroleum gas	1	1
Jet fuels	7	13
Gasoline	71	68
Distillate fuel	8	8
Residual fuel	8	5
Lubes and waxes	1	1
Electricity	nil	nil
Total	100	100

*Source: Report to the Office of Science and Technology, "Patterns of Energy Consumption in the United States," (Washington, D.C.: Stanford Research Institute, USGPO, January 1972), p. 145.

Table 5-17 Energy for Transportation in Q Btu*

Sector	*Use*	*(Q Btu)*	*Percent of net energy input*
Primary transport (propulsion)	Automobile	7.60	
	Truck and bus	3.06	
	Jet	1.63	
	Railroads	0.72	
	Marine	0.55	
	All other propulsion	1.59	
Subtotal		15.15	29
Secondary transport (related activitives)	Fuel refining, asphalt and road oil, energy	2.42	
	Primary metals used in transport manufacture	1.02	
	Manufacturing	0.53	
	All other secondary	1.05	
Subtotal		5.02	9
Total		20.17	38

*Source: Commoner, *op. cit.*

rels of oil. At 5.8 M Btu per barrel, this is 2.5 M barrels, 54.5 percent of the 4.5 B barrels consumed by all sectors that year. It is clear from these figures that transportation, more than any other sector, is leading the way into the Energy Crisis. It is, therefore, well worth our while to take a careful look at the various modes of transportation and at the efficiency of their operation.

There are two general functions of transportation: to move people and to move things. A second dimension is given by the distance moved and we can cover most of the variety by separately considering urban transportation and intercity transportation. All four of these dimensions are brought together in Fig. 5-14, which displays the record of total energy use (fuels only) in transportation since 1950 and also shows the contributions of the four categories: urban passenger transport

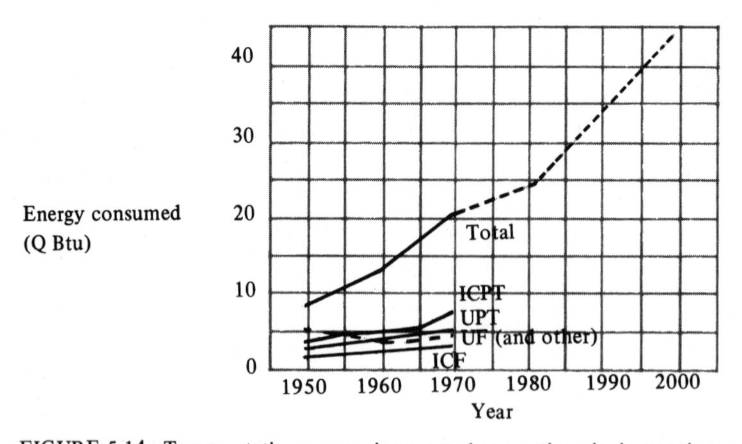

FIGURE 5-14 Transportation energy is expected to continue its impressive growth into the next century. Most of this growth can be attributed to passenger transport. [Source: E. Hirst, ORNL-NSF-EP-44, (Oak Ridge Nat. Laboratories, 1973).]

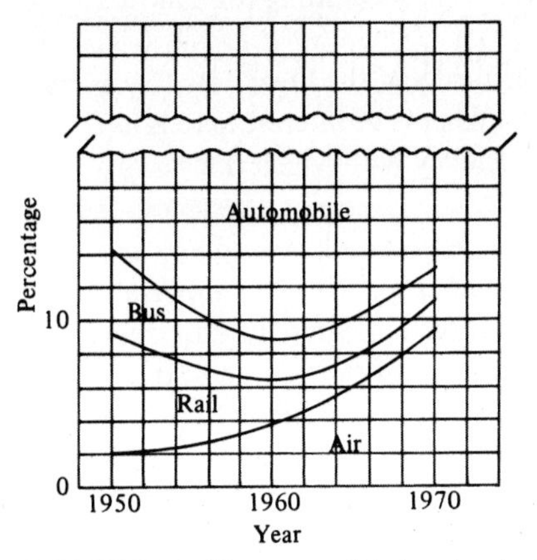

FIGURE 5-15 The automobile has dominated ICPT, accounting for about 90 percent of the total. Air transport is growing at the expense of rail and bus travel. (Source: Hirst, *op. cit.*)

FIGURE 5-16 The percentage of urban passengers carried by bus has declined from almost 20 percent in 1950 to less than 5 percent in 1970. (Source: Hirst, *op. cit.)*

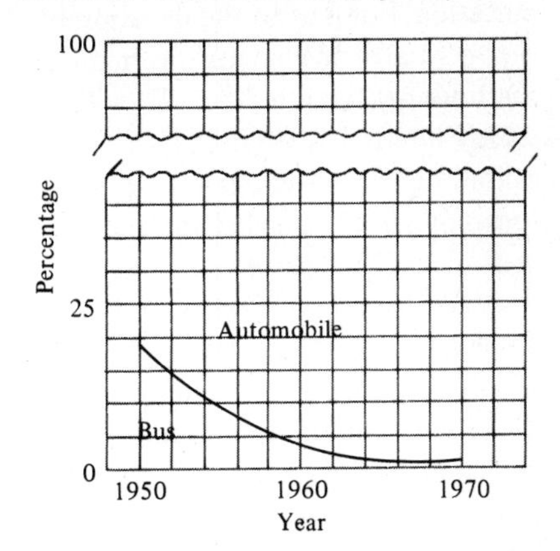

(UPT), intercity passenger transport (ICPT), intercity freight (ICF), and urban freight (UF). This last category, UF, includes all the other minor fuel uses, such as boats and private planes. All the curves of Fig. 5-14 are swinging upward with the exception of ICF, where the improved efficiency of diesel engines caused a drop over the first 15 years.

Figure 5-14 gives total energy consumption in the four categories. Equally important is the question of how the energy is expended within each transportation category. Figures 5-15, 5-16, and 5-17 display the percentage of passenger-miles or ton-miles carried by the various transport methods within each of the three categories, ICPT, UPT, and ICF. These three figures show the extent to which the internal combustion engine dominates transportation, accounting for, in 1970, 86

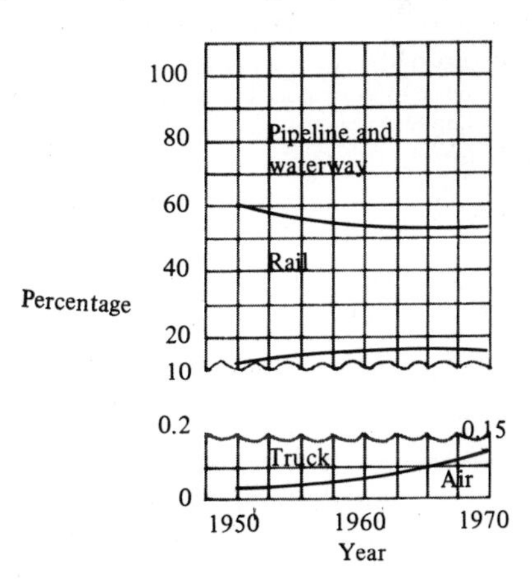

FIGURE 5-17 About 40 percent of ICF is carried by pipeline and waterways. The truck is taking over freight from railways and presently carries 15 percent as compared to railways 40 percent. Air freight is the fastest rising category. (Source: Hirst, *op. cit.)*

percent of ICPT, 95 percent of UPT, and in the truck, accounting for almost 20 percent of ICF.

These two sets of data, the total energy consumed and the breakdown by transportation, bring us to the question of efficiency. What is of interest here is not the thermal efficiency of the engine, but the efficiency with which the various methods get the job done. We will measure this by the inverse of efficiency, the *energy intensiveness* (EI)[7] in terms of passenger-miles per gallon or ton-miles per gallon. Figure 5-18 shows the great variation in the EIs for the most important forms of cargo transport and Fig. 5-19 displays the same measure of efficiency (or inefficiency) for passenger transport, including systems of the past and future.

A comparison of Figs. 5-18 and 5-19 with Figs. 5-15, 5-16, and 5-17 provides much insight into how we spend our transportation energy. In all cases the modes which are growing (automobile, truck, and air) are the energy-intensive ones, while the efficient buses and trains are being ignored.

A look back at Fig. 5-14 provides a projection into the future of the total energy consumed by transportation; it rises from 17 Q Btu in 1968 to 43 Q Btu by the year 2000. From that data we can calculate that the transportation sector alone will consume, in the next 30 years, 915 Q Btu of energy, or 150 billion barrels of oil, more than the total estimated "oil-in-place" in the Alaskan North Slope discovery. Furthermore, at present recovery rates, we will only get a third of this Alaskan oil. Put in other terms, the recoverable Alaskan oil will only last 14 years at the 1968 rate of use and 9 years at 1985 rates.

We will leave the story of transportation here, but we will return to it. It will clearly be one of the prime targets of our study of strategies for energy saving in Chap. 14.

[7]E. Hirst, "Energy Intensiveness of Passenger and Freight Transport Modes," ORNL-NSF-EP-44, Oak Ridge National Laboratories, April 1973.

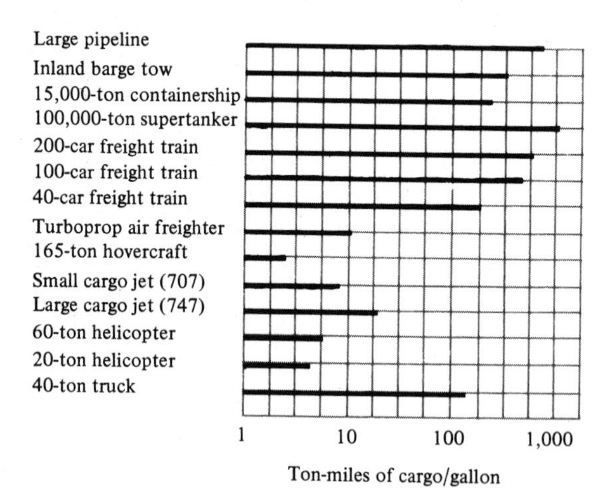

FIGURE 5-18 The most efficient cargo transport is by supertanker and is followed by pipeline. Truck and air freight are among the least efficient. (Source: Rue, A.E., "System Energy and Future Transport," *Technology Review*, January 1972.)

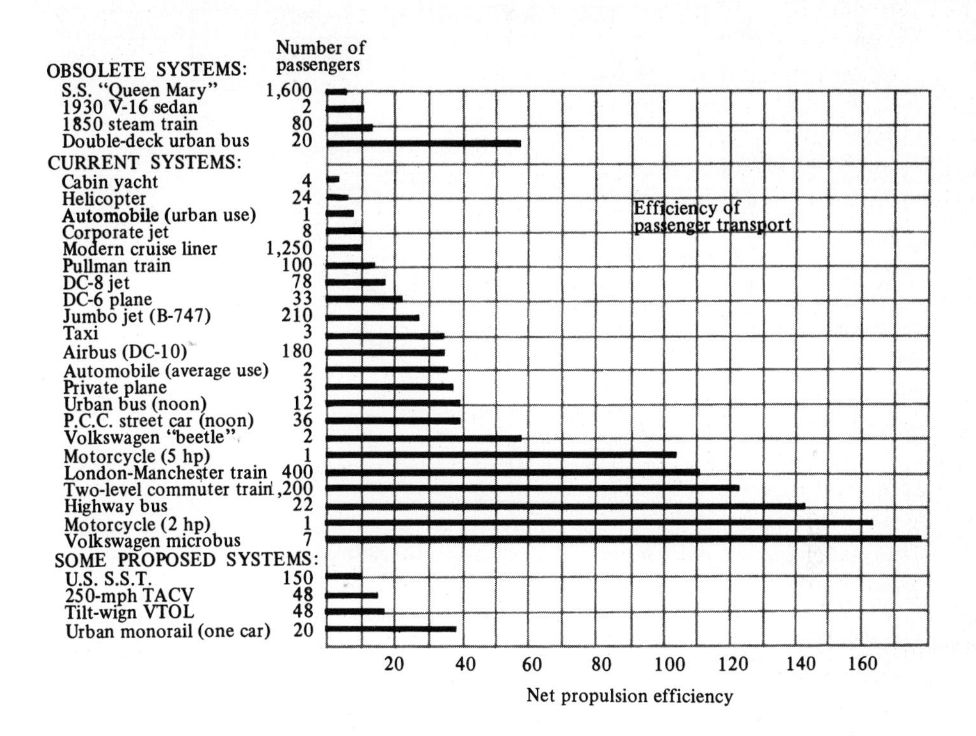

FIGURE 5-19 Net propulsion efficiency in passenger-miles per gallon. The most efficient passenger transport system of those shown is the Volkswagen microbus with seven passengers. Generally the fast, convenient, or luxurious systems are the least efficient. (Source: Rue, *op. cit.*)

THE UNDERPOWERED

We have focused our attention so far on the production and consumption of energy in the United States. Although justified in this by our nationality and that of most of our readers, we could hardly have chosen a less typical example. In 1970 the United States, with less than 6 percent of the world's population, consumed 36 percent of the total energy for that year. The per capita consumption of electric energy in this country (which was *not* the world's highest, as we shall see) was 7,715 kw-hr per person in 1968, 50 times larger than Ethiopia's reported 150 kw-hr, 76 times India's, and about six times the world average. To keep the uniqueness of our situation in perspective, we wish to look at some of the variation in amounts and patterns of consumption in other countries. We will also look a little more carefully at our own pattern of consumption, for it is obvious that the average figures we have been dealing with so far are compiled from a population whose consumption ranges between extremes. Are there underpowered on our own affluent island?

World Patterns

Figure 5-20 shows the growth of per capita consumption of commercial energy for the various continents. (The communist countries' centrally planned economies

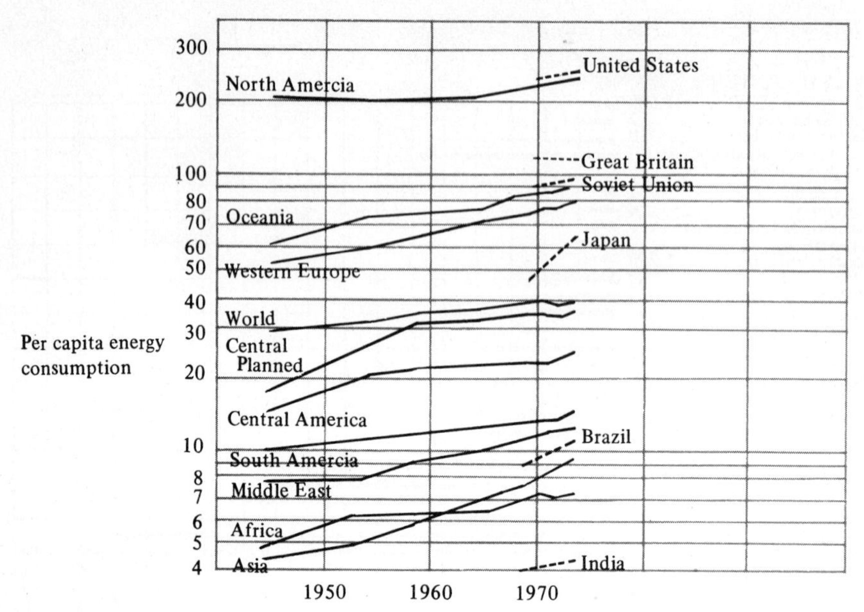

FIGURE 5-20 Per capita energy consumption varies by a factor of almost 60 between the U.S. and India with other countries and continents falling in between. Japan, and because of her influence, Asia, are showing the most rapid growth. (Source: *U.N. Statistical Yearbook*, 1969.)

are averaged separately.) The per capita consumption in North America is higher than in any other continent and is 35 times that of Africa which has the lowest average. The fact that commercial energy is plotted is partially responsible for the size of the differences, since Africa, Asia, and South America use significant amounts of noncommercial sources, for instance, wood in parts of Africa and South America, and cow dung in India. Total noncommercial fuel is estimated to be 10 percent of the world total and is used almost entirely for domestic purposes. The addition of these other fuels will lower the gaps between North America and the underdeveloped continents by 20 to 30 percent. When we consider, however, the probable efficiency with which these fuels are used, in makeshift stoves and open fireplaces, for example, their neglect in the comparison is not too misleading.

Although the general impression of all the curves in Fig. 5-20 is of similarity, a closer look shows regions with above average growth. Energy consumption in Asia, for all its enormous population, is rising at a rate equivalent to a 10-year doubling time as against 20 years for North America; the Middle East, after a bad period during the 1950s, grew rapidly during the 1960s. Africa shows an opposite trend. Figure 5-20 also shows records of 1965 and 1968 per capita consumption of the United States and five other countries, Great Britain, the Soviet Union, Japan, Brazil, and India. The first three are industrialized countries while the latter are representatives of the two underdeveloped continents. It has been estimated that 100 million tons of cow dung and vegetable waste is burned in India each year. This would raise her per capita figures by perhaps 30 percent. Similarly, the Brazilian figures are low since enormous quantities of wood are burned in that forested country. It is interesting to note that in 1895, when wood made up a third of our total energy, in the United States per capita consumption was already 110 M Btu—almost 25 times larger than present day India's (see Fig. 5-3). There are also striking differences between the growth rate of Great Britain and Japan. Japan's soaring energy consumption is all the more remarkable because that country imports 90 percent of its energy.

Even with the impressive rates of growth it will be a long time before the gaps are closed. If the United States level were to remain at its 1968 value, one can see that it will be many decades before any of the other lines will cross it and, for India and Africa, hundreds of years.

The differences in energy affluence among the various countries show up even more dramatically if we look at the growth of electrical energy consumption as displayed in Fig. 5-21. Most countries are turning to this intermediate form of energy; impressive rates of growth are shown, for instance, by Japan, whose consumption is doubling every six years compared to our ten-year doubling time, and by both India and China. India aims at producing 120 B kw-hr of electricity by 1975; but by then, its population will probably have reached 600 million and the per capita figure of 200 kw-hr will approximate the United States figure of 70 years ago.

There are great individual differences in per capita consumption of electricity. Figure 5-21 shows 1968 per capita data for selected countries. The highest per capita consumption of electricity, double that in the United States, is in Norway, where climate and geography have given it great hydroelectric resources; 99 percent of Norway's electricity comes from that source. Canada also exceeds the United States in per capita consumption of electric energy and its production is

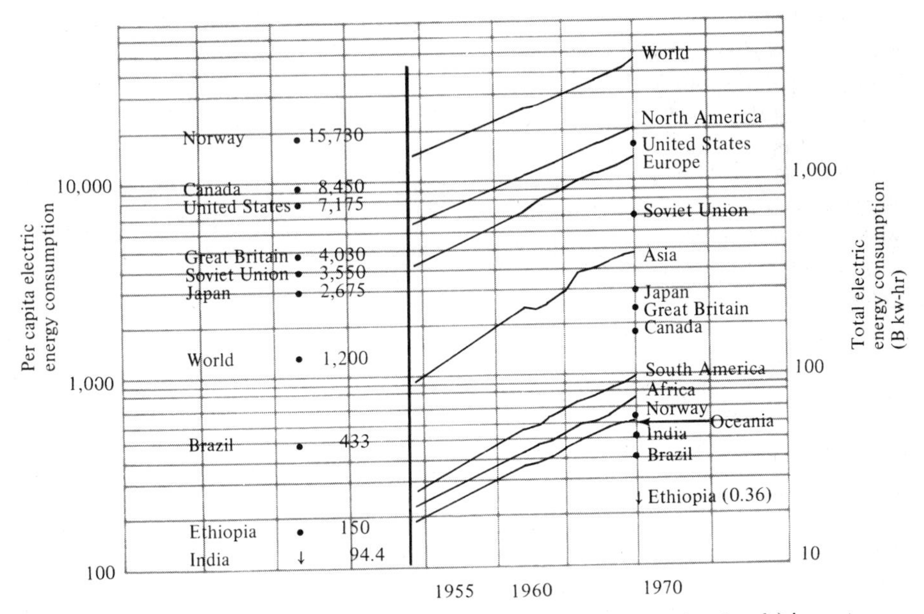

FIGURE 5-21 Although the range of total electricity production (right-hand scale) is great, all the continents show steeply rising growth curves. The per capita use of electricity (left-hand scale) is greatest in the countries with abundant hydroelectric resources. (Source: *U.N. Statistical Yearbook*, 1969.)

77 percent hydroelectric. At the other end stands Great Britain with no hydroelectric resources to speak of, and yet a respectable (by United States standards) per capita consumption. Ethiopia makes the smallest showing and that comes mostly from water power.

We can find these differences in quality in other areas also. Figure 5-22 compares the energy per capita spent on transportation, the domestic area (cooking and heating) and the industrial area for the five countries. We do not have this data from the Soviet Union. The countries hold their relative places but the effects of climate on domestic consumption shows up: Great Britain has a relatively large energy bill and Japan a small one. The inefficiency of our gasoline-powered transportation, as compared with Great Britain's and Japan's more efficient systems, is also indicated.

There are, therefore, not only differences in the amounts of energy available to citizens in the various parts of the world but differences in the quality, in the mix of modern convenient energy sources and ancient fuels. The potential of closing some of these gaps and the drain on the world's resources such ambitions will cause are the focus of Chap. 11.

Imbalance in the United States

Although the gaps are smaller, the same kind of inequities in energy per person as we have just seen in the worldwide data exist in the United States. Figure 5-23, for instance, presents data showing the distribution of electrical energy consumed by different income groups for selected conveniences which account for one third of all the electric power use. The 20 million households which make up the two wealthiest groups use twice the energy consumed by the 20 million households in the two poorest groups. The per household consumption in the wealthiest group is four times that in the poorest, a large portion accounted for by air conditioning. Only in television set ownership is there equity.

There are similar inequalities in the consumption of energy for transportation. The two- and three-car owners live in the suburbs and burn the gasoline; the inner-city dwellers use the available and energy-efficient urban transportation systems.

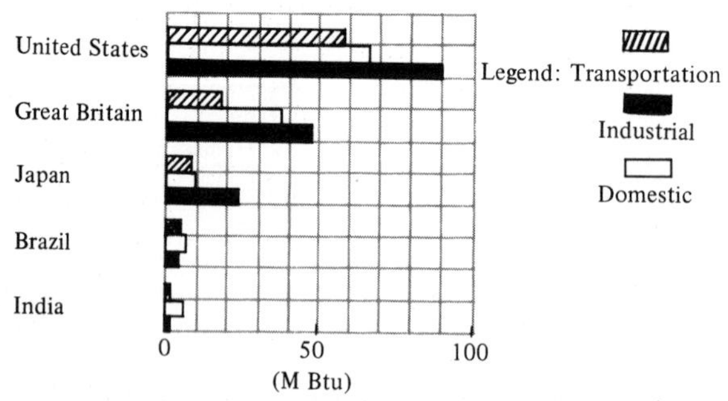

FIGURE 5-22 The dominance of domestic energy consumption in underdeveloped countries such as Brazil and India is apparent. The comparison also shows the large relative investment for transportation in the U.S. (Source: Guyol, N., *Energy in the Perspective of Geography*, (Englewood Cliffs, New Jersey: Prentice-Hall, 1971).

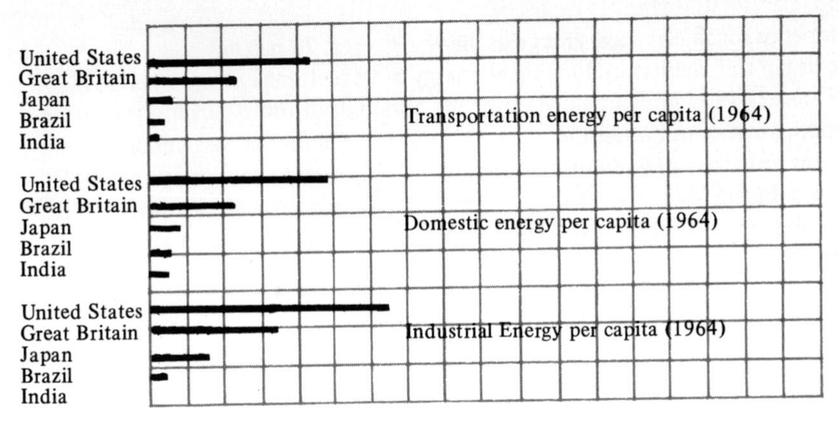

FIGURE 5-23 The inequities of distribution of wealth are reflected in unequal consumption of electrical energy among the various income groups. (Source: Commoner, *op. cit.*)

Projections of future demands and plans for energy conservation will have to face socio-economic questions as well as technological ones. Do we strive for energy parity? Will we reduce growth in consumption in higher income brackets and encourage it in the lower ones? Is equality of energy consumption one of the basic equalities of mankind? We will have to examine these questions and the effects of their answers again.

SUMMARY

This chapter has asked and answered two questions about energy in the service of man: Where does it come from? Where does it go? We found that 96 percent of our energy is obtained by burning the fossil fuels, those energy-rich molecules of carbon or carbon and hydrogen which store the solar energy of ancient life. The energy we obtain from their burning is used by industry (40 percent), transportation (25 percent), and in the residential and commercial sector (35 percent).

In dealing with the fuels and the end uses for which their energy is converted, we presented the themes of growth and change. We described the sequential replacement cycles in the fuel mix: wood replaced by coal, and coal, in turn, replaced by oil and natural gas. We also saw evidence of growth and change in the ways we use our energy. One of these examples, the growth of electrical energy consumption, will be examined further in the next chapter and will, along with transportation's consumption of energy, come in for more attention when we look for ways to save energy. Finally, we examined the imbalances in the way that energy is distributed throughout the world. We saw that the small United States population (6 percent of the world's) consumed 36 percent of the world's total energy and that, consequently, its per capita consumption was six times the world's average. We also saw a pattern of inequality within our own country where the wealthiest 20 percent of the United States households, for instance, consumed four times as much electrical energy as did the poorest 20 percent.

We began this chapter by looking at energy in its historical perspective: mankind's consumption growing slowly over the hundreds of thousands of years of his evolution. Most of our attention, however, was focused on the past hundred years of soaring growth, a period during which the world's energy consumption doubled five times. In the next chapter we pick out the most modern energy form, electricity, examine its production and consumption, and follow, again, the twin themes of growth and change.

6

Electricity: The All-American Energy

In Chapter 2 we followed energy through its many different forms, from the original "Big Bang" through the fern and moss jungle of Paleozoic earth, into the maw of an electric power plant furnace, and out through the wires to Chicago. The choice of electrical energy as the prototype form of energy for man was deliberate. Since its first commercial exploitation by Thomas Edison in 1882, electricity has become the preferred form of energy for an increasingly large number of uses. It is, as we have seen in Chap. 3, (particularly, Fig. 3-3) the most flexible form, since it can be converted to almost all other forms of energy. It is also the most rapidly growing one. Figure 6-1 compares the growth in the consumption of electrical energy with other rapidly growing energy uses and with total energy consumption.

The steepness of these growth curves is indicated quantitatively by the percentages which, as we explain in Appendix 1, can be interpreted in the same manner as the annual rate of interest for a savings account. Using the relationship shown in Appendix 1 we can also compute that the consumption of electrical energy is doubling every 8 to 10 years. This growth rate is expected to continue for at least the next 20 years. Thus, by 1990 the consumption is projected to be at least four times the 1.7 B mw-hr[1] of 1970.

[1] The mw-hr, which represents the product of power and time, is a unit of energy. It is related to the Btu by the following conversion: 1 mw-hr = 3.4 M Btu.

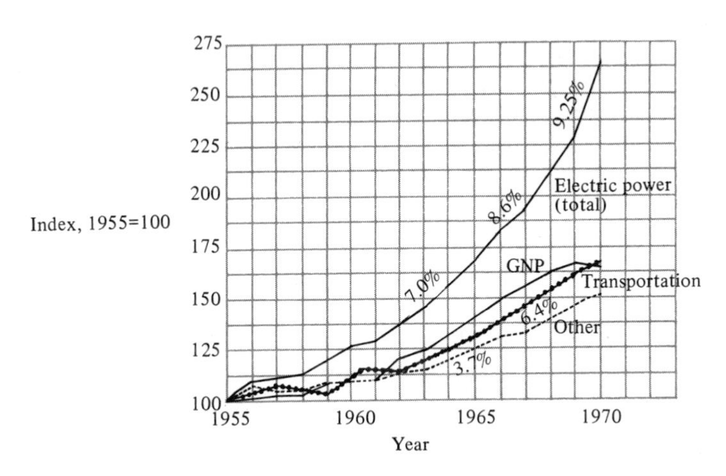

FIGURE 6-1 Comparative growth in types of energy consumption. Electrical energy consumption has the highest growth rate (yearly rates of increase are shown) and this dominance is expected to continue. (Source Earl Cook, "The Flow of Energy in an Industrial Society," Copyright ©September 1971 by Scientific American, Inc. All rights reserved.)

In many ways, the Energy Crisis is identified with electric power consumption; a great many of the tensions between the demands of our affluent, industrialized society and our increasing consciousness of our environment have developed in this area. The growth of this popular form of energy and its environmental impact, therefore, make it worthy of our close examination.

We took, as our example in Chap. 2, a coal-burning utility plant near Chicago and followed the electric energy produced from the coal out over the wires into that city. It is now time to look in more detail at its production, transport, and end use. What better place to begin that examination than Chicago?

Chicago's electric power is almost entirely supplied by the Commonwealth Edison Company. Electric power is big business and Commonwealth Edison is a big company: its net capital worth is 4.3 billion dollars, it employs 15,000 people, owns 12 fossil fuel generating stations, 2 nuclear plants, and 1 small hydroelectric plant, in addition to hundreds of small, special-purpose generators. The transport of electric power is accomplished by 4,000 miles of high voltage transmission lines and 112,000 miles of lower voltage distribution lines. Furthermore, like the rest of the American power industry, Commonwealth Edison must provide twice as much power by 1980.

The complexity of the power industry is well symbolized by Fig. 6-2. All the major steps from fuel delivery to end use are shown. Coal, the energy source in this example, is delivered at the lake shore or down one of the canals. It is specially processed before it is dumped into the furnace and burned to produce steam, which in turn powers the steam turbine. The conversion sequence is chemical energy to thermal energy to mechanical energy; then, at the end of the production line, the electric generator (see Appendix 4) makes the final conversion to electric energy. The mode of the future is also indicated in this process: the uranium-fueled reactor. This conversion pathway quickly joins the other and also ends with the steam turbine-electrical generator combination.

Electricity is kinetic energy; after it is generated, it must move out. We can follow its path in Fig. 6-2. The generated power, at 12,500 to 25,000 volts depending on the generator, is now converted by the transformers (see Appendix 4) to high voltages (as high as 765,000 volts in some parts of the country). This power is then sent out over the high-tension wires strung between the familiar metal towers, or, at somewhat lower voltages through underground cables.

Arriving at its destination, in this case the many transmission substations of the Chicago metropolitan area, the voltage is reduced to 12,500 volts (again by transformers) and sent out over the distribution system. Two paths are available: wooden poles with their insulator-dotted cross arms, or underground cables. The distribution system is the individualized performer. It must get power to where it is needed and provide it at the right voltage and quantity to the ultimate consumer. It accomplishes this through additional voltage transformation at the point of use.

In 1972 the Commonwealth Edison Company generated 53.9 M kw-hr of electrical energy. The sources of this enormous amount of energy (which was only 3 percent of the total electrical energy generated by all United States electric power companies) are shown in Table 6-1. Where did it go? Commonwealth Edison serves about 2.6 million customers, although not evenly. As shown in Table 6-2, the 704 large commercial and industrial consumers used more electric energy than did the

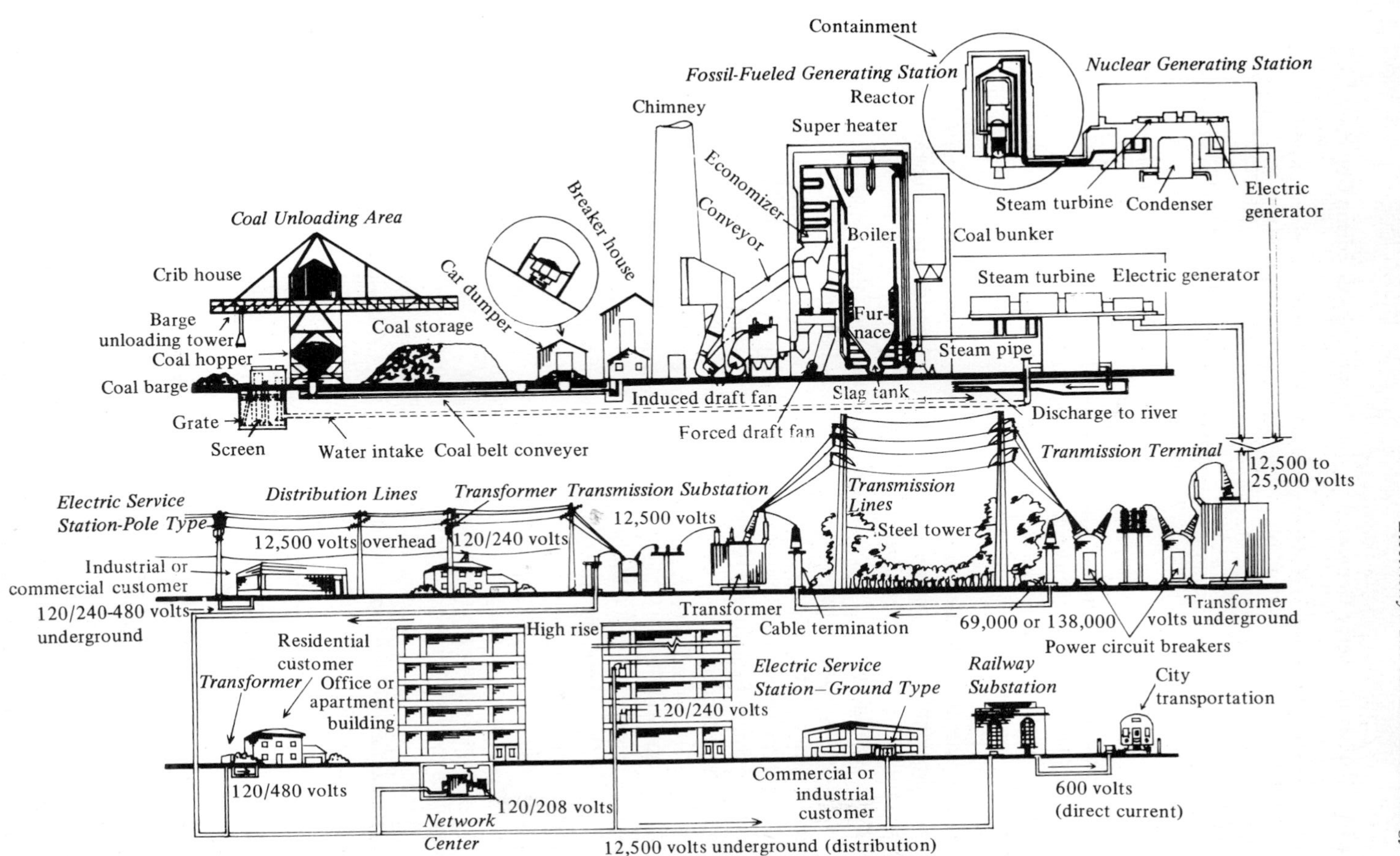

FIGURE 6-2 Major steps in the journey of electric power. The four steps–generation, transmission, distribution,end use–are shown with some indication of the variety which exists within each step.(Courtesy of Commonwealth Edison Co.,Chicago, Ill.)

Table 6-1 Sources of Energy Generated by the Commonwealth Edison Company (1972)*

Source	Amount (M kw-hr)
Fossil-fired steam	39,291
Nuclear steam	12,130
Diesel	8
Gas turbine peaking units	2,424
Hydro	16
Total	53,869†

*Data from the Commonwealth Edison Company, Chicago, Illinois.

†Difference in generation and utilization is due to power purchased and sold to other utilities, system electrical losses, and internal consumption.

2.4 million residential customers or the 221,000 small industries and commercial customers. Not only are the amounts of power per customer different, but the voltage, form, and rates of the delivered power are different.

For the residential customer and the commercial customer, the voltage is stepped down to either 120 or 240 volts (we often refer to it as 110 and 220 volts since the actual values differ a bit from company to company). On the average, the Chicago residential customer used 6,020 kw-hr of electrical energy per year, a bit less than the average United States figure of 6,530 kw-hr per year for residential customers in 1972. Table 6-3 gives some idea of how this energy was used. In the Commonwealth Edison territory, there are 2.4 million refrigerators and each uses about 1,000 kw-hr a year. There are 161,000 electric water heaters, each of which uses 4,000 kw-hr a year; 560,000 home freezers, at 1,200 kw-hr a year each. Even the 1.387 million electric can openers use their share; the American homeowner is truly plugged in.

The industrial picture is quite different. Some 917 M kw-hr of the 16,500 million total in that category went, for instance, to a large steel company outside of Chicago. This company takes its power directly from the transmission lines; the utility company steps the voltage down to either 34,000 or 13,200 volts, and the steel company does the rest. Huge amounts of power at 34,000 volts are used to heat electric arc furnaces for the production of steel. These furnaces, which are rapidly replacing the traditional coal-fired, open-hearth furnaces, use 400 kw-hr per ton of steel. In addition to the greedy consumption by the furnace, induction heaters are used for

Table 6-2 Utilization of Energy in the Chicago Area (1972)*

Sector	Amount (M kw-hr)	Number of customers
Residential	14,492	2,407,150
Small commercial and industrial	15,453	220,777
Large commercial and industrial	16,512	704
Public authorities	4,274	9,800
Railroads	391	2
Total	51,122†	2,638,433

*Data from the Commonwealth Edison Company, Chicago, Illinois.

†Difference in generation and utilization is due to power purchased and sold to other utilities, system electrical losses, and internal consumption.

Table 6-3 Commonwealth Edison Company Estimated Number of Electric Appliances in the Service Territory

Appliance	*Number*	*Annual consumption per unit (kw-hr)*
Air conditioning		
Central	273,000	3,600
Window units	1,505,000	1,375
Refrigerators	2,410,000	1,137*
Electric water heater	161,000	4,219
Electric heat	107,000	14,000
Home freezers	560,000	1,195†
Dehumidifiers	389,000	377
Electric ranges	560,000	1,175
Washers		
Automatic	1,266,000	103
Wringer and other	243,000	76
Electric dryers	351,000	993
Vacuum cleaners	2,518,000	46
Dishwashers	438,000	363
Food waste disposers	341,000	30
Television		
Black and white	2,263,000	362
Color	1,345,000	502

*For a standard 14-cubic foot model, a "frostless" model of the same size consumes, on the average, 1,829 kw-hr.

†A 15-cubic foot "frostless" model utilizes 1,761 kw-hr.

the preheating of steel ingots in rolling mills and there is the usual need of power for motors, lights, and so on.

Two other big customers are an oil refinery and an auto-assembly plant. Again, these are served directly from the transmission lines and the utility-owned transformers provide 13,200 volt power. The refinery uses its power for large and small motors, pumps, fans, and automatic flow control equipment. Total consumption was 427.4 M kw-hr in 1972. The auto-assembly plant used 137 M kw-hr to power arc welders, turn the myriad motors of the assembly line, and provide light and air conditioning to the plant.

A final example is the "railroad" customer of Table 6-2, the Illinois Central commuter line. Its needs are again different. The trains run on *direct current* (dc) power at a relatively low 600 volts. The utility provides power through the 12,500 volt distribution line shown in Fig. 6-2; the railroad must then provide for the stepdown to lower voltage and the conversion of *alternating current* (ac) power to dc. (The difference between direct current and alternating current is explained in Appendix 4.) This railroad used 310.5 M kw-hr in 1972.

This trip through the Commonwealth Edison service territory[2] serves as an intro-

[2]We wish to express our strong appreciation for the considerable assistance we received from Charles J. Jakubowski, Director of Economic Research, and Roland Kraatz at Commonwealth Edison Company, who provided most of the material on which this description is based.

duction to the four parts of power delivery—generation, transmission, distribution, and end use—and suggests some of the complexity and special difficulties in each area. Electric energy is not a product that can be stacked on a shelf waiting for use; it is kinetic energy of electrons in motion, from the generators around the long circuits back to their source. It must be used "on the fly." The examples of customer demand we have just given illustrate some of the utility's problems. Perhaps it is rather safe to project the demand of the steel and auto assembly company (unless there are strikes or recessions) but there are also those 1.75 million air conditioners used in this territory (see Table 6-3); when will they be turned on and for how long? The transient nature of electric energy makes it easily convertible but difficult to market and distribute. It also makes the electric power business an interesting one to study. With this introduction, we now turn to a more detailed description of electrical generation and transmission, including implications of its rapid growth over the next two decades.

THE COMMERCIAL GENERATION OF ELECTRIC POWER[3]

The four major steps in the journey of electric power are production, transmission, distribution, and use. Each step has its own characteristic problems and makes its own contribution to the national energy-environmental problem. Each step is, of course, related to the others, but we will discuss them separately in order to obtain a better perspective of their importance.

Electrical energy is produced commercially in only two ways: (1) by converting chemical energy directly to electric energy in the battery; and (2) by converting mechanical energy to electrical energy in the generator (see Appendix 4). The battery, while it accounts for only a miniscule fraction of the total electrical energy used in this country, is important because it can store energy until needed and produce it on demand. It is expensive energy; the 100 amp-hour automobile battery (12 volts) costs about 40 dollars and produces only 1.2 kw-hr of energy. This 33 dollars per kw-hr is in contrast to the average price of about 2 cents per kw-hr charged by the utility company. Automobile batteries have the advantage of being rechargeable; but even ten recharges does not make their energy output competitive in price. Our focus in this chapter is, therefore, on utility-generated, electric energy.

Appendix 4 describes how the conversion of mechanical energy to electric energy takes place and defines and points out some of the advantages which have caused ac power to dominate the power picture. The actual generators in use today are much more complicated devices than the one shown in Appendix 4, but they have the same features: a series of coils which turn in a magnetic field and sliding contacts (brushes) to bring out the generated current.

The generator is, of course, only part of the system; it requires something to turn it. It should be emphasized that the generator itself neither stores nor produces energy. What it accomplishes, and with efficiencies close to 100 percent, is the conversion of kinetic mechanical energy into kinetic electrical energy; sitting still, it is just a lump of metal.

[3]Most of the quantitative data in this and the following sections were obtained from the 1970 Report of the Federal Power Commission.

There are presently only two important sources of mechanical energy to turn the generators: the heat engines discussed in Chap. 5 and hydropower from the kinetic energy of falling water. The heat engines dominate the picture, providing, as shown in Table 6-4, motive power for 85 percent of the generated electricity, with hydropower providing the remaining 15 percent. Of the various heat engines, the most important is the steam turbine which generates 76 percent of the electricity. The efficiency of this conversion (thermal to mechanical to electrical) is thus inexorably limited by the efficiency limits on heat engines which we have discussed in Chap. 4; electric power has to come through the thermal bottleneck. The emergence of nuclear-fueled generating plants will not provide a bypass to this bottleneck, for, as we shall see in Chap. 10, nuclear energy is also converted to thermal energy and the electrical conversion in these plants will also be made by a steam turbine-generator set.

The total generating capacity in the contiguous United States in 1970 was 340,000 mw (see Table 6-4). This capacity (which is maximum power available, not the total energy) was in excess of the peak demand of 275,700 mw in 1970. Twenty-three percent of the generating capacity was held in reserve against unexpected demand or equipment breakdown. Table 6-4 also shows the same information projected to 1980 and 1990. The expected increase in nuclear power generation and the accompanying percentage decrease in the importance of hydroelectric power and electric power generated by fossil fuels are displayed. It should be remembered, however, that although electric power generated from fossil fuel plants will decrease on a percentage basis, dropping from 76 to 44 percent over the 1970-1990 period, the absolute capacity of fossil fuel generation is predicted to more than double from 259,100 to 558,000 mw. This means, of course, that the fuel requirements will double.

Let us look at these fuel requirements. Table 6-5 shows the amounts of the various fuels used in 1970 and the projections for 1980 and 1990. We can see the expected increase in the relative importance of nuclear fuel and the decrease in the fossil fuels, but the more important numbers are the actual fuel requirements. We see that the

Table 6-4 Generating Capacity by Types of Prime Mover, Peak Demands, and Reserves*

	1970†		*1980*		*1990*	
Capacity	*(mw)*	*(percent)*	*(mw)*	*(percent)*	*(mw)*	*(percent)*
Conventional hydro	51,600	15.2	68,000	10.4	82,000	6.5
Pumped-storage hydro	3,600	1.1	27,000	4.0	70,000	5.6
IC and gas turbines	19,200	5.7	40,000	6.0	75,000	5.9
Fossil steam	259,100	76.1	390,000	58.6	558,000	44.3
Nuclear	6,500	1.9	140,000	21.0	475,000	37.7
Total	340,000	100.0	665,000	100.0	1,260,000	100.0
Peak demand	275,700		555,000		1,050,000	
Indicated reserves	64,300		110,000		210,000	
Reserves in percent of peak	23		20		20	

*Source: *The 1970 National Power Survey*, Part I, (Washington, D.C.: Federal Power Commission, USGPO, December 1971), p. I-18-2.

†Year-end capacity is shown, some of which was installed after the date of peak demand. Therefore, actual reserves were somewhat less than indicated.

Table 6-5 Projected Annual Fuel Requirements for Thermal Power Generation*

Fuel	Year 1970	1980	1990
Coal	322 million tons	500 million tons	700 million tons
Natural gas	3,600 billion cu ft	3,800 billion cu ft	4,200 billion cu ft
Residual fuel oil	331 million barrels	640 million barrels	800 million barrels
Uranium ore†			
without plutonium recycle	7,500 tons	41,000 tons	127,000 tons
with plutonium recycle	7,500 tons	38,000 tons	108,000 tons

*Source: *The 1970 National Power Survey,* Part I (Washington, D.C.: Federal Power Commission. USGPO, December 1971), p. I-I-20.

†Short tons of U_3O_8 required to supply feed for diffusion plants to supply annual burnup and new reactor inventories.

amount of coal needed in 1990 is more than twice that burned in 1970, that one sixth more natural gas and almost two and one half times as much oil will also be needed. The demand for uranium, even with the breeder reactor, increases by almost a factor of 15.

Hydropower is expected to decrease in relative importance from 16 to 14 to 12 percent for the years 1970, 1980, and 1990, respectively. To achieve these levels, the industry is expected to add 35 percent more conventional generating capacity by 1980, and to increase the generating capacity of pumped storage units almost eightfold by 1980 and almost twentyfold by 1990. Pumped storage is a recent and somewhat controversial addition to the mix of generating systems which will be discussed in more detail later in this chapter.

More fuel, more capacity, more power—this is the story of the electric power industry.

Increasing Size

The electric utility industry, if considered a single entity, is the nation's largest industry with a present capital worth of 85 billion dollars in 1970. This investment in plant and equipment will have to be approximately quadrupled by 1990 if the industry is to meet expected demand for its product. The industry's total new capital requirements between 1970 and 1990 are estimated to be 400 to 500 billion dollars (based on 1970 dollar value), half the 1970 Gross National Product (GNP).

The utility industry is, of course, not one single industry but rather 3,500 separate and independent systems which are a complex mixture of investor-owned utilities, public agencies, cooperatives, and federal agencies. In spite of this large number, the trend has been toward consolidation and toward a smaller number of larger systems. There were 4,334 systems in 1927, half of them investor-owned, but the stock market crash wiped out many of these. In 1968 there were 405 investor-owned utilities, 2,075 public (nonfederal) ones, 960 cooperatives, and 5 federal utility agencies. The raw numbers are misleading, however. Seventy percent of these companies are engaged in distribution only; by type of ownership, 93 percent of the cooperatives, 66 percent of the public nonfederal, and 3 percent of the investor-owned companies are engaged in distribution only.

On the basis of annual sales or the total number of kilowatt-hours generated, the industry is clearly dominated by a few big investor-owned utilities. The 200 largest investor-owned systems own and operate more than 75 percent of the generating capacity and serve about 80 percent of the total customers of the electric power industry. The five large federal systems add 12 percent of the generating capacity and 13 percent of the customers. The trend is toward more consolidation; each year more and more of the small municipal generating plants are shut down and arrangements made to buy power from one of the large companies. The somewhat anomalous role of an investor-owned business in the monopolistic situation of a utility has been examined searchingly by Senator Lee Metcalf in *Overcharge.*[4]

No matter what the ownership, a common attribute has been the increases in size: the individual generators are getting bigger, the plants are getting bigger, and the interconnected systems are getting bigger. The reason behind all of these size increases is, of course, the efficiency of scale, the same type of reduction in product cost with increasing size that enables the supermarket to undersell the local grocery.

The cost of producing a kilowatt-hour decreases as the individual generator, or the utility plant, gets bigger. There is a reduction in capital costs and in operating personnel. In going from a 500 mw generating unit to a 1,000 mw unit, the capital investment per mw is reduced by 10 percent and the manpower requirements are reduced by 18 percent. The economies depending on plant size are similar; capital costs are lower (one set of offices will do for several generators); manpower, equipment for loading and storing fuel, and other basic costs are all reduced on a per mw basis.

Generating units are getting larger physically and, since the number of coils wound on the rotator determines the output power, their megawatt capacity has been increasing. In 1930, for instance, the largest unit on line had a generating capacity of 200 mw, and the average size was 20 mw. In 1970 the largest generator in service produced 1,150 mw for the Tennessee Valley Authority's (TVA) Paradise Steam Plant in Tennessee, and three 1,300 mw units were under construction. The average size of units under construction was 450 mw. Some appreciation of the size of these electrical monsters can be gained from Fig. 6-3 which shows the 809 mw generator at Commonwealth Edison's Dresden plant (a nuclear plant).

Not only are the sizes of individual generators increasing, but plant size is also increasing. In 1970 the largest plant was that same TVA Paradise Steam Plant with a total capacity of 2,558 mw of power. In 1973 the four-unit, 3,200 mw Monroe plant of the Detroit Edison Company came on line. The generating capacity of these huge plants can be best appreciated by comparing them with peak load demands of various urban areas. The peak load, or the largest amount of power needed during the year, was about 7,500 mw for New York in 1970, 3,000 mw for Washington, D.C., and 5,600 mw for the Detroit area. The one Detroit Edison plant can produce more than half the total needs of Detroit. This, of course, also means that the loss of one of the 800 mw generators at the Monroe plant takes out one seventh of Detroit's power. As of now, the large units have a poorer record of *forced outages,* or breakdowns, than do the smaller ones. This is one of the hidden prices of bigness.

[4] Metcalf and Reinewer, *Overcharge,* (New York: McKay, 1967).

FIGURE 6-3 The large, 809-mw generator at Commonwealth Edison's Dresden nuclear power station. The man is standing next to the generator. Three low-pressure turbines and a high-pressure one are to his left. Further down the room a second unit is under construction. (Courtesy of Commonwealth Edison Company, Chicago, Illinois.)

System size has also increased, largely through consolidation. Larger systems can offer more reliability, and can reduce costs for marketing and advertising. There are those who feel that the ideal system for this country would consist of a group of 15 to 20 closely integrated, large utility systems. Others point out that the problem of trying to regulate such large companies at the state level would be very difficult. The trend toward more and more concentration in the industry and to larger sizes of units, plants, and systems, however, will probably continue.

The Ups and Downs of the Industry

One of the greatest difficulties faced in planning and operating an electric power system is that of coordinating a series of generating devices, whose outputs are more or less fixed, with a demand profile that fluctuates rather considerably on a daily, weekly, monthly, and yearly basis. This mismatch between fixed output and fluctuating demand is exacerbated by the nature of electrical energy; it is kinetic energy and cannot be stored. In addition, it is rather awkward and inefficient to convert it to a potential energy form for storage, but we shall consider a few such conversions later in this section. Utility systems must, therefore, have a surplus of generating capacity available and must be able to turn various components off and on as the consumer's demand for power fluctuates. The amount of fluctuation in a typical week is shown in Fig. 6-4. There is not only a day-night change of about 40 percent but also, for instance, a 25 percent difference between the Wednesday and the Sunday peak.

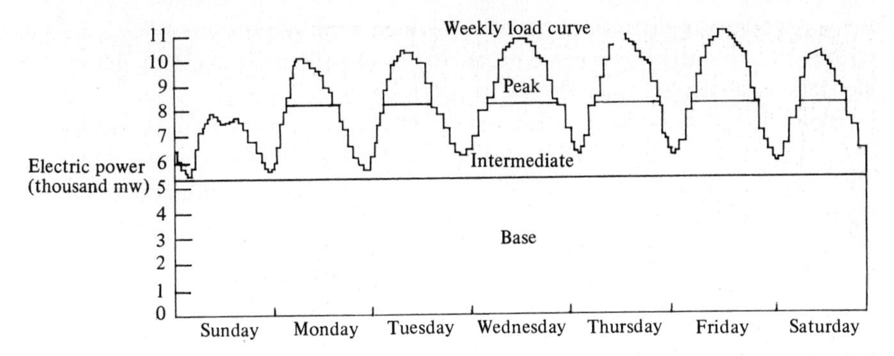

FIGURE 6-4 A typical weekly load curve. The demand for electric power varies from day to night and through the week. (Source: *The 1970 National Power Survey,* (Washington, D.C., Federal Power Commission, USGPO, 1971).

The large steam-driven generators, which provide about three fourths of the electric power, are designed to operate at maximum efficiency at their maximum power output. They are meant to run continuously; it is a waste of time and fuel to stop and start a coal fire; similarly, it is difficult and expensive to shut down a reactor. The big coal plants and the big nuclear plants which are being built are therefore designed to provide the steady "base load" shown in Fig. 6-4. To meet the intermediate load, gas- or oil-fueled steam plants are favored since they can be started and stopped relatively quickly and can be more precisely controlled by automatic equipment.

To take care of the small amounts of electric power at the very top of the peaks (the most difficult part of the load to predict), small gas turbines of the type described in Chap. 4, or diesel-driven generators are used. The turbine devices have a power capacity in the range of 10 to 60 mw; the diesels are somewhat smaller, in the range of 2 to 6 mw. Both the diesel engine and the gas turbine are inefficient, with only 25 to 30 percent efficiency, and burn costly fuel. Their advantages of automatic operation and quick turn-on time (newer units can be up to full power in 3 to 5 minutes) make them valuable as a standby reserve in case of unusual demand or breakdown of other equipment. The fact that they do not need cooling water makes their siting flexible. Surprisingly, the most flexible unit for meeting the fluctuating need, however, is the hydroelectric plant.

Hydroelectric Power

Historically, the first generating plant was thermal-powered; the second was a *hydroelectric* plant. The hydroelectric plants, as we saw in Table 6-4, are still trailing the thermal plants in importance and their percentage contribution to the total generating capacity is declining. In spite of this, however, the hydroelectric plant is a most important part of the generating system. The new development in this area, *pumped-storage,* is expected to further increase the importance of hydroelectric power.

Hydroelectric plants have some decided advantages over thermal plants: the fuel, gravitational potential energy supplied by the sun, is free and renewable. There is no emission of pollutants and no need for cooling water. The energy conversion does not go through the thermal bottleneck. It is a one-step conversion of mechanical energy to electric energy with greater than 90 percent efficiency and only the small but inevitable heat tax of friction losses.

Hydroelectric plants usually have long lives of about 100 years or so. (In some cases, however, heavy siltation may cause a shorter life, as in the Hoover Dam.) Their large, low-speed generators need only infrequent servicing. Most important, they can start and stop quickly and the speed is easily adjustable so that they provide a kind of *spinning reserve*; the turbines are turning and the generators only have to be engaged to respond quickly to power demand. Their disadvantages have more to do with their siting and what they do to the streams that power them. We will discuss some of those disadvantages in Chap. 9.

The most interesting hydroelectric plant from an energy point of view, is the pumped-storage plant. A sketch of such a plant proposed for the Allegheny River in Pennsylvania is shown in Fig. 6-5. These plants provide, at present, the only practical way for large-scale storage of electrical energy. They are made possible by the development of highly efficient, reversible pump-turbines. No other device illustrates the reversible nature of generators and motors so well. During the off-peak hours, electric power from steam plants is fed to the generators (now acting as motors) and they turn the turbines which force water up to the storage reservoir. This off-peak power is relatively cheap; there is no use for it and, yet, as we have said, it is wasteful to turn off the base load plants. When the demand increases, the water is allowed to run back down the tunnel; it turns the turbine which, in turn, operates the generator to produce the needed and now highly valued extra power. All efficiencies considered, it takes about 3 kw-hr of electric input energy to produce 2 kw-hr of output energy; the system is 67 percent efficient.

A further important advantage of pumped-storage systems is their ability to respond quickly to changes in demand. A spinning turbine can be fully loaded in one minute and, as tests have shown, some of the newer installations can change over

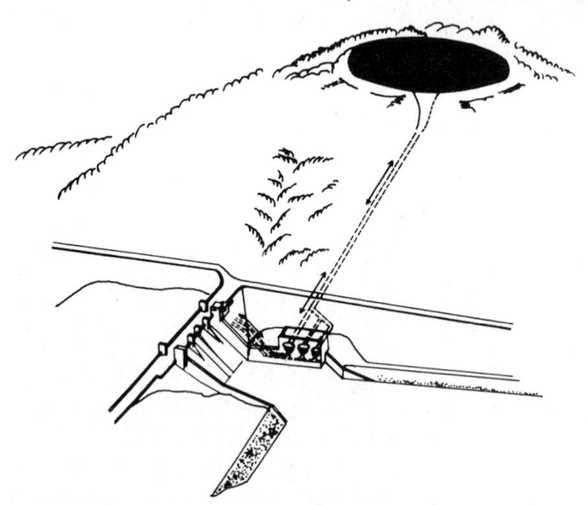

FIGURE 6-5 A sketch of the Kinzua pumped-storage project in Pennsylvania on the Allegheny River which will be able to generate 427 mw of power. (Source: *The 1970 National Power Survey, op. cit.)*

from pumping to generating in 5 to 10 minutes. The reconversion in response to power surplus is equally rapid.

Pumped-storage installations can either be combined units (such as the Kinzua project shown in Fig. 6-5) in which both conventional hydroelectric generation and storage capacity is present, or they can be pure storage units. The anticipated growth in pumped-storage generating capacity (3,600 mw in 1970, 27,000 mw by 1980, and 70,000 mw by 1990) is stimulated by the need for storage rather than for generation. For that purpose, all that is needed is a sufficiently large body of water and a nearby natural elevation; such locations are not difficult to find. That the growth of pumped-storage will be subjected to pressure from conservation groups is assured by the history of the famous Storm King project on the upper Hudson. There are also geographical considerations; transmission line lengths, for example, must not be so large as to counterbalance the expected savings with their own losses.

There may be future alternatives to pumped-storage. The competitive position of the battery is mentioned in Appendix 4. It is estimated that batteries with a cost of 6 to 10 dollars per kw-hr, instead of the present 20 to 70 dollars per kw-hr characteristic of automobile-type storage bateries, would be attractive for large-scale storage. There is considerable research under way to discover new battery materials and efficient ways to make them.

It is also likely that the eventual answer to the problem of electrical energy storage will be to use surplus electrical power to *electrolyse* water into hydrogen and oxygen and store these gases. In this process of *electrolysis* an applied voltage causes the water molecule to separate into hydrogen and oxygen. It takes energy to accomplish this and that energy can be reclaimed by burning the hydrogen to form water again (see Appendix 4). The hydrogen will have a variety of uses as fuel, including the regeneration of electric power in a fuel cell. We will discuss these future technologies further in Chap. 12.

The Generating Blend

One sees, from even the brief discussion of this section, that the electric power which flows so reliably from the wall outlet is provided for by a great variety of generating devices turned on and off in response to the varying demand. A typical response to the day-to-day variation shown in Fig. 6-5 is displayed in Fig. 6-6. The base load, and even some of the intermediate load, is carried by the "Atlases" of the industry; at present, these are the generators fired by fossil fuels, but the nuclear reactor will stand alongside these in the future. As Fig. 6-6 shows, even with these large units some increase and decrease of output must be obtained. This is preprogrammed, as is the use of their off-peak power to store energy by pumping water.

The remainder of the peaks are served by the spinning reserve of conventional hydroelectric power, by reclaiming the energy stored by pumping water up hill, and by turning on the diesels and gas turbines to fill out the tips of the peaks.

All in all, the system is marvelously complex and obviously requires careful monitoring and decision-making by man and computer. Until it is replaced by an entirely new technology, this system must continue to serve us and all its individual components must grow to meet our mounting demand for power.

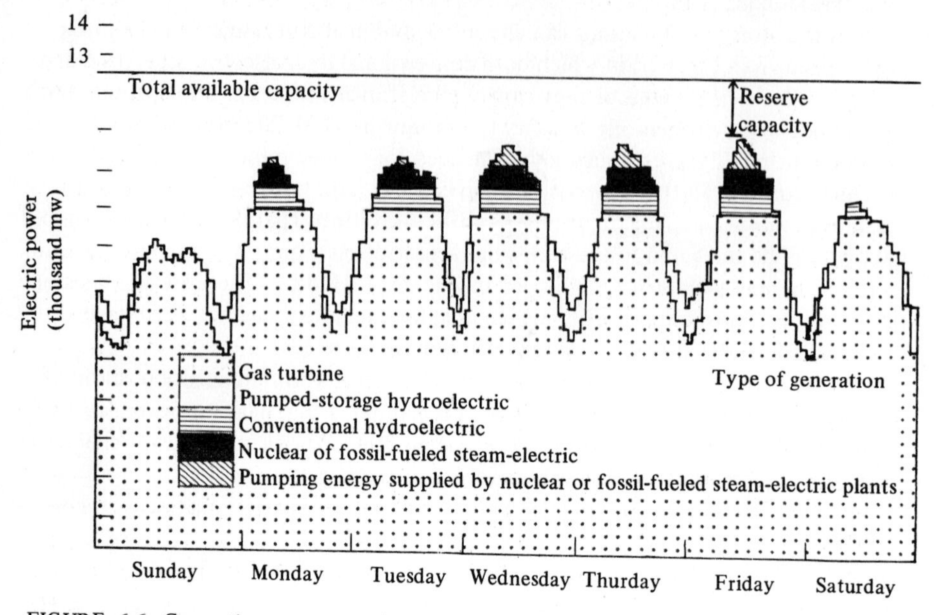

FIGURE 6-6 Generation to meet weekly loads. The contributions of the various types of generating units to provide for the fluctuating loads are shown. (Source: *The 1970 National Power Survey, op. cit.*)

TRANSMISSION AND DISTRIBUTION OF ELECTRIC POWER

There is more to the business of electric power than its generation; it has to be transmitted to the consuming area and then distributed to the individual users; in fact, for every dollar spent on electricity only about 40 to 50 cents represents the cost of generation. The transmission of the power (at high voltage) from distant generating plants adds another 15 to 20 cents and the distribution and delivery within the consuming area (at lower voltage) costs the final 35 to 45 cents.

As discussed in Appendix 4, the loss of electrical energy in transmission through a wire is given by the product of the square of the current i^2 with the resistance R of the wire. The power available from a generator, is the product of the voltage V times the current i. The same amount of power Vi can therefore be delivered at high current and low voltage, or at low current and high voltage. The energy losses in the transmission line, however, are much lower for the latter choice since i is small.

To minimize transmission losses, electric power is shipped at the highest possible voltages. In 1970, out of a total of 300,000 miles of electric power transmission lines in service, about 65,000 miles were extra-high voltage lines (EHV), with voltage greater than 200,000 volts; the lowest voltage was 69,000 volts. The progressive increase in the ultra-high voltage lines (UHV), greater than 530,000 volts, is also shown in Table 6-6 along with estimates of future growth. The 1990 total of all transmission lines is expected to be 500,000 miles.

The continued growth of EHV and UHV transmission is assured. Not only does the percentage of loss decrease, but the power-carrying capacity increases with voltage. One 765,000-volt line, for instance, has the carrying capacity of five 345,000-volt lines or thirty 138,000-volt lines. The 765,000-volt line needs only 200 feet of right-of-way as against 750 feet for the five 345,000-volt lines and about 3,000 feet for the thirty 138,000-volt lines. Increasing land prices, competition for space, and environmental and aesthetic objections to more lines provide strong incentives to increase the per line capacity even further.

The future trends in power-plant size and siting will also contribute to the growth in numbers and voltage of transmission lines. Not only are individual generators and plants growing larger, but future plans are for *energy parks,* groups of nuclear reactors on the shoreline or floating just offshore (see Figs. 1-2 and 1-6) and for concentrations of *mine-mouth* plants in the Western coal-producing regions. The six utility plants built or under construction in the Four Corners region (the controversy surrounding them is examined later) are tied by UHV lines to the cities of Arizona, Nevada, and to Los Angeles. The development of a similar generating complex on the North Central Plains of Montana, Wyoming, and the Dakotas will require UHV ties to St. Louis and other Midwestern cities.

EHV and UHV lines are also employed, to a growing extent, to interconnect formerly independent systems. This practice, which began in 1958 with a 345,000-volt tie between two Midwestern systems, has grown impressively: in 1968 there were 78 connections. The incentive for making interconnections is, of course, manyfold; the various systems provide reserve strength for each other, which has economic benefits; they also provide stabilization. The resulting interdependence of systems has its disadvantages, however, as the famous New York blackout of a few years ago demonstrated.

The shape of this growing web was displayed earlier in Chap. 1, where Figs. 1-3*a* and 1-3*b* contrasted the transmission line network in 1970 with that planned for 1990. Not only does the density of lines increase, but there is some change in the type of line; more UHV lines going up with greater than 1 million volts, more dc transmission, and so on. There will be still other contributions from technology.

New Transmission Technology

The increase of UHV voltages to 1 or even 1.5 million volts rests largely on improvements in insulation. One of the ironies of the present situation is that the sulphur oxides released by the coal-burning generating plants, dirty the insulators and allow the power to leak across their surfaces. Emission control, therefore, has immediate cost benefit.

Other new approaches to electric power transmission are a return to dc transmission, underground cables, and supercold transmission lines. Historically, dc power was the first to be generated and transmitted. These early systems, however, were, abandoned after the invention of the transformer, which allowed ac power to be easily stepped up or down in voltage. The recent development of convenient and high-powered electronic devices for converting ac to dc and vice versa, has returned dc to favor. The first such system was installed between Sweden and the island of Gotland in 1954. In the United States a 400,000-volt dc line, 846 miles long, connects gen-

Table 6-6 Transmission Line Mileages in the United States, 230 kv and Above*†

	230 kv	287 kv	345 kv	500 kv	765 kv	±400 kv (dc)	Total
1940	2,327	647					2,974
1950	7,383	791					8,174
1960	18,701	1,024	2,641	13			22,379
1970	40,600	1,020	15,180	7,220	500	850	65,370
1980	59,560	870	32,670	20,180	3,540	1,670	118,490
1990	67,180	560	47,450	33,400	8,940	1,670	159,200

*Source: *The 1970 National Power Survey,* Part I (Washington, D.C.: Federal Power Commission, USGPO, December 1971), p. I-13-4.

†By 1990 there may be significant applications of ac voltages higher than 765 kv and more extensive use of high-voltage dc than that shown in the table.

erating capacity in the Pacific Northwest with the consumer demand of Los Angeles.

Direct current transmission has several advantages. Some losses are reduced; the *corona losses*, or the discharge through air which causes the radio interference by transmission lines, is less troublesome because dc does not swing to the large peaks of ac; it can carry more power at less maximum voltage. There is less loss due to charging and recharging the line. (This charging current is also discussed in Appendix 4.) Direct current lines will also be used more and more to link systems together because this eliminates the necessity of precise frequency matching. Havoc results in linking two ac systems if their frequencies are different, if the peaks and valleys of the alternate voltages do not match up. If the power is converted and transmitted as dc and then converted back to ac by devices whose frequency is controlled by the receiving systems, no such problem occurs.

A second direction of new technology points underground. There is much pressure in urban and scenic areas to get the transmission lines out of sight. Underground UHV transmission is presently very expensive; the losses are greater, insulation is more difficult, and the ground does not do as good a job of carrying off the heat as does air. Cables overheat and destroy their insulated coverings. To counteract this latter problem, cable manufacturers are experimenting with forced cooling. Circulating cooling oil or gas in a pipe surrounding the transmission line, for instance, doubles the power that can be carried.

All these complications increase costs; it is estimated that for the same capacity of bulk power transmission, underground lines cost from 10 to 40 times as much as overhead ac lines. A cost study for an underground 345,000-volt line in a suburban area found that the cost per mw-mile for the underground cable was 1,600 dollars as against 110 dollars for overhead transmission. Since as much as 60 percent of the costs go for installation labor, there is not much hope of significantly reducing them. Overhead transmission will predominate for some years, except in dense urban areas where land, if it is available, is too expensive to buy for the necessary rights-of-way.

It may be possible in the future, however, to make underground power lines competitive by freezing them. As metal wires are cooled, their resistance drops. At the temperature of liquid nitrogen (−320°F), for instance, the resistance of aluminum (which will probably be used since it is lighter, stronger, and more plentiful than copper) is 10 times less than at normal outside temperatures. Since the power losses go as i^2R, this means that the current can be increased by $\sqrt{10}$ or about three times with the same i^2R loss. General Electric engineers have tested a *supercold* cable which they estimate could carry 3,500 mw of power, an amount about equal to the average power input to Philadelphia and more than double the highest power transmission levels to date.

If the temperature of metallic conductors are lowered even more, down to the temperature of liquid helium, a few degrees above absolute zero (-460°F), an even more amazing thing happens: the electrical resistance disappears altogether and the metal becomes *superconducting.* Since there are no i^2R losses for such conductors (this is actually true only for dc since for ac there are small losses which need further strategies of reduction) they can carry huge currents through very small amounts of metals. The problems which need to be solved before such cables become practical, however, are enormous. Refrigeration and insulation are the first ones. Another is the property of the superconducting materials (the rare metal, niobium, is the current choice) to become high resistance paths if the temperature slips above the critical value. If this were to happen, the i^2R heat would be sufficient to burn up the cable. The problem of terminating the cable, bringing the power into a distribution system at normal temperatures, must be faced; a continuous electrical path over which the temperature rises by almost 500°F must be invented.

With all these complications, however, superconducing cables are economically promising for some applications. At high-load levels, say 10,000 mw, a superconducting cable is projected to cost about 200 dollars per mw-mile as against the 1,600 dollars per mw-mile we quoted earlier for UHV underground transmission.

Present maximum line capacities are on the order of 1,000 to 1,500 mw. One superconducting cable would, therefore, have to carry the entire power load of Washington, D.C. (3,000 mw) or even New York (10,000 mw) to become commercially viable. The promise is bright enough to cause the power industry to invest heavily in research along these lines and the 1980s may see these superconducting lines carrying huge power loads from, for instance, offshore nuclear generating stations to their metropolitan clients.

Cost of Energy Transmission. With all these improvements the cost of shipping energy in electrical form will probably remain, as it is now, among the highest. A comparison of the cost of the various modes of energy transport is shown in Fig. 6-7. The cost ranges from the low figure of about 0.5 cent to ship one M Btu 100 miles using tankship transport of oil, to the high value of 10 cents shipping the same amount of electric energy the same distance by 220,000-volt transmission. It will always be necessary to ship electric power, but the final answer to the problem of shipping large quantities of electric energy long distances (from offshore generating stations, for instance) may be to convert it to another energy form such as hydrogen. We will consider this and other possibilities in Chap. 12, which deals with new conversion techniques.

The Distribution of Electricity

Electric power is produced at the generating plants at voltages from 2,000 volts up to 30,000 volts; it is shipped at voltages as high as 765,000 volts. The consumers,

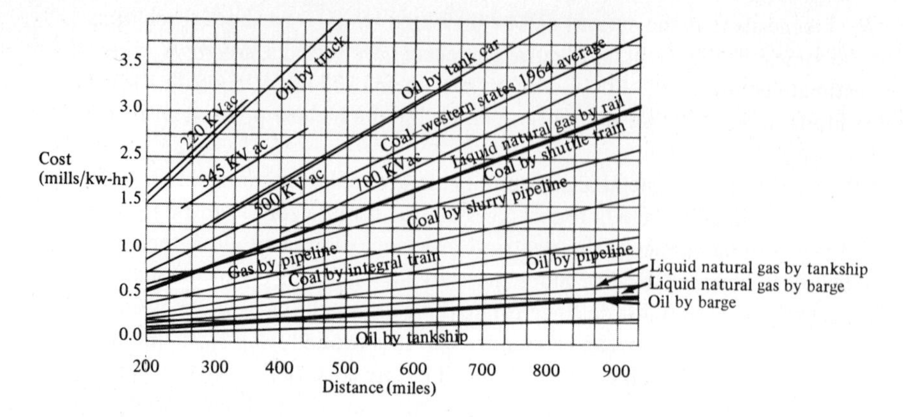

FIGURE 6-7 Transportation of electric energy is increasingly expensive as shipping distances increase. The shipment of ore by tanker is the least expensive form of energy transport. (Source: *The 1970 National Power Survey, op. cit.)*

however, have different needs, as we saw earlier in the examples from Chicago: 600 volts dc for the railroad, or 34,000 volts ac to the steel company. The homes and offices use much lower voltages, most of it at 110 volt ac but some at 220 volts ac. These lower voltages are more convenient; less insulation is needed, closer spacing of electrical parts is possible, and they are safer.

The present United States distribution system represents an investment of about 33 billion dollars, 40 percent of the total utility investment. As Fig. 6-2 shows, the two major components of the distribution system are the lines carrying power at 10,000 to 15,000 volts and the substations in which the voltages are stepped down and fed out to the many customers. Transformers at each customer location then complete the job.

We do not intend to discuss the many technical details of distribution and will only look at two subjects which are important for our considerations of energy and its related environmental problems.

Much of the pressure of environmental considerations is on the unsightliness of the overhead distribution lines. There are now almost 3.5 million miles of these lines and people are finding their cluttered silhouettes more and more objectionable. New techniques and materials have brought the cost of underground distribution down; it is now only 2 to 4 times as expensive (compared to 40 times more expensive for UHV lines) as overhead lines and is used more and more for extension to suburbs and in downtown areas. In 1968, 20 percent of the new distribution lines in the country were underground; this percentage is expected to be 70 percent by 1975 and 90 percent by 1990. But these are the new lines; an attempt to convert all existing overhead distribution lines to underground by 1990 would cost an estimated 170 to 200 billion dollars. Therefore, the overhead lines will, in all likelihood, be with us for some time so that we must be satisfied with new silhouettes and new colors and materials as environmental improvement.

An energy problem which will also demand attention is the increasing *power density* in our metropolitan areas. Power density, in units of kw per mile^2, for instance, is determined by customer density and power demand per customer. It varies greatly from the low demand in the spacious expensive suburbs to the high demand of the regions dominated by high-rise apartments or downtown business areas. In major metropolitan areas, densities of 100,000 to 300,000 kw per $mile^2$ exist now and densities of 350,000 to 1 million kw per $mile^2$ are projected. As a result, the use of electricity in cities, at least, is beginning to compete with the 500,000 or so kw per $mile^2$ energy-density we receive from the sun (averaged over 24 hours). It is a competition we must enter cautiously.

GROWTH AND THE CAUSES OF GROWTH

We have just examined what we could call the space dimension of electric power, following it through generation, transmission, and distribution. We must now look at the time dimension, the study of the rapid growth of this versatile form of energy, and some of the reasons for that rapid growth. We will want to identify, in particular, those end uses of electricity which represent either large consumption or are threatening to result in large consumption. Such end uses will be worthy of critical examination later, as we look at the potential for conserving energy.

The Past and Present

First let us briefly review the swift rise of electrical energy to its present importance. The first commercial electrical generating plant, the Pearl Street Station, opened for business in Manhattan on September 4, 1882. It was a coal-fired plant and produced power for 400 light bulbs of 83 watts each, a total of 33.2 kw. It came into operation just 26 days before the first hydroelectric station at Appleton, Wisconsin began to deliver its 12.5 kw of power. The thermal production of electricity has outranked hydropower ever since.

The demand for electric energy in the United States rose rapidly; by the year 1891 total United States consumption was at a level of 129 M kw-hr or 2 kw-hr per person. By 1900 the per capita consumption was 25 times greater, 49 kw-hr per person, and a total of 3.73 B kw-hr.

The complete record of the growth of electrical energy is shown in Fig. 6-8, with extrapolation to the year 1990. The doubling times for representative periods are indicated as are the growth rates in terms of yearly percentage increases. These curves, which group the data in 5-year intervals, show clearly the almost unbroken rise in electrical energy consumption, the total energy consumption doubling every 10 years, more or less, and the per capita energy doubling about every 14 years. They also demonstrate clearly that the increase in electrical energy consumption is not explained by attributing it to population growth; only 1.4 percent of the total annual growth rate (the difference between the total and per capita consumption) can be explained by population increase alone. The major contribution comes from the increasing per capita use.

In 1973 a total of 1.98 B mw-hr of electrical energy was consumed in this country as compared to 1.86 B mw-hr in 1972, an increase of 6.5 percent. On a per capita basis, with a 1973 population of 211.7 M, this is 9,400 kw-hr per person. The 1972

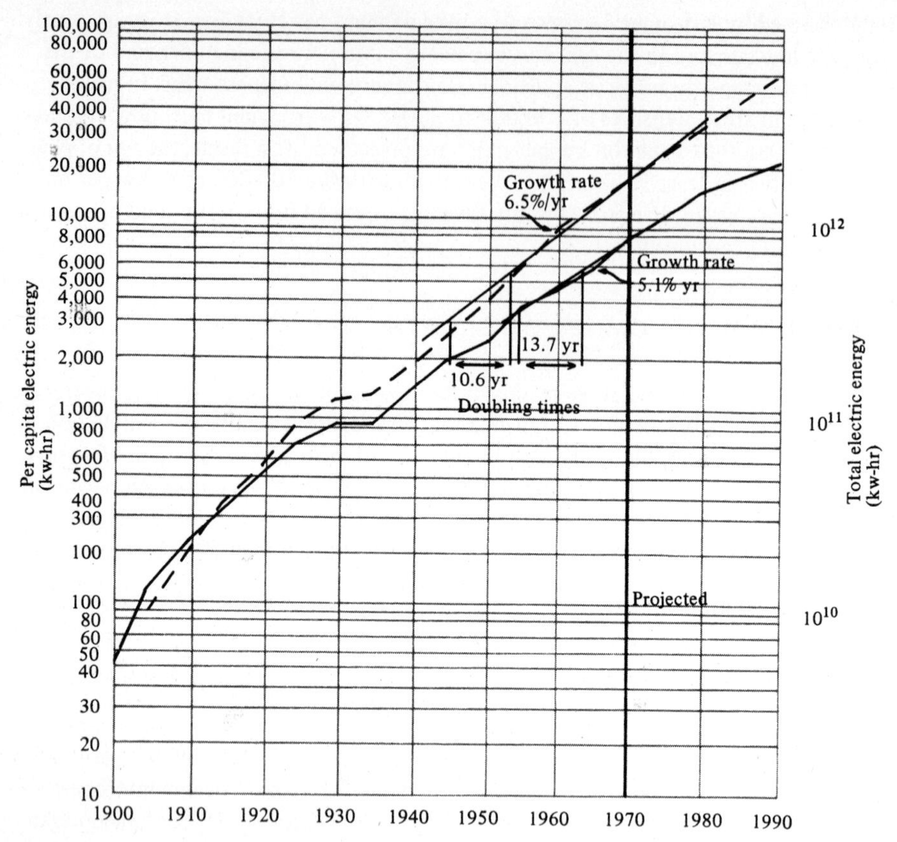

FIGURE 6-8 Total and per capita consumption of electric energy. The doubling times and growth rates during the present era are shown.

per capita consumption was 8,900 kw-hr per person. The growth rate of 5.6 percent was, therefore, greater than the per capita growth rate of 5.1 percent recorded in the 1960s. It appears that the projections of Fig. 6-8, which predict three doublings between 1970 and the year 2000, are not yet proven unrealistic. We will comment further on those projections in Chap. 11. For now the question is: Where was this energy used?

The total amounts of purchased electrical energy consumed in the three major sectors in 1973 are shown in Table 6-7 along with the percentage distributions. The

Table 6-7 Electrical Energy Consumption (1973)*

Sector	*Amount (M kw-hr)*	*Percent of total*	*Percent growth (1973-1974)*
Household and commercial	1,092	58.1	+5.2
Industrial	783	41.6	+4.2
Transportation	5	0.3	+3.8
Total	1,880		

*Purchased electricity, the total consumption is somewhat larger since some industries generate electricity for their own use.

Table 6-8 Percentage Consumption Electrical Energy*

Sector	*Amount (M kw-hr)* 1960-1968	1968	*Percent of total* 1960	1968	*Percent growth* 1960-1968
Residential	218	407	28.6	30.7	83
Commercial	153	316	20.2	23.8	106
Industrial	383	600	50.5	45.1	57
Transportation	5	5	0.7	0.4	0
Total	759	1328			75

*Source: "Patterns of Energy Consumption in the United States," Report to the Office of Science and Technology by the Stanford Research Institute, January 1972.

comparable data and percentages for 1960 and 1968 are shown in Table 6-8. Together, these two tables show that the residential-commercial consumption of electrical energy is growing more rapidly than industrial consumption. They also give conclusive evidence of the unimportance of electrical energy in transportation.

The Future

Table 6-9 provides a look into the future. It is, of course, roughly consistent with the projection of Fig. 6-8 and predicts, on the average, a quadrupling over the 25-

Table 6-9 Projected Increases in Electric Energy Requirements by Classification of Use Contiguous United States (1965-1990)*

Class of use	*Increase in energy requirements 1965-1990 (M mw-hr)*	*(percent)*	*Average annual rate of growth (percent)*	*Number of years to double consumption*
ELECTRIC UTILITIES				
Residential (nonfarm)	1,155	45.4	7.08	10
Irrigation and drainage pumping	23	20.9	4.62	15
Other farm†	70	25.9	5.24	14
Commercial	948	49.8	7.41	10
Industrial	1,950	44.7	7.04	10
Street and highway lighting	31	34.4	6.14	12
Electrified transport‡	3	6.0	1.90	37
Other uses	182	56.9	7.90	9
Losses and unaccounted for	410	44.6	7.04	10
Total utility	4,772	45.2	7.06	10
INDUSTRIAL ESTABLISHMENTS				
In-plant generation§	48	47.1	1.55	45
Total	4,820	416.2		

*Source: *The 1970 National Power Survey,* Part I, Federal Power Commission, (Washington, D.C.: USGPO, December 1971), p. I-3-18.

†Includes residential use on farms; other residential uses in rural areas included under "Residential."

‡Excludes electrification of automobiles.

§Excludes industry sales to electric utilities.

year period covered. We see from the data in Table 6-9 that residential and commercial consumption are expected to continue to lead the growth parade into the 1990s, followed closely by the industrial sector. Electricity is not expected to gain as rapidly down on the farm. In transportation, the one area where electric energy could be used more efficiently than the existing energy source, it makes the smallest gain.

The implications of Table 6-9, in concrete terms, are truly remarkable. We look around our house and wonder how can we use 4.5 times as much electric energy in 1990 as we now use. In the commercial area, what kind of expansion in consuming units and consumption per unit will be necessary to produce a 500 percent increase? There will be more people and more houses and buildings to shelter them, but this increase alone falls far short. Population growth from 1965 to 1990 will produce only about a 50 percent increase. Even in the industrial area, it is difficult to imagine that the production lines can increase the outpouring of goods enough to create the projected growth. These are, however, the predictions. If we equate increased electric power with more acres of land strip-mined for coal, more sulfur dioxide in the air, and more stored radioactive waste, it becomes clear that the predictions must be looked at more closely.

The Causes of Growth

We have neither the space nor the vision to follow all the many branches of our complex, consuming society into the future to find their contribution to the amazing and somewhat threatening picture of growth we have just presented. We will try to give concreteness to the projection by following a few of the dominant ones and by anticipating, here and there, the budding of new ones. In this selection the past will be our guide.

In Table 6-10 we summarize the 1968 consumption of electrical energy by end use in the two sectors, residential and commercial. We also show the percentage increase in electric energy use over the period 1960-1968. It is easy to spot the two uses that will help make Table 6-9 come true. In the residential sector the first four uses between them account for 57 percent of the consumption and three of these are growing rapidly. In the commercial sector we see two of the same end uses, refrigeration and air conditioning among the top three, joined there by lighting. Again, the big consumers are growing rapidly. By examining the reasons for some of these specific instances of growth we can, perhaps, gain insight into the overall phenomenon.

Electric Space Heating. It is evident in the data of Table 6-10 that *electric space heating* in the residential sector, is the most rapidly growing form. In the type of electric heating we are considering here, electric current is converted to heat by a high resistance wire located in a baseboard unit or in the walls, floor, and/or ceiling. In Table 6-11 we see that the growth in total energy consumed by residential heating comes from two trends: the number of houses having such heating has been increasing each year, and the total electric energy consumed per all-electric home has also been increasing. These homeowners are keeping their homes warmer (or cooler) and perhaps maintaining poorly insulated homes as well.

Table 6-11 also shows the Federal Power Commission's (FPC) projection of the growth of electrically heated homes. With such an explosive growth in homes with

Table 6-10 Electrical Energy Consumption by End Use (1968)*

End use	*Amount (M kw-hr)*	*Percent of total*	*Percent growth (1960-1968)*
RESIDENTIAL SECTOR			
Refrigerator	73.3	18	105
Water heating	65.2	16	44
Space heating	48.1	12	466
Air conditioner	45.2	11	221
Television	37.6	9	116
Cooking	28.1	7	32
Food freezer	23.5	6	167
Clothes dryer	15.0	4	120
Lighting	44.0	11	—
Other	27.3	7	20
Total	407.3		87
COMMERCIAL SECTOR			
Water heating	24.6	7.8	19
Refrigeration	71.5	22.6	36
Air conditioning	108.6	34.3	84
Cooking	2.3	0.7	57
Other	109.6	34.6	611
Lighting	59.0	18.7	
Advertising and display lighting	7.9	2.5	
Elevators	11.7	3.7	
Fans and air handling equipment	10.3	3.3	
Pump and motors	5.3	1.7	
Miscellaneous	15.2	4.7	
Total	316.5	100.0	108

*Source: Report to the Office of Science and Technology, "Patterns of Energy Consumption in the United States," (Washington, D.C.: Stanford Research Institute, USGPO, January 1972).

Table 6-11 Residential Consumption of Electric Heat*

Year	*Dwellings with electric heat (thousands)*	*Average annual consumption per dwelling (kw-hr)*	*Total consumption (M mw-hr)*
1960	719	11,908	8.6
1961	913	12,150	11.1
1962	1,058	12,397	13.1
1963	1,254	12,906	16.2
1964	1,910	12,460	23.8
1965	2,378	13,031	31.0
1966	2,698	12,652	34.1
1967	3,040	13,951	42.4
1968	3,388	14,153	48.0
1970	4,700		
1980	12,500		
1990	24,000		

*Source: Report to the Office of Science and Technology "Patterns of Energy Consumption in the United States," (Washington, D.C.: Stanford Research Institute, USGPO, January 1972), Table 18, p. 42.

electric heat and with the consumption per home increasing, the projection of Table 6-9 becomes a bit more believable.

We should not leave the subject of electrical heating without some comments on the reasons for its popularity. Since such heating systems consume great quantities of electricity they are quite naturally promoted heavily by the utility companies in their advertising, at high promotional rates. We will comment on this later. The builders also like this heating system; it reduces the initial cost of the house; no space is needed for boiler or furnace, nor is it necessary to build in heating ducts and pipes, or chimneys or storage tanks. It is also popular with apartment builders because of these same low construction costs and because the tenants usually pay the utility bills. The homeowner likes the simplicity and cleanliness. Electricity does not, as yet, play an important role in commercial space heating where the higher fuel costs make it less attractive. There is evidence, however, of increased interest even in this area. It is reported that electric heating was installed in 23 percent of the nonresidential buildings started in 1969.[5]

Since electricity competes with direct heat, gas, and oil, where overall system efficiency is greater, we will want to look again at electric space heating, with energy conservation in mind.

Appliances Unlimited. The remainder of the growth in the consumption of electrical energy will come from water heating, air conditioning, and from the appliances: televisions, dishwashers, cooking ranges, refrigerators, and the like. An intelligent guess at future growth in this kind of consumption is assisted by two kinds of information; how many more appliances will be needed and how effective is the competition from natural gas.

Some light is thrown on the first of these questions by data on market saturation, the percentage of households owning the various appliances. Table 6-12 shows the percentage saturation of the most important appliances in 1970. There is clearly room for significant growth in the numbers of air conditioners, food freezers, clothes dryers, dishwashers, and color television sets. The numbers of appliances will grow as each household buys one and as the number of households grow. This will push the energy consumption upwards. Adding to this, however, is the fact that the energy consumed per appliance is also growing. Evidence for this is shown in Table 6-13 which lists the ratio of the consumption of energy (gas or electric) per unit in 1968 to the 1960 per unit consumption. Most of those ratios are greater than one. The big contributors to this last category of growth are refrigerators, as the conversion to the more energy-expensive "frost-free" units continues; but there is also potential for contribution from food freezers, washing machines, and of course, color televisions.

We have so far only considered extensions of present use. If the past is any guide, we will also see new appliances for the home which will add to the electrical bill. Home-sized units for electric incineration and garbage processing are under development, and the electric automobile with home chargers is on the horizon. We will consider this latter possibility later on.

[5] Commoner, Boksenbaum, and Corr, *Energy and Human Welfare*, Vol. III (New York: Macmillan, 1974).

Table 6-12 Percentage Saturation of Various Electrical Appliances (1970)*

Appliance	*Saturation (percent)*
Cooking range†	56 (44)
Water heater†	32 (68)
Refrigerator	96
Clothes dryer†	27 (13)
Television	
Black and white	141
Color	30
Air conditioner	41
Clothes washer	92
Vacuum cleaner	92
Dishwasher	27
Waste disposal	26

*Source: *The 1970 National Power Survey*, Part I, (Washington, D.C.: Federal Power Commission, USGPO, December 1971), p. 1-3-8. Data on gas appliances from "Patterns of Energy Consumption in the United States," (Washington, D.C.: Stanford Research Institute, USGPO, January 1972).

†These appliances compete with gas appliances where percentage saturation is shown in parentheses.

The Expensive Substitutions. The nation's consumption of electrical energy is expanding by virtue of addition (more households, new appliances) but substitution is also making a major contribution. There are three different types of substitutions at work: the substitution of electric energy for other forms (electric space heating, for example), the substitution of energy-intensive materials for energy-economic ones, and the substitution of electric power for manpower. They are all worth brief comments.

Table 6-13 Growth of Gas and Electric Appliances*

	Energy consumption per unit 1968/1960 ratio	
Appliance	*Gas*	*Electric*
Water heater	1.07	1.05
Cooking range	1.00	0.96
Refrigerator	1.08	1.61
Clothes dryer	0.97	1.03
Freezer		1.67
Television		
Black and white		1.04
Color		1.09
Air conditioner		
Room		1.10
Central	1.00	1.00
Washing machine		1.51
Dishwasher		1.07

*Source: Report to the Office of Science and Technology, "Patterns of Energy Consumption in the United States," (Washington, D.C.: Stanford Research Institute, USGPO, January 1972).

Electricity for Other Fuels. We have already mentioned that the electric arc furnace is replacing the open-hearth furnace in steelmaking, and that induction heating of iron ingots is also growing in importance. Of the 26 M Btu per ton required to produce raw steel, 0.89 M Btu or 260 kw-hr were in electrical form (see Table 5-11). The new arc furnace takes about 400 kw-hr per ton; added to that will be about 360 kw-hr per ton of rolled steel. It is clear that the steel industry will make its contribution to electrical growth.

In other sectors of the primary metals industry, electric power is also growing in importance. The refining of copper and aluminum is accomplished in part (in the case of aluminum almost entirely) by electrolytic deposition. Copper is deposited in pure forms from crudely refined materials by electric current. Electric power is also used in the smelting of the ore. All in all, out of the 40 M Btu of energy to produce a ton of refined copper, 3.4 M Btu or 990 kw-hr are electric energy. This is three times the electric energy requirement of steel. In this industry also, the newer techniques are electrical: electric smelting, for instance, to replace the traditional smelting by natural gas heating.

The big consumer of electric energy is aluminum. Aluminum does not often occur in a pure state in ore, but in chemical combinations. Aluminum oxide (Al_2O_3) is a common one. The chemical potential energy of this form is lower than that of aluminum and oxygen separated, since it is formed spontaneously when aluminum is put in contact with oxygen. To separate the aluminum from the oxygen, therefore, requires an input of energy. The separation is accomplished by electrolysis in huge reduction cells in which aluminum oxide is dissolved in a molten metallic salt. Electrolysis, discussed previously in reference to water, is the reverse of the battery process; a potential difference (a voltage) is established across a metallic solution and the metallic atoms flow preferentially to one electrode (see Appendix 4). Large amounts of dc current, 80,000 to 100,000 amperes at low voltages (about 3 volts), are passed through the solution and the aluminum is deposited on an electrode. The power requirements are quite large. The total energy required is about 75 M Btu of which 55 M Btu, or about 16,000 kw-hr, are electrical. The interest in recycling aluminum is made clear by the comparable figures for scrap aluminum: 8.5 M Btu per ton, one tenth the energy cost of virgin aluminum.

The chemical industry is the largest industrial user of electrical energy. While electric motors and similar electromechanical devices consume most of this energy, electricity is playing a larger and larger role in the synthesis of new materials and especially in the cracking of oils in petroleum refining. There are also, on the horizon, a host of new energy-intensive, electrochemical production processes which will drive this industry's power demand on upward.

The one sector in which there has been little substitution has been transportation; moderate growth, 60 percent from 1965 to 1990, was forecast in Table 6-9. Some of that growth may come from electric cars. The FPC makes an estimate of 62 M mw-hr per year as the energy cost of charging 19 million electric cars each driving an estimated 6,500 miles per year. This electric load would be 3.8 percent of the 1970 total of 1,642 M mw-hr and 1 percent of the expected 1990 total.

Electric rail transport may also increase. If long-haul, high-speed railways become a reality, they could add 20 M mw-hr. Rapid transit will surely increase; the present electrified rapid transit systems in New York City, Chicago, Boston, Cleveland, Philadelphia, Camden-Philadelphia and Trans-Hudson consume about 3 M mw-hr of electric energy annually. The San Francisco Bay area is a late addition and Washington, D.C., Atlanta, and several others may be added by 1990. The FPS estimate is 7 M mw-hr by 1990. The total of electric cars, railways, and rapid transit, 89 M mw-hr, is only 1 percent of the 5.8 B mw-hr projected in Fig. 6-8.

Energy-Intensive Substitution. In this chapter and others we have indicated the role of electricity in the production of many of our modern materials such as aluminum, plastic, and synthetic fabrics. A substantial part of the growth in electrical demand as well as in total energy demand is caused by the substitution of these materials for wood, paper, cotton, and wool. A few examples will make this point.

Table 6-14 shows some energy data from the packaging industry which allow comparisons of the energy-intensiveness of the various materials used in making cans, bottles, and similar containers. The energy advantages of glass over plastic and steel over aluminum are demonstrated clearly. Table 6-14 also makes the surprising point that the styrofoam meat trays cost less energy than do the ones made from wood pulp. Even though styrofoam requires more energy than pulp to produce, the task of forming the tray is much more energy intensive for the latter. Another substitution which has already been mentioned is that of aluminum (16,000 kw-hr per ton) for steel (now 260 kw-hr per ton, but increasing). Simple mathematics shows the greater amount of energy required to produce aluminum. However, this substitution could actually help conserve energy. The aluminum industry has suggested that the use of more aluminum in automobiles could save energy by reducing car weight and thereby increasing gas mileage.

Several synthetic fabrics being used today are more energy-intensive than the materials they are replacing. Nylon, for example, formed from petroleum-derived raw materials, requires several energy-consuming steps in its manufacture, making it more energy-intensive than silk or cotton which it replaces. The same is true for the replacement of soap by detergents.

An interesting side effect of many of these synthetic substitutions, and part of the economic basis for their occurrence, is that they are energy-intensive rather than

Table 6-14 Energy Cost of Packaging Materials*

Type of container	*Production energy (kw-hr)*
Aluminum can (12 oz pop-top)	1.91
Steel can (16 oz)	1.17
Glass bottle (16 oz)	1.16
Glass jar (1 qt)	1.78
Plastic bottle (1 qt)–polypropylene	3.2
Plastic bottle (1 qt)–polyvinylchloride	3.8
Plastic bottle (1 qt)–polyethylene	2.90
Styrofoam tray (size 6)	0.25
Molded pulp tray (size 6)	0.45

*Source: H. Makino and R.S. Berry, "A Thermodynamic Analysis of Packaging, Transport, and Storage," (Chicago, Ill.: Illinois Institute of Environmental Quality, June 1973).

labor-intensive. They produce cheaper goods as well as fewer jobs. We will look at this type of substitution next.

Electric Power for Manpower. An interesting view of the causes of the growth of electric power demand in the industrial sector is presented in a paper by Barry Commoner and Michael Corr, "Power Consumption and Human Welfare in Industry, Commerce and the Home."[6] Commoner and Corr focus on what they call "electric power productivity," which they define as value added per kw-hr used in production, as compared to manpower productivity which is value added per man-hour. (Value added is the value of the goods produced minus the cost of material and energy used in their production.) What they found is shown in Fig. 6-9. Since 1947 manpower productivity has almost doubled; overall energy productivity rose strongly until about 1954, probably due to the improvements in overall efficiency (as we have seen in Fig. 5-3) but electric power productivity fell in the period 1947-1958 and has more or less stayed flat since then.

The overall data for the period 1947-1967 adjusted to constant 1958 dollars are displayed in Fig. 6-10. We see that while value added increased by a factor of 2.3 and electric energy input doubled, manpower only went up by 14 percent. Further analysis suggests that industries with low power productivity (primary metals, chemicals and petroleum, for example) have increased their consumption of electric power

[6] *Ibid*, Vol. III, Chap. 4.

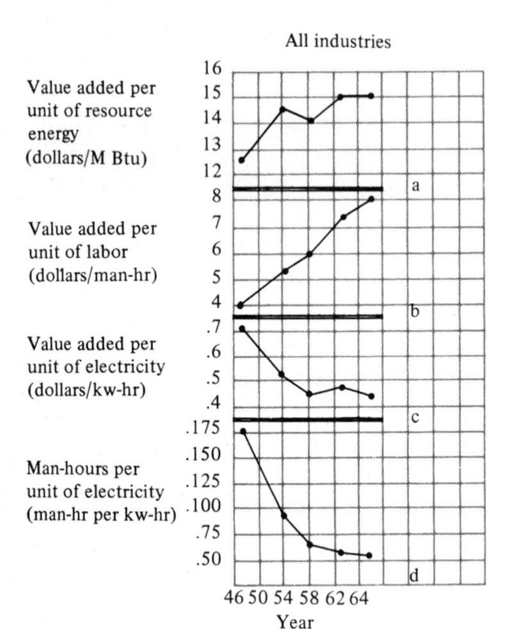

FIGURE 6-9 The drop in electric power productivity (*c*) and the dropping ratio of man-hours to kw-hr (*d*) suggests that electric power is replacing man power. [Source: *Energy and Human Welfare, A Critical Analysis,* (New York: Macmillian, 1974).]

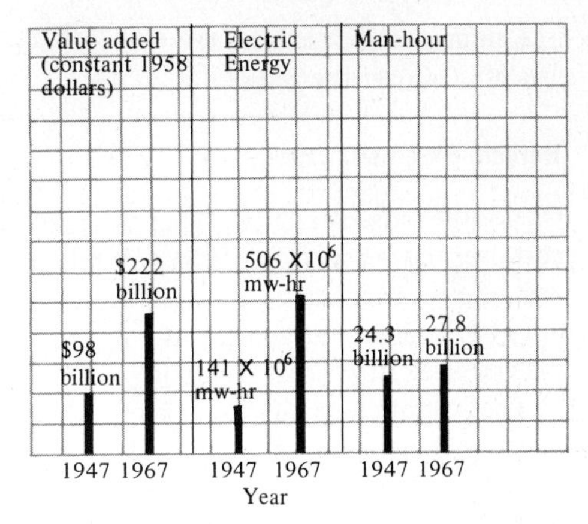

FIGURE 6-10 The comparisons of value added (in 1958 dollars), electric energy, and man-hours for the years 1947 and 1968 show striking differences.

more rapidly than industries with high power productivity. Commoner and Corr draw the following conclusion:

> On the basis of this analysis it becomes evident that the rapid growth in industrial power consumption *has not been accompanied by a comparable growth in the value of the goods produced by industry,* in large part because only a few industries, which contribute a relatively small share of the actual economic growth of industry, account for a good deal of the growth in power consumption. The rapid growth in power consumption is largely due to those industries which use power *least efficiently*—primary metals, chemicals and petroleum products in particular.

Comparing the use of power productivity to manpower productivity in these particular industries is really not justified. It is just these industries, as we have seen in earlier chapters, in which energy is used like an input raw material rather than like manpower. It is in other industries, however, where machines—particularly automated machines—have replaced or extended manpower, that the comparison is more informative. The tentative conclusion drawn from such a comparison is that, while the replacement of steam-driven prime movers by electric motors or human-operated lathes by automated ones increased the consumption of electric energy, it was also largely responsible for the *decrease* in total energy consumption per unit of goods produced since World War II. We have apparently reached the end of the period of large gains in efficiency of this type and must demand stricter energy accounting for any further replacements.

THE SELLING OF GROWTH

We have considered some of the causes of growth that have a physical base: growth from new appliances, new production methods, and substitution of materials, for example. We must also mention some of the other equally important causes which are more directly economic in nature and derive from the nature of electricity and the industry which sells it. The first of these is the bargain in energy that electricity has offered over the years.

Bargain-Basement Energy

One of the most attractive selling points of electric energy has been its decreasing cost over the years. The data base of this assertion is shown in Fig. 6-11, where the cost of electric energy in constant dollars is compared with the *consumer's price index* (CPI) which measures the average cost of other consumer items. Even in actual dollars, electricity's cost record is good.

The cost of electricity has fallen in spite of rising costs of fuel, labor, and construction materials because of increases in the efficiency of production. These increases are those of scale which we discussed earlier, and actual increases in the efficiencies of the steam turbines and, to a lesser extent, in the generators. In 1925, for instance, 25,000 Btu of fossil fuel energy was required by the average generating plant to produce 1 kw-hr of electricity. Since the actual equivalence of electrical energy is 3412 Btu per kw-hr, this conversion efficiency was 13.6 percent. In 1970 the average was 10,400 Btu per kw-hr and the best plant required only 8600 Btu per kw-hr; the corresponding efficiencies are 32.8 percent and 39.6 percent. The decline in the *heat rates,* as these Btu values are called, is shown in Fig. 6-12.

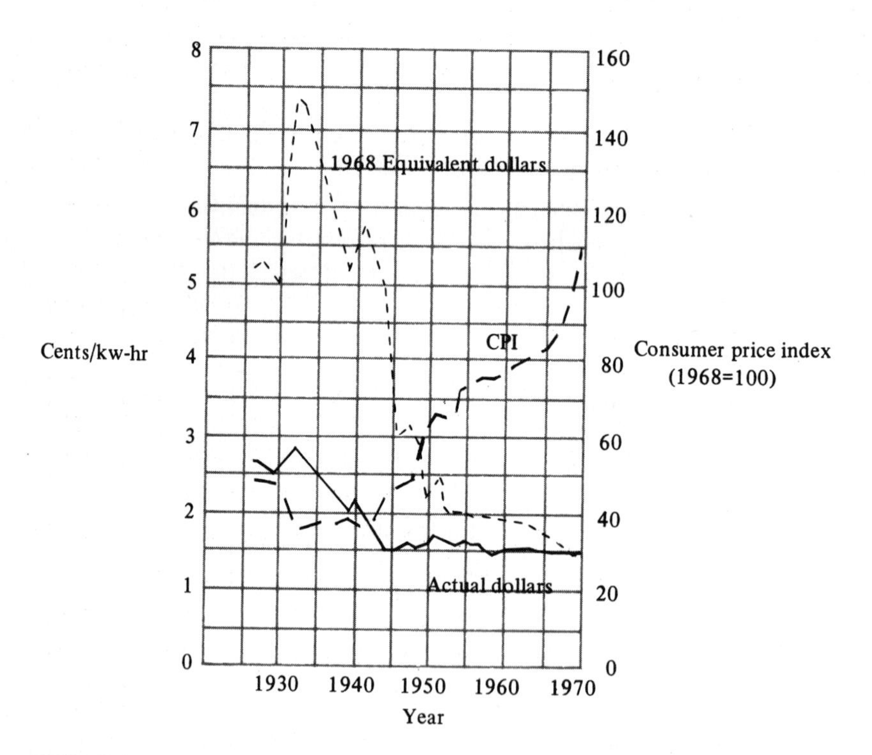

FIGURE 6-11 The cost of electric energy in equivalent 1968 dollars increased from 1926 to 1930 and has gone steadily downward since then in contrast to the rising CPI. The cost in actual dollars has also been decreasing. (Source: *Energy and Human Welfare, A Critical Analysis, op. cit.)*

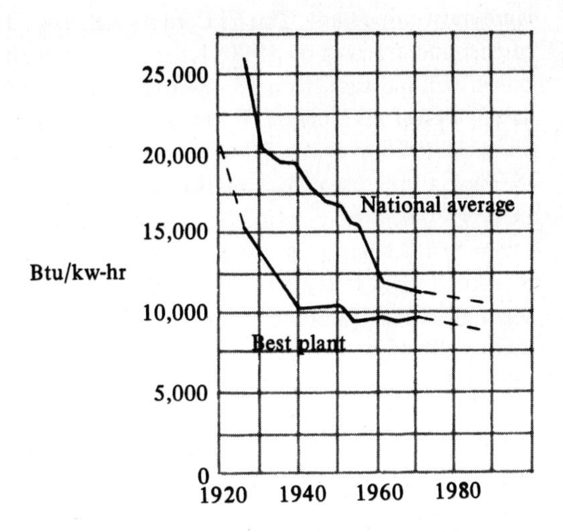

FIGURE 6-12 The heat rate, the amount of thermal energy (in Btu) required to produce a kilowatt-hour of electric energy, dropped sharply from 1920 to 1950 but has leveled out. At 100 percent efficiency the heat rate would be 3412 Btu. (Source: *Energy and Human Welfare, A Critical Analysis, op. cit.)*

The period of significant increase in production efficiency or *scale efficiency* seems to have come to an end. Figure 6-12 shows the leveling out of the production efficiency. The industry's costs, however, continue to rise, and, in 1971 we reached an historic turning point; for the first time, even constant dollar prices inched upwards while current dollar charges rose disturbingly; utility after utility went to its Public Service Commission to ask for rate increases. This upturn seems to have been brought about by the usual multitude of factors, but among these, rising fuel prices and the increases in costs brought about by more stringent air pollution standards

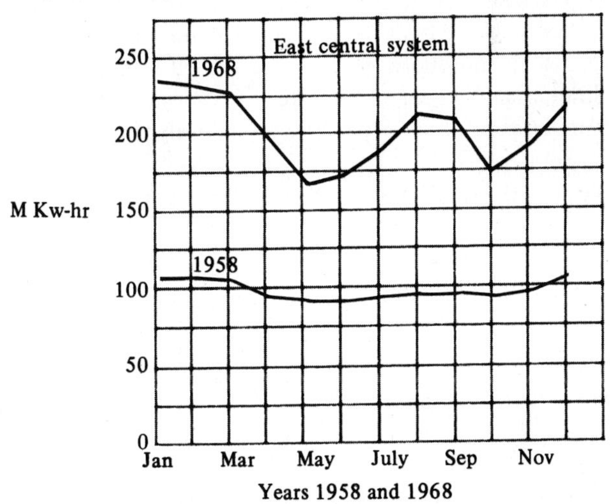

FIGURE 6-13 In 1958 there was a "valley" in the elctrical consumption curve for the East Central System. By 1968 the increase in air conditioning had created a summer peak. (Source: *Energy and Human Welfare, A Critical Analysis, op. cit.)*

were most important. The FPC, in its extensive 1970 report, has made estimates both of cost and demand by 1990. Its conclusion is that the total annual cost of electric power will increase from 18,484 million dollars in 1968 to 97,200 million dollars in 1990—a fivefold increase. It also anticipates that sales will increase from 1.198 T kw-hr in 1968 to 5,327 T kw-hr in 1990 and the average use per customer (which averages all the use categories) is expected to triple from 17,300 to 57,200 kw-hr per year. The average price per kw-hr is projected to increase from 1.54 cents per kw-hr to 1.83 cents per kw-hr, a 19 percent increase. (The comparison is in terms of 1968 dollars so that there is no inflationary input.) When inflation is taken into account, current dollar costs increase even more impressively (see Table 6-15). From these projections it appears that low cost as a cause of growth may lose some of its drive.

Perishable Goods

Another cause of growth comes from the transient nature of electrical energy and from the fluctuating demand we have mentioned; the seasonal demand variation, peaking in the summer air conditioning season in the South and in the winter heating (and dark) season in the North, for instance, as well as from the weekly variations shown in Fig. 6-5. These great variations put a burden on the utility industry. Sufficient generating capacity to supply the peaks has to be provided, but this capacity then lies idle during the off-peak times. Since fuel costs make up only one sixth or less of the total cost, unused generating capacity is expensive. Thus, there are powerful economic incentives to promote electrical energy uses that tend to even out the peaks.

The classic example of this promotion is provided by air conditioning. Figure 6-13 compares the seasonal demand data in the East Central region for the years 1958 and 1968. In the earlier year, the period from April to October was a valley, with 20 to 30 percent less demand than in the winter months. By 1968 a large air conditioner peak is occurring during the hot months and the valleys are in the spring and summer. Corroborating data is found when one looks at the percentage saturation of the residential air conditioning market; in 1959 it was 15 percent saturated for room air conditioners and 2 percent saturated for central air conditioners; in 1969 these figures were 36 percent and 9 percent, respectively. In fact, in certain parts of the country,

Table 6-15 Cost of Electricity to Ultimate Consumers 1968 and Projected 1990*

United States† average (year)	*Cents per kw-hr*	*1990 projected at various inflation rates*
1968	1.54	
1990	1.83	1968 equivalent dollars
1990	2.28	1%
1990	3.51	3%
1990	5.35	5%

*Source: *The 1970 National Power Survey*, Part I, Federal Power Commission, (Washington, D.C.: USGPO, December 1971), p. 1-19-2.

†Excluding Alaska and Hawaii.

the Southeast in particular, the air conditioning boom has led to such dominance of the summer peak over the winter peak that the promotional drive is to sell more electric heating to bring up the winter peak. That the utilities recognize this is shown by the following quotation made in 1955 by Phillip Sporn, a well-known industry spokesman:

> Consequently, with present trends holding, summer air conditioning will continue to be a difficult, if not an almost impossible load to service profitably. Besides its high demand and low annual load factor, its highest demands will occur on the relatively few extremely hot days when load carrying capacity of the system is considerably lower than in winter. Growth of air conditioning inevitably creates a system summer demand peak.
>
> As the most obvious solution to the utilities' problem is the development of winter load, the most logical means on the horizon is electric heating, either resistance heating or the heat pump. When wide acceptance of summer cooling materializes, as predicted, it will become an impressive percentage of the total residential load.[7]

This advice was taken. In the Southeast, for instance, in 1957 the ratio of the winter peak to the summer peak was 0.84 (more electric demand in the summer), in 1962 the same ratio was 1.24 (more electric demand in the winter). The 1968 ratio was about 1.05.

The most obvious tactic of the promotional effort to keep the generators turning is the "all-electric home" concept. The all-electric home medallions and the preferential rates which the utilities award owners of the all-electric home are a familiar sight in the South and are now creeping northward. This trend is expected to keep the demand growing; the all-electric home owner uses, on the average, three times as much electrical energy as other residential customers.

SELF-FULFILLING PROPHECY

We have tried, in this brief survey, to discuss at least representative examples of forces that will serve to make the projections of Table 6-9 become reality. One important force, which we have not yet discussed, is the projection itself.

In a certain sense all predictions influence results by serving, to some extent, as goals. Predictions of electric power consumption, however, are very likely to be self-fulfilling prophecies because of the nature of electric power, the system, and its rate of growth. In other words, the prediction almost guarantees the results. The two factors where interlocking action produces this guarantee, are the long times necessary to plan, build, and start up a big generating plant and the rapid growth of consumption. It requires 3 to 4 years of planning and another 4 to 6 years of building to get a generating plant on line. If power consumption were doubling at a leisurely pace, say, every 20 to 30 years, there would be some time to look ahead and see how the prediction was working out, to watch for new effects in the economy which might change it. With the short doubling time of 10 years, however, one has to begin now to build the 1980 plants. It is, therefore, necessary to accept the projection as accurate and to immediately begin to build the plants to double the generating ca-

[7] *Vistas in Electric Power,* Vol. 3, pp. 928-929.

pacity by 1980. When 1980 arrives the capacity is there and, for the reasons we discussed earlier, the utilities lose money if the capacity isn't used. The promoters then take over, lower rates are offered to bulk users, all-electric home owners are given preferential treatment, the full capacity is sold, the prediction comes true, and another round begins.

A specific example of this feedback between prediction and result appeared in the controversy-surrounded Calvert Cliffs nuclear plant on the Chesapeake Bay. A study showed the usual increasing need for electric power in the Washington, D.C.-Maryland area. This need overran the various environmentalist objections to the installation of the plant and two 845 mw nuclear-powered generators are scheduled to go into operation, one in 1973 and one in 1974. Then, in 1972 it became known that an aluminum plant, East Alcoa, was to be built in the region of the plant, attracted, in part surely, by the guaranteed low-cost electricity. Thus, we have a prediction of need and a generating plant is built; the existence of that power plant then attracts the consumers who make the prediction come true.

Projections are necessary, electric power is a necessity, and it is the responsibility of the utilities companies to provide this power. Nonetheless, the interaction between prediction and result must somehow be lessened and we must gain control of this always-rising curve of growth before resource scarcity and environmental deterioration exert their own irreversible control.

SUMMARY

In this chapter we have looked at one form of energy: electricity. This energy is unique in its flexibility and convertibility as well as in its rapid growth in popularity. Unfortunately, it is also unique in the low efficiency of its generation, as presently accomplished with heat engines supplying the motive power.

We described the four steps in the journey electric power makes from its source to its eventual reradiation as heat energy back into space. It is generated chiefly by fossil fuel powered plants, transmitted at high voltages to the consuming area, transformed to lower voltages and distributed to the customers where it is used in a great variety of ways.

A significant feature of generation is the increasing size of generators, plants, and systems. Modern generators now have capacities greater than 1,000 mw; the largest plant has a capacity greater than 3,000 mw, enough capacity to handle the power needs of Washington, D.C.

Transmission voltages, with present maximum values of 765,000 volts, are also increasing and will provably rise above 1 million volts. The next two decades will see the number of miles of these lines increase, from 300,000 in 1970 to 500,000 by 1990. Farther in the future lie supercold and superconducting lines to carry large power loads from energy parks to the metropolitan areas.

We looked at the end uses from two points of view: to see where the electric energy went and to gain some understanding of why its growth was so spectacular. The three major users are the residential sector (31 percent), the commercial sector (24 percent), and the industrial sector (45 percent). The first two sectors use most of this energy for comfort, heating, and air conditioning, although refrigeration and water heating are substantial consumers as well. In the industrial sector the primary metal and chemical industries are the big consumers.

The causes for the projected fourfold increase in consumption between 1970 and 1990 are varied. They come, for instance, from substitutions: electric space heating for gas and oil, or electric arc furnaces for coal-fired open-hearth furnaces, for example. Growth is also stimulated by the substitutions of energy-intensive materials for energy-economic ones (aluminum for steel or wood, plastic for paper, nylon for cotton) and the general substitution of electric power for manpower.

Of particular concern in the area of electric power demand, is that prophecies about it have a tendency to be self-fulfilling; that electricity, perishable product that it is, must be sold. This fact also is pushing the demand curves upward. As we consider, in the next three chapters, some of the environmental impact of the production and consumption of electric power, we will provide reinforcement to our belief that a tight rein on this growth is necessary. A strict accounting of its expenditure is needed if we are to have the many benefits of this flexible form of power and, at the same time, retain a healthy environment in which to enjoy those benefits.

7

What Goes Up: Energy and Air Pollution

September 16 in Piscataway was warm, humid and hazy. Early that morning the meteorologist at Newark Airport, 20 miles to the north, had noted in his log that the air condition was "HK"—haze and smoke. The wind was light and blowing from the east and south-east, unusual for September.

At the state air-monitoring post in a Newark parking lot, pollution gauges registered the 8 AM count of "oxidants" in the air at 0.022 parts per million—higher than usual, but no cause for alarm. "Oxidants," in the context of air pollution, are known to the rest of us as smog. They form in the atmosphere through a mixture of nitrogen oxides, hydrocarbons and sunlight. Ninety percent of oxidant pollution, the scientists agree, is produced by automobiles.

Quibbletown School (named for a colonial hamlet where residents couldn't agree whether the Sabbath fell on Saturday or Sunday) sits amid the Jersey slurb. All around it the megalopolis churns. Seven major state and federal highways pass within eight miles of the school. Busy boulevards link the highways. Dozens of factories squat beside the railroads and major roads, many of them the chemical manufacturers that make New Jersey the nation's leading chemical producer. New Jersey also leads in virtually every statistic measuring congestion—the highest population density, most roads per square mile, most cars per mile of road. All the wheels were turning as usual that day.

By early afternoon the temperature in Piscataway was 86°, the humidity 68 percent, the winds easterly at a feeble two knots. At 2:45 the Quibbletown football team trotted onto its practice field behind the school.

The practice was to be routine—15 minutes of passing, punting and limbering up, a couple of laps around the field, 20 minutes of calisthenics and a scrimmage. The 37 members of the team, known as the Braves, were all seventh- to ninth-graders, boys between 12 and 14. They were one day away from their opening game.

Coach Stuart Schnur ran the Braves through the two laps and the calisthenics. At 3:20 the scrimmage began, the starting offensive team running plays against the defense. The sky was still gray and hazy, the field damp from recent rains.

The two main ballcarriers were quarterback Tom McNeill and halfback Ray Jackson. After only five plays Schnur noticed that Jackson was holding his stomach. Jackson complained that he was having trouble breathing. Schnur took him out for a rest.

The offense ran a few more plays, and two more boys complained of feeling sick. They said they felt chest pains when they inhaled. Schnur halted the scrimmage and gathered the team around him. "I was mad," he recalled later. "I thought maybe they were just dogging it." Schnur suggested that another lap around the field might help, and the Braves trotted listlessly around the goalposts.

The scrimmage resumed, but the team looked even worse. "We huddle the

Kansas City Chiefs' way," Schnur said, "The linemen leaning foreward with their hands on their knees, the backs standing with their arms folded behind them. I noticed that the linemen were down on one knee, and the backs were leaning their elbows on the linemen. I train them to run back to the huddle—it looks sharp, you know. But they were dragging back."

The Braves sluggishly ran two or three more plays, but it was obvious they were weak and disorganized. "They looked like they were dead. They weren't even running the plays right," said Schnur. Three more boys were bent over now, holding their chests and coughing. Jackson and McNeill both seemed to be in considerable pain, half a dozen others were almost as bad. Schnur stopped the scrimmage again, this time for good. It was 3:45. One of the boys mentioned that he had seen a helicopter pass over the field earlier in the day. He thought it was spraying pesticide; helicopters frequently sprayed nearby marshes. Schnur, wondering if the boys had been inhaling pesticide from the turf, led his crippled band back to the locker room.

At 3 PM the Newark air-monitoring station showed an oxidant count of 0.080, an abnormally high reading. At another monitoring post in Bayonne the count was 0.096. The only place in the United States where oxidant levels are regularly that high is the Los Angeles basin.

In New York, meanwhile, the average oxidant level between 9 AM and 3 PM was 0.02. The wind was blowing more briskly in New York. There was a wide-ranging temperature "inversion" in the atmosphere, warm air clamped atop cooler air and trapping it, but it was not regarded by the Weather Service as serious. New York City's Air Resources Board rated that day's air quality as "acceptable."[1]

These Jersey high school football players were unpleasantly introduced to a consequence of the Energy Crisis that has not received the attention recently accorded the empty gasoline pumps. They were lucky; they were taken to local hospitals and after the doctors finally identified air pollution oxidants as the cause of their distress, they were given treatment and rest, and then released. There were no fatalities, and apparently, no long-term disabilities.

There is a tendency, in discussing this Energy Crisis, to focus on the problems of supply and demand, to ask: How long will oil supplies last? Will there be enough electricity for the air conditioners this summer? Will the nuclear reactors relieve the shortages? What happened to these kids in New Jersey reminds us that there is another set of questions we must ask which deal with the impact on the environment of the production and consumption of all this energy.

In this chapter and the two which follow we will consider some of these energy-related environmental effects. We begin with the one that is before our nose, so to speak, the fouling of the atmosphere. Air pollution is, of course, not exclusively due to energy production and consumption; industries, agriculture, paint sprayers, dry cleaners, and a host of other factors all contribute. As we shall see, however, energy conversion, in particular, combustion, is the major contributor. One can pinpoint the source of the New Jersey discomfort even more specifically, for we know that automobile exhaust is the major contributor to the photochemical smog described there.

In terms of mortality or money the Quibbletown episode seems rather minor: none of the boys was seriously affected, health insurance paid the costs. On a national scale, however, there is hard evidence of cost in both money and mortality.

[1] Jackson, "The Cloud Comes to Quibbletown," *Life*, December 19, 1971.

(We will look at some mortality-air pollution studies taken in New York City later in the chapter.) In 1968 the Environmental Protection Agency (EPA) estimated the cost of damage due to air pollution to be 16.1 billion dollars annually, or about 80 dollars for every person in this country.[2] Among these costs are:

6.1 billion dollars for health effects

4.8 billion dollars for effects on materials (rust and corrosion, repainting, damage to fabrics, rubber, galvanized metals, the cost of protective coatings)

5.2 billion dollars for damage to residential properties (reduction of values by pollution)

120 million dollars for damage to vegetation (damage to commercial agricultural crops)

One can easily think of costs that are not included such as damage to art, to precision instruments, the aesthetic costs of dirty living and working areas and the costs of cleaning them up, the cost of extra laundry and dry cleaning, the damage to residential flowers, shrubbery, and gardens, and so on. The 16.1 billion dollars is, of course, an estimate. It is not only difficult to directly identify air pollution damage but it is difficult to put a value on it. How does one put a dollar sign, for instance, on death, disease, and discomfort?

At the present time, most of these costs are being borne by the public, often by those who are not directly benefiting from the energy conversion itself, for instance, the Quibbletown football team. The effect of the more stringent controls which are emerging, however, is to shift some of the cost to the energy broker and, through him, to the consumer. The final breakdown is not yet in clear outline; what is clear is that the free ride is over; the true fare for energy will be collected.

The material which follows is designed to help the reader assess the causes and effects, in essence, the polluters and pollutants, and the complex ecological cycles and technological practices which affect and are affected. We begin by looking at the atmospheric dumping ground, the thin layer of air which sustains us and most of the rest of life.

AIR AS A "DISPOSAL"

From our new awareness we can see a dangerous shortsightedness in using air, that most precious necessity of life, as a "disposal" for our waste. In the past, however, when the amount of waste was small and its character fairly benign, this "disposal" was more than adequate, its periods of breakdown hardly noticed.

As our species rode the exponential growth curves (see Chap. 5) into this century, the amount and nature of the junk we put into the atmospheric "disposal" changed drastically. It is now operating, around our cities at least, near overload, so that even temporary breakdown is both costly and painful. We can no longer take its workings for granted; we have been forced to understand the "disposal" in detail, to find out how it works, when and why it doesn't work, and how much of what it can eliminate.

[2] L. Barrett and N. Waddell, "Cost of Air Pollution Damage: A Status Report," United States Environmental Protection Agency, February 1973.

When It Works

Basically we look to air for two functions in the handling of gaseous waste: dilution and dispersion. We count on small-scale circulation in the enormous ocean of air to take the unwanted or dangerous material we release and mix it so thoroughly that the final concentration is no longer dangerous to life or growth. We count on large-scale movement of the air, the solar-powered winds, to carry the wastes away from the city's smokestacks and highways and spread it, in diluted form, over the whole countryside. These two processes are then followed by cleansing mechanisms. The alien particles either serve as the nuclei of raindrops or are swept into raindrops and fall to the earth; the large particles settle to earth, drawn by its gravity. Absorption in the oceans, and, to a lesser degree, filtering by plants and other living things, help complete the task.

Dilution, the first function, depends on quantity and, fortunately for us, there is a lot of atmosphere on our planet, an estimated 5.6 Q tons (5.6×10^{15} tons). In contrast to this it is estimated that the worldwide production of carbon dioxide from the burning of fossil fuels in 1967 was 1.5 B tons (1.5×10^9 tons). Sulphur, particularly in sulphur dioxide, is one of the most serious of the man-made pollutants; it is added to the atmosphere at a rate of only 70 million tons per year. It is unlikely therefore that man will drastically alter the composition of the atmosphere; his contributions, on the average, will not rise above small fractions of a percent. But man is not spread around the earth "on the average" and, therefore, he manages to seriously pollute the air wherever he is concentrated.

The chief components of the atmosphere are nitrogen and oxygen; by volume air is about three quarters nitrogen and one fifth oxygen. In addition to these major constituents, there are traces of other gases; argon, neon, and helium are the next most plentiful. The relative amounts of the various *permanent constitutents,* as these are called, are shown in Table 7-1. As the label "permanent" implies, these percentages remain about the same from place to place or time to time.

In addition to the permanent constituents, the *variable* gases where relative contribution changes from place to place and time to time are also shown in Table 7-1. The most important of these is water vapor, and its concentration, the humidity, shows the most variation. It is followed in importance by carbon dioxide, which like water, plays a major role in life processes. It is *recycled,* or taken from the air by plants which, in turn, produce oxygen; animals reverse the cycle. Although man, largely through his burning of fossil fuels, is beginning to make a measurable contribution to the carbon dioxide content of the atmosphere, we will postpone discussion of this pollutant until Chap. 8 which deals with thermal pollution and climatic change, for that is where its major effect is expected.

The other variable gases such as methane (CH_4), carbon monoxide (CO), and ozone (O_3) are *pollutants.* In our sense of the word, this means that they can cause unwanted effects on living things. While all of them can be produced by man's activity and most of them are associated specifically with the production and consumption of energy, they are also injected into the atmosphere by natural processes. Some of the methane, for instance, is from swamp gas produced by organic materials decaying in water. Carbon monoxide is produced by plants and marine organisms as well as by

Table 7-1 Constituents of the Atmosphere*

Constituent	Chemical formula	Percent by volume	Parts per million by volume
PERMANENT GASES			
Nitrogen	N_2	78.084	
Oxygen	O_2	20.946	
Argon	Ar	0.934	
Neon	Ne		18.2
Helium	He		5.2
Krypton	Kr		1.1
Hydrogen	H_2		0.5
Nitrous oxide	N_2O		0.3
Xenon	Xe		0.09
VARIABLE GASES			
Water vapor	H_2O	0.7	
Carbon dioxide	CO_2	0.032	1.5
Methane	CH_4		0.1
Carbon monoxide	CO		0.02
Ozone	O_3		0.01
Ammonia	NH_3		0.001
Nitrogen dioxide	NO_2		0.0002
Sulfur dioxide	SO_2		0.0002
Hydrogen sulfide	H_2S		

*Source: Sanuel J. Williamson, *Fundamentals of Air Pollution,* (Reading, Mass.: Addison-Wesley, 1973), p. 56.

atmospheric reactions with methane. Ozone is also produced in large amounts at the top of the atmosphere by the action of ultraviolet radiation on oxygen and, in fact, shields us from this radiation. Ammonia (NH_3), nitrogen dioxide (NO_2), and the sulphur compounds are also naturally produced in important ecological cycles. Table 7-2 summarizes and compares the natural and man-made contributions of these latter gases (excluding water) and provides a perspective on the amount of time that it takes them to leave the atmosphere. In some instances the total contributions from natural processes far outweigh man's additions but man still manages to foul those parcels of air in which he lives.

Form and Motion of the Atmosphere. The atmosphere in which all these gases find their temporary or permanent home is but a thin film around the earth. Most of its mass, 99.999 percent, is found below a height of 80 kilometers, or 50 miles. This thin film, however, is *stratified* and these strata have significance to the disposal of pollution.

The most important layer, the lower layer, is called the *troposphere*; it is here that weather is formed and that pollutants are dumped, diluted, and dispersed. The upper boundary of the troposphere, the *tropopause,* is a real boundary: there is little movement of air and, therefore, of pollutants through it. The height of the tropopause varies with latitude, longitude, and the seasons, but it is roughly about 7.5 miles above the equator sloping down to perhaps 5 miles above the poles.

Under normal conditions the temperature of the atmosphere drops with distance away from the warm surface of the earth; the instances when it doesn't are the tem-

Table 7-2 Constituents of The Atmosphere and Common Pollutants*

Gas	*Concentration (ppm)*	*Combustion emission*	*Natural emissions*†	*Half-life*‡
CO_2	320	13	160	2-5 yr
CO	0.08	0.27	4.0	0.2 yr
SO_2	0.0002	0.13	0.044	4 days
Hydrocarbons (except CH_4)	<0.001	0.1	0.2	–
NO and NO_2	0.001	0.05	0.5	5 days
H_2S	0.0002	0.003	0.1	2 days
CH_4	1.5	–	2.0	1.5 yr
N_2O	0.3	–	0.6	4 yr
NH_3	0.01	–	1.2	7 yr
O_2	2.1×10^5	-10	100	1,000 yr

*Source: S. J. Williamson, *Fundamentals of Air Pollution,* (Reading, Mass.: Addison-Wesley, 1973).

†In billions of metric tons (1000 kilograms) per year.

‡The estimated time for half of an emitted amount of a pollutant to be remined by natural means.

perature inversions so important in air pollution. The tropopause, however, is always cold, averaging about 220°K, or -53°C. Because of this coldness, water vapor condenses into droplets which then freeze at this boundary and are prevented from being carried higher. This is indeed fortunate, for if the water were to be exposed to the ultraviolet radiation in the stratosphere instead, it would dissociate into hydrogen and oxygen, and the light hydrogen would escape from the earth. Thus, the tropopause is a lid which preserves the earth's water.

There are layers above the troposphere: the *stratosphere* where the temperature rises with height, the *mesophere* in which it again drops, as well as others. For our story, however, we can confine our attention to the lowest region, the troposphere, which is the working part of the "disposal."

Winds. The great quantity of air can dilute even the large amounts of gaseous waste that our energy conversions are producing. Quantity alone is not enough, however; we count heavily on turbulence to mix the wastes into the atmosphere and vertical and horizontal air currents, or *winds,* to carry them away. That this is so, is painfully obvious on those days when air currents refuse to rise, the winds are stilled, and we are forced to breathe our refuse.

The energy which runs our pollutant disposal comes from the sun. That it works for us is due to the fortunate circumstance, which we have noted earlier: the atmosphere is virtualy transparent to the sun's radiation, and roughly 50 percent of the solar energy is absorbed at the earth's surface. The troposphere is, therefore, warmed from the bottom. It is this condition that gives rise to the vertical and horizontal air currents which disperse our gaseous waste. We see the effects of the opposite condition, surface warming, in the ocean. The depths of the ocean are quite stable so that there are no appreciable currents for mixing and dispersal. Inhabitants of the ocean floor have a most difficult time getting rid of their wastes.

The mechanisms by which solar energy (kinetic energy in the form of light) is transformed into wind (kinetic energy of moving air) are most interesting, as are the descriptions of the massive movements of air masses which form our weather. We will be forced to bypass discussion of most of this and concentrate on those mechanisms which are important for the dispersal of pollutants.[3]

Atmospheric motions, large and small, are caused by the warming of the lower layers of air by the sun-warmed ground. Hot air is less dense than cold air and rises: wind and weather start from there. The temperature difference between the equator and the poles causes the large-scale motions: warm air rises and is replaced by cooler air from the polar region. This north-south motion is then broken into large swirls of air by the rotation of the earth. The "cells" which form are shown in Fig. 7-1. It is these "cells" which account for the unidirectional *trade-winds.*

Out of this "cell" structure are formed the large masses of high and low pressure air that move across the temperate zone, the so-called *cyclones* and *anticyclones.* The cyclones are low-pressure regions; the air is warmer so that it expands and rises and, due to the earth's rotation (the *Coriolis force*), spirals inward horizontally, counterclockwise in the Northern Hemisphere and clockwise in the Southern Hemisphere (see Fig. 7-2). The anticyclones are high-pressure regions with the cooler, heavier air subsiding as the spiral unwinds (see Fig. 7-3).

The spiraling masses of air, whose dimensions may be a hundred or so miles, have a major impact on weather and therefore on pollution. Some of them are more or less fixed in location. The *semipermanent anticyclone* over the Pacific is, as we will see, one of the causes of the pollution distress of Los Angeles. What is most important for the working of the "disposal," however, are the migratory cyclones and anticyclones which cross the country.

The low-pressure cyclones bring cloudy skies and rain. The air rises and cools; the water vapor condenses and soon overloads the carrying capacity of the air. The opposite occurs in the anticyclones. The subsiding air is warmed as it is compressed

[3] A thorough treatment of the pertinent meterology can be found in Miller, *Meteorology,* (Columbus, Ohio: Merrill Books, 1966).

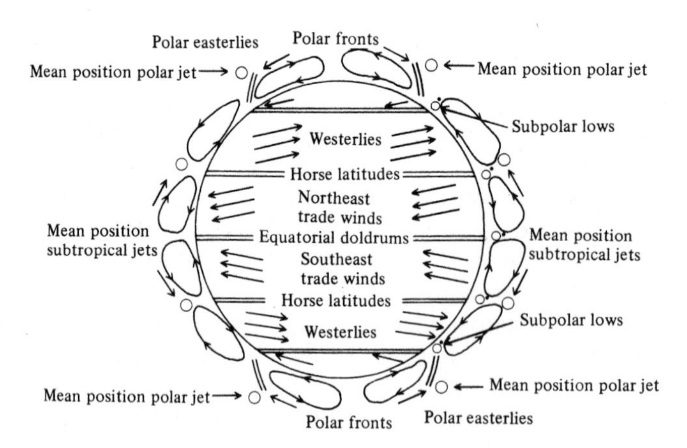

FIGURE 7-1 The circulation of air from the warm equator to the poles is broken up into cells by the effect of the earth's rotation.

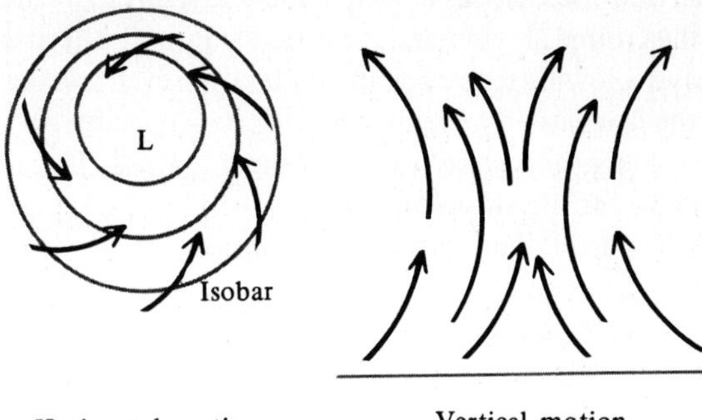

FIGURE 7-2 In a cyclone the winds spiral inward (counterclockwise in the Northern Hemisphere) and produce a rising vertical current.

and can carry more water. As a result, anticyclones are usually characterized by clear, calm air while the winds and rising air of the cyclones produce excellent pollution dispersal.

The anticyclone often creates a *temperature inversion.* The subsiding air is warmed by compression as it descends and can become warmer than the air underneath it. Thus, the normal *decrease* in temperature with height above earth's surface is *inverted* and, for a while, the temperature *increases* with height. These inversions, as we shall explain later, put a lid on the lower regions of the atmosphere and suppress the vertical currents which are so necessary for dispersal. They are a prime cause of air stagnation, one of the conditions that causes the "disposal" to fail.

When It Doesn't Work

By now temperature inversions are part of our daily vocabulary; they mean, for the urban dweller, hazy skies and noxious air. To describe their meteorological and geographical causes as well as their effects we must say a little bit more about the vertical motion of air.

As we have already noted, these motions begin because warm air is less dense than

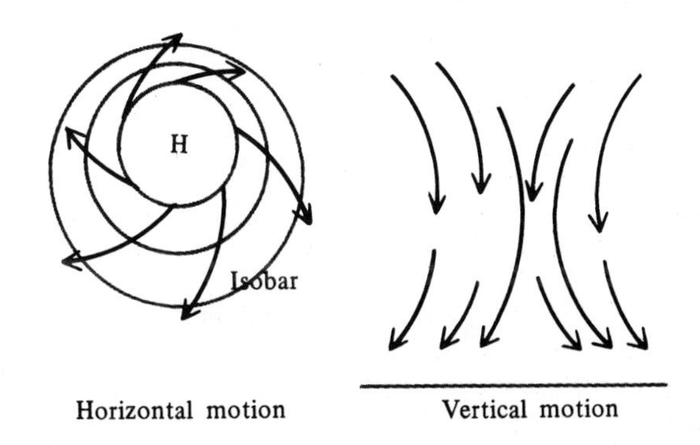

FIGURE 7-3 The subsiding vertical movement of air in an anticyclone is accompanied by winds that spiral outward (clockwise in the Northern Hemisphere).

cool air and rises. Let us consider what happens to a parcel of air warmed by contact with the ground. It rises and, as it floats upward, the atmospheric pressure around it decreases allowing it to expand and become even less dense. At the same time, however, the decrease in pressure cools it (see Appendix 3). This, in turn, increases its density and weakens the force pushing it upward. The crucial comparison that determines how far the air parcel will rise, therefore, is the comparison between the cooling of the parcel of air due to its expansion and the drop (with altitude) of the temperature of the air around it.

We have indicated earlier that the temperature in the troposphere normally decreases with altitude. This *temperature profile,* as it could be called, depends upon many things: the prevailing winds, the intensity of the sunlight, the humidity, to name some. If the temperature of the surrounding air decreases with altitude *more* rapidly than does the temperature of the parcel, then the difference of temperature and, consequently, the difference in pressure, increases with altitude. The parcel will move upwards more and more rapidly until it is either stopped by the tropopause or, more likely, it rises into a region of different temperature-altitude characteristics. This is a condition of instability; any small increase in temperature of an air parcel under these atmospheric conditions will cause it to rise immediately.

Instability of this sort is rare, although it sometimes occurs in the winter. The more likely cause of instability is the moisture content of the air parcel. If the air has high humidity, the cooling we have referred to may cause it to condense thus releasing a rather large amount of heat energy: 1050 Btu per pound of water. This released heat then slows down the rate of cooling and can keep the parcel warmer than the surrounding air, allowing it to continue rising. It is this kind of instability that produces the summer thermals on which buzzards and gliders ride.

Unstable air produces a good day; the smoke and exhaust is carried upward by rising air, mixed thoroughly, and carried away by the winds which blow at the higher altitudes. It is the condition of stable air which makes the "disposal" break down.

Air is stable if the temperature of the surrounding air drops *less rapidly* with altitude than does the temperature of the ascending parcel. Under these conditions the temperature, and, therefore, the pressure difference between the parcel and its surroundings, become less and less until finally they are the same and the parcel rises no higher. Such a condition can hold pollutants near the earth's surface. Stability can be brought about by horizontal air movements, warm air moving in over cooler air, by cooling at the surface (over the ocean, for instance), or by the presence of clouds that absorb energy radiated from the ground and warm the air at upper levels.

Temperature Inversions. What is much more serious than the condition just described is a temperature inversion, an *increase* of temperature with altitude. There is no way an expanding, cooling parcel of air can make its way through such air. This condition is usually responsible for holding the polluted air down in the nose of a city.

There are several atmospheric conditions which can bring about a temperature inversion. We have already mentioned one: the anticyclone, or high-pressure air mass in which subsiding air is compressed and warmed, producing a region of warmer air above the cooler surface layer (see Fig. 7-4). If this mass stops its cross-country move-

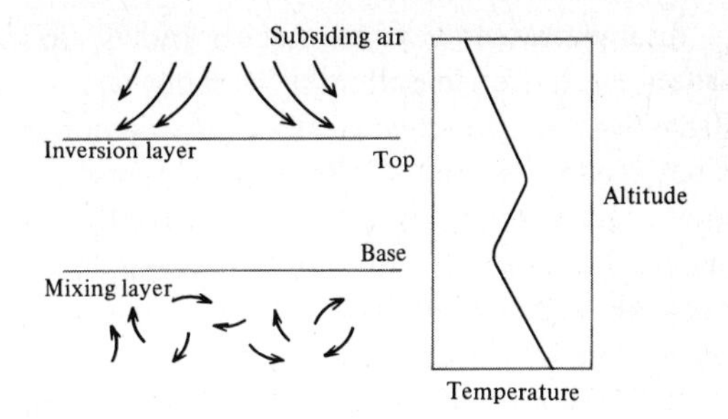

FIGURE 7-4 Subsiding air warms by compression and creates an inversion layer (a region in which the temperature rises with altitude) which prevents vertical circulation.

ment, then the area underneath it experiences stable, in fact stagnant, air. In the good old days, these periods of stagnant air posed no threat; they brought the "Dog Days" of summer picnic weather, calm, lazy days with clear skies. It is not so now. Most of the very serious air pollution episodes, such as in Donora, Pennsylvania in 1948, London in 1952, and New York in 1962 were due to such stagnation.

There are many regions in the United States where this kind of stagnation occurs quite frequently. The Great Basin between the Sierras and the Rockies is one, the Central Basin of California another. In the Eastern region, the Southern Appalachians experience these periods of stagnation with above-average frequency and, it is believed, this contributes to the accumulation of natural hydrocarbon emissions where haze gives the name to the Blue Ridge and Great Smoky Mountains.

While these stationary anticyclone inversions are the most massive and most serious, there are other mechanisms which produce temperature inversions. There are, for instance, inversions associated with the movement of hot and cold *fronts.* A front is the interface between two large air masses of different temperature and usually, different humidity and pressure. If the front consists of warm air moving into a colder region, it is called a warm front and vice versa. These fronts have the distinctive shapes shown in Fig. 7-5. In both cases the warm air is forced up over the cold air, since the warmer air is lighter. When the cold air is moving (a cold front) friction with the ground curves it over while the friction drags out the cold air pushed back by a warm front (see Fig. 7-5).

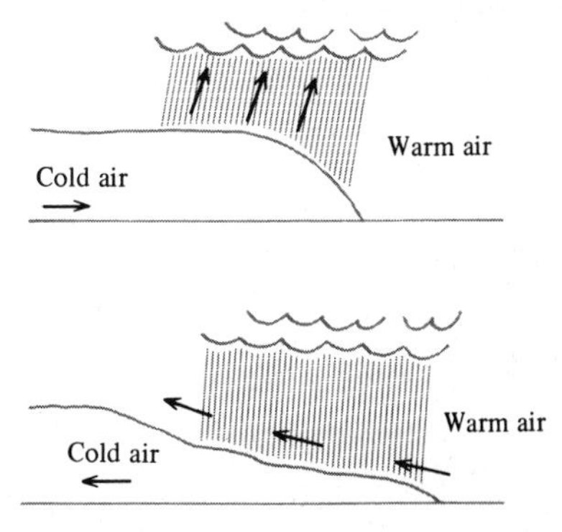

FIGURE 7-5 Schematic representations of a cold front (top) and a warm front (bottom).

As any amateur weather-watcher knows, clouds and rain are associated with fronts, caused by the condensation in the moist warm air which is lifted and cooled. From Fig. 7-5 we see that such weather precedes a warm front and follows a cold front. There is an inversion in both cases, of course, for there is warm air above cold and the temperature does not decrease with altitude. Pollutants are thus trapped in the underlying cooler air and, should the front stop moving over an urban area, there are bad times below. Fortunately, the turbulence and horizontal winds usually make up for the lack of vertical movement.

A second kind of inversion occurs when warm air moves over a cold surface; its lower layer is cooled and *ground-based inversion* is produced. The seriousness of this depends on its depth; a tall smokestack, for instance, may puncture it and release its pollutants above the inversion. Warm air blowing over a mountain and riding out across cold air produces a similar inversion. Such conditions often occur at Denver, Colorado during winter months.

The most common and least troublesome type of inversion comes at night. On a clear night, the earth's surface radiates energy away and cools rapidly: the lower atmosphere is then cooled while upper regions remain warm. These *radiative inversions* are frequent and their existence is shown by the ground-hugging nature of smoke and exhaust in the early morning hours. Since anticyclones (high-pressure centers) are usually clear, this effect can increase the problem of nighttime pollution during a period of high-pressure stagnation. Fortunately, in most instances the sun quickly warms the ground and destroys the inversion—unless fog or pollution shields the ground and keeps it cool.

Geographical Causes of Inversion. There are certain geographical conditions that, combined with meteorological features, cause frequent inversions. It is such an occurrence that is partially responsible for the sad plight of Los Angeles. We say *partially,* because Los Angeles has many conditions which conspire to cause its smog and we can use this to illustrate several of the ones we have already described, as well as some new ones.

The dominant influence in the workings of Los Angeles' inefficient pollutant disposal is the semipermanent inversion which sits over the region most of the time. As one can see, referring back to Fig. 7-1, at about 30° north latitude the two circulation cells are both sending air down toward the earth's surface. This subsiding air, driving against the lower air, is warmed by compression. This warm-inversion layer produces stagnation; 30° north is the Horse Latitudes, the scene of frequent becalmings of the ships which sailed for the New World. Los Angeles, at 32° north latitude gets both the clear, dry weather that this semipermanent anticyclone produces and the lid on pollution dispersal. It is reported that Los Angeles experiences, on the average, 300 days of temperature inversion in a year.

This effect is added to by geographical factors. Los Angeles is surrounded on the landward side by mountains which hinder horizontal wind flow. Cyclic breezes are a feature of seashores. During the day, hot air rises from the land and is replaced by cool air from the water; at night, the land cools faster than the sea and the breeze reverses blowing back toward the sea. This land breeze carries back over Los Angeles, the pollutants carried inland during the day; at times the pollutants circulate back and forth across the basin.

The land-sea breeze cycle produces an inversion over the land at night, as the pressure from the subsiding air warms the upper layers. Mountains and valleys show similar effects. During the day, the air along the mountainsides is warmer than air at a similar height but further away, and there is an upslope wind. At night, the situation is reversed: the ground cools more rapidly by radiation; the adjacent air is cooled, producing a downslope wind and the inversion always associated with subsiding air.

Land-sea breezes and upslope-downslope winds with their day-night cycles create serious pollution problems bringing the polluted daytime air back over a city at night. Los Angeles boasts both types; the land-sea breeze cycle is troublesome to other coastal cities and even to Chicago. The upslope-downslope cycle adds to the woes of Denver.

Fortunately, the winds and weather are still uncontrollable; we take what we can get. Where man can exert his control is at the other end: he can decide what and how much is dumped into the "disposal." We will now look in detail at the major pollutants and the ways in which they cause damage and finally, at the polluters and the methods for their control.

AIR POLLUTION AND MAN

It is difficult to avoid writing down, at some stage, a definition of a pollutant. Perhaps we should adopt the age-old definition of beauty and admit that "pollution is in the nose of the inhaler." That definition would have done for Queen Elizabeth I who, it was reported: "findeth hersealfe greately greved and anoyed with the taste and smoke of sea-cooles." However, while the pollutants of today include the sharp and easily recognized odor of ozone as well as an appreciable quantity of visible smoke and soot, they also include much more subtle waste gases, some of which, like carbon monoxide, are only detected by their effects.

We will define a pollutant as something that is not a natural constituent of the atmosphere, and add the qualification that it adversely effects something that man values (his health or his crops, for example). In the discussion which follows we will additionally limit ourselves to pollutants which are released into the atmosphere as a consequence of the production or conversion of energy. This latter qualification does not significantly reduce our scope, for as we shall see, energy conversions are the major sources of air pollution.

One other distinction which must be made in discussing pollutants is that between primary and secondary pollutants. Primary pollutants are those discharged directly into the atmosphere. Once in the atmosphere, triggered by sunlight or by mixture with other atmospheric materials, these primary pollutants may undergo chemical reactions and change their form. These new forms are called secondary pollutants. They are not necessarily secondary in their effects or importance; some of them, like the irritating peroxyacyl nitrates (PAN) caused by sunlight falling on automobile exhaust, are more troublesome than many of the primary pollutants.

The five most important primary pollutants, in terms of amounts emitted, are: (1) the *particulates* (fine ashes, soot, and various droplets); (2) the *oxides of sulphur* (which we will label SO_x); (3) *carbon monoxide* (CO); (4) the various *hydrocarbons* (HC); and (5) the *oxides of nitrogen* (which we will label NO_x). In addition to these five which are now routinely measured, we will consider lead, which is released in the burning of leaded gasoline. We will postpone discussion of carbon dioxide until Chap. 9, for its effects are on the climate rather than directly on man and his belongings. These pollutants, which come predominantly from energy conversions, produce several types of damage. Before we discuss the various adverse effects of specific pollutants, however, there are some important general statements to make about air pollution and man.

Effects on Humans

We will deal with health effects of specific pollutants later on in this chapter. Before such detail, however, we wish to set the stage with some background comments. The first of these concerns the respiratory system, the point of attack for airborne pollutants.

The Defense. The respiratory system in man is a most complex structure which has the basic functions of inhaling and exhaling air as well as providing for the diffusion of oxygen from the air into the blood and for removing carbon dioxide from the blood. We can divide it roughly into two parts: (1) the upper system which includes the nasal cavity and the passages connecting it to the *windpipe* or *trachea*; and (2) the lower system which includes the branching air passages of the *lung*, the *bronchi*, which divide rootlike until they end in the functional unit of the lung, the *alveoli*. These are sacs, shaped like small clustered balloons with a diameter about 0.2 mm. Their smallness allows the transfer of oxygen through their thin walls to the blood-carrying capillaries.

The average adult respiratory system has a capacity of 500 cm^3, about half a quart of air. Two thirds to three fourths of this is held in the lower system. With this air comes all the waste material that the great "disposal" has not gotten rid of. Fortunately, the respiratory system has several lines of defense to protect itself from foreign materials in the air.

The first line of defense is the filtering nasal hairs, or *cilia*; particles which penetrate this line are caught by the mucus-lined walls of the winding nasal passages and expelled with the mucus. Soluble gases such as sulphur dioxide are absorbed in the passage walls. The upper system removes most particles larger than 10 microns (10^{-3} cm) in diameter.

In the lungs themselves, the smaller particles settle out from the now slowly moving air onto the mucus-lined walls of the bronchi and are expelled. This mechanism removes particles down to perhaps 1 micron (10^{-4} cm) in diameter. Between 1 micron and 0.1 micron diameter the particles are likely to stay in the air long enough to be exhaled, but below that size the probability of their capture again rises. For these very small particles the bombardment by air molecules, the so-called *Brownian movement*, drives them into the walls.

To take care of particles which either settle or are driven to the walls of the alveoli there is an ingenious defense; cells from the irritated tissues attach themselves to the foreign body and engulf it. This mass is then either taken through the wall into the bloodstream or a lymph node, or it works its way up the system until it is expelled.

The Damage. Ingenious as the defenses are, they are not perfect and there is mounting evidence that pollution can cause disease. It is important to understand the nature of this evidence. In the case of pollution by inhaled air, as with deliberate personal pollution by inhaled cigarette smoke, the evidence is statistical. It is not possible to deal in cases of individual cause-and-effect; instead, we must look at probabilities.

There are two types of statistical evidence that convince most medical experts that air pollution is harmful to human health. One is the type of study which shows that death rates in an urban area experiencing a severe pollution episode are strongly correlated with the intensity of that pollution. The kind of data that exists in this area is illustrated by Fig. 7-6. We see that the peak in deaths appears strongly connected to the peak in air pollution.

We should make it clear that data such as that in Fig. 7-6 do not *prove* that air pollution is a cause of death; something else, such as a change in temperature or pressure may have been the operative factor. This kind of correlation is also not sufficient for the setting of safety levels for air pollution. The suggestion of a connection is too strong to be ignored, however, and there exist other studies that reinforce the belief in a connection. Greenburg, *et. al.*[4] in a study of the correlations between deaths and amounts of air pollution in New York City from 1963 to 1968 estimate that the pollution, which during that period ranged from mild to severe, led to 10,000 deaths a year. These deaths were at least pollution-related: they were perhaps only triggered by the stresses of pollution. It is not possible to say that some of them

[4]Greenburg and Schimmel, "A Study of the Relation of Pollution to Mortality," *Journal of the Air Pollution Control Association,* **22**: 607, August 1972.

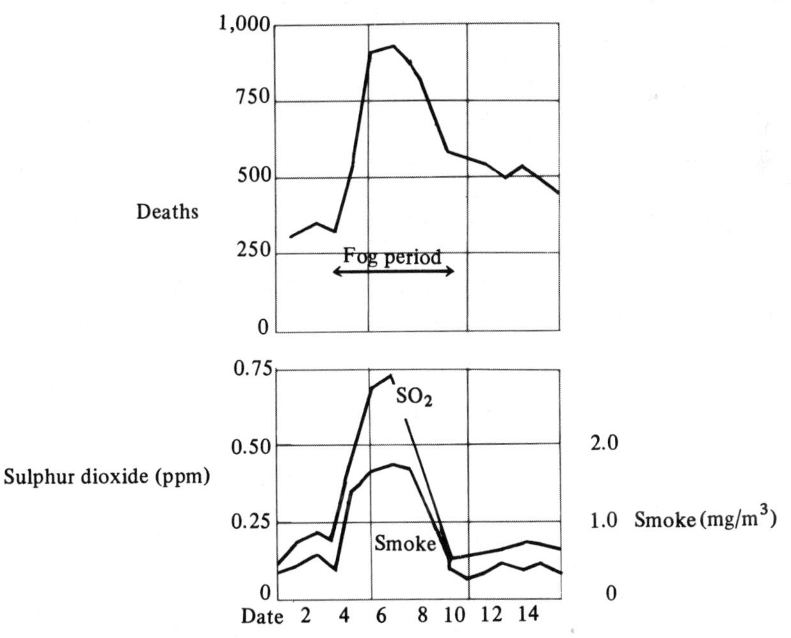

FIGURE 7-6 Deaths are shown to be related to sulphur dioxide concentration and smoke during the December 1952 "killer smog" in London. (Source: Wilkins, *Journal of the Royal Sanitation Institute,* **74**: 1, 1954).

might not have occurred anyway at some later date; but pollution seems, at least, to have brought them on earlier.

The second kind of data emphasizes this point: pollutants usually don't act alone in causing disease and death. In the air pollution disasters in, for instance, Donora, Pennsylvania in 1948 and London in 1952, it was the aged and those with heart trouble or respiratory diseases such as emphysema and asthma who died. Air pollution served as a last straw on an already-stressed human organism. This effect in combination, the so-called *synergistic effect,* has also apparently shown up in studies of the joint effects of smoking and urban air pollution. In several studies of mortality from lung cancer, urban smokers had death rates from lung cancer three times that of smokers in nonurban environments.[5] Again, the increased death rates of urban smokers over nonsmokers suggests, but does not prove, that air pollution makes smokers more susceptible to lung cancer. Similarly, it does not prove that air pollution causes lung cancer any more than does the comparison of smokers and nonsmokers prove that smoking causes it. The cause may be some other factor in the urban environment. Since known cancer-causing compounds are found in both polluted air and in tobacco smoke, prudence suggests a reduction of urban pollution levels (as well as a reduction in smoking).

With this explanation, we can present a list of the various kinds of respiratory and heart diseases that appear to be worsened by air pollution exposure (see Table 7-3). As we study these relations more deeply, the list lengthens.

Synergy. We need to make a few additional comments on the phenomenon of *synergy,* the alteration of the effect of one agent by the presences of one or more other agents. This is a common phenomenon in air pollution, and greatly complicates its study. We have provided one example from the combined effect of air pollution and smoking. A more direct example is provided by the combination of particulate matter and sulphur dioxide. We mentioned earlier that sulphur dioxide is removed in the upper respiratory system. If, however, particulate matter is present, sulphur dioxide may combine with water and create an absorbed coating of sulphuric acid on the particles. That particulate matter, if it is of the right size, may be drawn into the lung

[5] Summarized by Williamson, "Fundamentals of Air Pollution," (Reading, Mass.: Addison-Wesley, 1973). See Table 2.2, p. 46.

Table 7-3 Respiratory and Heart Diseases Affectedly Air Pollution*

Respiratory diseases	*Heart diseases*
Tuberculosis of respiratory system	Arteriosclerotic heart disease, including coronary disease
Malignant neoplasm of respiratory system	Hypertensive heart disease
Asthma	Rheumatic fever and chronic rheumatic heart disease
Influenza	Other diseases of the heart, arteries, veins
Pneumonia	Certain types of nephritis and nephrosis of the kidneys
Bronchitis	
Pneumonia of newborn	

*T. A. Hodgson, "Short-Term Effects of Air Pollution on Mortality in New York City," *Environmental Science and Technology,* **4**: 589, July 1970.

where the sulphuric acid can cause serious damage. There are other synergistic effects which increase the danger of other pollutants, making this area of medicine a very complex and unsatisfactorily understood one.

We have tried to give the impression that it is extremely difficult to make definitive statements of a causal nature about disease and air pollution. It is, therefore, difficult to identify thresholds below which damage is not expected and to set standards. It is particularly unsatisfactory to set standards for isolated pollutants since the effects are so dependent on synergistic reactions. We will, however, in the sections which follow, try to summarize the major effects expected or observed from the most prevalent forms of combustion-related air pollution.

The Particles of Pollution

The particles in the air, the aerosols, smoke, dust, droplets, and so on are the most visible pollutants. Aerosols are formed in two ways: (1) by the break-up of larger chunks of material (through grinding, plowing, or spraying) and (2) by the gathering together (condensation or agglomeration) of smaller pieces of material. In addition, there is biological material such as fungi and pollen among them.

The combustion-related aersols of our concern are particles of tar (heavy hydrocarbons) and soot (carbon) which escape unburned in the exhaust. With flyash (mineral residue from the fuel) these particles make up half the weight of the urban particulate burden; much of the rest are droplets such as the sulphuric acid formed from sulphur dioxide.

The physical and biological effects of these particles depend on their size. The larger particles reduce visibility as well as the amount of sunlight reaching the ground. They carry damaging materials such as sulphuric acid to the surfaces they strike, increasing cleaning and painting bills for both cities and individuals. The magnitude of their effect can be seen from dust-fall data. In urban areas this ranges from 10 to 100 tons per square mile.

As we will discuss in Chap. 9, there is some evidence that particulate pollution is also having an effect on the weather. These small particles floating in the air make ideal nuclei for the formation of rain droplets and may be affecting rainfall patterns. By scattering the incoming solar radiation they may change the energy balance of the atmosphere.

Finally, we must be reminded that man is not the only producer of aerosols, that volcanoes, ocean spray, and erosion contribute significantly. There seems to be a minimum value of 5 to 10 micrograms per cubic meter (a microgram is 10^{-6} gm) or about 200 particles per cubic cm. This low value can be measured, for instance, over the ocean. This minimum or background value is compared with other measurements including those over American cities (see Fig. 7-7). It is clear that the urban air is far more cluttered than that above those few areas where man is an exotic.

The Classical Smog

The most distinctive components of the *classical smog,* the foul-looking and foul-smelling blanket associated with the catastrophies in London, Donora, Pennsylvania, and the Meuse Valley in Belgium, are the various sulphur compounds. This is now

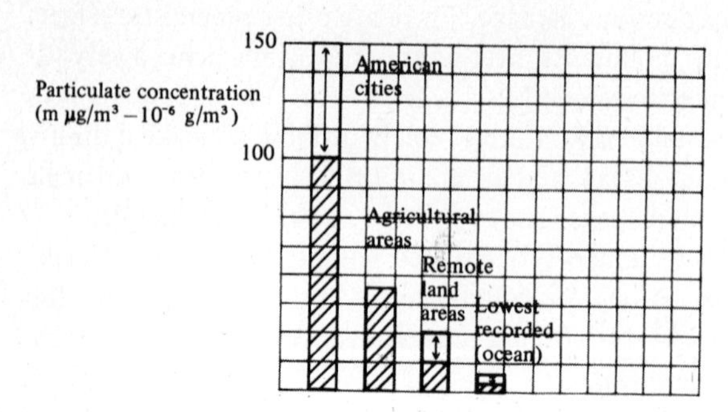

FIGURE 7-7 Particulate concentrations over cities are about five times greater than over remote areas. (Source: Williamson, Samuel, *Fundamentals of Air Pollution,* Reading, Mass.: Addison-Wesley, 1973).

called classical smog to distinguish it from the *photochemical smog* which Los Angeles has made famous (or infamous) and which derives from modern automobile exhaust.

Sulphurous smog can come from nonenergy-related sources; in certain regions the smelters for copper, lead, and zinc are the primary contributors. We will, however, restrict our discussion to sulphur compounds released in the combustion of fuels, and coal is the main offender.

In the combustion of fuel many reactions take place at different temperatures and in different environments. Most fuels have sulphur in them and, heated in the presence of oxygen, sulphur dioxide is formed. Some 98 percent of the sulphur content is emitted in this form. The sulphur dioxide is not in its equilibrium state and it oxidizes in contact with air to form sulphur trioxide (SO_3), a *hydroscopic,* or water-attracting, molecule which, with water, forms sulphuric acid.

Sulphuric acid is the most easily identified damaging component of the classical smog. It produces irritation in the upper respiratory system and in the lungs, when carried there by particles of a size that allows them to escape the defense mechanisms of the respiratory system. One of its more subtle, but nonetheless serious effects, is to increase the *airway resistance,* that is, to make it harder for air to pass through all the many passages of the respiratory system. This may be only troublesome for a healthy person, but can be fatal to someone already afflicted with chronic bronchitis or emphysema.

For the reasons we have mentioned the health effects of sulphur dioxide, as with all pollutants, are very difficult to give a quantitative measurement. There are synergistic effects, dependencies on the length of exposure, and other factors not easily measured. The best that can be done to sum up our knowledge of the health effects of sulphur dioxide is shown by the graphic presentation in Fig. 7-8. It is clear that even at concentrations of a few parts per million, sulphur dioxide is something to be avoided.

Another observed effect of sulphuric acid pollution is on quite a different scale. The aerosols containing this acid have a fairly short residence time in the troposphere: two days to a week. It is often washed out by rain. In the past few years it has been noticed that rain is becoming more and more acidic. Acidity is measured by chemists according to *pH.* A pH of 7 is neutral; numbers below that are acidic, above are alkaline. The scale is logarithmic. A pH of 4 is 10 times more acidic than 5, and so on. Rains in Scandinavia have been measured with a pH as low as 3. Figure 7-9 shows

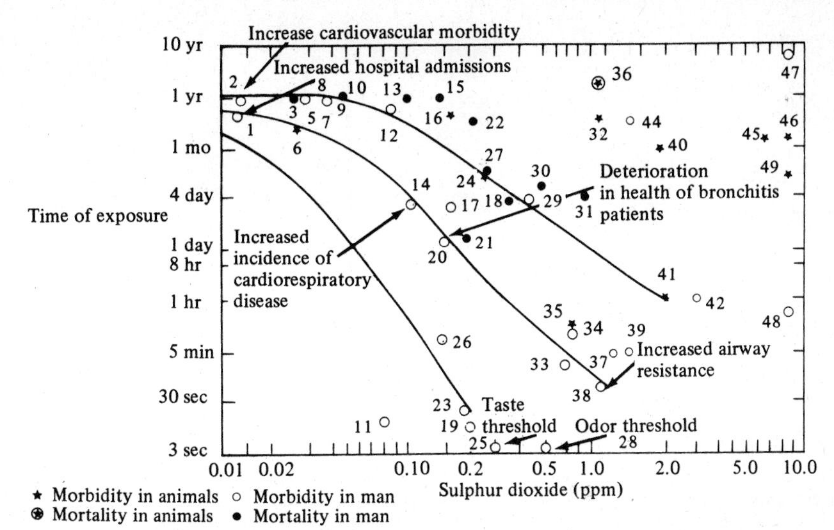

FIGURE 7-8 Rough thresholds of the incidence of health effects for various concentrations and exposures of sulphur dioxide are shown. (Source: "Air Quality for Sulfur Oxides," National Center For Air Pollution Control, U.S. Department of Health, Education and Welfare, 1967.)

the historical record of the increasing acidity of rain over Northern Europe. The increasing acidity of water can have serious consequences for aquatic life. At a pH of 5 the salmon disappear, at 4.5 the trout disappear. Serious effects on fish life in Norwegian lakes and streams have been reported. Changes in the pH of soil are also serious. Plants grow best at a pH between 6.0 and 7.0 and can be damaged by a lower pH as well as by direct exposure to sulphur dioxide.

It is this same sulphuric acid that is causing serious damage to buildings of concrete, limestone, marble, and to art objects the world over. In sufficient concentrations it can discolor paint, corrode metal, and weaken or destroy clothing, especially nylon hose and leather. Again, there are synergistic effects; a soot particle with absorbed sulphuric acid, for instance, is very damaging to the local area in which it lands.

This, then, is most of the story of the sulphurous smog; we have but to pinpoint the sources, assign the blame, and discuss control mechanisms. Before we do that, however, we turn to the aother smog, the photochemical smog, with which the Angelenos have to live.

Photochemical Smog

While records of classical smog go all the way back to Elizabethean England, to the first heavy dependence on coal as a fuel, photochemical smog is a modern development, a mark of a society which depends on the automobile.

Classical smog is a mixture of smoke and fog, or at least moisture. It most often occurs in winter months when much fuel is being consumed. It has an *oxidizing* effect, that is, it takes oxygen from the materials it interacts with, chiefly by the reactions of $SO_2 \longrightarrow SO_3$ and $SO_3 \longrightarrow H_2SO_4$. Photochemical smog, on the other hand, occurs in regions of bright sunlight, usually in the summer, and it does not depend on moisture. Furthermore, the primary pollutants in photochemical smog have a

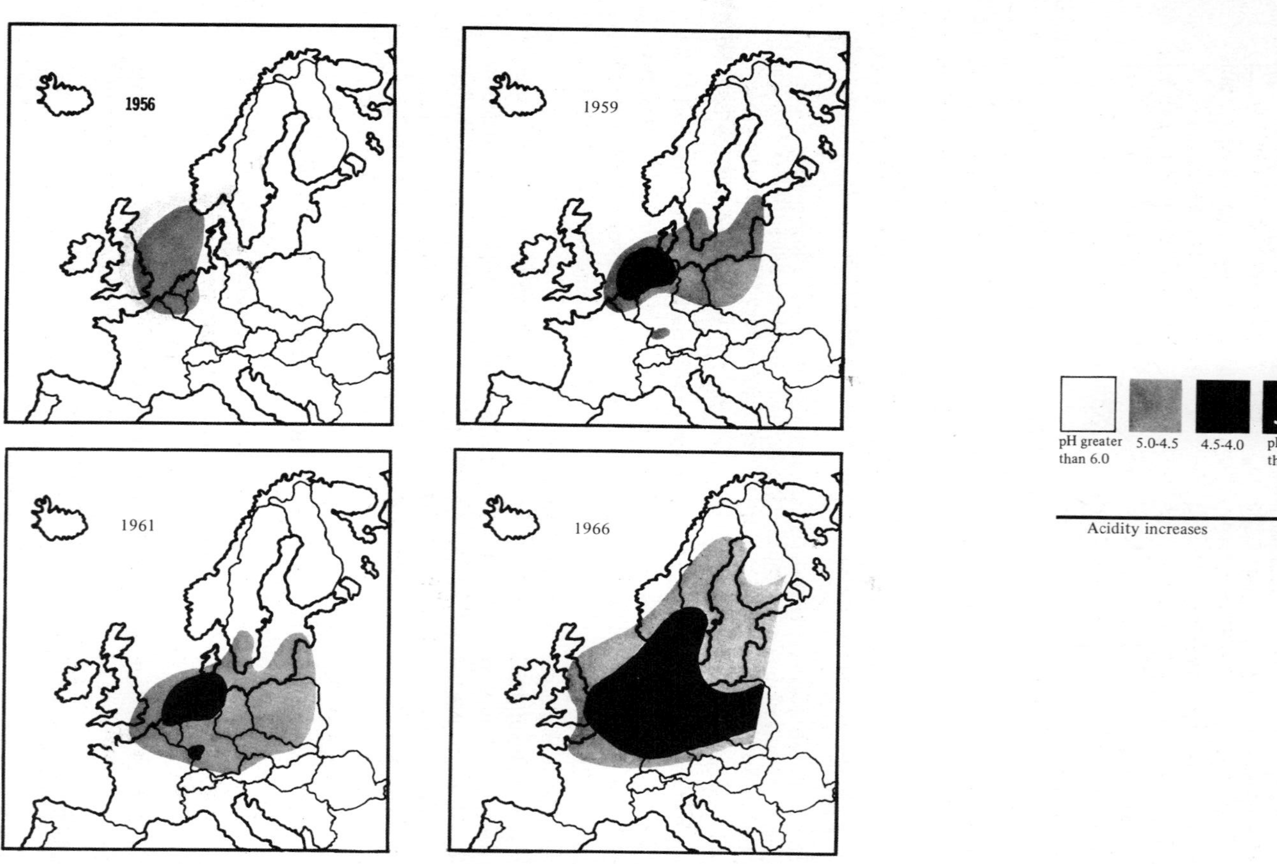

FIGURE 7-9 Acid rainfall, highest in the Netherlands, is spreading over Europe. (Source: "Acid Precipitation in Europe," *Environment,* **14**:2, p. 36.

reducing effect, that is, they give up oxygen to other compounds. The oxidizing effect of photochemical smogs come from secondary pollutants.

The characteristic effects of photochemical smog are eye irritation, a brownish cast to the sky (usually with a sharp upper limit caused by an inversion layer), respiratory distress, and more severe plant damage than is experienced with the sulphurous smog. The major components of this smog are the various compounds of nitrogen and oxygen, the nitrogen oxides, and the hydrocarbons, which are compounds of hydrogen and carbon. Both of these are produced by the burning of fuels, gasoline being the major contributor.

The damaging components of photochemical smog are, for the most part, secondary pollutants, produced both by the action of sunlight on NO_x and HC, and by reactions between the primary and secondary pollutants and the products formed. It is a complex chemical soup only superficially understood in either chemical or biological terms.

Two of the typical chains of photochemical reactions are the production of ozone from nitrogen dioxide and the production of a host of organic irritants labelled the PANs from hydrocarbons. Nitrogen oxide (NO) is produced whenever air, containing as it does both nitrogen and oxygen molecules, is heated to temperatures above 1000°C. Its formation does not depend on the nature of the fuel being burned. During combustion at these high temperatures, the oxygen and nitrogen molecules are dissociated and then recombine to form nitrogen oxide. Exhausted from automobile engines or electric power plants, the nitrogen oxide is oxidized by oxygen in the air to nitrogen dioxide (NO_2), a toxic gas at sufficiently high levels (10 parts per million).

Nitrogen dioxide immediately makes its presence known. While nitrogen oxide is colorless, nitrogen dioxide absorbs sunlight strongly at short wavelengths, taking out the blues and greens and leaving the brownish color typical of this kind of smog. More importantly, the high-energy, very short wavelengths of sunlight break nitrogen dioxide apart, forming nitrogen oxide and oxygen. The oxygen atom quickly joins an oxygen molecule to make ozone.

Ozone is one of the most obvious and most troublesome secondary pollutants of photochemical smog, responsible for the strong and unpleasant odor. It is detectable in concentrations of 0.02 parts per million and commonly exists at ten times that level in smog. It is a strong oxidizing agent and is responsible for much of the damage to plants and to materials.

The most noticeable feature of ozone in this smog is that there is too much of it, ten times too much, to have been produced by nitrogen dioxide alone. This leads us to the other component, the hydrocarbons, and their role.

Hydrocarbons come from automobile fuels for the most part, although they can be released from any combustion process. In the photochemical smog mix, they compete for the oxygen atoms released by the dissociation of nitrogen dioxide and are oxidized to forms of the general type HCO. Most of the oxidized HCO compounds are very reactive; they react, for instance, with more oxygen molecules to form HCO_3, a chemical species called a *radical* which is even more reactive. The HCO_3 radicals can then react again with oxygen to form more ozone or with nitrogen dioxide to form the PANs. These compounds are also strong oxidizing agents and, along with ozone, are the most troublesome components of photochemical smog. Other hydrocarbon reactions produce aerosols which scatter light and reduce visibility.

It is the oxidizing agents, ozone and PAN acting alone or synergistically, that cause most of the damage. PAN, for instance, causes most of the eye irritation, even at levels of less than one part per million. (Effects have been detected at 0.7 part per million.) Ozone is not as potent in the eyes but, like the sulphur oxides of the classical smog, can increase the airway resistance and make breathing difficult even, in some subjects, at the level of 0.1 part per million found commonly in urban smog. Although no permanent effects of ozone exposure have been identified in humans as yet, animal experiments suggest that they are to be expected from long exposures. Based on animal experiments there is also evidence of increased susceptibility to bacterial infection in the ozone-irritated lungs.

The major difference between the sulphurous classical smog and the photochemical smog is that statistical studies of the latter have not shown, so far, the correlations with increased mortality that has been observed with the classical smog. It may be that the rapidly changing character of this smog and the scarcity of pollution-measuring devices (one or two per city) conspire to conceal these correlations; it may be that deaths follow after a long time interval, or that, there have not yet been any deaths due to photochemical smog. The American dependence on the automobile will, no doubt, give us plenty of time to carry out the necessary studies.

Plants are not as lucky; the photochemical smog with its oxidizing agents is very damaging: levels of PAN as small as 0.01 to 0.05 part per million can produce damage within an hour. Leafy vegetables are the most sensitive, but even trees can be damaged. Ozone is almost as bad. It is no wonder that the Los Angeles freeways have plastic trees (and ozone will probably affect them eventually).

Carbon Monoxide

Carbon monoxide is a component of photochemical smog, although it remains largely separate chemically. It is worthy of separate treatment. In terms of mass emitted it is the most important pollutant released at six times the rate of nitrogen dioxide, for instance, and is found wherever there are automobiles.

Carbon monoxide results from the incomplete burning of fuel, or combustion in the presence of insufficient oxygen; proper burning conditions produce carbon dioxide. It is unfortunately, odorless, and poisonous in high concentrations. It works by putting the red blood cells out of action by taking the place of oxygen in hemoglobin. The affinity of carbon monoxide for the iron atom in hemoglobin is about 200 times greater than is that of oxygen. It not only replaces an oxygen molecule, but by means not yet understood, makes the blood cell hold tightly to its other oxygen molecules, thus preventing the use of the oxygen in the normal oxidizing functions of blood.

The effects of carbon monoxide depend not only on the concentration in parts per million, but also on the exposure times. What happens at various combinations of concentrations and exposure times is summarized in Fig. 7-10. The levels at which serious effects appear are higher than for some of the other smog components, but not out of reach for our automobile-hooked culture; Table 7-4 presents, for perspective, some measured levels.

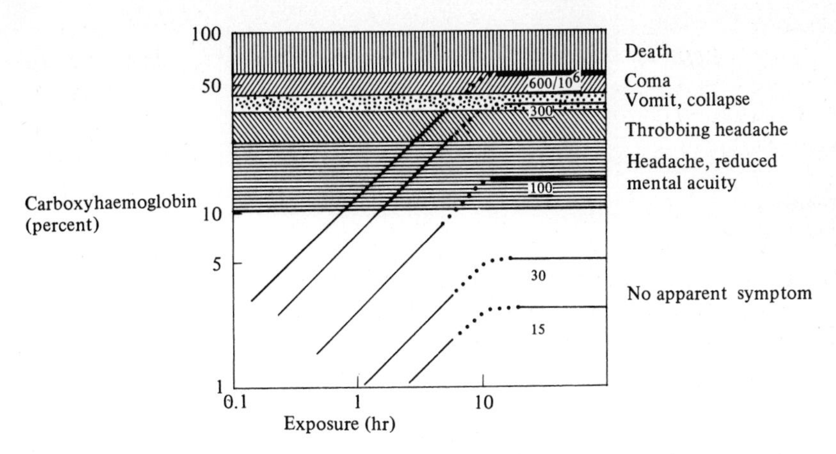

FIGURE 7-10 The general response of humans to various concentrations and exposure times of carbon monoxide is shown. The percentages of hemoglobin with attached carbon monoxide are shown on the left-hand vertical scale. (Courtesy of General Motors Corporation.)

Lead

Lead is the last member of the poisonous tribe of pollutants which our energy-consuming society discharges into air. Although there are several industrial sources, the main culprit, again, is the automobile. Lead atoms hooked on to a hydrocarbon in gasoline form an *additive* which reduces *knocking*, premature firing of the compressed fuel mixture. There are about two thirds of an ounce of lead in a gallon of premium gasoline and half that in regular. There were, therefore, 560,000 tons of lead in the 27 billion gallons of gasoline sold in 1968. Three quarters of this 420,000 tons, was released to the atmosphere.

This additive is justified from the point of view of efficiency; the increase in the miles-per-gallon performance of automobiles during the late 1930s and early 1940s was due to its invention (see Fig. 4-3). Now, however, it is an additive to the atmos-

Table 7-4 Carbon Monoxide Levels at Various Locations*

Location	*CO Levels (average ppm)*
Los Angeles Freeways	37
Los Angeles Freeways, slow, heavy traffic	54
Los Angeles, severe inversion	30 for over 8 hr
Parking garage	59
Cincinnati intersection	20
Detroit, short peak	100
Detroit, residential area	2
Detroit, shopping area	10
Manhattan, intersection	15 all day long
Allowed industrial exposure for 8 hr (for comparison)	50 recently lowered from 100

*Source: Virginia Brodine, Ed., *Air Pollution: A Scientist's Institute for Public Information Workbook*, (New York: SIPI, 1970, p. 13.)

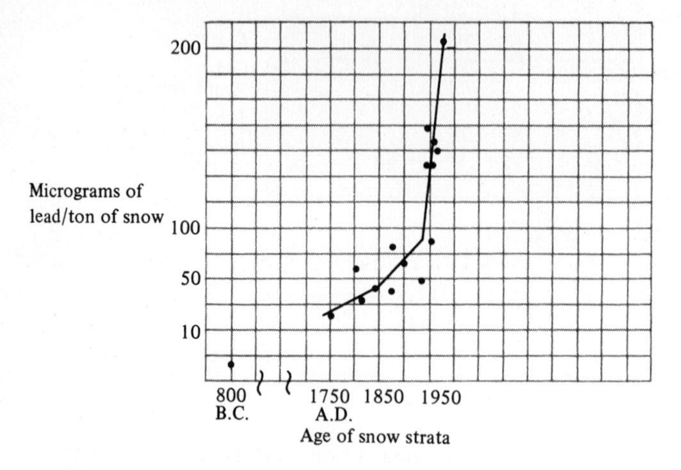

FIGURE 7-11 Measurements of lead concentrations for layers of different age in the Greenland ice sheet show the effects of industrial civilization and the automobile. (Source: "Science and Citizen," *Environment,* April 1968, P. 72.)

phere. It is found everywhere, even in the Greenland ice sheets. Lead concentrations in the annual ice layers in northern Greenland are shown in Fig. 7-11. The steep growth since the 1950s is obvious. We can rather convincingly identify the source of lead in the air by data such as that of Fig. 7-12. The connection between the amount of gasoline sold and the lead concentration seems to be a strong one.

The same correlations are shown in measurements made in man: Boston tunnel employees, for instance, have about twice the body burdens of farm workers.

We take in about 20 to 30 micrograms (10^{-6} grams) of lead per day by breathing and with food and drink. At this low level the effects are not well understood. There is evidence of mild brain damage at 60 to 80 micrograms of lead per 100 grams of blood. Lead is also deposited in bone tissue and inhibits the action of the bone marrow. Since it accumulates, its increasing environmental levels are worrisome. Again, since the large-scale experiment is now under way, we will, in the following years, learn more about the damage it causes.

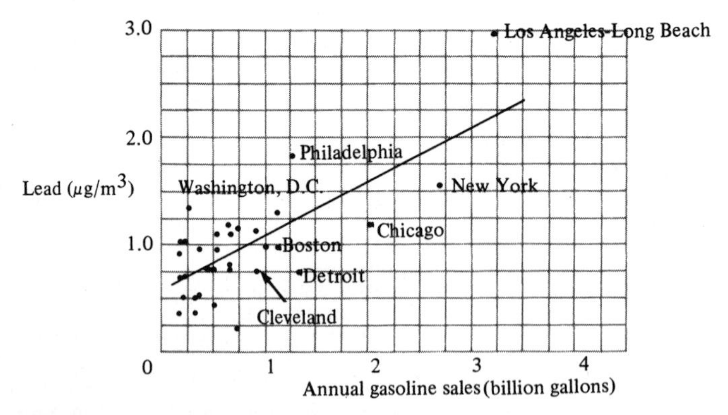

FIGURE 7-12 The fact that the cities with larger volumes of gasoline sales also show larger lead contamination provides strong evidence of its source. (Source: The Automobile and Air Pollution Part II, U.S. Department of Commerce, December 1967.)

THE POLLUTERS AND THEIR CONTROL

We have so far discussed the atmospheric effects that interfere with efficient operation of our "disposal," and looked, in some detail, at the dangerous materials we dump into it. We will now turn our attention to the sources of the pollutants and to the amounts that are being dumped into the "disposal" year by year.

Obtaining the desired numbers is not easy since we have no nationwide network sufficient to measure total pollution levels. We must, therefore, rely on a knowledge of the amounts of pollutant emitted by known quantities of the various fuels, by a knowledge of how much of each kind of fuel is burned, what typical furnace and automobile engine efficiencies are, and similar information. A rough idea of what comes from where is given in Figs. 7-13 and 7-14 which present the uncontrolled emission components from the burning of fuels for transportation or the generation of electricity. We can anticipate from these data that transportation will be the dominant contributor to carbon monoxide and hydrocarbon emission, while the fuels used to produce electricity will make the major contribution to sulphur oxides and the particulates.

This anticipation is realized by the data of Table 7-5 which indicates for each of the five major pollutants, the amount and percentage from each source. The data in Table 7-5 is summarized graphically in Fig. 7-15. Only in the production of aerosols is the monopoly of production by the energy-converting sources broken.

The damages from the pollutants are, of course, not directly measured by the total mass emitted. Some, like the PANs, cause serious eye irritation at less than 1 part per million, while the damage threshold for carbon monoxide is perhaps ten times as high.

A pollutant ranking that takes into account both the amount of material emitted and the level of concentration at which they begin to cause damage has been suggested. In his article, "Rating of the Major Air Pollutants and Their Sources by Effect,"[6] E.G. Walther computes what he calls an *effect factor.* He determines from the National

[6]E.G. Walther, "Rating of the Major Air Pollutants and Their Sources by Effect," *Journal of Air Pollution Control Association,* **22**: 352, May 1972.

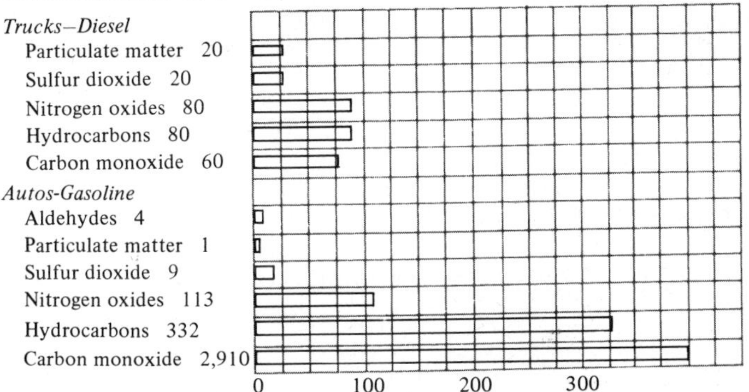

FIGURE 7-13 The automobile is the most important source of the nitrogen oxides and hydrocarbons of photochemical smog as well as of carbon monoxide. (Source: The Automobile and Air Pollution Part II, *op. cit.)*

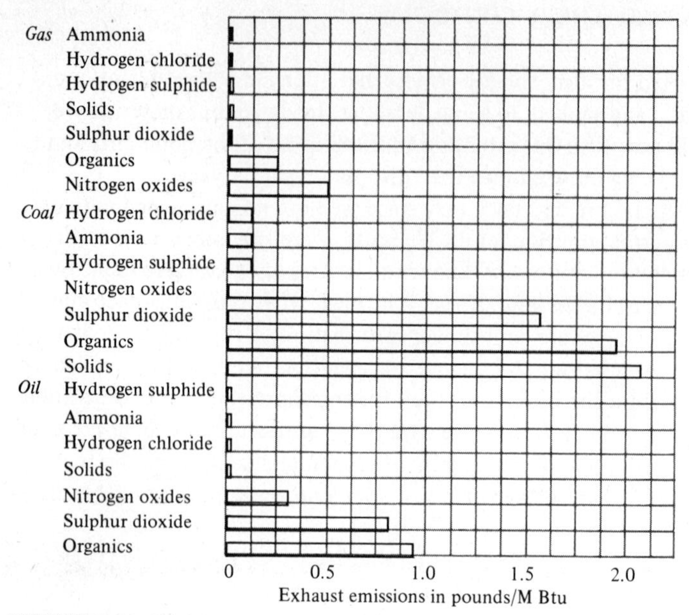

FIGURE 7-14 Both in terms of overall amounts and the specific pollutants, gas is the cleanest fuel and coal the dirtiest. (Source: The Automobile and Air Pollution Part II, *op. cit.)*

Ambient Air Quality Standards, the maximum allowable concentration (mass per unit volume of air) during a full day for each pollutant. Setting the effect factor for carbon monoxide, the largest of these, at 1, the effect factor of each of the other pollutants is taken to be the ratio of the maximum allowable one day concentration of carbon

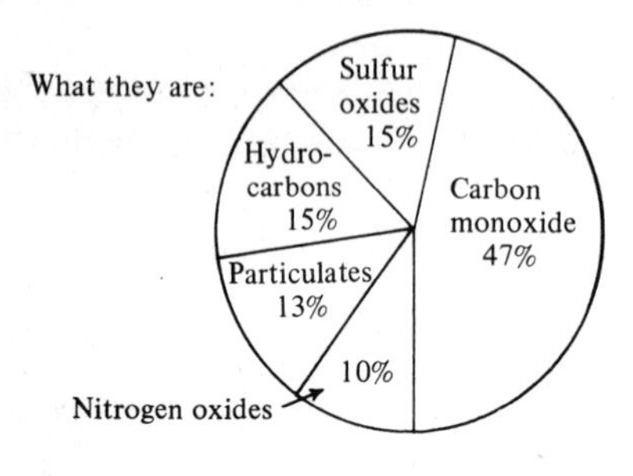

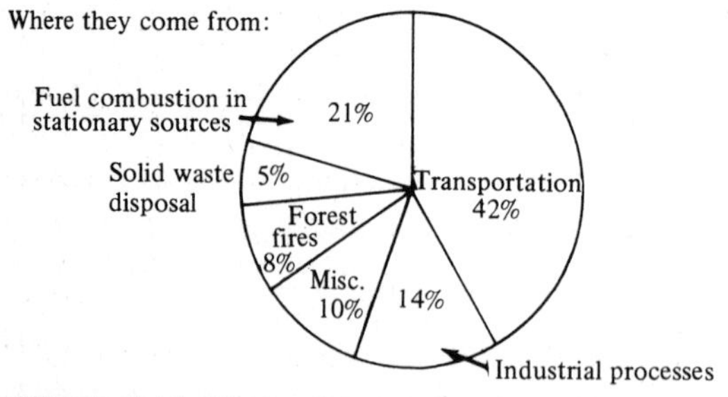

FIGURE 7-15 The breakdowns of pollutants by weight and by source are shown. As Table 7-6 shows, the effects of these pollutants are not necessarily proportional to their weight. (Source: The Automobile and Air Pollution Part II, *op. cit.)*

Table 7-5 Annual Air Pollution Emissions Classified by Pollutant and Ranked by Mass*

Pollutant	Source	Annual quantity (10^6 tons)		Annual quantity (percent)	
CO			151.4		53.3
	Transportation	111.5		73.7	
	Miscellaneous	18.2		12.0	
	Industry	12.0		7.9	
	Solid waste disposal	7.9		5.2	
	Stationary fuel combustion	1.8		1.2	
				100.0	
HC			37.4		13.3
	Transportation	19.8		53.0	
	Miscellaneous	9.2		24.6	
	Industry	5.5		14.7	
	Solid waste disposal	2.0		5.3	
	Stationary fuel combustion	0.9		2.4	
				100.0	
Aerosol			35.2		12.5
	Industry	14.4		40.8	
	Miscellaneous	11.4		32.4	
	Stationary fuel combustion	7.2		20.5	
	Solid waste disposal	1.4		4.0	
	Transportation	0.8		2.3	
				100.0	
SO_x			33.4		11.9
	Stationary fuel combustion	24.4		73.0	
	Industry	7.5		22.5	
	Transportation	1.1		3.3	
	Miscellaneous	0.2		0.6	
	Solid waste disposal	0.2		0.6	
				100.0	
NO_x			23.8		8.5
	Transportation	11.2		47.1	
	Stationary fuel combustion	10.0		42.0	
	Miscellaneous	2.0		8.4	
	Solid waste disposal	0.4		1.7	
	Industry	0.2		0.8	
				100.0	
Total			281.2		100

*Source: E. G. Walther, "A Rating of the Major Air Pollutants and Their Sources by Effect," *Journal of the Air Pollution Control Association,* 22: 352, May 1972.

monoxide to maximum allowable one day concentration of each of the others.

There are two sets of national standards: (1) primary standards based on the protection of health; and (2) secondary standards based on protection of public welfare (protection against adverse effects on property, climate, vegetation, and comfort, for example). Thus, in some instance Walther computed two sets of effect factors. These are shown in Table 7-6. From this point of view, the hydrocarbons are the

Table 7-6 "Effect Factors" of Pollutants*

Parameter	Effect factor CO	HC	SO_x	NO_x	Aerosol
Primary standards	1	125	15.3	22.4	21.5
Secondary standards	1	125	21.5	22.4	37.3

*Source: E.G. Walther, "Rating of the Major Air Pollutants and Their Sources by Effect," *Journal of Air Pollution Control Association*, **22**:352, May 1972.

most troublesome, 125 times more potent than carbon monoxide, followed by the nitrogen oxides and the aerosols (particulates) which are about 20 times as potent.

The five pollutants can be ranked by their overall effect, which is the mass of the pollutant (see Table 7-5) multiplied by this effect factor. This ranking for the primary standards is shown for each pollutant in Table 7-7; their percentage contribution to the total effect is also shown. In Table 7-8 the effects are displayed, classified by their source and the percentage effect contributed by each source is shown.

From these two tables we can conclude that transportation is the major contributor to the effect of pollution; it also contributes the largest mass of pollutants. It is followed, however, by miscellaneous factors such as forest fires, agricultural burning, and coal dump fires, and industry. Stationary fuel combustion (power plants) drops to fourth instead of the second place occupied in the mass ranking of Table 7-5. The hydrocarbons, largely from automotive fuel combustion, are the most important pollutants, contributing 71 percent of the effect compared to their 15 percent of the mass.

With this somewhat more pertinent ranking in mind, let us now look at our number one pollutor, the automobile, and examine both the ways in which pollutants are emitted and the means and determination available for their control.

The Dirty Automobile

As we have seen, the automobile is the most important source of the nitrogen oxides and hydrocarbons which provide the raw materials for the photochemical smog. It is also responsible for two other energy-related pollutants we have discussed, carbon monoxide and lead.

Table 7-7 Annual Air Pollution Emission Classified by Pollutant and Ranked by Effect*

Pollutant	Annual amount (million tons)	Effect factor†	Effect	Percent of total effect
HC	37.4	125	4681	70.6
Aerosol	35.2	21.5	757	11.4
NO_x	23.8	22.4	533	8.0
SO_x	33.4	15.3	511	7.7
CO	151.4	1	152	2.3
Total	281.2			100

*Source: Walther, *op. cit.*

†Effect factors are based on primary air quality standards. Amounts are for 1968.

Table 7-8 Annual Air Pollution Emissions Classified by Source and Ranked by Effect*

Source	Annual amount (million tons)	Effect†	Percent of total effect
Transportation	144.4	2877	43.3
Miscellaneous‡	41.0	1461	22.1
Industry	40.0	1129	17.0
Electric utilities	44.3	867	13.1
Waste disposal	11.9	300	4.5
Total	281.2		100

*Source: E. G. Walther, "Rating of the Major Air Pollutants and their sources by Effect," Journal of Air Pollution Control Association **22**, 352, (May 1972).

†Effect factors are based on primary air quality standards. Amounts are for 1968.

‡"Miscellaneous" refers to forest fires, agricultural burning, and coal waste fires.

We have earlier summarized what is presently known about the effects of carbon monoxide and lead. There is little to add to what we have already said about the production of carbon monoxide; it will be produced whenever fuels are burned with insufficient oxygen. We will discuss means for its control along with other emission control practices.

The control prognosis for lead is somewhat different. It was introduced into gasoline in the late 1920s and sold as tetraethyl lead by the former General Motors subsidiary, the Ethel Corporation. As much as 6 billion pounds of lead have been released into the atmosphere since then. We may be coming to the end of the lead era, however. Cars can run on nonleaded gasoline; Amoco gasoline has always achieved its antiknock properties by adding a suitably altered hydrocarbon rather than by adding tetraethyl lead. Most of the oil companies, now under federal pressure, are offering nonleaded gasoline. The most effective pressure for lead removal has come, however, from its effect on the emission control devices. Lead quickly deactivates the catalytic reactors favored by the major automobile companies. This may produce the first major break in the history of close cooperation between the motor and oil companies.

Emission Controls. The sources of the various pollutions from different parts of the car's anatomy are shown in Fig. 7-16. We see that 50 percent of the hydrocarbon emission, and practically all of the nitrogen oxides and carbon monoxide, are in the exhaust; the remaining half is produced almost equally by *blowby* at the crankcase, gas forced past the piston rings, and by evaporation from the carburetor and the gas tank. Under stricter controls the automakers now handle these last two satisfactorily by better sealing and by feeding blowby gases back to the engine. The exhausted gases remain a perplexing problem.

The early attack in automotive pollution focused on carbon monoxide and hydrocarbons. Since the presence of either of these is an indication of the same engine problem, incomplete combustion caused by a deficiency of oxygen, the same cure worked for both. By increasing the air/fuel ratio, the production of either of these pollutants can be minimized. In the 1960s the carmakers turned to "leaner" mixtures (increased ratio of air to fuel) and reduced the carbon monoxide and hydrocarbon emission.

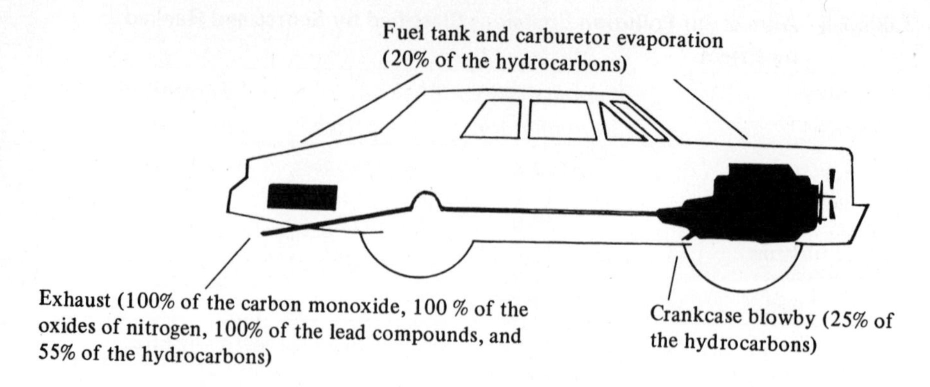

FIGURE 7-16 The sources of the various pollutants in the automobile. (Source: The Automobile and Air Pollution Part II, *op. cit.)*

Unfortunately, the lean mixture burns at a higher temperature, and this increases the emission of the nitrogen oxides. The experience in California, the first state to enforce strong emission controls, was that although the controls did cause design changes which reduced the emission of carbon monoxide and hydrocarbon, emission of nitrogen oxides from these cars increased by 40 percent. In 1971 the restrictions were extended to include emission of nitrogen oxides.

The issue of clean air standards and the automobile has now been joined at the national level. The Clean Air Act of 1970 specified that the 1975 model car must emit no more than 10 percent of the carbon monoxide and hydrocarbon allowable in 1970 models and that 1976 model cars must emit no more than 10 percent of the nitrogen oxides allowed in 1971 models. These standards have produced near-panic (or perhaps near-rebellion is the better term) in the auto industry.

Emission control can be approached either through engine redesign or by add-on features. So far, the United States auto industry has chosen the add-on route, with much publicity being released about the reductions in gas mileage and increases in price which will result. The basket in which all their eggs seem to be resting is the catalytic converter, a device which fits into the car's exhaust system and oxidizes the hydrocarbons and carbon monoxide to water and carbon dioxide.

The major problems of these devices are erosion and deactivation due to the impinging of high velocity lead particles on the inner surfaces, the need for new materials able to stand the high temperatures (up to 1700°F), the cost of the catalytic metals (platinum is one of them) and the long warm-up times and balkly starting that they cause. The life expectancy of these converters is unsatisfactory; so far, they are seriously impaired after 5,000 miles, while the EPA requires them to go 50,000 miles. The need for, and availability of, maintenance and untraditional engine adjustment are also uncertainties in their future.

While these converters can take care of the hydrocarbons and carbon monoxide, they cannot, at the same time, reduce the nitrogen oxides. That problem is being approached from several different directions. One is to produce a dual-catalyst converter, or two catalysts: one to handle the hydrocarbon and the carbon monoxide and the other for the nitrogen oxides. This solution suffers from all of the problems we have just summarized. The other is to make changes in the engine operation to reduce nitrogen oxides at the expense of an increase in hydrocarbon and carbon monoxide by making the mixture "richer" and then to take the increased hydrocarbon and carbon monoxide out with the converter. Such an arrangement would obviously reduce gasoline mileage.

New Engines. A more drastic, but potentially more satisfactory, response to the need to reduce engine emissions is to completely redesign the engine. This approach is much less popular with the automakers. There are, however, several interesting developments at this time. The Wankel engine, which we discussed briefly in Chap. 4 offers some hope. The Mazda, with this rotary engine, has officially met the 1975 standards. As we said earlier, the Wankel is not cleaner than the piston engine; it is, in fact, dirtier. It is just smaller for the same horsepower, so there is room for plenty of control devices under the hood.

There is much excitement over the *stratified charge* engine. In such an engine, the amount of charge, fuel plus air, is adjusted to engine conditions, directed to the area where it will burn best and fired at just the precise instant. The EPA has tested a Honda with a moditied version of this engine and it produced carbon monoxide and hydrocarbon emission 50 percent *less* than the 1975 standards and nitrogen oxides emissions 70 percent less than the 1976 standards. In the Honda engine there is a prechamber in which an extra-rich mixture (low air to fuel ratio) is fired and generates very hot combustion gases; these hot combustion gases then ignite a very lean mixture (high air to fuel ratio) in the main chamber. The lean mixture, with its supply of oxygen, reduces carbon monoxide and unburned hydrocarbons, while the fact that the high temperatures occur in a rich mixture reduces nitrogen oxides. Detroit, however, has so far been less than eager to turn to Japan for help; only Ford is working seriously on stratified charge engines and it estimates that its programmed combustion (PROCO) engine might be ready for limited use in 1979.

The diesel is an existing engine with low emission; Daimler-Benz (Mercedes Benz) is the only company other than Honda and Toyo-Kogy (Mazda) to produce a car that meets the 1975 standards. In the diesel, fuel is sprayed into air that has been heated by compression; the fuel burns much more completely and burns before it reaches the cylinder walls. The pollutants associated with incomplete combustion, carbon monoxide and hydrocarbon, are thus, greatly reduced. The nitrogen oxides will be a problem due to the lean mixture and an additional exhaust control may be needed to meet 1976 standards. Also in its favor is the higher thermal efficiency of the diesel. The major disadvantages are increases of particulates, noise, and odor over the gasoline engine. There is hope that these last three environmental hazards can be reduced.

Further in the future are other engines such as the gas turbine and improved steam engines. There are also the possibilities of new fuels. Alcohol, which can be made from organic waste or grain surpluses (if we ever have these again) and added in amounts up to 25 percent of the gasoline, produces significant pollutant reduction. Engines can be converted to run on gaseous fuels such as propane and methane. The latter is derived from decaying chicken manure in a much-publicized British car. Hydrogen may be a fuel of the future. We will discuss the disadvantages and advantages of hydrogen as a fuel and look at the other nonpolluter of the future, the electric car, in a later chapter. For the 1970s, however, we seem stuck with the piston engine and with the gasoline fuel which it requires.

There has been strong pressure on the EPA to relax its insistence that the carmakers meet the 1975 and 1976 standards. The substantive arguments have been based on lowered fuel economy, the increased cost, and the danger of going into full-scale assembly-line commitment to an experimental, and presently unsatisfactory, system. The mileage per gallon argument can be answered by the comparisons which were made by Ruckleshaus in Chap. 4: that air conditioners, automatic transmission, and especially increasing car weight cause equal or greater penalties. The Chevrolet Impala, for instance, has become almost 1,000 pounds heavier since the early 1960s. The cost argument is also weakened by comparisons; Table 7-9 shows the estimated costs of meeting the 1975-1976 standards and compares them to other costs that consumers have accepted. The benefits of reduced emission do not seem to carry a pricetag outside of consumer range.

The most telling argument turned out to be the last one, that the auto companies just weren't ready, and rather than try to force them to put the catalytic converters on all 1975 cars, William D. Ruckelshaus, EPA Administrator at the time, announced in April 1973, less stringent 1975 levels of carbon monoxide and hydrocarbon. Following this decision, it was inevitable that the 1976 requirements would be relaxed also, and in late July 1973, Will W. Fri, Acting Administrator of the EPA, set less stringent nitrogen oxides levels. He cited not only the implementation problems, but preliminary evidence that the nitrogen oxides standard was set too low in the light of recent health data. The pressure continues, however, and the implementation of the standards may be postponed, again, until the middle of the 1980s.

The results of this controversy are not all in. The enforcement of even the interim standards will, it is clear, reduce emissions. What is not clear is whether the pressure from the EPA and foreign competition will stimulate the American auto companies to save their dinosaur, the piston engine, from extinction. There have been some net gains: the 1975 cars will, we are now told, have better gasoline mileage than the 1973s. Given the record over the past 30 years, that is no small victory.

Table 7-9 Comparison of Costs to the Consumer of 1975 and 1976 Emission Controls with Other Automotive Costs

Automotive costs	*Sticker price*	*Annual O and M*	*1975-1985 National cost*
Engine Modification	$100	$ 5	$10.8 billion
1975 standards	185	10	40.9 billion
1975-1976 standards	205	42*	70.8 billion
Vinyl roofs (46%)†	92	—	6.9 billion‡
Radio (80%)	59	—	7.7 billion
Air conditioning (69%)	350	36§	65.1 billion

*Gasoline prices are assumed to be 36 cents per gallon for leaded regular.

†Percentage figures indicate the percentage of cars sold with this option.

‡Calculation is adjusted for percent of sales figure.

§Air conditioning is estimated to have an average fuel penalty of 9% in urban driving. Maintenance costs were not included because of lack of data.

Stationary Fuel Combustion

The generation of electricity produces 78 percent of the sulphur oxides, 27 percent of the particulates, and 44 percent of the nitrogen oxides that go to make up the total yearly pollutant load of the atmosphere (see Table 7-7). Its contributions to carbon monoxide and hydrocarbon pollution are neglible. At present, our major concern focuses on the sulphur oxides since the electric utilities plants are important but not the sole contributors to the classical (sulphurous) smogs. Before discussing the source of this sulphur and the means for controlling it, let us say one word of reminder about the nitrogen oxides. The absolute contribution of the utilities plants to this raw material for photochemical smog is not as large as that of the automobile. We must remember, however, that nitrogen oxides are formed whenever there is a fire in air and the amount produced increases with increasing temperature. Two facts that we emphasized in Chap. 6 return to haunt us here: (1) the trend in the generation of electricity is to go to higher and higher temperatures; and (2) the number of generating plants is expected to double in the next ten years. The generating stations may soon by able to get into the photochemical smog business themselves.

For the present however, the sulphur oxides are the major culprit. They are formed by the oxidization of sulphur. The amount released into the air thus depends greatly on the amount of sulphur in the various fuels. As was shown in Fig. 7-14 there is a definite hierarchy of fuels as far as sulphur is concerned; natural gas is practically free of this element since it is removed from it before combustion. From the point of view of the sulphur oxides, natural gas is "Mr. Clean."

Oil is of varying but generally low sulphur content; it ranges from as high as 2 percent down to 0.5 percent and it is believed that Alaskan oil will be low in sulphur. The average sulphur concentration in the fuel oil used in power plants is about 1.5 percent. Part of the reason for the fuel oil shortage in the winter of 1972 was a shortage of low sulphur oil which would meet state air pollution standards.

Coal is the major offender. Its sulphur content also depends on its geographical source and can vary from 0.2 percent up to 7 percent by weight. Coal-fired power plants produce 90 percent of both the sulphur oxides and the particulates which are charged to the generation of electricity.

Air quality standards in many urban areas will require coal of less than 1 percent sulphur for power plant use in the next few years. Today 90 percent of the coal does not meet that standard. We have large, low-sulphur coal reserves, but they are not near the metropolitan areas. The enforcement of this standard will cause major changes in coal mining and transportation, but here also, the present Energy Crisis is being used to mount an attack on the Clean Air Act regulation. Congress and other legislative groups have given approval to the relaxation of many clean air laws allowing high sulphur coal to be used as a fuel again. Major amendments are being proposed to further weaken many of the controls on emission.

From an environmental point of view the more reasonable approach would be to remove the sulphur from the coal. Unfortunately, most of the sulphur is chemically part of the coal and precombustion removal is difficult. With the improved combustion technology to be discussed in Chap. 12, the fluidized bed, for instance, removal is quite simple, but this requires major conversion of power plant boilers.

The present approach is to tackle the more difficult job of dealing with the sulphur oxides in the exhaust gases. The first response was to increase the height of the smokestacks. The average height for stacks built in 1969 was about 600 feet compared to 240 feet in 1960. Stacks as high as 1,000 feet tall are under construction. For example, Washington Monument is 555 feet tall. These stacks disperse the pollutants into regions of the troposphere where dispersion is more likely to take place. Since these taller stacks are designed to discharge very large amounts of pollutants, their overall effect on pollution control is doubtful. They do reduce the concentration in the immediate area of the plant, but under stable air conditions they may cause a large region of concentrated pollutant downwind.

A number of processes for the removal of the sulphur oxides from stack gases are being studied. Ground limestone injected into the furnace, for instance, causes the production of solid sulphates ($CaSO_4$) which are then taken out, along with the other particulates, by electrostatic precipitators. In these devices, dust is attracted to charged collectors in the same manner that bits of paper are attracted to a comb charged with static electricity by a passage through the hair. There is a potential market for sulphuric acid, which can be made from those sulphates in the fertilizer industry for example.

A catalytic process, using aluminum oxide and producing either sulphuric acid or elemental sulphur is being tested in a pilot plant by the United States Bureau of Mines. It is calculated that it would produce 180 tons of sulphur per day from an 800 mw power plant burning 3 percent sulphur coal. Again, it is hoped that sale of the sulphur will reduce the costs of removal.

The magnitude of the sulphur released by the burning of the fossil fuels may work against the recovery of some of the cost of removal by sale. On a worldwide basis about 70 million tons of sulphur are released into the atmosphere each year, largely by the burning of fossil fuels; about 38 million tons are released in the United States alone. On the other hand, total world production of mined sulphur is 30 million tons. If a significant fraction of the fossil fuel sulphur is recovered, it could produce a great effect on the sulphur market changing the price and reducing the amount of cost recovered.

The sulphur oxides problem will clearly be with us for awhile. Even with the advent of nuclear power, as we saw in Chap. 6, the amount of coal burned each year is expected to increase, due to the increasing demand for electric power. We will pay more through pollution, or through control, for these future kilowatt-hours.

SUMMARY

The swiftly rising curves of population, production, and consumption have brought us into the age of shortage; we are running short on space, raw materials, and energy, and now our air is threatened.

Air pollution is still essentially a local phenomenon, as it has been since man first lit a fire in a poorly ventilated cave. The localities threatened now, however, are cities and even countries.

We have historically counted on the massiveness of the atmosphere to dilute our gaseous waste and the almost constant motion to carry it away. The atmospheric "disposal" worked well most of the time, and its failures, even over nineteenth-century London, seemed not too serious. We have, however, not only greatly increased the amount of material going into the "disposal," but built our cities without regard for its geographical variation in efficiency. We find, to our great distress, that many of them are located at places where the disposal action is often faulty, and that all of them suffer from some increasing number of breakdowns.

The major breakdown of the "disposal" occurs, as we learned in this chapter, when there is a temperature inversion: when the temperature increases with altitude, and the lower, polluted air is prevented from rising. Such conditions, in the winter over a coal-burning city, lead to the classical (sulphurous) smog, and to increased death rates. In the sunlit summer, an inversion over any automobile-clogged city gives us the photochemical smog with its eye searing PANs and crop damaging ozone.

In this chapter we also weighed both the contributions to air pollution from the various types of combustion and the body's defenses against them. We examined the cost of the failure of these defenses. As a final comment on the status of our air, we will summarize the federal air quality standards, the goal against which we judge our progress, and also look at the records of this progress over the past few years. The primary (health protection) standards and secondary standards (protection of crops and materials) are given in Table 7-10. The standards are stated in terms of concentrations (fractions of a gram per cubic meter of air, for instance) and exposure times; longer exposures require lower concentrations. With these standards as reference, Fig. 7-17 shows the score; it presents average measurements of several of the pollutants over the past decade or so. We can make several comments. First, the trend of the urban data is down; things are getting better. One notices that the nonurban total suspended particles (TSP) level is, however, rising. Power plants are moving away from the cities. We also see from these data that there is need for improvement. Comparison with Table 7-10 shows that while the average value of carbon monoxide is well under the standard, the average values of urban TSP and sulphur oxides are dangerously near their standard limits.

Table 7-10 National Ambient Air Quality Standards

Parameter	*CO*	*HC*	*SO_x*	*NO_x*	*Aerosol*
Primary	10 mg/m^3 (8 hr)* 40 mg/m^3 (1 hr)* (32 ppm)	160 μg/m^3 (3 hr)*	80 μg/m^3 (1 yr) 365 μg/m^3 (1 day)	10^2 μg/m^3 (1 yr) (250 μg/m^3, 1 day, recommended but deleted)	75 μg/m^3 (1 yr) 260 μg/m^3 (1 day)^3
Secondary	10 mg/m^3 (8 hr)* 40 mg/m^3 (1 hr)* (32 ppm)	160 μg/m^3 (3 hr)*	60 μg/m^3 (1 yr) 260 μg/m^3 (1 day)	10^2 μg/m^3 (1 yr) (250 μg/m^3, 1 day, recommended but deleted)	60 μg/m^3 (1 hr) 150 μg/m^3 (1 day)^3

*Maximum concentration not to be exceeded more than once a year.
mg/m^3 = 10^{-3} grams/cubic meter
μg/m^3 = 10^{-6} grams/cubic meter
ppm = parts per million

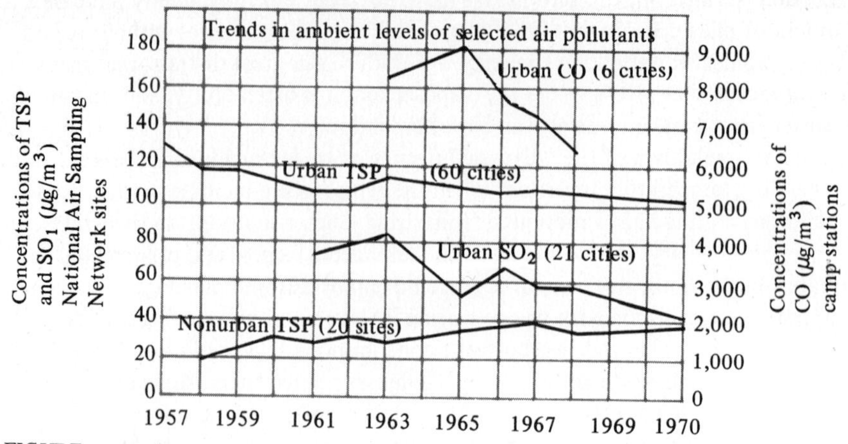

FIGURE 7-17 There is some encourageing decrease in the urban pollutants, but nonurban pollutants are increasing. (Source: "Environmental Quality 1972," Council on Environmental Quality, August 1972, p. 7.)

We call attention, however, to the footnote to Table 7-10, where it is stated that most of these levels are "not to be exceeded more than once a year." Is that requirement being met? In Table 7-11 we present the record of air pollution measurements in several large American cities. The so-called *MAQI Index* in column 1 of this table is a complicated ratio which is greater than 1 if the standards have been exceeded; all of the MAQI Indexes are considerably larger than 1. The other columns give even more easily interpretable data: the ratio of the annual averages of concentrations of the various pollutants to their standards. This ratio in close to 1 for certain pollutants, with several examples of ratios which are greater than 1.

One can read all this data in various ways; a middle-of-the-road interpretation is that we have slowed the ascent of air pollution curves and caused them to turn over and fall here and there. There has been some clearing of the air.

The future, however, remains cloudy. The continued growth in the consumption of energy will keep amounts of pollutants growing. Therefore, the achievements of control technology, and the investment in it, must grow more rapidly. Unfortunately, from the commercial point of view, investment in pollutant control is not a money-making venture. The emission control device does not make it easier to sell an automobile, while the air conditioner does. The natural inertia of industry to change is now being bolstered by the suddenly apparent energy shortages. Our media are suddenly flooded with subtle and not-so-subtle warnings that we are going too far, too fast, with environmental protection. The legislative victories of the environmentalists, won in the late 1960s, are seriously threatened. The EPA and its congressional backers are fighting determinedly to hold on to reasonably strict standards. It is not, at this moment, possible to predict the outcome; in fact, we may not even notice a victory for the environmental protectors. There will, however, be a certain smell to failure.

Table 7-11 Air Pollution Data for Selected Cities*

Year	*MAQI*	SO_2	*TSP*	NO_x
ATLANTA				
1968	2.83	0.39	1.07	2.20
1969	2.51	0.33	1.03	1.80
1970	2.60	0.24	1.20	1.65
DETROIT				
1968	4.01	0.83	1.79	2.44
1969	3.68	0.69	1.55	2.25
1970	3.39	0.46	1.51	1.69
ST. LOUIS				
1968	3.82	1.14	2.16	2.17
1969	5.35	0.91	2.45	2.02
1970	4.41	0.73	2.05	1.72
MILWAUKEE				
1968	4.27	0.48	1.85	2.05
1969	3.17	0.20	1.47	1.69
1970	2.69	0.19	1.21	1.21
NEW YORK				
1968	6.07	3.00†	1.59	2.87†
1969	5.01	1.69†	1.40	2.93†
1970	3.48	0.91†	1.64	1.49†
BALTIMORE				
1968	3.51	1.07	1.31	1.78
1969	3.28	0.75	1.47	1.86
1970	4.17	0.68	1.51	1.80
BIRMINGHAM				
1968	5.09	0.21	2.44	2.46
1969	4.25	0.21	2.16	1.75
1970	5.05	0.10	2.07	1.10
BOSTON				
1968	2.35	0.65†	1.24	1.04†
1969	2.51	0.80†	1.13	0.75†
1970	2.12	0.59†	1.07	0.96
WASHINGTON, D.C.				
1968	NA	NA	1.15	NA
1969	1.51	0.36	0.97	1.29
1970	1.52	0.34	1.03	1.15†
PHILADELPHIA				
1968	3.84	1.13	1.49	1.65†
1969	2.72	0.88	1.69	NA
1970	0.99	1.05	1.80	2.28

*Source: Third Annual Report of the Council on Environmental Quality, *Environmental Quality*, (Washington, D.C.: Council on Environmental Quality, 1972), Tables 6, A-5, pp. 10, 44.

†Data invalid for at least one quarter because of inadequate number of samples.

8

When You're Hot, You're Hot: Energy and Thermal Pollution

The "smogs," the cause, curse, and cure of which we explored in the previous chapter, are largely a by-product of our primitive methods of energy conversion. The major source of their raw materials is the burning of fuel, the only significant technique we have mastered to convert the chemical energy of fuel to thermal energy. In the future we see hope to supplant this reaction by replacing chemical fuels with nuclear fuels in the fission and fusion reactions, or by direct conversion of chemical energy to electric energy in, for instance, the fuel cell (which we will discuss in Chap. 12).

The hope that new techniques will do away with the pollutant, cannot sustain us, however, in the present chapter. Heat, which we consider a pollutant here, will be with us as long as we use energy; it is wedded to energy conversion by the unbreakable bonds of the Second Law of Thermodynamics. In every conversion process some available energy, chemical or gravitational potential energy, for instance, or mechanical kinetic energy, is converted to low temperature heat, and cannot be reconverted to useful work. As the flow diagram in Fig. 5-12 showed us, heat itself is the desired output of more than half of United States energy conversion; transportation and the production of electric energy account for most of the rest. We have already seen that in both these latter processes, large amounts of heat energy (three quarters of the gasoline engine's input and two thirds of the steam turbine's) are thrown away. More waste heat is produced by, for instance, friction in the moving automobile and by i^2R heating in electrical transmission lines. In the end, of course, all the energy coming in at the left of Fig. 5-12 becomes heat energy, is dispersed into the environment, and is, eventually radiated out into space. Therefore, thermal pollution differs in two major ways from the smog components we have just discussed: (1) it is an unavoidable consequence of all energy conversions; and (2) it does not accumulate indefinitely, but is radiated away at a fixed rate.

The vastness of space easily accepts the puny amount of heat our fossil fuel fires produce; it is unnoticed alongside the blazing stars. On earth, however, the heat must be transported through air or water, and, in its passage it has the capacity to disrupt delicately balanced natural processes. It is in this sense that we call it a pollutant.

WASTE HEAT AND WARM WATER

All energy conversion processes produce heat, and if this heat energy is not desired to warm a building, drive a chemical reaction, melt steel, or take part in some other similar function, it is waste heat. In this sense our large and densely populated urban areas with power consumptions as high as 300,000 kw per mile^2 are important heat sources. Later we will discuss some of the changes in the climate of the cities that are influenced by these spread-out sources. At the present time, however, the major sources of concern for thermal pollution are the stationary power sources: the electric generating plants.

The efficiency of a steam engine, as we have seen in Eq. (4-3*b*), is given by $(1 - T_{in}/T_{out})$ where T_{in}/T_{out} is the ratio of the output and input steam temperatures. To minimize this ratio, the input temperature must be as high as possible. It is limited, however, to about 1000°F by the strength of the pipes and the boilers. The output temperature must be as low as possible; this is accomplished by forced cooling. The heat energy remaining at the end of a cycle must be taken away and, in a steam engine, cooling water in the *condenser* allows the steam to condense back to water so that it can be recirculated in the boiler. We can understand the necessity of this step both from the relationships in the efficiency equation and on a physical basis. If the steam pressure is not reduced by cooling and condensing, then the piston will have to do work against this pressure in the exhaust stroke and this work will be wasted.

As we have seen earlier, the efficiency of the best present-day fossil fuel burning-generating plants is about 40 percent. Thus, for every kilowatt-hour of energy produced, about 5200 Btu of heat energy must be disposed of. Some 15 percent of the input heat energy is lost from the boilers and pipes or goes up the stack with the exhaust gases, leaving about 3900 Btu per kw-hr to be disposed of in other ways. To explain this more mathematically, since 1 kw-hr = 3400 Btu, the total energy required is about 8600 Btu of which 0.6 or 5200 Btu are wasted; 1300 Btu (15 percent of 8600 Btu) go up the stack, leaving 3900 Btu.

While the nuclear reactor, which is expected to rapidly come into its own as a source of electrical energy, will reduce the demand for fossil fuel, it will not relieve us of this waste heat problem. In comparison to the modern fossil plants, the present-day reactors operate at much lower steam temperatures and have an efficiency of about 33 percent. Thus, the reactor needs 10,000 Btu or more of heat energy to produce 1 kw-hr of electric energy. The Vermont Yankee nuclear plant in New England, for instance, needs 10,500 Btu per kw-hr. The nuclear reactor does not discharge appreciable heat energy up the stack so it must get rid of about 6500 to 7000 Btu of waste heat, almost half again as much as the fuel-burning plant. We are being promised by their proponents that the *Breeder Reactors* now under development will operate at higher steam temperatures and thus have efficiencies comparable to the modern fuel-burning plants. Figure 8-1 compares the useful and wasted energies from representative present and projected plants. It appears unavoidable that the switch to nuclear power will increase the amount of waste heat for disposal.

There is not much choice as to where to dump this heat energy; it can go into the air or it can be carried away by water. Air does not have a high heat capacity but, as

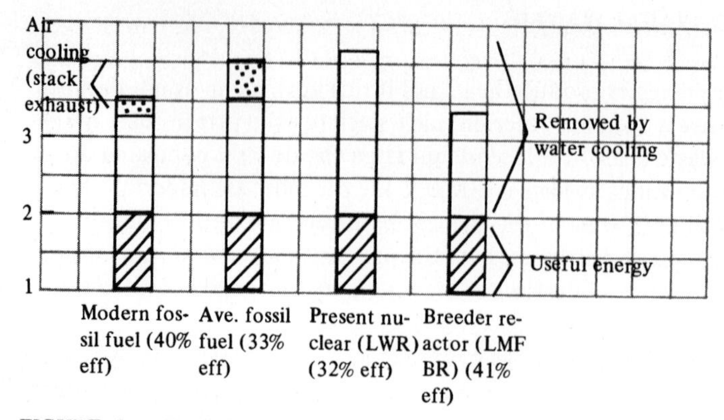

FIGURE 8-1 Useful and wasted energy of various generating plants. Since the fuel-burning plants release 15 percent of their input heat energy to the atmosphere their demand for cooling water is significantly lower than the nuclear plants.

we have seen in the previous chapter, there is much of it, and the heat itself stimulates the vertical currents which help carry it away. Water is the preferred cooling medium, however, for it has good heat-transfer properties, a large heat capacity, and, (in rivers, at least) a natural flow to carry the heat away. We will examine the use and the problems of these heat disposal techniques.

Water as a Heat Sink

In some ways water seems to have been designed as a cooling medium. It has the unusually high heat capacity of 1 Btu per °F pound (see Table A3-2 for comparisons). Water can absorb energy, not only in the random motion of the molecules, but also because of the structure of the molecule itself; it can absorb energy in vibratory motion within the molecule. It has the advantage of a high heat conductivity and it makes good thermal contact with a hot object, thus allowing quick transfer of heat energy. Finally, being a fluid, *convection currents,* which are caused by hot, less dense water rising so that cold water replaces it, carry heat away.

Water, however, is such a generally useful substance that we must make some priority judgments among its uses. It serves as a cooling fluid, as a cleaning fluid, as a medium for waste disposal, a medium for recreation, and as a necessity of life for most organisms. We will point out some of the conflicts among these uses as we proceed. Let us now look at its use as a cooling fluid.

Electric generating plants are, of course, not the only industries which require water for cooling. Steel plants need water to cool the ingots; chemical plants, oil refineries, paper mills and food processors use their share. The electric industry, however, is the major user, accounting for about 80 percent of the total.

The easiest, and, therefore, the traditional way to cool a power plant, is to pump water from a nearby large source (a lake, river, or bay), circulate it through the hot system, and return it, now at a higher temperature, to its source. A considerable volume of water is needed for this *once-through cooling,* as it is called. As we have just computed, a modern fuel-burning generating plant produces with each kilowatt hour of electric energy, 3900 Btu of heat energy which needs to be carried away. Thus, a

1,000-mw generating plant of 40 percent efficiency releases 3.9×10^9 Btu per hour or 1.1×10^6 Btu per second. A gallon of water weighs about 8 pounds; remembering the definition of a Btu (it raises 1 pound of water 1°F) a gallon of water can remove about 120 Btu if its temperature is allowed to increase 15°F. Cooling for this 1,000-mw, 40 percent efficient plant, therefore, requires 9,000 gallons of water per second. In other terms, each kilowatt hour of electrical energy generated requires 33 gallons of water. In a 33 percent efficient fuel-burning plant of 1,000 mw (the present national average), the requirements would be 12,000 gallons per second or about 45 gallon per kw-hr. A 1,000-mw nuclear plant operating at present efficiencies needs 15,500 gallon per second or 56 gallon per kw-hr.

From these calculations we see that generating plants, which, as we have mentioned, are getting larger and larger, require enormous amounts of cooling water. The 3,200-mw Monroe plant of the Detroit Edison Company requires, if the allowable rise in temperature is 15°F, at least 28,800 gallons of water per second from Lake Erie.

Since nuclear plants will not reduce the electric industry's needs for cooling water, but will, in fact, increase them, we must look ahead to the demands for water which the rapid growth (doubling by 1980, quadrupling by 1990) of that industry will make. The FPC has estimated that the United States will have 390,000 mw of fossil fuel steam-generating plants and 140,000 mw of nuclear plants in 1980 (see Table 6-4). This capacity will increase, it is predicted, to 558,000 mw of fossil steam and 475,000 mw of nuclear by 1990. From these estimates, we can compute the cooling water demand for those years. Table 8-1 displays the result of this calculation under the assumption that fuel-burning plants and nuclear plants are, on the average, 35 percent efficient in 1980 and 40 percent efficient in 1990. Fortunately, the real demand on the United States supplies of running fresh water, which average 1,200 billion gallons per day during a year, will not be that large. Already 30 percent of the cooling water needs are filled by salt water at coastal or bay generating sites, and this percentage will probably grow. An additional reduction in the estimate can come from the assumption that many of the plants will have part or all of the cooling provided by the cooling towers whose operation we will discuss later. A recent FPC estimate which takes these factors into account is shown in Fig. 8-2. Even with these reductions in the estimate, it is still predicted that 200 billion gallons of water a day, one sixth of the total average fresh water runoff in 1990, will be needed to cool electric power plants.

Table 8-1 Maximum Cooling Water Needs 1980, 1990

	*1980**		*1990†*	
Type of plant	*Capacity (mw)*	*Cooling water (10^6 gal/sec)*	*Capacity (mw)*	*Cooling water (10^6 gal/sec)*
Fossil fuel	390,000	4.4	558,000	5.6
Nuclear	140,000	2.1	475,000	5.6
Total (in 10^6 gal/sec)		6.5		11.2
(in billion gal/day)		610		970

*All plants 35 percent efficiency.

†All plants 40 percent efficiency.

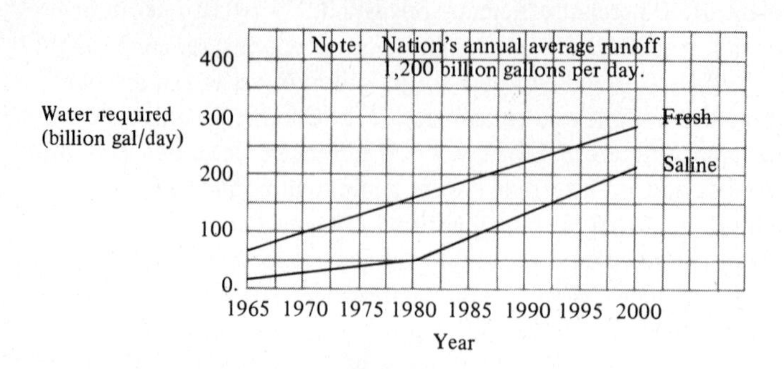

FIGURE 8-2 Present and future cooling water requirements for electric power plants. The fresh water demand in 1990 is one sixth the total average daily runoff.

These data should not be interpreted as predicting that the water from one sixth of the United States streams will run through power plants. In reality, the water in a given river may be used over and over again by plants along it with some (but not all) of the heat being lost from the water's surface to the atmosphere. On the other hand, the 1,200 billion gallons per second of United States runoff is an average over a year; river flow however, is not steady. Flood conditions which exist for perhaps one third of the year account for two thirds of the runoff. The Hudson River, for instance, at flood has a runoff of 310,000 gallons per second which is reduced to as low as 15,500 gallon per second during dry periods. This lower flow of water could only service one 1,000-mw present-day reactor. It is reported that already during such periods of low flow more than 100 percent of the waters in certain rivers in the Northeast are used for cooling.

The worst recorded case is that of the Mahoniny River at Youngstown, Ohio, whose waters were used many times, leaving them at a temperature of 140°F in the summer. In fairness to the utility companies it must be admitted that the major offenders were steel companies, and that the situation of that unfortunate river is now improving under regulatory control.

The Ohio case reminds us, however, that it is in the Midwestern and the Eastern Central States that the water problems will be most acute. Figure 8-3, prepared by the same FPC study group, shows the capabilities of the continental rivers to supply cooling water for fossil fuel plants. The capability would be 40 to 60 percent less for nuclear plants. With an assumption that the large plants will be separated by 25 miles to allow for cooling of the river water and, allowing a 5°F rise at minimum flow, the study predicts that the Ohio could supply cooling water for 60,000 mw of fuel-burning plants or 40,000 mw of nuclear plants. There are already 40 steam electric plants with a capacity of 18,000 mw on the Ohio. It is clear that the doubling by 1980 and quadrupling by 1990 will severely strain the Ohio's cooling capacity.

The problems of coastal sites are serious also, as evidenced by the recent battle over the plans of the Florida Electric Company to build a nuclear generating plant on Biscayne Bay, south of Miami. Their plans called for a discharge of about 2 million gallons a minute into this bay. This amounts to the equivalent of 1,800 acres of water 4 feet deep, 1 to 2 percent of the total volume of this shallow bay (averaging 7

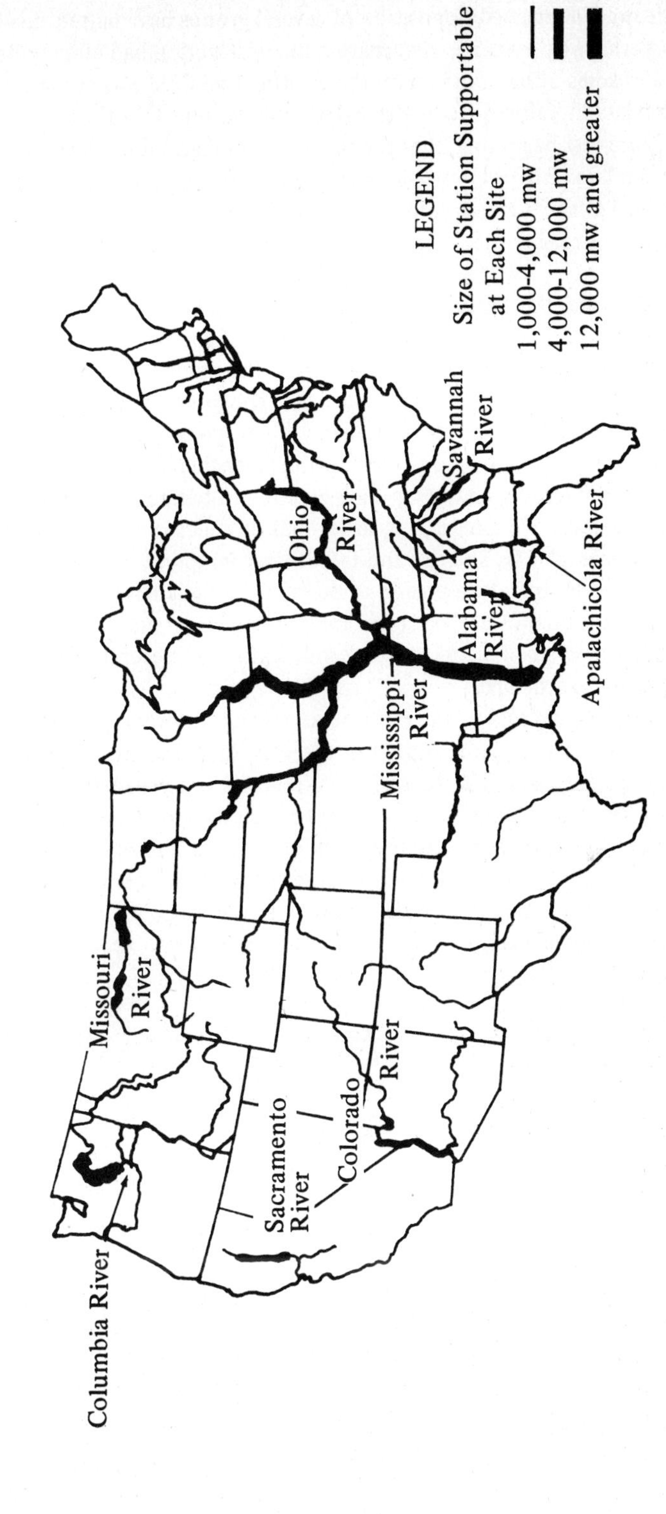

FIGURE 8-3 The generating capacity which will be allowed by the cooling capacity of U.S. rivers.

feet deep). The massed opposition of several groups have halted this construction, aided perhaps by evidence that marine life destruction had already been caused in over 450 acres of bay bottom by the existing fuel-fired power plants.

The data in Table 8-1 and Fig. 8-1 should warn us that these battles over "water rights" are just beginning, that the increasing demand for "clear, cool water" will cause similar confrontations along most of the rivers in the populated regions of this country. To take a constructive role in such controversies, we must examine the consequences of raising the temperature of a river.

Dump It Down the Drain. What happens to this hot water after it is dumped into a river or a lake? Since it is hotter than the rest of the river by 10 to 30°F, it is less dense and stays on top. It spreads out from the point of discharge, like the plume of smoke from a smokestack spreads into air. The hot layer rides on the surface of the river for a while but loses its heat energy by evaporation and through radiation, conduction, and convection, to the air, to the layers beneath it, and to the banks.

In a few miles, depending on the amount and temperature, it will cool back to the normal level. We mentioned earlier that 25 miles is needed for this to occur for the 1,000-mw plants. The rise and fall of river temperature that this causes is shown in Fig. 8-4, a temperature profile of the Monongehela River in 1960. The peaks caused by cooling water discharges are quite apparent and the extent of the river affected by these small plants is 1 or 2 miles, although several long stretches of elevated temperature are visible.

In a lake or a reservoir the result is somewhat different. These static bodies of water are normally stratified, that is, there are layers at different temperatures which do not normally mix. The bottom layer remains at a constant lower temperature most of the year. Power plants use this to their advantage, pumping the cooler water from the lake bottom and discharging it at the top. This has the effect of increasing the stratification of the lake and, because the bottom waters are usually richer in nutrients, it may increase the algae growth in the warm surface layer.

Heat energy is not the only pollutant discharged into a river or a lake by once-

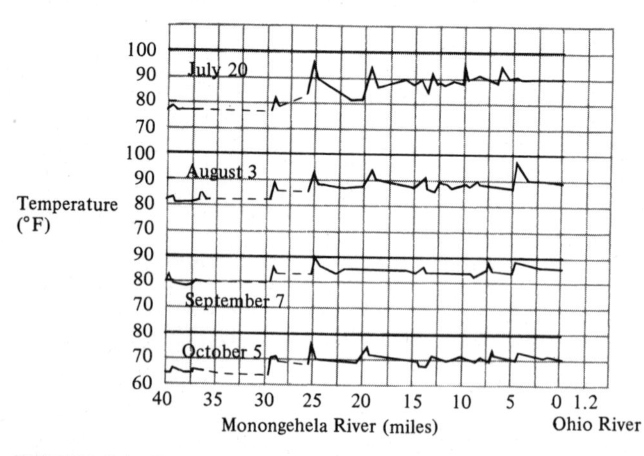

FIGURE 8-4 Temperature profiles of Pennsylvania's Monongehela River, summer 1960. Heat discharges by the industrial plants along the river are shown by the peaks.

through cooling. The water, on intake, is usually treated with chemicals to reduce the growth of slime, and kill certain spores and larvae. The discharged water carries this and any corrosion products it has picked up with it back to the water source.

These corrosion products have turned out to be more important than anticipated. They were, for example, finally identified as the villain in the case of Maryland's "green oysters." Shortly after installation of a big steam electric plant on the Patuxent River, it was noticed that the oysters below the discharge canal were beginning to turn green and die, and that the area affected was spreading and growing. Although certain algae do turn oysters green, laboratory analysis connected this greenness with increased copper uptake, the copper coming apparently from the condenser tubes of the steam plant. At least it was determined that oysters set at the intake of the plant thrived while oysters at the outlet turned green and died. It is still not completely understood how much copper is corroding away, and why it is.

The problem of chemical, and especially metallic, water pollutants is important, but not specifically an energy problem. Heat, however, is energy, so let's look more carefully at our warmed-up river.

Effects of Hot Water. The water in a river or a lake presents a different view to different viewers: it is a sewer and a source of tap water to a city manager, a swimming hole to a hot passerby, a fishing ground to a fisherman. Increasing its temperature affects all these uses in different ways. Two major changes take place. As the temperature rises, the water's ability to hold dissolved oxygen decreases; an increase from 55° to 68°F, for example, reduces the dissolved oxygen by 13 percent. This only occurs if the water is saturated with oxygen; studies at plant sites so far do not show much change in oxygen levels.

A second change with rising temperature is an increase of the rate at which chemical reactions take place. This is a gain for the water supply engineer; it becomes cheaper to purify water as it becomes warmed, an estimated savings of 30 to 50 cents per million gallons for each 10°F rise in temperature. As water is warmed, however, the growth of taste- and odor-producing organisms, for instance, the blue-green algae, is stimulated, making the purification task more difficult. The water will eventually have to be cooled since temperatures near 50°F seem to be preferred for drinking.

Temperature is also an important factor in determining the water's ability to assimilate the organic wastes dumped into it. Since waste assimilation depends on oxidation, it is affected directly by the oxygen content. It is also indirectly affected by the stimulation of the blue-green algae which compete for oxygen in decay and many of which are toxic to fish, livestock, and humans. The overall effect of a temperature increase seems to be a lowering of the water's ability to satisfactorily process its sewage load.

The use of river water for cooling threatens aquatic life by means other than temperature changes. A trip through the condenser system is usually lethal to any organism, and fish are killed by the pumps, the screens, and other mechanisms. One of the larger fish kills, at Consolidated Edison's Indian Point Station in 1969, for instance, was not a thermal kill. In the period from December to February of that year, some 150,000 fish were killed. It was at first believed that thermal shock was the culprit, but after study, it was determined that they were probably attracted to the intake region by warm discharge water carried upstream by the tide. There they were trapped in some of the screens and entrance channels and died of exhaustion. The fish, however, whatever the culprit, were dead by reason of the plant's cooling demands.

This complexity of results is found when one looks at the effects of increased temperature on aquatic life in general. Temperature is important in several ways; too high or too low a temperature can kill. Small changes have their affects also; metabolic processes and oxygen demand increase with temperature, but dissolved oxygen decreases. Temperature regulates many of the life processes of aquatic organisms, affecting growth rates, reproduction, feeding, digestion, respiration, migration, and other important processes. It seems clear that any temperature change will cause a change in the overall aquatic ecology of a stream, encouraging certain organisms and discouraging others. Its effects are amplified in shallow water where important spawning and food-producing activities take place. Faced with this complexity, and lacking much of the necessary knowledge of the aquatic ecology, the most prudent course is obviously a cautious one.

There are some data which show the results of laboratory experiments on the tolerance of fish to maximum and minimum temperatures (see Table 8-2). Such data are available on only 5 percent of the 1,900 or so identified North American fish species. One can see from this table that the effects of temperature vary from specie to specie. This temperature dependence is shown even more dramatically in Fig. 8-5, which relates the "well-being," (measured by size) to temperature. One can see that in 81°F water, a trout would be dead, a pike very unhappy, while a largemouth bass

Table 8-2 Results of Laboratory Investigation of Minimum and Maximum Temperatures for Certain Freshwater Fishes*

Species	*Acclimated to °F*	*Minimum temperature†* (°F)	*Time (hr)*	*Maximum temperature†* (°F)	*Time (hr)*
Largemouth bass	68.0	41.0	24	89.6	72
	86.0	51.8	24	93.2	72
Bluegill	59.0	37.4	24	87.8	60
	86.0	51.8	24	93.2	60
Channel catfish	59.0	32.0	24	86.0	24
	77.0	42.8	24	93.2	24
Yellow perch	41.0			69.8	96
Winter	77.0	39.2	24	86.0	96
Summer	77.0	48.2	24	89.6	95
Gizzard shad	77.0	51.8	24	93.2	48
	95.0	68.0	24	98.6	48
Common shiner	41.0			80.6	133
	77.0	39.2	24	87.8	133
	86.0	46.4	24	87.8	133
Brook trout	37.4			73.4	133
	68.0			77.0	133

*Source: Pacific Northwest Water Laboratory, "Industrial Waste Guide on Thermal Pollution," Federal Water Pollution Control Administration, September 1968.

†These are the water temperatures survived by 50 percent of the test fish.

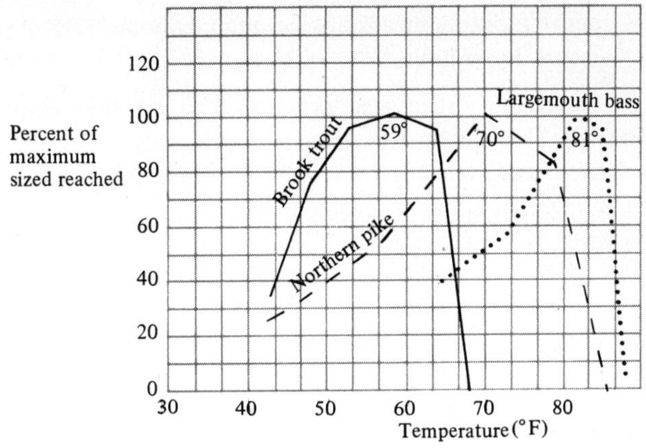

FIGURE 8-5 "Well-being" (growth rate) of trout, pike, and bass as a function of temperature shows the great differences which exist.

would be at home. Not only is the temperature dependence different from fish to fish, but the liveable range for a given fish changes during the various stages of a fish's life. During certain stages variation by more than 5°F can be fatal.

One could also correctly guess that the optimum temperature varies with the season. Fish are known to congregate near the warm discharge outlets in northern climates in the winter. In summer many of the rivers are already near their limits; and most of the recorded thermal fish kills have been in this season.

There is evidence that fish use water temperatures to guide and control their migrations, and that hot water can block their passage up a stream. Increased susceptibility to disease has also been reported; there is particular concern over the Columbia River salmon; increased water temperature seems to have brought about an increase in a heretofore rare disease, and a heightened susceptibility to death from nitrogen bubbles in the blood, similar to the "bends" of deep sea divers. On the other hand, one sees reports of more nesting activity and increased salmon size.

Faced with this sparse and offtimes conflicting information, the recommendation is to change the temperature as little as possible. More specifically, the 1968 Report of the National Technical Advisory Committee on Water Quality Criteria made the following recommendations to the Secretary of the Interior:

1 Temperatures should not be raised more than 3°F (epilimnion[1] of lakes and reservoirs) or 5°F (streams) during any month of the year, and normal daily and seasonal temperature variations should be maintained if warm-water fish populations are to be maintained.

2 Inland trout streams, headwaters of salmon streams, trout and salmon lakes and reservoirs, and the hypolimnion[2] of lakes and reservoirs containing salmonids should not be warmed.

3 Discharges of heated wastes to coastal or estuarine waters should be closely managed to avoid increases of over 4°F (fall through spring) or 1.5°F (summer)

[1] Top layer.

[2] Bottom layer.

above monthly means of maximum daily temperatures and rates of change over 1°F per hour.[3]

These are fairly strict standards, much stricter than most presently applicable state standards. The aquatic ecosystem is so delicately balanced, however, that changes will no doubt occur even if water temperatures are kept within these limits. As we have seen in Fig. 8-4, temperature excursions of 20°F were not unusual for the Monongehela River of a few years ago. In the present and future the bind will be tighter as we try to achieve the conflicting goals of discarding larger and larger amounts of waste heat, and keeping our rivers and lakes viable. There seems only one way out of this bind; if we are to be denied the use of water as a "heat sink" we must turn to air.

OTHER HEAT-DISPOSAL SYSTEMS

The disposal of waste heat is a stepwise process. In the once-through cooling technique, river water is used to carry the heat energy away from the plant. The air then picks it up by the three methods of heat transfer (conduction, convection, and radiation) and by evaporation. After joining the other components of atmospheric heat energy and following the complicated radiation and absorption cycles that exist in the atmosphere, it is ultimately radiated away into deep space.

Since we, as a society, have until now, made no charges for the use of our rivers, once-through cooling is the cheapest and most widely used technique. The water quality requirements we have just discussed, however, are forcing the utilities companies to look to other means for getting the heat energy out of the condenser and into the atmosphere. In contrast to some of the problems posed by air pollution, the alternate technology for this purpose exists.

Cooling Ponds

We have mentioned that natural lakes or reservoirs are presently used as heat sinks. In regions of low land-cost it has been feasible to build a cooling pond to achieve the same results. Hot water is discharged into the pond, spreads over the surface, is cooled largely by evaporation, and cooling water is pumped from the pond. The obvious advantage to the utilities is that the pond is dedicated to this use and its temperature characteristics are under their control and supervision.

Since it is the pond's surface area that provides for the interchange of heat with the atmosphere, it must be large. Current estimates are that 1,000 to 3,000 acres per 1,000 mw are needed; ponds with areas of 1,000 acres per 1,000 mw can suffice only in areas of rapid evaporation. Pond size is also affected by the temperature of the discharge water: higher temperatures cause more rapid evaporation and more rapid heat dispersal by convection and radiation. If provision is made for spraying the discharge water into the air to increase evaporation the size of the pond can be reduced by a factor of 20. We should point out that these requirements are not really differ-

[3] "Water Quality Criteria," National Technical Advisory Committee, Federal Water Pollution Control Admin., USDI, Washington, D.C. (1968).

ent from the size requirements of the cooling area of a river's surface. In that case however the 1,000 to 3,000 acres moves downstream.

Since evaporation is a major mechanism for cooling, it is clear that some means for "make-up" of the water lost from the pond must be provided, usually through a small feeder-stream or by pumping from some large source. The amount of water needed for a 1,000-mw plant is in the range of 10 to 20 million gallons per day.

Several power companies are experimenting with the use of treated liquid sewage effluent as "make-up water" in the ponds or in the wet cooling towers to be discussed next. This use, while it has complications, is at least in the direction of attacking two major problems at once.

Cooling Towers—Wet

When suitable sites for ponds are unavailable and once-through cooling is not practical, *cooling towers* are needed. These are structures in which the water is circulated, brought into contact with the air, and the heat exchange accomplished. There are two types: (1) *wet towers* in which the water is brought into direct contact with the air and the heat is transferred principally by evaporation; and (2) *dry towers* which are constructed much like an automobile radiator (but on a large scale) and in which the heat transfer is mostly by conduction or convection.

Wet towers can be used in any region where the humidity is low enough to allow a reasonable amount of evaporation. The heated water is either sprayed into the air in droplets or allowed to flow in a film over a lattice work. The cooled water is collected at the base and pumped back to the condenser. Brackish or even salt water can be used if precautions are taken to reduce the amount that blows out over the countryside. This problem of chemical fallout must be faced anyway, since chemicals are added to the water to reduce slime growth.

Most of the wet towers in this country use *forced-draft* cooling, in which large fans blow air either along or through the falling water. Such a tower is pictured in Fig. 8-6. Since water is continuously removed by evaporation or *blow-down* (forcing the water out through the nozzles, for instance, to remove the accumulation of solids) it must be replaced. The requirements, however, are only 2 to 4 percent of the water required for once-through cooling.

The use of *natural-draft* towers is beginning to interest utilities companies. These towers are essentially a chimney, open at the bottom as well as the top (see Fig. 8-7). They provide an upward current of air heated by contact with the hot water which then rises up the chimney and is replaced by a flow of cooler air at the bottom. These structures are huge, on the order of 400 feet high and 400 feet in diameter, and designed in an interesting geometrical shape as indicated in Fig. 8-7. They require a much larger initial investment than the forced-draft towers, but have much lower operating and maintenance costs.

Disadvantages of Wet Towers. The disadvantages of the evaporative coolers are all connected with the fact that they discharge large amounts of humid air out over the countryside. At a minimum, the towers will be a constant emitter of clouds, changing the average humidity in the local area. To prevent corrosion and fouling by organic

FIGURE 8-6 A forced draft cooling tower system located at the Utah Power and Light Company's 200-mw plant near Kemmerer, Wyoming. It is approximately 360 feet long, 48 feet wide, and 31 feet high.

growths, many different kinds of chemicals are added to the water and, as they are blown out over the countryside, added to the environment.

Dry Cooling Towers

In dry cooling towers the water circulates in a closed system and is cooled by heat exchange with the air through conduction or convection. The dry tower, while it has the smallest environmental impact, is the most expensive mechanism and, therefore, is not put to much use by the utilities. Again, there are both forced-draft and natural-draft systems. In forced-draft systems, the hot condenser water is circulated through the usual finned cooling tubes and air is blown over them by fans. In very hot weather the fans, called on for steady, intense use, can reduce the plant's overall efficiency by 6 to 8 percent because of the energy needed to drive them.

The natural-draft towers would have less energy loss. In both wet and dry towers, however, the cooling water enters the condenser at a higher temperature, on the average, than when taken from a lake or river, therefore reducing the plant efficiency by increasing the output temperature.

The cooling effect of dry towers depends on heat exchange with air and, since the heat conductivity of air is poor, these towers must be significantly larger; consequently, they are more expensive than evaporative coolers. The increased cost is balanced by their low environmental impact. Since they do not require a supply of make-up water they are most suited to dry, semiarid areas.

FIGURE 8-7 The natural draft cooling tower for the Portland General Electric Company's Trojan Plant on the Columbia River. This tower is 499 feet high and the facility includes a recreation lake, boating facilities, a reflecting pool, and a whistling swan lagoon.

Dump It in the Ocean

The impressive size of the ocean has inspired many thoughts in man down through the ages. Among the least attractive ones is the thought that it is so big you can dump anything into it without effect. The now visible evidence of small clumps of congealed oil and bits of styrofoam floating everywhere suggests this theory is erroneous. Nonetheless, we look to the oceans now as a heat sink for waste heat; in fact, as Fig. 8-2 showed, in 1970 we obtained about 30 percent of the cooling with salt water. This percentage is expected to increase.

Salt water has its own difficulties as a cooling fluid. It demands a much more corrosion-resistant system than fresh water and, if used in evaporative cooling, there are serious difficulties associated with the spray of salt water over the area.

If the coastal waters deepen rapidly, cooler waters from the unmixed deep layers of the ocean could be used and then discharged at temperatures very near the surface temperature, with minimal ecological effect. The cold bottom water is usually rich in nutrients which have settled from the surface. The world's choice fishing areas, such as the anchovy harvesting waters off the coast of Peru, are coastal areas where ocean currents force the cold deep waters to the surface. There are suggestions that pumping for sea water cooling of power stations might achieve similar effects.

Floating Nuclear Plants. The availability of abundant sea water for straight-through cooling is one of the major attractions of the concept of floating nuclear plants (FNP). An artist's conception of one of these planned for the coast of New Jersey, was shown earlier as Fig. 1-1. These plants which are to be assembled on huge barges 400 feet square, will have a 32-foot draft and a rise of 177 feet above the waterline. Each FNP will have a reactor capable of generating 1,150 mw and the two or more of them will be moored offshore, in water 40 to 70 feet deep and protected by a breakwater. The power will be transmitted to land by cable—an obvious situation for the newer transmission technology discussed in Chap. 6.

A production facility, Offshore Power Systems (OPS), is a joint venture of Westinghouse and Tenneco Company, now under construction on the St. John's River at Jacksonville, Florida. Four FNPs, including the two to be installed off the Jersey coast (2.8 miles offshore and 12 miles northeast of Atlantic City) have been ordered by the Public Service Electric and Gas Company of New Jersey.

The idea of floating the power plants has attractive features; not only is the cooling water problem simplified, but less land is used and the reactor is isolated from population areas in case of accidents. The possible consequences of an accident with the resulting contamination of sea water is still being studied. It seems likely, however, that several of these plants will be floating off our coasts by the mid-1980s.

Be Careful of Bays and Estuaries. A major concern of environmentalists has been to keep power plant discharges away from shallow water bays and estuaries because of the importance of these habitats as spawning and food-producing areas for salt-water life. One of the worries over the FNPs is their closeness to shore since they must now be within the territorial limits of 3 miles. It is expected that the desirability of locating them further offshore (with floating breakwaters) will be an important argument for increasing the territorial limit.

The fragility of the shallow shoreline areas is a more important concern for coastal plants. In Biscayne Bay, south of Miami, we have already mentioned that thermal kills of sea organisms had been recorded even before the proposed tripling of the plants' size. The projected damage to that shallow bay, which is a most important fishing area, rallied many groups to its defense.

The opening round in a similar battle over the use of Maryland's Chesapeake Bay was won by the power company. The now famous Calvert Cliffs nuclear plant (2 units of 890 mw each) is situated near the headwaters of the Bay. It is located, in fact, at a "bottleneck" for migrating fish and in an important spawning area of the Bay's famous blue crabs. There was concern that the hot discharge water would cause unforeseen problems.

Cooling towers were finally ruled out on economic terms and because of the damage from salt spray of evaporative coolers. Once-through cooling using Bay waters was the choice. The specifications of thermal releases are as follows:

1. The mean surface excess temperature would be 0.4 degrees F in the tidal segment adjacent to the Calvert Cliffs plant.

2 The area having a mean surface excess temperature of 3 degrees F or greater would be 23 acres.

3 The area having a mean surface excess temperature of 1.5 degrees F or greater would be 225 acres.

4 The Bay water having an excess temperature exceeding 1 degree F would occupy less than 10 percent of the cross-section of the Bay adjacent to the plant and about 20 percent of the volume of the tidal segment adjacent to the plant site.[4]

The temperature changes and areas affected seem innocuous enough, until one reads in another section of the report:

> The demand for electric energy in the United States is doubled approximately every ten years. Based upon the estimated rate of growth in the Chesapeake Bay region, Section IIIA of this report shows that there will be a threefold to eightfold increase in power demand in this region by the year 2000. If we assume that this demand is satisfied by building individual 3000 megawatt power plants (approximately twice the size of that proposed for Calvert Cliffs), then ten such plants will have to be built in this region during the next three decades.[5]

The Bay can stand one such plant. Further on in the report, one sees a computation that the Calvert Cliffs plant uses only 3 percent of the evaporated cooling capacity of the Bay (based on an allowable 1.5°F rise in temperature), but ten big plants would use 60 percent. At some "point in time," between now and the year 2000, the citizens of Maryland will have to decide whether to pay more for their electricity or use the Bay as a cooling pond. It is a decision facing most of us in slightly altered form.

COSTS AND BENEFITS

If we are to have clean air, then the automobiles will need emission control devices which we, the consumers, will pay for; the utilities will need electrostatic precipitators and sulphur dioxide removal systems; consequently, the cost of electricity to the consumer will increase. We will also have to pay for clear, cool water.

As we have said, the technology exists to remove the heat safely, even to the construction of the nonimpacting dry towers. On an absolute basis it is expensive technology; Table 8-3 shows some current estimates of the capital costs of various cooling systems. These costs are in dollars per kilowatt and range, therefore, for a 1,000-mw plant, from 2 million dollars for once-through cooling a fossil fuel-plant, to perhaps 30 million dollars for a dry tower on a present-day reactor. These are large numbers, but the construction cost of that nuclear plant is about 250 million dollars without the tower. The fuel-burning plant, by comparison, costs 150 million dollars. The fuel costs are, respectively 1.6 million and 3 million dollars, the operating and maintenance costs are 300,000 dollars and 200,000 dollars per year. As percentages of the total, even dry towers are not prohibitively expensive.

[4]Governor's Task Force on Nuclear Power Plants, Dr. William W. Eaton, Chairman, "Nuclear Power Plants In Maryland," (Annapolis, Maryland, December 1969), p. 50.

[5]*Ibid.*

Table 8-3 Comparative Costs of Cooling Water Systems for Steam-Electric Plants*

Type of system	Investment cost, (dollars/kw)†	
	Fossil-fueled plant‡	Nuclear-fueled plant‡
Once-through	2.00-3.00	3.00-5.00
Cooling ponds	4.00-6.00	6.00-9.00
Evaporative cooling towers:		
Mechanical draft	5.00-8.00	8.00-11.00
Natural draft	6.00-9.00	9.00-13.00
Nonevaporative cooling towers:		
Mechanical draft	18.00-20.00	26.00-28.00
Natural draft	20.00-24.00	28.00-32.00

* *The 1970 National Power Survey,* Part I, (Washington, D.C., Federal Power Commission, USGPO, December 1971), p. I-10-8.

† These investment costs represent ranges derived as of the year 1969. Future construction costs will probably be higher.

‡ Based on unit sizes of 600 mw and larger

The cost increases are even less impressive when expressed as the increase in cost to a consumer (see Table 8-4). We see that the largest increase is 0.15 mills[6] for a wet mechanical draft tower. Even a dry tower, whose cost we estimate from data in Table 8-3, to be perhaps 10 to 15 times the cost of once-through seawater, would only add 0.3 to 0.5 mills per kw-hr to the average 20 mills per kw-hr cost of electric energy. It would appear from these data that costs will not stand in the way of cooling our waters. It is, in fact, the necessity of raising those extra millions of investment money that is the real barrier, and public pressure is cracking through this. The second plant scheduled to be built on the Chesapeake Bay, for example, is scheduled to have cooling towers.

[6] A mill is a thousandth of a dollar or a tenth of a cent.

Table 8-4 Cost of Cooling Above that Required for Once-Through Freshwater Systems*

Cooling system	Range of estimates (mills/kw-hr)
Once-through seawater	0.023-0.034
Cooling pond	0.033-0.092
Wet mechanical draft tower	0.15-0.30
Wet natural draft tower	0.14-0.15

* Source: Adapted from Christianson and Tichinor, "Economic Aspects of Thermal Pollution Control in the Electric Power Industry," Table 1. Reprinted in "Environmental Effects of Producing Electric Power," Part 1, Hearings before the Joint Committee on Atomic Energy, (Washington, D.C., 1969).

Let's Call It Thermal Enrichment

The most satisfactory solution to the waste heat problem would be to stop wasting it. There is a fundamental paradox in a society which uses half of its energy for heating purposes and throws away almost 20 percent of the available energy as heat (a figure that will rise as electrical energy use outstrips direct fuel use). The basic problem is in bringing need and source together.

The needs are in the cities, where there are homes and offices to be heated, and around them, where industries (oil refineries and steel mills) need process heat. The large fossil fuel plants are dirty and sprawling; a 3,000-mw plant needs 900 to 1,200 acres of land for coal storage and ash disposal and must have facilities for handling special trains with 75 to 120, 100-ton coal-cars. A plant of this size uses 25,000 to 30,000 tons of coal a day. Such plants are not welcome neighbors. Nuclear plants occupy less room, 300 to 400 acres for a similar size plant, and fuel handling is no problem as far as volume is concerned. But the danger of accidents and the radioactivity of the shipments cause most communities to shy away from them also.

There are other problems; it is economically difficult to ship hot water any great distance. The heat energy is discharged in a degraded form at low temperatures (85 to 105°F). Therefore, it is not of any real use to a steel mill or a petroleum refinery. In addition, it is always available, whether it is needed or not, and the commercial building which used it for space heating in the winter would have no use for it in the summer (until an air conditioner using thermal energy is perfected).

In spite of these drawbacks, some waste heat is used now. There are buildings in New York, St. Louis and other cities heated by small urban-located generating stations. There is an oil company in New Jersey which uses waste heat from a neighboring power plant. In this case the power plant deliberately runs at a lower efficiency so as to discharge the water at a higher temperature. The combined efficiency of the two plants is of course significantly improved.

A most optimistic view of the future would have power plants located in "energy parks" along with industries which need the electric power and the heat. Each Btu that serves another purpose reduces the overall environmental impact of discharged heat and saves energy as well.

Multipurpose plants are also being suggested: generating plants near the ocean where discharged heat energy is used for distillation desalting of ocean water. High-quality desalted water could then be mixed with less satisfactory water (local brackish water, for instance) for irrigation.

It has also been suggested that new towns be built around large energy plants; cheap, available electric power could provide for mass transportation, part of the solid waste could be burned as fuel, liquid sewage effluent would be used as cooling water, waste heat would provide space heating and snow-free roads in the winter. Although all those activities exist separately in experimental projects, no one has yet put them together. Again, a missing ingredient is thermal-driven air conditioning.

There are other potential uses: warm water irrigation might extend growing seasons and allow the growing of tropical foodstuffs in greenhouses; the heat could be used in aquaculture, extending the range and hastening the growth of shellfish in the Puget Sound and oysters and lobsters off the coast of Maine. These experiments are in progress. It has also been proposed to use this waste heat, for instance, to keep northern rivers or, at least ports, ice-free for longer periods.

One notices that most of the suggestions for the uses of waste heat point northward and winterward. Biscayne Bay does not need warm water; nor do summer crops. It may well be, if electrical transmission can be made less expensive, that power stations will migrate northward where their heat is needed for a greater portion of the year. But even (or perhaps especially) the North is ecologically delicate, and one must look very carefully at ecological and climatic effects before allowing such a migration.

SUMMARY

The steam-driven turbines, even in the nuclear reactor age, will continue to monopolize the conversion of thermal energy to mechanical energy for the generation of electricity. To get their maximum efficiency (to obtain the lowest output temperature possible) they must be cooled.

At present, this cooling is accomplished by immersing the steam pipes of the condenser in running water pumped in from a river, lake, or cooling pond: 12,000 gallons of water per second for a 33 percent efficient, 1000-mw plant. At the projected rate of increase of electric power generating capacity, we will need to provide almost 1 trillion (1,000 billion) gallons of water a day to cool the 1990 plants. Even with some of this demand shared by sea water, the FPC projection is for 200 billion gallons of freshwater per day in 1990, one sixth the total United States freshwater runoff.

Several alternatives to straight-through cooling are possible; wet towers in which the water is cooled by evaporation or dry towers in which the water is cooled by air circulation. These techniques are expensive, adding millions, or in the case of the most expensive dry tower, tens of millions of dollars to the plant cost. They add, however, only small percentage costs to the price per kw-hr of the electricity.

We also described some of the biological cost of using our rivers as heat sinks. A change in temperature changes the ecological acquatic balance, killing some fish and changing migration and spawning patterns. Unfortunately, it also makes the rivers' role in processing sewage and the other waste material we are loading them with, more difficult.

To put the use of rivers for cooling our electric power plants in proper perspective, we must look not only at the cost of cooling, but also at the various other uses of water and put criteria and prices on them.

Among the competing uses of water are the domestic uses, the industrial uses for cleaning, cooling, mixing, the life-supporting uses for fish, irrigation for many food crops, and recreational uses for swimming, fishing, boating, and so on. One can put temperature criteria on some of these. An example of this is given in Table 8-5. If we look again at the temperature profile of the Monongehela River shown in Fig. 8-4, we see that already in 1960 the high temperature peaks were moving it out of most of these uses.

In each of these uses water also has a monetary value of some sort. Although this is obviously difficult to determine, some crude comparisons are given in Table 8-6. These figures were arrived at by estimating "the maximum amount people would be

Table 8-5 Resume of Numerical Temperature Criteria for Water Use*

Use	Change from ambient (°F)	Upper limit (°F)	Rate of change (°F/hr)
Swimming		85†	1
Water supply	5.0	85	
Fish and squatic life:			
Flowing streams:			
Warm water	5.0	80-93	
Cold water	3.0	68	
Lakes	3.0		
Marine:			
Winter	4.0		1
Summer	1.5		1
Irrigation		85‡	
Industrial process		95-100§	
Cooling		100-120§	

*Source: Office of Science and Technology, "Considerations Affecting Steam Power Plant Site Selection," (Washington, D.C., December 1968).

†Desirable. ‡Also, not less than 55°F

§For most uses, temperature is considered acceptable as received.

willing to pay for the use of the water . . . rather than forego the use or amount entirely."[6] The values for sport fishing and cooling water are added to the original table. The cost of cooling water was estimated by valuing the 30 to 50 gallons needed per kw-hr at the 0.04 to 0.4 mills estimated additional cost that cooling adds to the cost of electric energy (see Table 8-4). Its value for cooling may be underestimated, but

[6] E. F. Renshaw, "Value of an Acre Foot of Water," *Journal of American Water Works Association*, **50**: 3: 303, March 1958.

Table 8-6 Value of an Acre-foot* of Water

Use	Value (ave)	Value (Max)
Domestic†	$100	250
Industrial†	40	160
Sport fishing‡	20	
Irrigation†	2	27
Hydropower†	0.75	6
Waste disposal†	0.60	2.50
Cooling§	0.50	4.50
Navigation†	0.05	1.20
Commercial fishing†	0.03	1.10

*A volume of water 1 acre in surface area and 1 foot deep; it equals 325,000 gallons.

†E. F. Renshaw, "Value of An Acre-foot of Water," *Journal of American Water Works Association*, **50**: 3: 303, March 1958.

‡R. D. Hook, paper at Pennsylvania Water Pollution Control Association, University Park, Pennsylvania. August 10-12, 1960. Reprinted in *Thermal Pollution*, Part 1, p. 70.

§Assuming 30 to 50 gallons per kw-hr and an additional 0.04-0.4 mils per kw-hr as the cost of cooling towers.

even with that concession, it is not one of the expensive uses of water. It should be clear, however, that these different values for the various uses of water lead to conflicts which will, as free water becomes more and more scarce, have to be resolved in a marketplace where profit and loss are broadened to include as profits, such pleasures as swimming in clean, cool water or landing a game fish, and as losses, the sight of a once proud river serving an industrial master.

9

Energy and the Environment: The Rest of the Bill

The Energy Crisis, which we see through the eyes of the news media, focuses on shortages of natural gas, gasoline, heating oil, and the like, and on the exponentially increasing demands that are bringing us to shortages. In the last two chapters, however, we have been describing surpluses, not of energy, but of the dangerous by-products of its conversion. There are no shortages here; our reliance on the gasoline-fueled automobile for transportation and the steam electric plant for industrial, commercial, and residential power are overloading the natural disposal systems. Urban air is becoming full of airborne pollutants and our continental rivers are threatened with overheating.

In this chapter we will look to other effects on the environment of the production, transport, and consumption of energy. Some of these effects are spectacular: coal mine accidents and oil spills become national news. Others are only visible close-up; the barren, strip-mined Appalachian mountainsides and the subtle damages of pumped-storage plants, are of most concern to those whose lives and livelihoods are involved. Also alarming because of their potential, are the small but global influences, such as the changes of climate which the buildup of carbon dioxide portends.

None of these potential problem areas can be overlooked. If we are to survive our Energy Crisis we must do much more than find more resources; we must diminish environmental impact, choose between different energy conversion techniques, and set priorities among end uses. In this exercise of judgment we need the entire picture. We will begin putting the final touches on that picture by looking at some of the hidden costs of producing coal, that dependable fuel to which we are once again looking for large amounts of energy.

ENVIRONMENTAL COSTS OF COAL MINING

Coal is a fuel of the past and the future. As we saw in Fig. 5-6, use of coal has twice reached a peak, in 1920 and 1945, and then declined to lesser importance. The projections for 1980, however, call for 850 million tons of coal to be used, almost 200 million tons more than the previous peak of 660 million tons in 1920.

Coal is mined in a variety of ways. The old method of tunneling into an underground seam, blasting, and loading by hand, has almost disappeared; in fact, mining from the surface with power shovels is now a major technique, accounting for almost half the production. Figure 9-1 shows the percentage of the total produced by the various methods. Both strip mining and auger mining,[1] which are forms of surface mining, accounted for 50 percent of the total in 1971. Most of the underground coal is now at least cut by machine and the use of *continuous mining machines* is growing rapidly. These machines not only tear into the coal seam but also load the coal onto conveyors in one operation. They can mine as much as 8 tons of coal per minute.

Continuous mining machines of this type are employed in the most modern underground mining technique, *long-wall mining.* This technique has been used in Europe for some time, but is only just beginning to be experimented with in this country. It offers a major increase in the productivity of mines. Currently about 50 percent of the coal must be left in the form of pillars to hold up the mine roof. In long-wall mining a continuous mining machine, protected by powered roof supports and served by an armoured conveyor system, slowly grinds its way into a coal seam, allowing the roof to fall behind it. Although this increases the land subsidence (unless the area is filled in with debris or ashes) it allows the removal of 70 percent or so of the coal and is, as well, inherently safer. The employment of such advanced techniques will be necessary to enable underground mining to compete with surface mining.

The net impact of mechanization, of course, is to increase the productivity of manpower. Figure 9-2 shows the rapid increase in tons of coal produced per man-hour over the past several decades. Although each of these techniques poses hazards to the environment and to the men who use them, we will concentrate our discussion on the two general categories, underground mining and strip mining.

Strip Mining

Strip mining is the more apparent environmental evil because its damage is out in the open. We shall see later, however, that underground mining also makes its pres-

[1] An auger is a huge drill, several feet in diameter, which is used on coal seams accessible from the surface, but too deep for strip-mining.

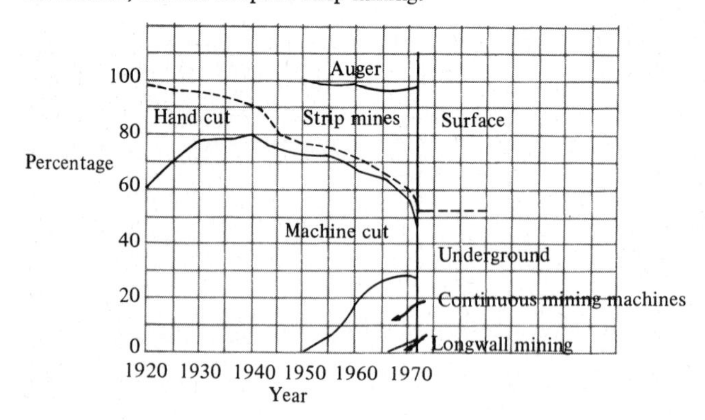

FIGURE 9-1 Coal mining by machine has been steadily increasing and mining from the surface by power shovels (strip mining) and by augers now accounts for about half of the total coal produced. (Source: U.S. Bureau of Mines.)

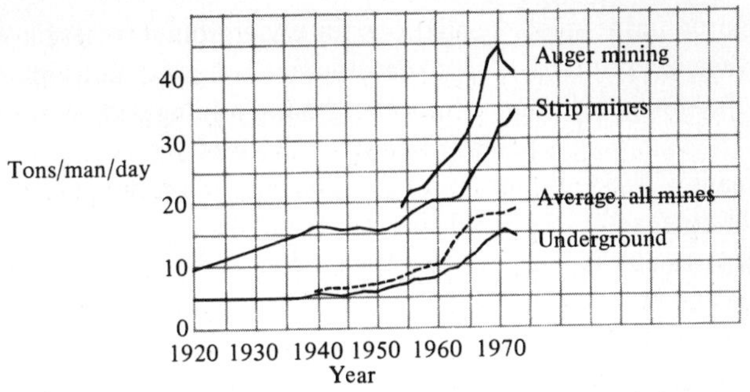

FIGURE 9-2 The increased reliance on automatic mining machinery has quadrupled the man-hour productivity in coal mining.

ence felt. In strip mining a huge power shovel removes the soil above the coal seam (the overburden) and then digs out the coal. Some 47 percent of the coal produced in this country in 1971 was mined in this fashion. The swing to this technique since World War II is due, of course, to its economic advantage: hand labor is replaced by the labor of huge machines. "Big Muskie," the current champion, shown in operation in Fig. 9-3, has a "bucket" which holds 220 cubic yards of earth, 325 tons per scoop (an ordinary dumptruck holds 10 to 15 cubic yards). With such a machine going after the coal (and huge trucks to carry the earth and coal at 200 tons a load), strip-mined coal sold at the mine for an average of $3.75 per ton in 1968 while coal from underground mines sold for $5.22 per ton. Increasing labor costs and the costs of the new

FIGURE 9-3 Big Muskie, a "dragline" used for strip mining is the largest mobile land machine in the world. It's bucket holds 220 cubic yards, 325 tons of earth and rock. The huge spool in front of the dragline is its power cord.

rigorous mine-safety regulations for underground mines have increased the gap between the prices. The rapid development of continuous mining and similar automatic bulk-mining techniques for underground mining will be needed to keep it competitive.

Strip mining is practiced on both flat land and on mountainsides: in the Appalachians of West Virginia and Pennsylvania and on the prairies of Illinois and Ohio. The total amount of land that has been "disturbed" (in the mild phaseology of the Department of the Interior's 1965-1966 survey) by all types of strip mining was estimated at 3.2 million acres, 5,000 square miles, about the area of the state of Connecticut. Coal mining accounted for about 40 percent of this land, or 1.3 million acres. Allowing 300,000 acres for the additional land strip-mined between 1965 and 1972 brings the total to almost 2 million acres. The Soil Conservation Service claims that no more than one third of this strip-mined land has been adequately reclaimed by either natural means or man's efforts. Thus, part of our price for coal (which was not borne by those who profited from it or burned it) is almost 1.5 million acres of barren land.

Area Strip-Mining. What does stripped land look like? It depends on the topography. In flat terrain a power shovel removes the soil above the coal, the *overburden,* in a long trench. The coal is then dug out and carried away. A parallel trench is then dug, the overburden from this one dumped in the old cut, the coal removed, and a new parallel trench dug. The depth at which coal can be profitably reached has been increasing as the equipment has become larger. The maximum overburden which could be handled in 1955 was about 70 feet; in 1965 it was 125 feet; and "Big Muskie" can now go after coal at 185 feet.

In *area mining* the flat farmland that might once have existed before stripping is replaced by a series of ridges and gullies; Figure 9-4 shows a particularly bad example. One does not need to be a soil expert to see that the value of this land for crops, recreation, or anything is practically zero. Area mining is now common in the Midwest, and the huge, untapped coal reserves that underlie much of Montana, Wyoming, the Dakotas, and the Southwest also lend themselves to this technique. An estimated 135 billion tons of this coal (which is only 10 percent of the total coal reserve) is in beds less than 100 feet below the surface and thus suitable for stripping. Two thirds of this strippable coal is west of the Mississippi. Thus, the promise of wealth and the vision of ugliness which faced the Easterners and Midwesterners in the 1930s and 1940s now haunts the Indian owners of the Black Mesa and the farmers of the Dakota grass lands. Before we look at the implications of the utilities' determination to "win the West," however, let us complete the cost accounting for coal mining.

Contour Strip-mining and Appalachia. *Contour strip-mining* of mountainside coal deposits is very different from the area strip-mining we have just described and in many ways more destructive. The power shovel begins a cut at the outcropping where the coal seam reaches the surface. The overburden is thrown down the mountainside and the coal in the long *contour trench* is removed in a manner much like paring an apple. The shovel then moves in toward the center of the mountain, each cut removing more overburden until the *highwall,* now 100 feet or so high, is too

FIGURE 9-4 Strip mining of the flat terrain leaves a tumbled, sterile landscape.

thick to remove. An auger may be used here to drill out more coal.

The discarded overburden, stacked on the edge or thrown down the slope, is the major cause of damage. With no trees or grass to anchor it, it is easily eroded and washed into the flatlands and streams below. Erosion rates as high as 400 tons per year per acre from Appalachian *spoil banks,* as the piled debris is called, have been measured. Adding to the spoil-bank erosion is that from the access roads, where rates of 472 tons per acre per year have been observed. An Appalachian natural forest, on the other hand, erodes at about 1 ton per acre per year.

Erosion is not the only threat from the tumbled, unstable land. Falling boulders and frequent landslides contribute their steady and sometimes spectacular damage. Tragedies of recent memory, one in Wales and one in West Virginia resulted from a spoil bank sliding down on the coal town below it. The highwalls themselves circle many mountain summits making them inaccessible. A United States Department of Interior survey of Appalachia in 1966 measured 740,000 disturbed acres, and estimated an additional 31,000 acres per year. By now one can be sure the total is over a million acres, almost 1 percent of Appalachia's 117 million acres.

Acid Water. Where water reaches the remains of the coal seams and the sulphur-containing pyrites which contaminate them, a second destructive pollutant, sulphuric acid, is formed. As we discussed in Chap. 7, acidity, which is commonly indicated by the pH number, is a determinant of aquatic life. Most fish cannot survive at a pH less than 5. It is also important to vegetation; no vegetation can live at a pH less than 3 and most vegetation needs at least a pH of 5 to survive. In the Appalachia survey, 4 percent of the spoil banks had a pH less than 3 and 79 percent were between 3 and 5. It is not land which is hospitable to life.

Because the Appalachians are a land of heavy rainfall, the acid does not stay at the site but leaches out into the streams, making them too acidic to support fish. By 1963 some 6,000 miles of Appalachian streams and 15,000 acres of impoundments had been damaged by mine acids; the United States total at that time was two or three times that figure.

Contour mining at its worst is shown in Fig. 9-5. All mountainsides, fortunately, are not this bad. The Department of Interior survey of Appalachia reported that of the spoil banks it examined, 15 percent were protected by sufficient vegetation, 15 percent had fair to good coverage, 20 percent required seeds and fertilizer, 30 percent had no cover and required extensive treatment, and 20 percent (including, one would suppose, that in Fig. 9-5) could not support vegetation. The implied goal here was modest: stabilize and protect the site and stop the erosion. To return the area to its natural prestripped state was beyond hope.

Reclamation

It is possible to reclaim stripped land. The Germans have had great success in this. In the Rhineland coal fields they remove and store both the top soil and the subsoil, and replace it when the coal has been extracted. The coal seams they are working, however, are hundreds of feet thick, instead of the 5- or 6-foot depths of our strip mines, and the profit per acre is so high that they can afford to spend 8,000 to 10,000 dollars an acre to restore it to crop and forest land.

We could do much, however, without pricing coal out of the market. The contour mines can be refilled and graded and then the graded areas fertilized and seeded. Sometimes, it is also necessary to create a drain with a culvert to carry the water away quickly before it can form sulphuric acid. It is also possible (but expensive) to neutralize the acid with limestone. Interestingly enough, the acidic condition can also be neutralized with ashes. This technique has a certain neatness to it, for one of the

FIGURE 9-5 The combination of contour strip mining and erosion has left this wasted mountain top in the Applachians.

most serious solid waste problems we now face is that of the disposal of the millions of tons of coal ashes removed from the electric utility furnaces each year (31.5 million tons in 1970).

Backfilling and grading contour mines is expensive. Three different techniques and results are shown in Fig. 9-6, along with costs in dollars per acre. We have also indicated the costs in cents per ton of mined coal. Even at the prices quoted earlier of $3.75 per ton of stripped coal (which is much less than today's prices), these costs do not seem a heavy burden for consumers to pay.

After the backfilling and grading, the land, often acidic and devoid of humus or any organic nutrients, must be induced to grow vegetation. Seeding is not enough; the land must be heavily fertilized. One is attracted, in the "two birds with one stone sense," by experiments in Illinois where sewage residue (and even liquid sewage) is being spread over graded, strip-mined areas. Perhaps some of these attempts to close ecological cycles, even if it is the "long way round," will work. At the present, they are only small-scale experiments.

Seeding and fertilizing, of course, adds to the cost of reclamation. The Department

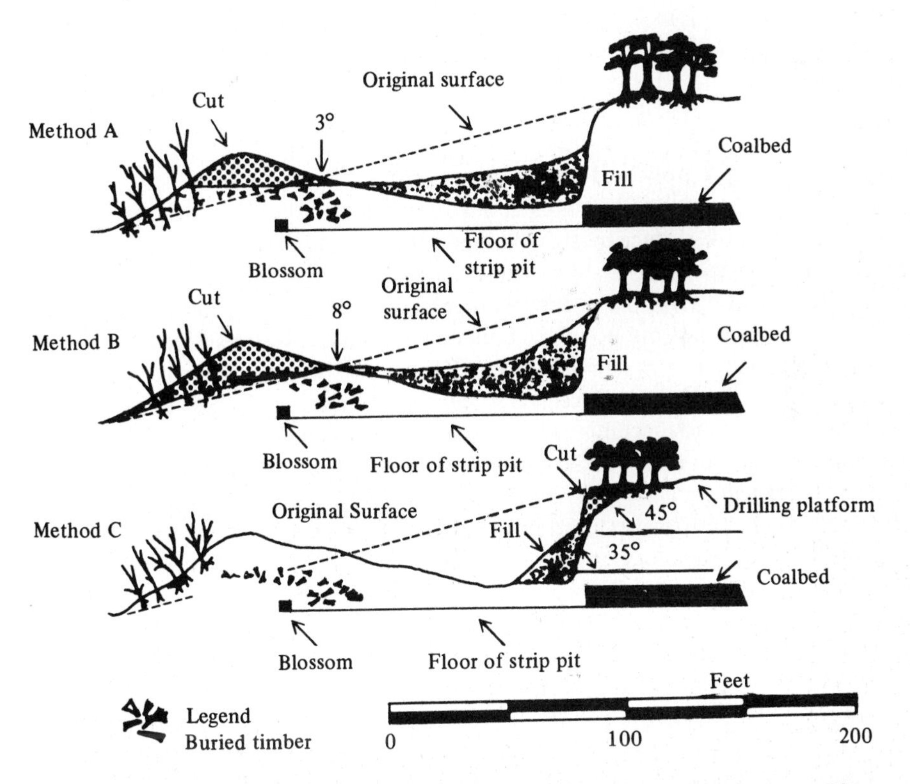

FIGURE 9-6 Three different types of contour mining reclamation. In Method A a bulldozer fills in the cuts; Method B is similar but the restoration of the original slope requires the cutting of buried timbers; Method C is accomplished by blasting rather than bulldozing. The cost per acre and per ton (at 1970 prices) are:

Method A:		$250/acre	$0.06/ton
Method B:	(If preplanned)	$400/acre	$0.10/ton
Method C:		$460/acre	$0.11/ton

of the Interior estimates that 55 percent of all the strip-mined land that needs replanting could be revegetated at 232 dollars an acre, adding only 3.2 cents per ton to the coal mined from these areas. Prices range upwards and downwards from this. There is a company which offers airplane seeding at 35 dollars per acre. Toward the other end of the range, the Bureau of Mines reclaimed a strip-mined area in Pennsylvania for 814 dollars per acre (11 cents per ton of coal) by growing grass only and 1,448 dollars per acre (20 cents per ton of coal) for land also planted to trees. That land is now a state park. Even the state park is not in the condition of the original land, but it is now at least usable and attractive, and is not itself a source of further damage to the environment.

The costs of reclamation, even 20 cents per ton of coal, do not seem prohibitive. Unfortunately, there are not as yet rigorous state or federal requirements to assure even minimal efforts. Most states require a coal-mining company to post a small bond and a certain amount per acre, the average being 350 dollars. Since this is near the minimal range of reclamation costs, it is frequently economically advantageous for an operator to leave the land as is and forfeit the bond. One hopes that our newly environmentally conscious Congress will provide leadership and legislation that will force coal companies to exercise more enlightened stewardship. Surely the consumer will be willing to pay a little more for his coal in order to save his land.

Underground Mining

In spite of the rapid growth of surface mining, half of the coal mined in 1971 came from underground mines. Since only about 10 percent of our total coal reserves can be reached from the surface by strip-mining techniques, underground mining will continue to be important.

To the casual observer the underground mines do not show massive environmental effects; they are messy, but no more so than many other industries. Unfortunately, closer inspection uncovers serious damage. The most troublesome and extensive damage comes from abandoned mines. Water seeping into and out of these mines interacts with the sulphur which accompanies coal (and is the source of its other pollutant, the sulphur oxides of smog) and forms sulphuric acid. We have already discussed the effects of this acid water; aquatic life, for instance, is very sensitive to the acidity of its environment. Acidity also interferes with the other uses of water we have considered in the previous chapter: industrial, municipal supplies, cooling, navigation, to name a few. The added bill that mine acid adds to these uses is put at 3.5 million dollars. The losses from potential recreation are much more difficult to quantify, but this loss is also in the millions of dollars.

There are several means available to reduce this pollution. It can be controlled at the source or treated after it gets into the streams. The source-control techniques attempt to prevent the interaction of water and sulphur by sealing the mines to prevent water from entering, by locating mine outlets above ground-water drainage, and by rapidly removing any water which does enter. These techniques are expensive: it costs from 1,000 to 20,000 dollars to seal a mine, and while new water quality standards may improve the situation in the future, it is difficult to force remedial efforts on the abandoned mines which are the major source of acid water drainage in the Appalachians.

The Department of the Interior has estimated that the cost of controlling acid mine drainage for the entire country would be about 6.5 billion dollars. The estimate by the State of Pennsylvania of the cost to remove the acid from its own streams is 1 to 2 billion dollars. Since there are not very many direct savings to be applied to these costs (the damage being loss of recreation and industrial potential, for the most part) it is not likely that an effort of this magnitude will be undertaken. Pennsylvania has begun a 150-million dollar 10-year program for acid-mine water abatement which will clean up certain important rivers and streams and, perhaps, provide research results that will allow more effective and inexpensive future efforts.

Land Subsidence and Fires. Two other troublesome environmental residues of underground mining are land subsidence (cave-ins) and fires. The Department of the Interior estimates that coal mines have undermined some 8 million acres and that 2 million acres have subsided. Most of this is in rural areas where the damage is not catastrophic, although it can make the land unusable as crop land, change drainage patterns, and cause other problems. The 7 percent of subsidence that occurs in the cities can have much more obvious effects as Fig. 9-7 shows.

Another hidden cost of underground mining is fire. Mine fires begin from accidents in operating mines or are started in abandoned mines. These fires deplete our coal reserves; their heat destroys much vegetation and they often release dangerous gases. Since 1949 the United States Bureau of Mines has had a mine-fire control program and has brought 150 such fires under control. There are, however, 200 more fires still burning. One of them, at New Straightsville, Ohio, has been burning since 1884 in spite of about 1 million dollars spent trying to control it. A fire near Scranton, Pennsylvania has cost 5.5 million dollars in unsuccessful expenditures.

FIGURE 9-7 Cave-ins from underground mining has caused considerable damage in urban areas such as these in eastern Pennsylvania.

Mine Waste. Coal mining is also the third-ranking industrial producer of mineral waste. Ninety percent of this comes from the washing of coal to remove impurities. This material is piled up in the unsightly waste heaps one sees throughout the coal-mining areas, almost 2 billion tons of it by now. The environmental damage associated with these banks is more than aesthetic; they are especially troublesome when they are near populated areas, as are 60 percent of them. Not only do wind and water erosion spread their dirtiness, but many of them, (about 500 in 1964), are on fire, adding to local air pollution problems.

The refuse banks burn because of their relatively high carbon content. The most desirable solution, therefore, would be to burn them to create electricity. Some experiments with this waste as fuel are being conducted by utilities plants. It is also being tried as an additive for concrete or cinder blocks and as road construction material. Since the total amount of this waste will be more than doubled by the year 2000, widespread use of it in constructive ways must be encouraged.

Warning: Coal Mining May Be Detrimental to Your Health

Although Congress passed the Coal Mine Health and Safety Act in 1969, coal mining continues to be one of the most dangerous–if not the most dangerous–of occupations. Table 9-1 shows the recent data on fatalities and injuries. What is more meaningful from the point of view of energy production, are the death and injury rates per man-hour and per ton of coal mined. These are also shown in Table 9-1. The historical record of total deaths, deaths per million man-hours and deaths per million tons, is presented in Fig. 9-8. We see that the number of deaths per year has been reduced by a factor of 10 since the early twentieth century. The death rate per million tons of coal has been dropping due to the increasing productivity of machines, but the recent data in Table 9-1 shows that even now, on the average, one coal miner dies for each 2 or 3 million tons of coal produced.

Most of the deaths occur in underground mining. In 1969, for example, there were 155 underground fatalities and 35 in strip and auger mining. Thus, 19.4 percent of the fatalities occurred in surface mining which produced almost 40 percent of the coal that year; in fact, one of the reasons for the dropping death toll during the past decade has been the switch to strip mining.

Again we are gored by an energy dilemma; the 20 billion tons of coal to be mined between 1970 and 2000 will take an additional 6,400 lives at 1971 death rates (the lowest recorded. At 1970 death rates this would be 8,600, which give some idea

Table 9-1 Deaths and Injuries from Coal Mining*

Rate	*1967*	*1968*	*1969*	*1970*	*1971*
Total deaths	222	311	203	260	180
Total injuries	10,120	9,640	9,920	11,500	11,350
Deaths per million man-hours	0.92	1.33	0.85	1.00	0.71
Injuries per million man-hours	41.8	41.1	41.8	44.4	45.1
Deaths per million tons of coal	0.40	0.56	0.36	0.43	0.32
Injuries per million tons of coal	18.04	17.42	17.42	18.8	20.2

*Source: "Injury Experience in Coal Mining," Bureau of Mines Information Circular 8599, Washington, D.C., 1973.

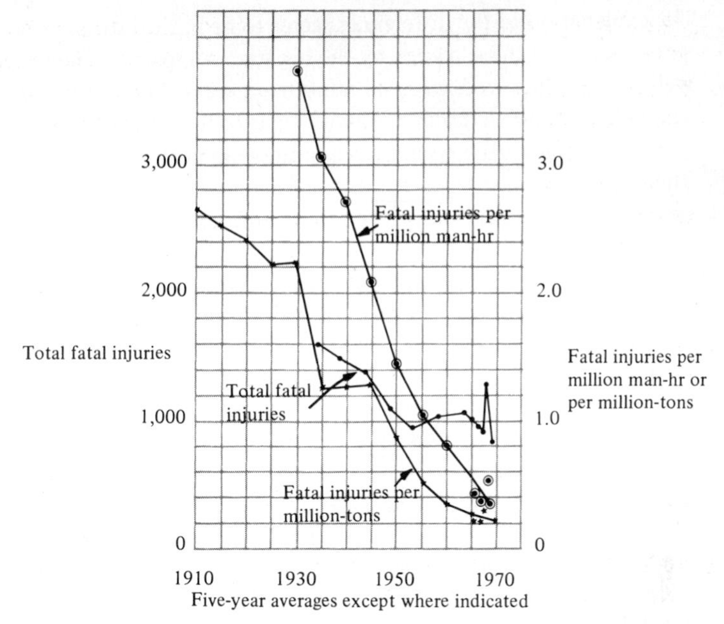

FIGURE 9-8 Total fatalities and fatal injuries per million tons of coal have dropped because of the increasing mechanization while deaths per man-hour remain at about one per million.

of the uncertainity of the estimate. If a larger percentage of this coal is strip-mined, the deaths will go down—but the environmental damage will go up.

There is convincing evidence that underground coal mining, even with present equipment, does not need to be so hazardous. If the company-by-company statistics are examined, great variation is found (see Table 9-2). Some of this is due to surface mining. Amax coal gets almost all of its coal from surface mining. United States Steel, however, with the lowest fatality record, is almost entirely an underground mining operation, while Peabody Coal (80 percent surface mining) has one of the poorest safety records by this measure.

Table 9-2 Underground Coal Mining Fatalities—by Companies*

Company	*Fatalities per million man-hours*
U.S. Steel mines (4)†	0.28
Amax Coal (8)	0.35
Bethlehem Steel mines (10)	0.44
Eastern Associated Coal (7)	0.53
General Dynamics mines (6)	0.72
Island Creek Coal (3)	0.85
Peabody Coal (1)	0.94
Old Ben Coal (9)	1.07
Pittston (5)	1.10
Consolidated Coal (2)	1.52

*Source: "Safety Underground—Coal Mine Study Shows Record Can by Improved When Firms Really Try," *Wall Street Journal,* January 18, 1973. Reprinted in *Energy Conservation,* Part I, Hearings before the Committee on Interior and Insular Affairs of the U.S. Senate, March 22 and 23, 1973. Serial No. 93-7.

† Rank in parenthesis based on annual tonages

The explanation of the differences seems to be in the attitude toward safety. The safest mines are those owned by the big steel companies. They have a tradition of well-designed and well-enforced safety regulations in the steel plants and carry the same attitude and determination into their coal-mining operations.

We must eventually go underground for our coal; that is where most of it is. Therefore, in human terms it will pay to study these successful operations and enforce similar regulations in the other companies.

Black Lung. The heaviest cost of coal mining in human misery and in money, comes from the disease known as *black lung.* The number of miners with this debilitating disease may be as high as 100,000.

Black lung is caused by inhalation of coal dust in quantities that overpower the lungs' cleansing mechanisms. It leads to cell damage that produces a form of emphysema and increases the miner's susceptibility to other respiratory diseases. The incidence of such disease among coal miners is 8 times higher than in the total United States male population. Mechanization, and unfortunately, the continuous-mining machine in particular, may increase the incidence of black lung due to its greater dust production, unless preventive techniques are vigorously employed.

Black lung is not only costly in terms of health but also in terms of money. Table 9-3 gives estimates of some of the annual costs. One must now add to this account, benefits paid to the disabled workers and their families since the enactment of the Coal Mine Health and Safety Law. During 1970, 250,000 claims were filed, and the cost of the 170,000 which were eventually processed was 151 million dollars. The amount appropriated for 1972 is about 385 million dollars plus an additional 260 million dollars for 1970 and 1971 obligations. If these charges were made to the coal industry, it would increase the price of coal 64 cents per ton. This is an 11 percent

Table 9-3 Annual Economic Losses from Coal Worker's Pneumoconiosis*

Losses:	*Annual*
TANGIBLE	
Cost to the industry	
Loss of production due to worker morbidity	$20,000,000
Workmen's compensation	11,000,000
Training employees to replace disabled workers	1,000,000
Cost to the miners	
Lost income	23,000,000
Medical expenses	13,000,000
Cost to the government	
Welfare	7,000,000
Total	$75,000,000
INTANGIBLE	
Loss of family cohesion	
Frustration of disabled workers	
Children and workers are unlikely to be educated	
Debilitated community development	
Poor regions are unable to attract new industries	

*Source: Commoner, Boksenbaum, and Corr, *Energy and Human Welfare,* Vol. I, (Riverside, N.J.: MacMillian Information, 1974).

increase over the $7.07 average price of coal in 1971; an increase less than the cost of food increases we have been lately experiencing.

It would seem that the matter of environmental "diseconomies" is clearer in the coal industry than in some of the other energy industries we have or will look at. Strip-mined land, and even acid-water damage, can be reclaimed at a price which adds a small percentage (5 to 10 percent) to the price of coal. The medical bill for black lung could be picked up by coal consumers. It is true that, due to the pervasiveness of the use of electricity, all of us would share in the bill ultimately. We would share, however, in proportion to our use of coal. This more direct billing would put more pressure on the coal companies to introduce reclamation and safety measures and reduce these costs. With a resurgence of coal as a source of energy, we must find ways to reduce its environmental costs.

OTHER COSTS OF ELECTRIC POWER PRODUCTION

To complete the record we should mention some of the other environmental costs that we will have to pay because of our growing dependence on electric power. In Chap. 1 (Figs. 1-3*a* and 1-3*b*) we showed the expected increase in the transmission line network between 1970 and 1990. In 1970 those high-voltage lines occupied 4 million acres of land. By 1990 they will occupy about 7 million acres. It is clear from Fig. 1-3*b* that acreage alone will become a factor to those whose idea of scenic beauty does not include a vision of high-voltage towers. There will be few places without them. The industry, in cooperation with the federal and state governments, is considering multiple use of highway rights-of-ways as one way to meet the problem.

Power plants themselves occupy large tracts of land, 900 to 1,200 acres per 3000 mw for coal-fired plants. Between the years 1970 and 1990, the utilities companies are expected to add 382,000 mw of new capacity and to replace 85 mw of old capacity. Minimal needs will be for some 155 of these 3,000-mw plants occupying an additional 140,000 to 190,000 acres. If cooling towers and cooling-pond use become important, as they should, these acreage needs will at least double.

Nuclear plants make smaller demands for land, 200 to 400 acres per 3,000 mw. The 494,000 mw of nuclear capacity expected to be added between 1970 and 1990 will require 165 of these large plants and 33,000 to 66,000 acres of land. In both these cases the advantages of locating generating plants on rivers, lakes, bays, or seacoasts often brings about sharp confrontations with those who value these areas for their recreation or scenic values. This choice is another one that must be put before the consumer when he turns on his electric appliance.

Environmental Effects of Hydroelectric Plants

Only 17 percent of the total electric power was produced by hydroelectric plants in 1970, and this percentage is expected to fall to 13 percent in 1980 and 12 percent in 1990. The number of plants, however, will grow. The FPC projections call for additions of 39,000 mw of this type of generating capacity by 1980 and 58,000 mw more by 1990. Included in these additions are 23,000 mw of pumped storage to be added by 1980 to the 4,000 mw available in 1970, and 44,000 mw more between 1980 and 1990.

Since the war between conservationists and the utilities originally began over the damaging of scenic rivers and the construction of storage reservoirs, it is a good idea to examine the environmental costs of these facilities.

Damming rivers to create reservoirs is one way in which man is counteracting his other activities that are leading to the premature aging of many natural lakes. If you exclude Alaska and the Great Lakes, we have already produced more square miles of artificial reservoirs than there are of natural lakes in this country.

The creation of a reservoir has many long- and short-range ecological effects. It causes local changes in ground-water levels and in atmospheric moisture content. It changes water temperature and chemistry. Perhaps the most drastic changes from the point of view of aquatic life are the interference with migration and spawning. Upstream migration is assisted by fish ladders, and other mechanisms, but the experience with these is that the runs are reduced. Downstream migration is more seriously affected as the fish have to go through the turbines.

The elimination of seasonal flooding also has important consequences. The floods are not only counted on to bring nutrients downstream, but many species spawn and their young mature in the shallow-flooded areas that lie along streams and rivers in the spring.

The effects of dams have not been adequately studied; rarely are ecologists consulted in the early planning. The effects, however, are not all bad; flood elimination is a net good on a human measurement. The fishing below the dam where nutrient-rich bottom waters are spilled out and aerated is often very good.

The classic example of the disasterous effects of ignoring ecology in the planning of dams is the Aswan Dam on the Nile. The dam has provided water for irrigation and power generation, and, by raising the ground water levels, has improved agriculture above the dam in general. The elimination of the seasonal floods and the reduction of the nutrients flow, however, has all but wiped out the sardine-fishing industry at the mouth of the Nile. The creation of many thousands of square miles of shallow water in flooded areas and irrigation ditches has brought on an epidemic of schistosomiasis, a debilitating parasitic disease carried by the snails which live in such waters.

With all these other damaging effects on the environment we must not become so practical that we forget the most obvious one. As we have mentioned, hydroelectric dams are best built in narrow canyons with swift streams. These spots are also among our most scenic. Glen Canyon in Utah is now Lake Powell. The Bureau of Reclamation even proposed two 1,000-mw dams in Grand Canyon. The public outcry seems to have stopped that unbelievable desecration and is successfully holding up the Hell's Canyon Dam in the deepest gorge in North America. The opposition and the scarcity of good sites is causing the power companies to turn to the newest form of hydroelectric power, pumped storage.

Pumped Storage. The qualifications for a pumped-storage hydroelectric site are much less stringent than for conventional dams. All that is needed is a modest amount of water (which can be recycled) and a modest hill to build the storage reservoir on. Even here, however, one of the earliest pumped-storage development projects set, "nature lovers" and "power producers" at each others throats. This bitter confrontation came over Consolidated Edison's plans to build a pumped-storage plant at Storm King Mountain on the Hudson River. Building the storage reservoir would have required the leveling of that impressive summit, the flooding of 240 acres of wilderness, and the general destruction of "one of the grandest passages of river scenery in the world." The opposition to this has been loud and successful. The application has been under litigation of various sorts for about eight years and the outcome is still in doubt.

Pumped storage has important engineering advantages. From an energy-economic point of view it is a good way to store cheap offpeak power and make it available for peak use. It gives a power system needed flexibility. Pumped-storage plants can be built in places where the flow of the river is not enough for conventional hydroelectric plants, since the water can be stored and recycled.

The pumping produces unusual effects. On the Connecticut River at the Northfield Mountain plant, the 10,000 feet^3 per second to be pumped out of the main channel is greater than the normal flow of that river for most of the year. It is expected that the river will flow backwards during the storage operation. The effect of this changed flow on fish activities is not known. What does seem to be clear, however, is that it could be very confusing during migrations when fish use the current to guide them. One other effect on fish is their entrapment and travel through the pumping cycle. Some fish apparently make the round trip and survive. Although some are killed by the turbine pump blades, the increased living space afforded by the upper reservoirs seems to compensate for losses by encouraging population expansion. There are in addition, examples of recreational purposes such as camping and fishing which are being served by the upper reservoirs.

Hydroelectric Plants as Peaking Plants. As we noted in Chap. 6, hydroelectric plants (both conventional and pumped-storage) are being used more and more as *peaking plants.* They are kept on reserve to be drawn on when demand is high. This in itself has environmental backlash. It means that water levels fluctuate greatly, doing damage to shallow water and bottom-dwelling aquatic life. The storage aspects are disturbing. In Consolidated Edison's case, for instance, the cheap off-peak power that would be used to pump water uphill would come at night. It would mean that the fuel-burning plants, most of which are in urban New York, would be burning all night, discharging their pollution into the usually stagnant nighttime skies.

Peaking needs must be met. It is not clear, however, that our rivers should be called upon to do it. The invention of a less destructive alternative is another challenge for the future power engineer.

EFFECTS ON CLIMATE

While we must carefully scrutinize the local impact of each new production facility, we must not forget the subtle global effects which may, in the end, be the most important environmental consequences. The most important global effect which we presently anticipate is the increase in the carbon dioxide content of the atmosphere due to the burning of fossil fuels. In considering the effect of this carbon dioxide on the atmospheric temperature, we will also look at the relative contribution made by the direct input of heat from our conversion of other forms of energy into thermal energy. In this same vein we will look at microclimatic changes (changes in the climates of cities) brought about, in some part, by their massive thermal energy release.

Carbon Dioxide, Dust, and the "Greenhouse Effect"

Carbon dioxide is a small but important constituent of the atmosphere. It plays an important role as a fuel for photosynthesis and also influences the radiation balance of the earth.

The burning of fossil fuels now releases some 14 B tons of carbon dioxide into the atmosphere each year and the rate is increasing with increasing fuel consumption. For comparison, photosynthesis annually consumes about 100 B tons of carbon dioxide; the man-made addition is now about one seventh of the photosynthetic consumption.

It is difficult to determine how much of this carbon dioxide accumulates in the atmosphere and how much is stored elsewhere. No careful measurements of carbon dioxide concentrations in the atmosphere were available until the late 1950s. Measurements since then show a steady increase and an accumulation over the estimated 290 parts per million of the nineteenth century (see Fig. 9-9). In these measurements we see firm evidence for a present rate of increase of about 0.2 percent per year in atmospheric carbon dioxide and a total increase of perhaps 10 percent in this century.

Figure 9-10 summarizes the carbon cycles of importance to this discussion. We see that there are two important reservoirs, in addition to the atmosphere, in which the carbon dioxide or the carbon is stored: the biosphere and the ocean.

If we are to estimate the effects of an increase in atmospheric carbon dioxide, we need to know how the uptake in these two reservoirs will be affected by the atmospheric concentration. The biosphere (all the living things) probably has in it one or two times as much carbon as there is in atmospheric carbon dioxide. If atmospheric

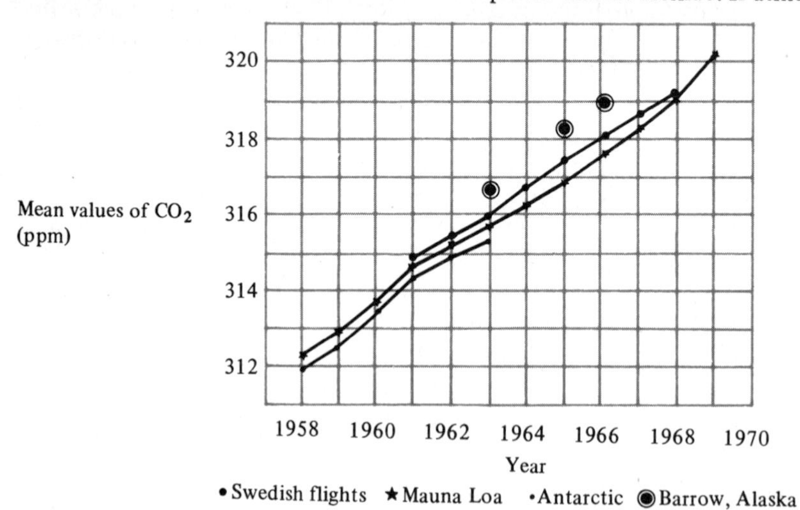

FIGURE 9-9 Measurements at several locations show a steady rise in the carbon dioxide content of the atmosphere. (Source: Report of the Study of Critical Environmental Problems, MIT Press, Mass.)

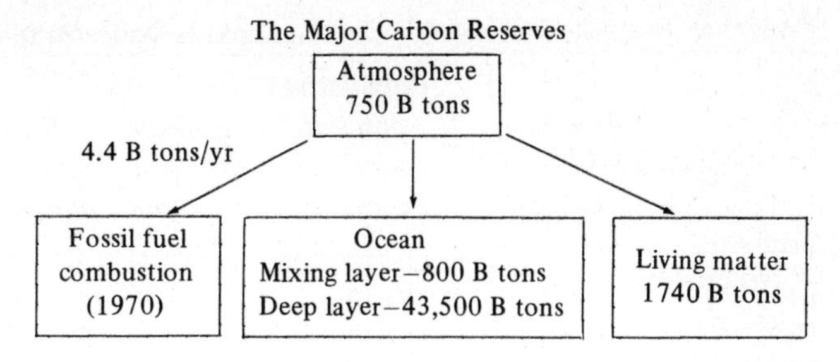

FIGURE 9-10 The content and interchange between the various carbon reservoirs shows that most of the carbon is stored in the ocean at deep levels where it is not available for mixing with the atmospheric carbon dioxide.

carbon dioxide increases, the biospheric mass might respond by growing faster, thus absorbing some of the increase. This would only be temporary storage, however, since the eventual decay returns the carbon dioxide to the atmosphere. Only about 1 part in 10,000 of the carbon in photosynthesis goes into fossil form, or creates new fossil fuels.

The ocean, on the other hand, has about 60 times as much carbon stored in it as is stored in the atmosphere. Most of this is in the deep layers and has a "turnover time" of perhaps 500 to 1,000 years. Changes in the atmospheric concentration affect only the top layer.

From the data in Fig. 9-10 and computations of carbon dioxide releases from the burning of fuel, we can estimate that one third to one half of the carbon dioxide introduced into the atmosphere stays there, while the other half or so is absorbed in these other reservoirs. It is important to note the feedback effects; forests can increase their rate of growth if carbon dioxide levels increase (and if they are not all cut down). In the other direction, an increase in the atmospheric and, therefore, the oceanic temperature which might be caused by increased atmospheric carbon dioxide could reduce the ocean surface layer's ability to retain carbon dioxide, thus increasing the concentration of carbon dioxide further.

These qualifying statements emphasize the complexity of the relationships we are discussing, and the uncertainty in our understanding of them. With these reservations, we can estimate that the total carbon dioxide added to the atmoshpere will be twice the 1970 levels by 1990 and almost three times that by 2000 and, that during that period, the total atmospheric carbon dioxide will have increased over the nineteenth-century levels by 25 to 30 percent. These projections are summarized in Table 9-4.

Using the same assumption we should note with some concern that by the year 2000, we will only have consumed between 2 percent and 12 percent of our total fossil fuel reserves (the uncertainty comes from uncertainty in the amount of reserve). Burning the rest of it could increase the carbon dioxide content of the atmosphere by as much as a factor of 4.

The Greenhouse Effect. What is, of course, more important to us than the amount of carbon dioxide are its effects. At this stage we can only speculate as to what might happen. It is important, however, that we make this speculation; by the time we be-

Table 9-4 Possible Atmospheric Carbon Dioxide Concentrations*

Year	*Amount added from fossil fuel (Mt/yr)*†	*Cumulative amount added over previous decade (Mt)*	*Concentration by volume (ppm)*	*Total amount in atmosphere (Mt)*	*Percentage of annual addition remaining in atmosphere*
1970	15,400	126,500	321	2.50×10^6	52
1980	22,800	185,000	334	2.61×10^6	52
1990	32,200	268,000	353	2.75×10^6	52
2000	45,500	378,000	379	2.95×10^6	51

*Source: *Mans Impact on the Global Environment,* (Cambridge: MIT Press, 1970).

† *Note:* 1 million, metric tonnes is equal to 1.1 tons or 2,200 lbs.

gin to gather field data from the real world it may be too late to avoid serious and even catastrophic change.

The effects we expect are alterations of atmospheric temperature. As we have pointed out in earlier chapters, the earth is in radiative balance with the sun. It receives energy at short wavelengths, bounces it around, changes its form a few times (see, for instance, Fig. 2-2), and then reradiates it at long (infrared) wavelengths.

Carbon dioxide is essentially transparent to the incoming short wavelengths, so the increase we anticipate will not change the incoming flux of energy. It is strongly absorbing, however, in certain parts of the infrared wavelength region, and will therefore capture a percentage of the energy radiated out from the earth. The net effect of this increased atmospheric absorption will be to increase the temperature of the atmosphere. This transparency to incoming radiation and opacity to outgoing is called the *Greenhouse Effect* because it is similar to the effects of the windows in a greenhouse.

Our present understanding of the atmosphere and the interrelated effects of humidity, cloudiness, dust, and carbon dioxide is not very precise. The best that can be done is to create simple models and base our predictions on them. The most sophisticated of the present calculations predicts about a 2°C increase in atmospheric temperature if the carbon dioxide content is doubled, or about a 0.2°C increase for each 10 percent increase in atmospheric carbon dioxide.

Observed Changes. Precise temperature measurements only exist for the past century. It is possible, however, to deduce earlier atmospheric temperature with some accuracy from fossil records of certain kinds. It appears, from this record, that there have been rather large changes over long periods of time (the various "ice ages," for example) and that from the late nineteenth century until about 1940 there was a gradual warming trend amounting to perhaps 0.6°C. In the last 25 years what has been observed is a 0.2°C cooling.

These seem like very small changes. The atmospheric balance is so delicate, however, that even these small changes have had great effects on the climate. During the warming period of the early twentieth century, the frost and ice boundaries moved northward, glaciers retreated, and rainfall was reduced in many parts of the world. The cooling trend since 1940 has reversed this. Sea-ice coverage in the North Atlantic in 1968, for instance, was the most extensive in the past 60 years, resulting in serious losses to the Icelandic fisherman.

It is considered possible that some of the pre-1940 heating is due to a buildup of carbon dioxide. Since most of the 10 percent increase in atmospheric carbon dioxide has come since then, it is difficult to assign more than 0.1°C or 0.2°C to this cause. The remaining drop in temperature during the time of heavier carbon dioxide additions is puzzling.

In an effort to assign the cause of this temperature decrease some meteorologists have suggested that another pollutant associated with man's activities, dust, may be to blame. The amount of aerosols in the atmosphere has been increasing, as we have reported. It is suggested that this dust is scattering solar energy back into space, thus reducing the incoming energy, and, therefore, the heating, at the earth's surface. However, there is also a strong possibility that some of the man-made aerosols, especially those containing carbon, iron, and other metallic oxides, will absorb the incoming radiation rather than scatter it. This would warm the atmosphere. It will take a considerable program of research and observation to determine what is happening, to explain the cooling, and to predict with confidence what the ultimate effect of the increasing load of carbon dioxide will be.

We can summarize this discussion of carbon dioxide in the following way. There is no question that the carbon dioxide concentration is increasing due to the burning of fossil fuels. There is also no question that the earth's energy balance is so delicate that small temperature changes can cause large effects. Witness the changes in ice coverage associated with the 0.2°C cooling. A change of 1 or 2 percent in the energy available at the surface can bring in a new ice-age or melt the polar caps. Thus, atmospheric temperature has an amplified effect and changes, inadvertent or deliberate, should be allowed only after a thorough understanding of their effects is obtained. We are presently far from that state of knowledge.

Direct Heating

In Chap. 3 we discussed the energy balance and traced the incoming solar energy as it is reflected, scattered, absorbed, and reradiated. Averaged over its surface over a period of 24 hours, the earth receives power from the sun equivalent to 17 watts per square foot. By his burning of fossil fuels man produces, at present, 7.8×10^{-4} watt per square foot, averaged over earth's surface. On the basis of such a global average, man's thermal effect is miniscule.

It is not enough, however, to deal with averages; man does not consume energy and produce waste heat evenly over the globe, but concentrates it in urban areas. In the 4,000 square miles of the Los Angeles basin, for instance, it has been calculated that the heat generated by power plants, automobiles, and other sources amounts to more than 5 percent of the solar energy absorbed at the earth's surface in that area. By the year 2000 it is predicted that the heat generated will be 18 percent. A similar calculation for the Northeastern United States, an area of 350,000 square miles, sets the thermal waste at 1 percent and predicts a rise to 5 percent by the year 2000.[2]

[2] "Man's Impact on the Global Environment," (MIT Press, Cambridge, Mass. 1970).

We need to look at this thermal pollution problem in terms of the radiation balance of the earth. Most of the incoming solar energy is eventually radiated away so that the earth remains at an even temperature. A careful accounting, however, shows that there is a difference between the incoming short-wave radiation and the outgoing long-wave radiation of about 6.3 watts per foot^2. This amount of energy is used up in the evaporation of water and in the kinetic energy of air movement. A small fraction (about 1 percent) of it goes into chemical energy through photosynthesis.

At present, man's thermal production is only 1/10,000 of this 6.3 watts per foot^2. This *radiation balance* is, however, a most sensitive determinant of earth's condition. It is argued that an increase of only a few tenths of a percent would melt all the polar ice. It is with some concern, therefore, that we compute the waste heat of future societies, all enjoying the present United States level of energy consumption. If the present annual rate of increase of energy consumption (4 percent) is continued for 200 years, man's thermal input will be one third the radiation budget. A 10 percent rate of increase will reach one third in 100 years.

This long-range concern over energy consumption goes beyond the myopic worries of the United States Energy Crisis of the 1970s. The threat is, in fact, a threat from success, from plentiful fusion power, for instance. It is not too early to begin the detailed studies that are needed to tell us just how much energy man can afford to release on this finite globe.

Cities as Heat Islands

It has been apparent for some time that cities were hotter than the surrounding countryside. The local weather forecaster usually points this out: "mid-30s in the city and 10 degrees colder in the suburbs tonight." What is not as well known is that the entire climate of the city is different. Table 9-5 summarizes most of the observed differences.

The causes of these climatic changes are manyfold and not all energy-related. An excellent popular summary by William Lowry appears in *Scientific American*.[3] There are two major causes of the *heat island* over the cities. The absorption of solar energy is increased by the concrete buildings, by the black asphalt, and by the very geometry of the city in which the tall buildings trap the sunlight in multiple reflections. The usual cooling mechanism, evaporation, is also reduced; the water which falls on a city quickly drains away; there is no grass and sod to hold it and allow it to evaporate.

Man-made energy release does not yet play an important role in the summer. In the winter, however, it is already quite significant. At the mid-latitudes, where most of our large cities are located, the winter sun comes in at an angle and the solar input is reduced. The pollutant cloud that hangs over a city also reduces the incoming energy. In winter man heats his cities. In Manhattan, for instance, the heat produced by combustion of fuel during the winter is reported to be 2.5 times the solar energy that reaches the ground, while it is only one sixth of this in the summer.

One of the most troublesome effects of this heat island is to form a minicyclone over the city, which is hot air rising up over the urban area and cooler air coming in

[3]William Lowry, "The Climate of Cities," *Scientific American*, **217**: 15, August 1967.

Table 9-5 Climate Changes Produced by Cities (after Landsberg, 1962)*

Element	*Comparison with rural environs*
Temperature	
Annual mean	1.0°-1.5°F higher
Winter minima	2.0°-3.0°F higher
Relative humidity	
Annual mean	6% lower
Winter	2% lower
Summer	8% lower
Dust particles	10 times more
Cloudiness	
Clouds	5%-10% more
Fog, winter	100% more
Fog, summer	30% more
Radiation	
Total on horizontal surface	15%-20% less
Ultraviolet, winter	30% less
Ultraviolet, summer	5% less
Wind speed	
Annual mean	20%-30% lower
Extreme gusts	10%-20% lower
Calms	5%-20% more
Precipitation	
Amounts	5%-10% more
Days with <0.2 in	10% more

*Source: H. E. Landsberg, "Man-Made Climate Changes" *Science* **170**, 1265 (1970).

at low levels from the countryside. As much as this ventilation is appreciated, it also recirculates the pollutants of the city. They are carried up and out over the countryside only to be returned through the suburbs by the incoming winds.

As the energy consumption of metropolitan areas increase, we can expect stronger interactions. One can, for instance, speculate on the effects of more and more air conditioning. The electric energy consumed in moving the heat from indoors out, heats the city, calling for more and more air conditioning. Here, too, as in so many other situations we have discussed, we look into the future toward a problem whose solution is not in sight; again, we can only call for moderation in growth and intensification of research.

THE WESTERN "COAL RUSH"

So far in this chapter we have spent our words describing the often hidden costs of the energy we have used in the past. We have concentrated our attention on coal, in part, because it has been an important fuel, but largely because of its role in the future. All signs of the future say "electricity" and, for the next two decades at least, most of the electricity will come from coal. The environmental costs which we have been describing will continue to mount, but with a difference: the mining operations are moving westward. Figure 9-11 shows the major coal reserves of this country. The

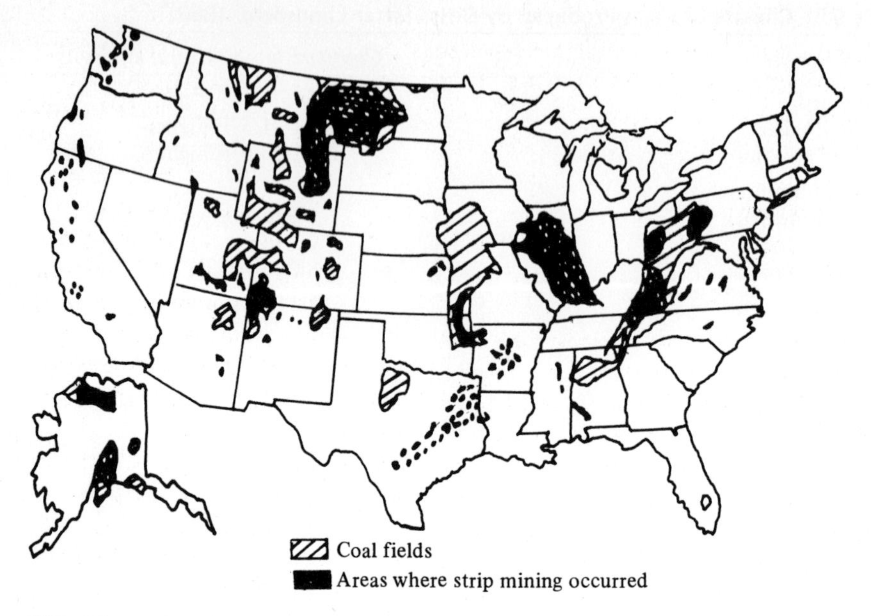

FIGURE 9-11 Coal fields of the U.S. Attention is turning to the deposits in the Western states from Montana down to New Mexico and Arizona where much of the coal can be strip-mined.

huge deposits of the East and Midwest are still formidable, but most of the easily mined coal has gone. The "coal rush" has now moved onto the next big black area on that map, the deposits which stretch from Montana, the Dakotas, and Wyoming down into New Mexico. The causes of this coal rush are clear: the booming West Coast needs electric power, electric power needs cheap strip-mined coal, and this western region has vast reserves (100 billion tons or so) of coal which can be reached from the surface. The region is also sparsely populated so the voices of protest should be few. A perfect set-up—or so it appeared.

In this final section we will look at this western coal rush in two different locations. In one of them, the Four Corners area, the rush is well under way. In the other, the Wyoming, Montana, and Dakotas area, the claims are just being staked. Again, we will try to assess the costs and to put them down under the headings these last three chapters have established: air pollution, thermal degradation of water, land abuse, and so on. But here we will remind you of another cost, the losses which come to those who love and live on land that has suddenly become the "overburden" of a strip mine.

Mining the Black Mesa

In the 1950s a consortium of 23 Western power companies banded together to try and meet their region's growing need for power. With nuclear reactors unproven, and faced by the strict air pollution regulations of their major urban areas, they not only looked to the New Mexican coal for energy but decided also to build the power plants there and ship the electricity to where it was needed. They chose, for these mining and generating complexes, the desert area of Four Corners, where New Mexico, Arizona, Utah, and Colorado join. This area seemed to be ideal: it had huge reserves of coal near the surface, ripe for inexpensive strip mining. It is water-poor, but the Colorado and its tributaries could, it appeared, provide enough cooling. Perhaps more important, it is sparsely settled and the poverty stricken Navajo and Hopi Indians who make up most of the population should, it seemed, welcome the economic well-being the utilities developments would bring.

The plans of this utility consortium, which called itself WEST for Western Energy Supply and Transmission, called for the construction of 36,000 mw of generating capacity in that region by 1990. In the first phase of construction, 13,895 mw by 1980, six plants were planned: Four Corners (2,220 mw) and San Juan (990 mw) on the San Juan River in New Mexico; Navajo (2,310 mw) and Karparowits (5,000 mw) on the Arizona and Utah sides of Lake Powell; and Mohave (1,580 mw) on the Colorado below Lake Powell (a man-made reservoir formed by damming the Colorado). The last of these super six would be Huntington Canyon (2,000 mw) on Huntington Creek in Utah. The locations of these plants are shown in Fig. 9-12.

Four Corners became operational in 1963 and quickly put its shadow over the area. The designers had apparently not gone in for much in the way of pollution abatement devices. The plant spewed 200 to 250 tons of particulate matter per day in the black plume out over the desert—more, in fact, than the combined daily particulate emission of New York and Los Angeles. The Gemini 12 moon voyages marked this plume as the only visible human artifact of their early view of earth.[4] This was too much for New Mexico. Regulatory agencies forced the plant to install much more effective emission control apparatus, but even now particulate emission is still 60 tons a day, to which is added 350 tons a day of sulphur oxides and nitrogen oxides.

The Four Corners plant is in Fruitland, New Mexico. It uses New Mexico's coal, almost 7 million tons in 1972. It uses scarce New Mexican water; 40 million gallons per day are pumped from the San Juan River as "make-up" for evaporative losses from its almost 4,000-acre cooling pond, Lake Morgan. It is New Mexico's air which

[4]This identification has since been challenged. There is some evidence that the photograph showed jet contrails.

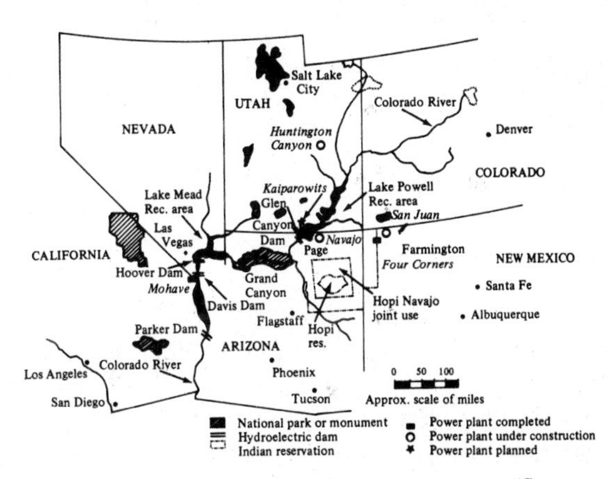

FIGURE 9-12 Locations of the six power plants proposed for the Four Corners area. (Source: The Conservation Foundation, Washington, D.C.)

is polluted, yet only 10 percent of the electric power produced there goes to New Mexico. The big user is southern California. New Mexico is learning the frustrations of the "underdeveloped countries."

A deeper controversy surrounds the next plant, Mohave, which became operational in 1970. It was not the anticipated output which troubled here: the plant's emissions were to be 97 percent controlled. It was the input that caused concern. The 16,000 tons of coal Mohave needs per day (as well as the 23,000 tons per day needed by the nearby Navajo plant) is to be strip-mined from the Black Mesa, that 3,300 square mile plateau which is the ancestral home of the Hopi and Navajo.

Both the promise and the nightmare is there for the Indians. The promise is the 3 million dollars annual royalties which the Peabody Coal Company is paying for the leased mineral rights on almost 65,000 acres of the Black Mesa. The nightmare is the vision of the 400 acres per year which will be stripped during the 35 years of anticipated operation. The coal company will gross about 750 million dollars during this period.

The potential for damage goes beyond the strip mining for coal. To get the coal to the Mohave plant, 275 miles away, the company plans to grind it, mix it with water, and pump it as a "slurry" through an underground pipeline 18 inches in diameter. Water is scarce on the Mesa and the slurry pipe will, at full capacity, use 4,500 gallons per minute.

The coal company, of course, does not propose out-and-out rape of the land. To get the water they need they are drilling wells 3,000 feet deep reaching water which, their geologists claim, is completely separated from the shallow Indian wells. Furthermore, they have lined their wells with concrete to prevent the draining of the Indian water sources. A federal study, however forecasts a lowering of the water table in parts of the Black Mesa, by as much as 100 feet.

The Indian owners have also been promised that the stripped land will be recontoured and revegetated. Their reactions are mixed; some believe that the money is payment enough for the damage and danger to their land. Others, instilled with the traditional Indian "land ethic," are still vigorously fighting the coal contracts; many environmental and conservation groups have rallied to their aid. The battle still rages; its early stages are well told by Phillip Herrera in Part Two of *Energy*, a Sierra Club Battlebook;[5] and "Showdown at Four Corners" by Anthony Wolff in the *Saturday Review*.[6] The conclusion has yet to be told.

The water demand seems at this time to be the real crunch. The Mohave plant has recently requested water for a 3,000-mw installation, but is permitted only half that amount. The future of the Karparowits Plant is also uncertain because of lack of water, and the Department of the Interior recently imposed a one-year moratorium on its construction.

For all the outcry and sympathy for the Indians, the demand for electricity remains. Table 9-6 compares the 1970 peak demands in some of that region's cities with 1980 and 1990 projections. We see that Los Angeles by itself could probably account for all the projected 36,000 mw of 1990 capacity of the six Four Corners plants. The Black Mesa is in deep trouble.

[5] John Holdren and Phillip Herrera, *Energy*, (San Francisco, New York: Sierra Club, 1971).

[6] A. Wolff, "Showdown at Four Corners," *Saturday Review of Science*, **55**: 29, June 3, 1972.

Table 9-6 Peak Power Demands (mw) of Some Cities Served by the Four Corners Plants*

City	*1970*	*1980*	*1990*
Los Angeles	9,150	19,100	37,800
Las Vegas	688	1,470	3,160
Phoenix	2,000	4,260	9,180
Tucson	688	1,420	3,060

*Source: *The 1970 National Power Survey,* Part III (Washington, D.C.: Federal Power Commision, December 1970).

Coal in Cowboy Country

The Indians are not alone on the horns of this dilemma; their mythical adversaries, the cowboys, are there with them, or at least the ranchers of the Northern plain states are. In a 250,000 square mile area of Wyoming, eastern Montana, and western North and South Dakota, a similar "coal rush" and "power play" is under way. It has its source in the same twin need for electric power and coal that led to the Four Corners development. These plans are even more awesome. There are an estimated 1.5 trillion tons of coal in the area, about 100 billion tons of which are near enough to the surface and in thick enough seams (20 to 750 feet thick) to be profitably strip-mined. The region contains 40 percent of the total United States coal reserves. It has not been heavily mined so far because of the lower heat content of its coal, and its great distance from urban areas which need the energy. It has low sulphur coal, however, and this fact, plus the inexorable demand for electric power, has turned the coal rush in that direction.

The development plan, put forward as the North Central Power Study by the United States Bureau of Reclamation, calls for the eventual establishment of 42 mine-mouth power plants in the area. The general location of these plants is shown in Fig. 9-13. The power goals are 50,000 mw of capacity by 1980 and 200,000 mw by 2000. The capacity projected for 1980 will be greater than the present total generating capacity of most of the countries of the world. Japan's total capacity in 1968 was 52,650 mw; the Soviet Union and England are the only other countries whose 1968 capacity exceeded 50,000 mw.

Even more impressive, 13 of the plants are proposed as 10,000-mw stations, three times larger than the largest 1970 plant. One notices from the map in Fig. 9-13 that most of the giant plants are targeted for Wyoming and southern Montana between the Yellowstone and Little Missouri Rivers.

At the 50,000-mw level the plants need 855,000 acre feet of water per year. The study proposes a system of dams, reservoirs, and pipelines in the Yellowstone Basin to provide water for the power plants and the additional water needed by the planned coal gasification and petrochemical companies to be located near the power plants. Critics have pointed out that 855,000 acre-feet is half the flow of the Yellowstone in a dry year and that the total flow in the Yellowstone Basin is only 1.8 million acre-feet per year.

We can read down the list and see the environmental bill mounting: at the 50,000-mw level air pollution will be 100,000 tons of particulate matter per year even with 99.5 percent ash removal; sulphur oxides, 2.1 million tons; nitrogen oxides, 1.9 mil-

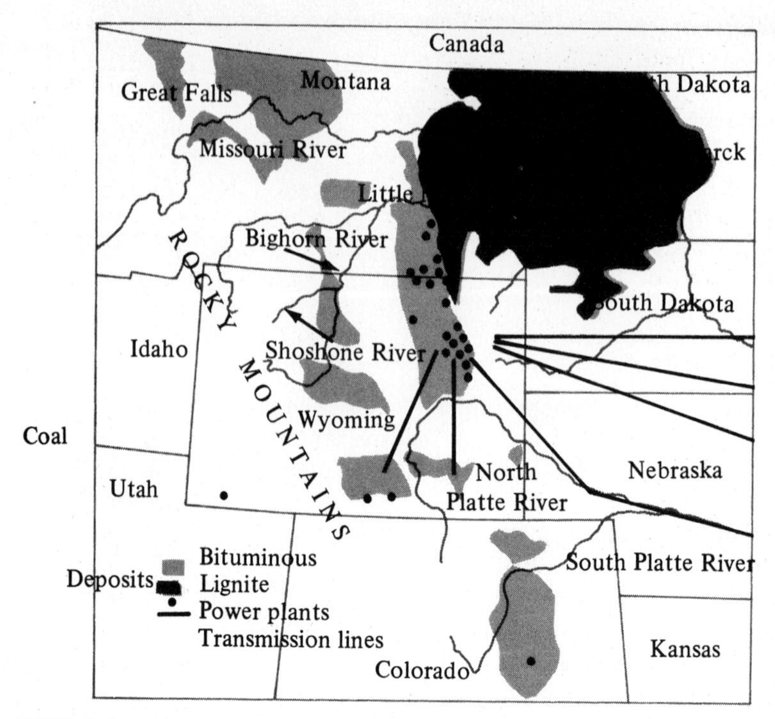

FIGURE 9-13 Coal deposits and proposed power plant location in the Montana-Dakota area.

lion tons. Land abuse presents additional problems: 8 percent of the tri-state area strip-mined, 8,000 miles of transmission lines, and a 500 percent increase in population in the coal-mining and power-plant areas.

The North Central Power Study was planned to bring about an orderly development of the region. It has not succeeded, and the local residents are caught in the squeeze. Ranchers are losing their rights to leased land; land owners are finding that they own the surface, but not the coal.

The Bureau of Indian Affairs has acted as an agent for some of the Indian tribes. It persuaded the Northern Cheyennes to lease out over half of its reservation acreage, 292,680 acres in all. It has been estimated that more than 5.5 million acres of Indian and federally owned land has already been leased.

The whole situation screams for an "environmental impact statement," an assessment of the environmental cost of these developments and a listing of such options as are available to hold these costs to a minimum. So far, the impact is showing but not the statement. It is not even clear that the tremendous momentum of the project can be controlled.

The Applachian Mountains, the Midwestern plains, the Southwestern desert and now the Northern plains, all ripped or to be ripped apart for coal. What is striking, as we review this at the beginning of the 1970s, is not that it has happened, but that we have not yet learned how to handle it. This rush in Wyoming shows few learned lessons from Appalachia.

Perhaps it is not too late. It is our hope that by presenting the Four Corners and

North Central picture in some detail we will cause the consumers in Los Angeles or Kansas City or St. Louis to reflect on the environmental costs of their electricity and to insist that they be paid. It is not certain that these arid lands can ever be completely restored. They will produce, however, at 20,000 tons per acre and 5 dollars per ton, 100,000 dollars per acre. The proposed expenditure of 500 dollars per acre (2.5 cents per ton) could easily be doubled, it would seem, without becoming a distortion of perspective. We cannot, however, count on the good intentions of the coal men, their feelings were expressed by one of the power shovel operators: "Hell its desert, strip it." Some counter pressure must come from the metropolitan consumer.

THERE'S MORE

In this chapter we have singled out some of the ways in which the production and consumption of energy has environmental impact. Not only is the list incomplete, but it grows constantly. We have not, for instance, discussed oil spills, one of the damages of energy transportation. This hazard, however, must be faced as we turn to the Mideast for oil or ship it by tanker from Alaska. It is not the major catastrophies which cause the most harm. In the celebrated Torrey Canyon disaster off the English coast in 1967, 100,000 tons of crude oil were lost, but an estimated 3.5 million tons of oil are routinely discharged into the ocean each year. The total pollution from exploration, pipeline breaks, oil in sewage, and other processes related to energy production is estimated at 5 to 10 million tons per year. Most of this pollution occurs in the biologically rich and sensitive coastal waters. We are only now beginning to study, in detail, the effects of petroleum products on marine ecology; and the prospect of deep ocean ports and increased offshore oil drilling make urgent the knowledge to be gained from these studies.

The newspapers are beginning to call to our attention the dangers of the growing web of natural gas distribution pipes beneath our cities, as accidental breakage and leaks cause an increasing number of explosions. The most spectacular recent episode occured in Michigan where drilling for gas changed the stability of a large underground (natural) concentration which then migrated across the countryside sporadically exploding large craters in the soil.

There are other hazards to come: there is some concern over the effects of cloud cover from high flying jets (not yet a significant factor but something to be examined). The opposition to the Supersonic Transport (SST) was, in part, due to concern over the effects of its pollutants on the stratosphere, in particular on the blanket of ozone which protects us from the sun's ultraviolet rays. We have yet to mention the shadow of radioactivity that hangs over nuclear power. We will look at that in the following chapter.

SUMMARY

In this chapter we totaled up the rest of the environmental bill for the production and conversion of energy. At the production end we looked at some of the environmental costs of coal mining. The most obvious damage comes from strip mining which now produces 50 percent of our coal. It is estimated that 2 million acres of land have been strip-mined so far and that only one third of these disturbed acres have been adequately reclaimed. The focus of strip mining has shifted west and we can keep track of the progress of the huge shovels across the plains by remembering that a 10,000-mw plant uses about 100,000 tons of coal per day. The Western lands will, in 1980, be supporting 50,000 mw of generating capacity. Since the coal mines will produce perhaps 20,000 tons per acre, the plains will be disappearing at a rate of 25 acres per day.

Underground mining will also continue; most of our coal is too deep to be strip-mined. There are both environmental costs such as drainage of acid water into streams, land subsidence, fires, and human costs to this technique, if we take into account the one fatality per million man-hours and the few hundred thousand cases of black lung already recorded.

There are smaller but more pervasive changes under way: the carbon dioxide released to the atmosphere has the capability of causing a temperature rise. For reasons that are not yet understood, this rise has not yet taken place. The increasing amount of heat, the final form of energy, is becoming a factor in the climate of our cities and may provide the final stop sign to unlimited energy growth.

We also added some other costs to the total: the loss of scenic beauty and threat to aquatic life of the hydroelectric dams and pumped-storage plants. The latter, by making it profitable to keep the coal-fired plants burning, add to air pollution woes. The transmission of energy by wire impacts more and more on our land; its shipment by sea is adding to the oil which is now an observable component of sea water.

We should underline the message of these last three chapters. We have need of energy and it is legitimate to focus attention, as we will in succeeding chapters, on new conversion techniques and new sources. We must insist, however, that each increase in production or consumption be accompanied by a detailed study and report on the environmental cost. We have used up the environmental credit of this planet; from now on it will be pay as you go.

10

The Mixed Blessing of Nuclear Energy

The plume of smoke above its factories was once the pride of a growing community; now it is a signal for alarm. Cheap motor fuel, the mass-produced automobile, and millions of miles of concrete have tamed this vast country and made Americans the most mobile people the world has known, traveling distances for recreation that would have awed even the hardiest nomads of a few centuries ago. Now the exhaust from millions of gasoline engines keeps many of our countrymen and women indoors on the hot stagnant days of air pollution alerts.

The story of energy, as we have developed it so far, contrasts great benefit with increasingly visible cost; it is a story of mixed blessings. This chapter, which will deal with the latest addition to the energy converters, will not break that pattern. Nuclear energy (energy from fissioning nuclei), in its present realization, is the most controversial form of energy with which we have ever had to deal. We can see the range of controversy in the words of its proponents and antagonists.

Dr. Glenn Seaborg, noted nuclear chemist and Chairman of the United States Atomic Energy Commission (AEC) from 1961 until 1971 sums up his view:

> By the year 2000 we will see about a 1,000 million kilowatts of electricity generated by about 1000 nuclear power reactors, sharing about half of the nation's power load, with highly improved fossil-fueled plants carrying the other half.

And later in the same article:

> Long before the year 2000, it will have become routine for the high level waste produced in reprocessing fuel from these plants to be converted to solids and buried or stored where it cannot reach the biosphere.
>
> Located according to a national electric power siting plan, and equipped with the latest cooling technologies, these nuclear plants will not be allowed to produce harmful thermal effects on their local environment. And the waste heat from many of them will be diverted to beneficial uses.[1]

Dr. John W. Gofman, an AEC radiation research scientist testifying before the State Senate of Pennsylvania in August 1970, made this recommendation:

> I should like to recommend that the Pennsylvania Legislature take the important and constructive step of declaring a 5-year moratorium on the planning, construction, and operation of nuclear power plants above ground anywhere in Pennsylvania. This would represent a first step toward the safe consideration of nuclear energy in the future.

[1] G. Seaborg, "On Misunderstanding the Atom," *Bull. of the At. Sci.*, **27**: 46, September 1971.

He went on to summarize his reasons for this recommendation:

1 Electric power requirements have not been rationally discussed in relationship to the intrusion upon a livable environment.

2 Nuclear electric power development has proceeded with the most grave failure of appreciation of the radiation hazard to the population.

3 Nuclear power development represents an anti-democratic disfranchisement concerning which citizens are properly becoming aware and incensed.

4 Nuclear electric power plants represent a gigantic experiment conducted upon, and at the possibly serious expense of, the citizens of your State. No one knows *at all* the risk of catastrophic accidents.

5 The creation of fast breeder reactors and its concomitant "plutonium economy" may well represent man's greatest immoral act.[2]

These two quotations present the mix of the blessing. In many ways the fission reactor seems to be an energy source that has arrived in the nick of time, saving us both from the shortage of fossil fuels and from the insult to our atmosphere which comes from burning them. As John Gofman and many others remind us, however, the nuclear age will be a very different sort of age than any we have known. There are dark clouds behind the silver linings.

In this chapter we will make a serious attempt to analyze both the reasons for seeing darkness in the clouds and the reasons for the brightness to the linings. Our failure to reach the safety of neutrality, however, is preordained. We will be too cautious for the "nuclearphiles," and not critical enough for the "nuclearphobes." To limit the grounds for disagreement we will try, for the most part, to deal with the facts of nuclear energy and leave their interpretation to you.

The fact which must be dealt with at the beginning is that nuclear reactors are already on the scene. At the most recent count (January 1974) there were 42 nuclear electric power stations in operation with a generating capacity of 25,000 mw. These stations ranged in size from the 53-mw plant at Genoa, Wisconsin to the 1,065-mw Browns Ferry Plant just put into operation at Decatur, Alabama. Their ages are also of some interest; the first commercial nuclear power plant was the 90-mw Shippingport Plant in Pennsylvania which came on line in 1957; the most recent one probably came on line as this book was published.

Figure 10-1 shows the geographical locations of nuclear reactors which are already in operation, under construction, and on order. It is only in the Rocky Mountain region of the country that a visit to such a plant is more than an easy automobile drive for anyone.

Nuclear energy has, therefore, arrived; 3 percent of this nation's electric power came from that source in 1972. It would take a major catastrophe to convince the electric utility companies (to convince the public ultimately) that the existing plants should be shut down. We should expect them to live out their 30 to 50 years of productive life. There does not, in fact, seem to be enough realistic public pressure, at this stage, to stop the continuing growth in the importance of nuclear generation.

[2] J. Gofman, *Testimony Before the Select Committee on Nuclear Energy,* State Senate of Pennsylvania, August 20, 1970.

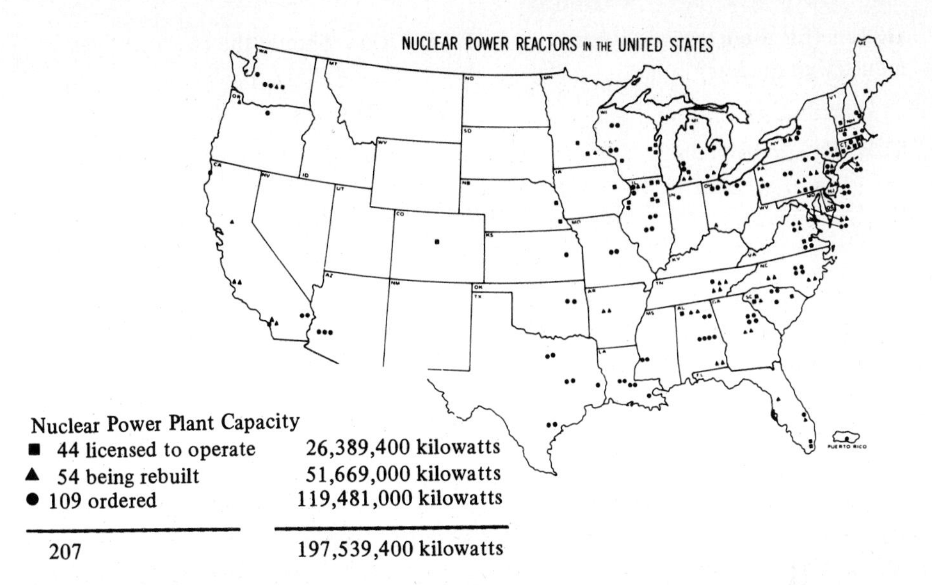

FIGURE 10-1 The location of reactors in service, under construction, and ordered as of January 1974.

Again, only a proven major threat to health and happiness, or the sudden emergence of a better source of energy will stunt their rapid growth.

Projections of future growth vary in their optimism, for instance, from a "high" AEC estimate of 602,000 mw by 1990 to the FPC estimate of 475,000 mw for the same target year. Many reasonable people, however, are worried by these projections; some because of the "all eggs in one basket" approach which has starved out research on other potentially important energy substitutes; others because they are deeply concerned with the irreversibility of the release of the nuclear genie and the by-products of his work. Concern is especially sharp over the next generation of nuclear reactors, the breeder reactors. This new reactor type, which received the presidential seal of approval in President Nixon's June 1971 energy message, promises much more energy and more trouble per pound of nuclear fuel than the present models.

In this chapter we will dissect some of these worries, look at them, and at the background of information and uncertainty against which they must be judged. In the process of this analysis, we will discuss, in general terms, the process of converting nuclear potential energy into heat energy by the fission chain reaction, and the important features of the different types of reactors. We will try to give an objective description of the advantages and disadvantages of this new source of electrical power, and include among the disadvantages, not only the air and water pollution, but also the longer-range and potentially deeper problems of radioactive waste and accidents. We will also look at this energy converter of the future, the breeder reactor, and point out, to the extent that we can at this early date, its real potential for both creating and solving problems.

As with many of our previous topics, there is room here for more physics; in this case, for an interesting glimpse into the submicroscopic world of the nucleus. We deal with some of these basic ideas in Appendix 5. In what follows we concentrate on answers (for there are more than one) to this question: Should the electric power industry go nuclear?

BURNING URANIUM

We will briefly describe the fission reaction and the techniques for controlling it in the section which follows. For a more complete discussion the reader is referred to Appendix 5.

The success of the conversion of nuclear energy depends (as does the success of the burning of coal) on the establishment of a *chain reaction.* We have to be able to start the fission reaction (ignite it, so to speak) and then obtain enough energy from ignition to cause the reactions to not only continue, but also to grow; in a manner analogous to using the damper on a coal furnace, we also must be able to control our nuclear chain reaction.

The fuels for the fission reactions are the nuclei at the heavy end of the periodic table, uranium (U), plutonium (Pu), and thorium (Th). When these heavy nuclei split, energy is released; this process is known as *fission.* During fission the energy is, in fact, carried away as kinetic energy by the recoiling products, nuclear fragments and particles, of the fission event. Fission is stimulated by neutron bombardment of these nuclei, and it is the release of neutrons in the fission event that produces the chain reaction. Two or three (2.5, on the average) neutrons are released from each fissioning nuclei; they can, therefore, cause two or three more nuclei to fission, and each of those two or three can release two or three more, and so on. Since the time between one set of fission events and those caused by it (between generations, so to speak) is only 10^{-4} seconds, the reaction, once started, quickly involves the entire fuel assembly.

While any heavy nucleus can be caused to fission if struck by a high-energy neutron (or, in fact, any high-energy particle), the probability of this occurring is highest when the neutrons are moving slowly. These neutrons are known as *thermal neutrons* in the jargon of nuclear physics. There is only one naturally occurring nucleus, the rare isotope[3] of uranium, U^{235}, which can capture these thermal neutrons and undergo fission. Two other nuclei, U^{233} and Pu^{239}, can be produced in a reactor and will also serve as nuclear fuel. It is the capacity to produce these fuels that creates most of the interest in the breeder reactor which we will discuss later. The reactors we are presently depending on, however, use U^{235}, which makes up only 0.7 percent (1 nucleus out of 140) of natural uranium ore. U^{238} is the dominant isotope.

Briefly, the chain reaction depends on the release of more than one neutron in each fission event. These neutrons must be slowed down to thermal energies; that is the job of the *moderator* (water in the present United States reactors). The reaction is also controlled through these neutrons by controlling the number of them allowed to cause fission. Neutrons are taken out of the flux by "capturing" them in boron containing *control rods.* If the average number of neutrons available to cause fission

[3]The "isotopes" of elements are all chemically identical since they have the same number of protons in the nucleus. Their nuclear properties differ, however, as they have different numbers of neutrons. The superscript (235 in this case) indicates the total number of nuclear particles (neutrons and protons) in the nucleus.

is reduced to about 1 per fission, the growth of the chain reaction is stopped and the reactor stays at a constant level of power. An important feature of this control, as we explain in Appendix 5, is the small fraction of the neutrons which are delayed, and which therefore allow some time to make fine adjustment of the neutron intensity.

These are the major features of the fission chain reaction. Let us now consider the structure of a nuclear reactor.

NUCLEAR POWER PLANTS

From the preceding there emerges a crude picture of a nuclear reactor as consisting of three main elements: fuel, moderator, and controls. The energy is released by the fuel, the moderator feeds low-energy neutrons back into the fuel to sustain the fission reaction, and the neutron-absorbing controls take the neutrons out of the chain reaction in order to slow it down or stop it. Let us look now at the reactor structure in more detail.

Fuel

Most present-day reactors must have U^{235} for the chain reaction. It is possible to use natural uranium ore with its 1-in-140 U^{235} nuclei, as a fuel, but there are problems. The U^{238} nuclei capture some of the neutrons and take them out of the flux available to cause the chain reaction. If natural uranium ore is the fuel, then it is necessary to use a moderator which does not capture slow neutrons. Reactors using natural uranium fuel, therefore, have either "heavy water" or carbon (graphite) as moderators. The British reactors are mostly of this style.

Most United States reactors use ordinary water as a moderator. Since the protons (hydrogen nuclei) in water can capture a neutron and form the heavy hydrogen isotope, deuterium (which is present in "heavy water"), it is necessary to compensate for their loss to the water and the U^{238} and increase the concentration of the fissionable nucleus, U^{235}, in the fuel. This is called *enriched* fuel.

The process of enrichment is no easy task. It is no great problem to separate elements from one another as they are chemically different and one can use their chemical properties to sort them. Isotopes lack this handle since they are chemically the same. The only difference is in their mass and size, and so the separation techniques must take advantage of that. U^{235} enrichment is achieved by converting the uranium to a gas, uranium hexaflouride (UF_6) and allowing it to leak through a series of porous walls. The $U^{235}F_6$, molecules are a little lighter, move a little faster, are a little smaller, and have a higher probability of getting through the walls than the do $U^{238}F_6$ molecules. Thus, at each step of the way, the mixture on the other side of the porous wall is a little bit enriched in U^{235}. The fuel for most United States reactors is enriched to about 3 percent or, perhaps 3 or 4 times its natural abundance.

This process is expensive and energy-consuming; in fact, it is only in the past year or so that the nuclear industry has become a net producer of electrical energy. Even in the late 1960s more electrical energy was consumed in the enrichment process and in the manufacture of reactor components than the operating reactors produced.

Reactor Construction

The working components of a fission reactor are shown schematically in Fig. 10-2. The fuel, usually in the form of a *sintered*[4] pellet of uranium dioxide (UO_2) is contained in a series of *fuel rods.* These rods are about 0.5 inches in diameter and 12 feet long. The fuel is contained inside a *cladding* of some noncorrosive, light, strong metal such as zirconium.

These fuel rods are then arranged in a bundle of 49 rods and assembled with the control rods to form a fuel assembly such as the one shown in Fig. 10-3. A commercial reactor contains several hundred individual fuel rods; in Niagra Mohawk's Nine-Mile Point generating unit, with 625-mw capacity, there are 532 bundles of 49 rods containing 115 tons of uranium oxide pellets.

In United States reactors the rods are surrounded by water, which serves a dual purpose. It acts as a moderator as we have discussed, slowing down the fast neutrons released by fission, and it also serves as a coolant. This latter function is equally important. As we have said, the fissioning U^{235} splits into two *fission products,* which fly apart with considerable kinetic energy. This is quickly converted, by collisions with the remaining material in the rod, to heat energy. Thus, the fuel rods get very hot and would quickly melt if the circulating water did not continually remove the heat energy. This is also the mechanism by which the energy of the reactor is retrieved.

Interspaced with the fuel rods are control rods containing boron. These rods, which are also shown in Fig. 10-3, can be pulled in or out of the reactor core. The number which are inserted and the depth to which they are inserted give the operat-

[4]Heated and formed into a ceramic.

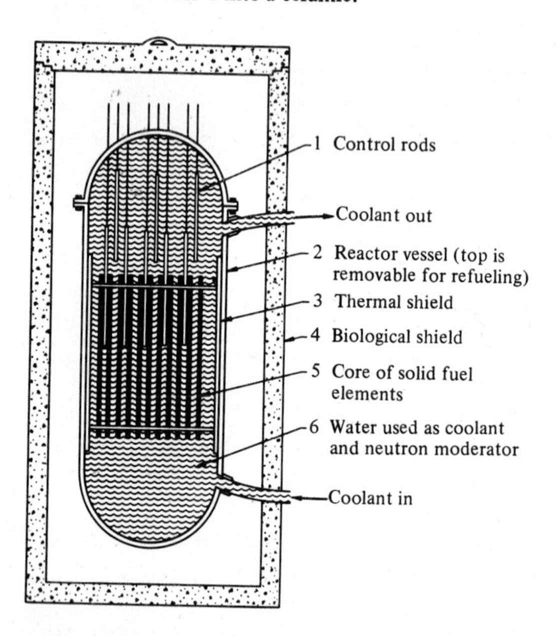

FIGURE 10-2 The general features of a reactor are shown. The water serves as coolant and neutron moderator. The pressure vessel is surrounded by a strong concrete outer shell for further protection.

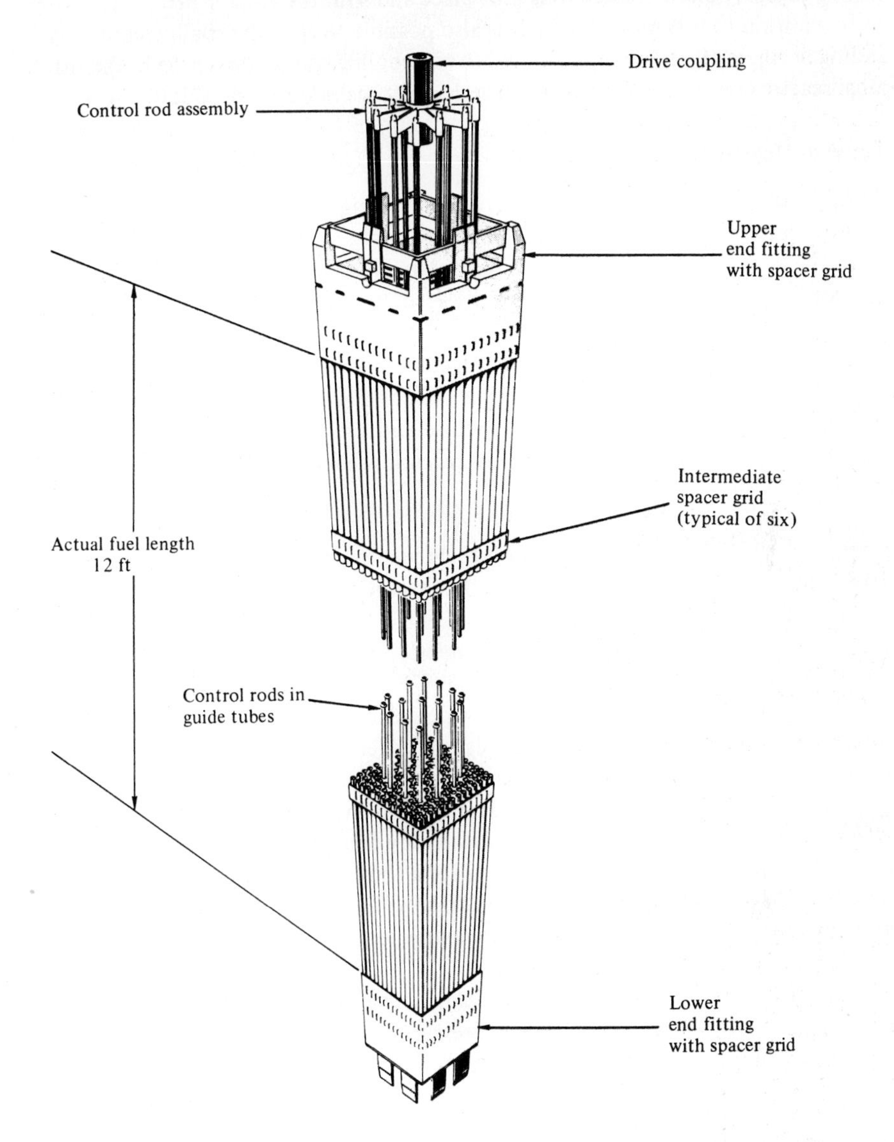

FIGURE 10-3 The assembly of fuel and control rods which make up a fuel assembly. There are typically 49 rods in an assembly and hundreds of these bunches in a reactor core. (Source: *The Environmental Impact of Electric Power Generation: Nuclear and Fossil,* Pennsylvania Department of Education, 1973.)

ors fine control over the number of neutrons available, and thus, over the number of nuclei fissioning per second. In this way the power level at which the reactor operates is determined.

Since it is the number of neutrons flying about that is critical, electronic *neutron counters* can be used to control the setting of the control rods. These same devices can be used to *scram* the reactor when the neutron count raises to a dangerous level, that is, to shoot extra control rods into place and shut the reactor off.

In addition to the control rods, it is also possible to stop the chain reaction by adding boron in the form of boric acid to the cooling water circulating in the reactor. Most reactors have both these control and safety techniques available.

Types of Reactors

With a core of fuel and control rods, with circulating water as coolant and moderator, we now have the basic requirements of an operating nuclear energy conversion device. The next step is to get this heat energy to a heat engine, convert it to mechanical work, and turn a generator. It is in the accomplishment of this heat transfer that differentiation between various reactor types takes place.

Table 10-1 lists the various types of reactors in commercial use or under development. In this country, at the present time, all commercial reactors are thermal neutron reactors, light water-moderated, and use enriched uranium as fuel. There are two basic types now in use, the *boiling water reactors,* (BWR), and the *pressurized water reactors* (PWR).

PWR. In the PWR there are two separate circulating water systems as shown in Fig. 10-4. The water that circulates through the reactor core is under very high pressure, about 2,000 pounds per square inch. At this high pressure, water does not boil at the normal 212°F, but at much higher temperature. Thus, the water in this first system circulates at a temperature as high as 600°F and remains a liquid. It is then circulated through a heat exchanger, and the heat energy is transferred by a set of coils immersed in the water of a second circulating system. This water is under much lower pressure; it is caused to boil by the heat energy transferred to it from the hot coils and produces steam which goes to a steam turbine and turns a generator.

From the turbine on, the operating machinery is basically the same as in a fossil fuel plant. In particular, there is still need for a third water-circulating system to condense the steam back to water. In contrast to the first two systems which are essentially closed, this last one is open and connected to a body of water or a cooling tower.

In the PWR it is easier to contain the radioactive materials that inevitably leak through the fuel rod cladding. The first pressurized system is completely separated from the other parts of the system and in particular, from the turbine.

We might mention one other feature of the PWR here which is of importance

Table 10-1 Types of Reactors

Neutrons	*Fuel*	*Examples*
Slow or thermal	Natural U; carbon moderated	British
	Natural U; D_2O moderated	Savannah River and research reactors
	Enriched U or Pu^{239}; H_2O moderated	PWR, BWR
	Enriched U or Pu^{239}; carbon moderated; He cooled	HTGR
Fast	Enriched U or Pu^{239}; unmoderated, sodium cooled	LMFBR

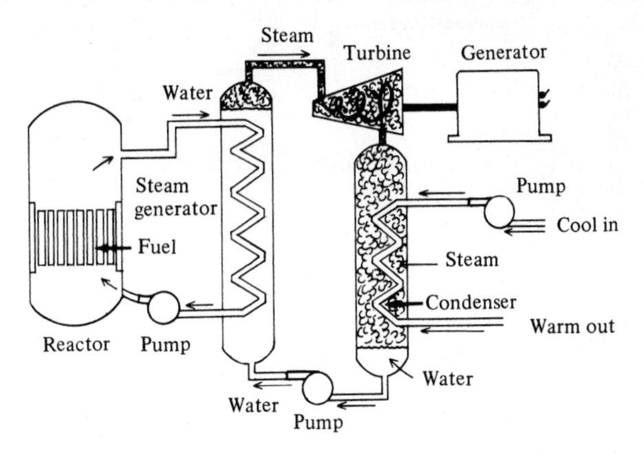

FIGURE 10-4 The PWR has two separate circulating coolant systems. In the system in contact with the core, the water is under such high pressure that it doesn't boil.

when we discuss its radioactive wastes. As the fuel is gradually used up in a reactor and neutron-absorbing impurities accumulate, the reactivity of the core decreases. To compensate for this in a more uniform manner than is possible with control rods, some boron is added to the cooling water at the beginning of the reactor's operating period. As the overall power level gradually lessens, the boron, which has been absorbing neutrons, is gradually removed from the coolant allowing more neutrons to take part in the reaction. Unfortunately, the reaction between boron and neutrons produces, in addition to Li^7 as we describe in Appendix 5, a radioactive form of hydrogen, tritium. It is difficult to prevent escape of this material into the environment. We will look at this potential danger in a later section.

The PWR was the first reactor to be put into commercial service, at the Shippingport Plant in Pennsylvania. It is also the type used in the nuclear submarines.

BWR. The other reactor type in commercial use is the BWR whose first commercial prototype was installed at Commonwealth Edison's Dresden, Illinois plant in 1960. In the BWR, the first pressurized water-circulating system is eliminated. The water in contact with the hot fuel rods is at a lower pressure, 1,000 pounds per square inch (psi), and is allowed to boil. As shown in Fig. 10-5, the steam generated by this boiling rises to the top of the core region, and, after passing through steam separators and dryers, is piped directly to the turbogenerators.

In this reactor there is a possibility of radioactive materials getting to the turbine and leaking from there to the environment. The circulating water and steam picks up radioactive material in several ways. The cladding on the fuel rods is so thin (0.02 inches) that some of the fission products diffuse through it. The cladding also occasionally ruptures, and, since it is not economically feasible to shut down the reactor to remove one fuel rod, the reactor is allowed to operate with up to 1 percent of the rods leaking. Some fission products thus escape directly into the water. There are also inevitable impurities in the water and these are made radioactive by the intense flux of neutrons in the reactor. Therefore, it is of extreme importance to make

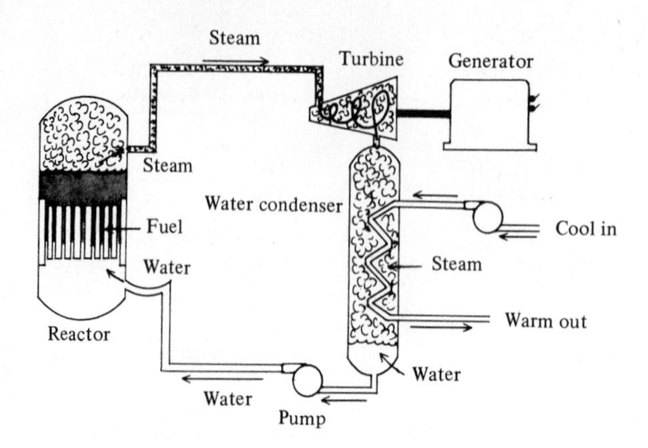

FIGURE 10-5 In the BWR the circulating water is boiled to generate the steam for the turbine. The turbine must then be sealed so that the radioactive materials in the steam do not escape.

certain that the turbines in the BWR are sealed so that none of the radioactive steam or water escapes directly to the environment. In compensation for this disadvantage, the BWR has, as far as safety is concerned, one rather significant advantage over the PWR. Boron is not added to the water as it is in the PWR; it would foul up the turbine. There is a special feature to the BWR that gives the same type of control.

In the water around the hot fuel rods, steam bubbles are formed. Neutrons are not as effectively moderated in the bubbles (because of the decreased density) as in the liquid water. Thus, as the chain reaction grows in the reactor, the rods get hotter and more steam bubbles are formed. The increase in the space occupied by the steam bubbles, however, reduces the moderating effectiveness of the water, reduces the number of slow neutrons, and thus slows the growth of the chain reaction. This is a situation which contributes to reactor stability. It operates in the other direction as the fuel rods lose their reactivity; bubble size decreases and more slow neutrons are available to compensate for the aging of the fuel.

HTGR. There is some interest in a new type of reactor, the high temperature gas-cooled reactor (HTGR). This reactor, while it is a converter (like the BWR and PWR) and not a breeder, does produce new fissionable material. It differs from the light water converters in having graphite as a moderator and helium as a coolant. The helium circulating through the core reaches a temperature of 1400°F and through a heat exchanger (coils immersed in water) produces steam at 1000°F. Thus, the HTGR offers the possibility of a thermal efficiency of 40 percent which is comparable to the best fossil fuel plants. Studies are also underway on the feasibility of using the hot helium directly in gas turbines, which would increase the efficiency even more.

In the HTGR it is possible to add Thorium, Th^{232}, to the reactor and create more fuel. The thorium nucleus can capture a slow neutron and be converted to U^{233} which, like U^{235}, is a good nuclear fuel. This reactor is called an *advanced converter* since it does create some fuel for itself out of the nonfissioning component.

An experimental HTGR of 40 mw is in operation at Philadelphia Electric Company's Peach Bottom site, and another of 330 mw is being constructed at Fort St. Vrain, Colorado. Whether this type will become important in the future depends on the experience with these plants and on the success of the breeder reactors which they to some extent anticipate.

These three types account for all the reactor sitings identified on the map of Fig. 10-1. More detail is given in Table 10-2 in which the type, the generating capacity, and the expected date of commercial operation of plants now "on order" are given. One can see from the dates that Table 10-2 gives a pretty complete preview of 1980.

BREEDER REACTORS–A GLIMPSE OF THE FUTURE?

The reactors which we have discussed so far are classified as converters: their major purpose is to convert the fissionable fuel–almost exclusively U^{235}–into energy. Although the energy these reactors produce per ton of fuel is quite impressive when compared with fossil fuel plants (200 tons of uranium per year for a 1,000-mw reactor as against 2,000,000 tons of coal for a coal-fired plant of the same capacity) their fuel conversion efficiency is not really very impressive.

Of the 200 tons of uranium used as fuel, only 6 tons (in a 3 percent enriched fuel) is U^{235}. Even this U^{235} is not totally converted. One ton of U^{235} should theoretically be able to produce about 10^7 mw-hr (10 M mw-hr) of energy.[5] A 1,000-mw generator operating at 80 percent capacity will produce 7 M mw-hr. Thus, the actual efficiency of conversion of U^{235} is given by (7 M mw-hr)/(6 tons $\times$ 10 M mw-hr per ton) = 11 percent. Since uranium itself is a rather scarce resource, attention from the beginning of reactor development has turned to ways of obtaining energy from the much more abundant U^{238}.

Breeding Nuclear Fuel

In Appendix 5 the process by which a nonfissionable nucleus is made into a fissionable one is described. This is called a *breeding reaction* and the nuclei which can be converted in this way are called *fertile* materials. The two fertile materials of importance to the nuclear reactor program are Th^{232}, which can be converted into a fissionable isotope of uranium, U^{233}, and the abundant isotope of natural uranium, U^{238}, which can be converted into the fissionable isotope Pu^{239}

The important parameters in the evaluation of breeder reactors are the *breeding ratio,* the ratio of fissile material (U^{233} or Pu^{239}) produced, to the fissile material consumed, and the *doubling time,* which is the number of years it will take for a breeder reactor to produce enough extra fissile material to fuel a second reactor.

Breeder reactors can operate either with low-energy neutrons, *thermal breeders,* or with high-energy neutrons, *fast breeders.* We will briefly mention the first of these types and spend most of our time talking about the fast breeders which have the highest priority for development.

[5] The energy equivalent of 1 ton of U^{235} is 3.3 $\times$ 10^7 mw-hr. At a thermal efficiency of 33 percent we can expect to get about 10^7 mw-hr of electrical energy.

Table 10-2 Nuclear Power Reactors in the United States

Site	Plant name	Capacity (net kw)	Utility	Commercial operation
ALABAMA				
Decatur	Browns Ferry Nuclear Power Plant: Unit 1	1,065,000	Tennessee Valley Authority	1973
Decatur	Browns Ferry Nuclear Power Plant: Unit 2	1,065,000	Tennessee Valley Authority	1974
Decatur	Browns Ferry Nuclear Power Plant: Unit 3	1,065,000	Tennessee Valley Authority	1975
Dothan	Joseph M. Farley Nuclear Plant: Unit 1	829,000	Alabama Power Co.	1975
Dothan	Joseph M. Farley Nuclear Plant: Unit 2	829,000	Alabama Power Co.	1977
Chilton County	Central Alabama Nuclear Plant: Unit 1	1,200,000	Alabama Power Co.	1982
Chilton County	Central Alabama Nuclear Plant: Unit 2	1,200,000	Alabama Power Co.	1983
Elmore County	Central Alabama Nuclear Plant: Unit 3	1,200,000	Alabama Power Co.	1984
Elmore County	Central Alabama Nuclear Plant: Unit 4	1,200,000	Alabama Power Co.	1985
Scottsboro	Bellefonte Nuclear Plant: Unit 1	1,189,000	Tennessee Valley Authority	1979
Scottsboro	Bellefonte Nuclear Plant: Unit 2	1,189,000	Tennessee Valley Authority	1980
ARIZONA				
Wintersburg	Palo Verde Nuclear Generation Station: Unit 1	1,270,000	Arizona Public Service	1981
Wintersburg	Palo Verde Nuclear Generating Station: Unit 2	1,270,000	Arizona Public Service	1982
Wintersburg	Palo Verde Nuclear Generating Station: Unit 3	1,270,000	Arizona Public Service	1984
ARKANSAS				
Russellville	Arkansas Nuclear One: Unit 1	850,000	Arkansas Power & Light Co.	1974
Russellville	Arkansas Nuclear One: Unit 2	912,000	Arkansas Power & Light Co.	1976
CALIFORNIA				
Humboldt Bay	Humboldt Bay Power Plant: Unit 3	65,000	Pacific Gas and Electric Co.	1963
San Clemente	San Onofre Nuclear Generating Station: Unit 1	430,000	So. Calif. Ed. & San Diego Gas & Electric Co.	1968
San Clemente	San Onofre Nuclear Generating Station: Unit 2	1,140,000	So. Calif. Ed. & San Diego Gas & Electric Co.	1979
San Clemente	San Onofre Nuclear Generating Station: Unit 3	1,140,000	So. Calif. Ed. & San Diego Gas & Electric Co.	1980
Diablo Canyon	Diablo Canyon Nuclear Power Plant: Unit 1	1,084,000	Pacific Gas and Electric Co.	1975
Diablo Canyon	Diablo Canyon Nuclear Power Plant: Unit 2	1,106,000	Pacific Gas and Electric Co.	1976
Clay Station	Rancho Seco Nuclear Generating Station	913,000	Sacramento Municipal Utility District	1974
*	–	1,128,000	Pacific Gas & Electric Co.	1981
*	–	1,128,000	Pacific Gas & Electric Co.	1982
Vidal	Vidal Generating Station: Unit 1	770,000	Southern California Edison Co.	1981
Vidal	Vidal Generation Station: Unit 2	770,000	Southern California Edison Co.	1982

Table 10-2 Nuclear Power Reactors in the United States (continued)

Site	*Plant name*	*Capacity (net kw)*	*Utility*	*Commercial operation*
COLORADO				
Platteville	Ft. St. Vrain Nuclear Generating Station	330,000	Public Service Co. of Colorado	1974
CONNECTICUT				
Haddam Neck	Haddam Neck Plant	575,000	Conn. Yankee Atomic Power Co.	1968
Waterford	Millstone Nuclear Power Station: Unit 1	652,100	Northeast Utilities	1971
Waterford	Millstone Nuclear Power Station: Unit 2	828,000	Northeast Utilities	1974
Waterford	Millstone Nuclear Power Station: Unit 3	1,150,000	Northeast Utilities	1979
DELAWARE				
Summit	Summit Power Station: Unit 1	770,000	Delmarva Power & Light Co.	1980
Summit	Summit Power Station: Unit 2	770,000	Delmarva Power & Light Co.	1982
FLORIDA				
Florida City	Turkey Point Station: Unit 3	693,000	Florida Power & Light Co.	1972
Florida City	Turkey Point Station: Unit 4	693,000	Florida Power & Light Co.	1973
Red Level	Crystal River Plant: Unit 3	825,000	Florida Power Corp.	1974
Ft. Pierce	St. Lucie Plant: Unit 1	801,000	Florida Power & Light Co.	1975
Ft. Pierce	St. Lucie Plant: Unit 2	801,000	Florida Power & Light Co.	1979
*	–	1,300,000	Florida Power Corp.	1983
*	–	1,300,000	Florida Power Corp.	1986
GEORGIA				
Baxley	Edwin I. Hatch Nuclear Plant: Unit 1	786,000	Georgia Power Co.	1974
Baxley	Edwin I. Hatch Nuclear Plant: Unit 2	795,000	Georgia Power Co.	1978
Waynesboro	Alvin W. Vogtle, Jr. Plant: Unit 1	1,121,000	Georgia Power Co.	1980
Waynesboro	Alvin W. Vogtle, Jr. Plant: Unit 2	1,121,000	Georgia Power Co.	1981
Waynesboro	Alvin W. Vogtle, Jr. Plant: Unit 3	1,121,000	Georgia Power Co.	1982
Waynesboro	Alvin W. Vogtle, Jr. Plant: Unit 4	1,121,000	Georgia Power Co.	1983
ILLINOIS				
Morris	Dresden Nuclear Power Station: Unit 1	200,000	Commonwealth Edison Co.	1960
Morris	Dresden Nuclear Power Station: Unit 2	809,000	Commonwealth Edison Co.	1970
Morris	Dresden Nuclear Power Station: Unit 3	809,000	Commonwealth Edison Co.	1971
Zion	Zion Nuclear Plant: Unit 1	1,050,000	Commonwealth Edison Co.	1973
Zion	Zion Nuclear Plant: Unit 2	1,050,000	Commonwealth Edison Co.	1974

Table 10-2 Nuclear Power Reactors in the United States (continued)

Site	*Plant name*	*Capacity (net kw)*	*Utility*	*Commercial operation*
Cordova	Quad-Cities Station: Unit 1	800,000	Commonwealth Ed. Co.-Ia.-Ill. Gas & Elec. Co.	1972
Cordova	Quad-Cities Station: Unit 2	800,000	Commonwealth Ed. Co.-Ia.-Ill. Gas & Elec. Co.	1972
Seneca	LaSalle County Nuclear Station: Unit 1	1,078,000	Commonwealth Ed. Co.-Ia.	1978
Seneca	LaSalle County Nuclear Station: Unit 2	1,078,000	Commonwealth Ed. Co.-Ia.	1979
Bryon	Byron Station: Unit 1	1,120,000	Commonwealth Edison Co.	1980
Bryon	Byron Station: Unit 2	1,120,000	Commonwealth Edison Co.	1981
Braidwood	Braidwood: Unit 1	1,200,000	Commonwealth Edison Co.	1980
Braidwood	Braidwood: Unit 2	1,200,000	Commonwealth Edison Co.	1981
Clinton	Clinton Nuclear Power Plant: Unit 1	955,000	Illinois Power Co.	1980
Clinton	Clinton Nuclear Power Plant: Unit 2	955,000	Illinois Power Co.	1982
INDIANA				
Porter County	Bailly Generating Station	660,000	Northern Indiana Public Service Co.	1979
IOWA				
Palo	Duane Arnold Energy Center: Unit 1	569,000	Iowa Electric Light and Power Co.	1974
KANSAS				
Burlington	Wolf Creek Generation Station: Unit 1	1,150,000	Kansas Gas & Electric–Kansas City P & L	1981
LOUISIANA				
Taft	Waterford Generating Station	1,113,000	Louisiana Power & Light Co.	1977
St. Francisville	River Bend Station: Unit 1	934,000	Gulf States Utilities Co.	1980
St. Francisville	River Bend Station: Unit 2	934,000	Gulf States Utilities Co.	1981
St. Rosalie	–	1,200,000	Louisiana Power & Light Co.	1982
St. Rosalie	–	1,200,000	Louisiana Power & Light Co.	1984
MAINE				
Wiscasset	Maine Yankee Atomic Power Plant	790,000	Maine Yankee Atomic Power Co.	1972
MARYLAND				
Lusby	Calvert Cliffs Nuclear Power Plant: Unit 1	845,000	Baltimore Gas and Electric Co.	1974
Lusby	Calvert Cliffs Nuclear Power Plant: Unit 2	845,000	Baltimore Gas and Electric Co.	1975
Nanjemoy	Douglas Point Project: Unit 1	1,178,000	Potomac Electric Power Co.	1980
Nanjemoy	Douglas Point Project: Unit 2	1,178,000	Potomac Electric Power Co.	1981

Table 10-2 Nuclear Power Reactors in the United States (continued)

Site	*Plant name*	*Capacity (net kw)*	*Utility*	*Commercial operation*
MASSACHUSETTS				
Rowe	Yankee Nuclear Power Station	175,000	Yankee Atomic Electric Co.	1961
Plymouth	Pilgrim Station: Unit 1	664,000	Boston Edison Co.	1972
Plymouth	Pilgrim Station: Unit 2	1,180,000	Boston Edison Co.	1980
MICHIGAN				
Big Rock Point	Big Rock Point Nuclear Plant	75,000	Consumers Power Co.	1965
South Haven	Palisades Nuclear Power Station	700,000	Consumers Power Co.	1971
Lagoona Beach	Enrico Fermi Atomic Power Plant: Unit 2	1,123,000	Detroit Edison Co.	1976
Lagoona Beach	Enrico Fermi Atomic Power Plant: Unit 3	1,172,000	Detroit Edison Co.	1981
Bridgman	Donald C. Cook Plant: Unit 1	1,060,000	Indiana & Michigan Electric Co.	1974
Bridgman	Donald C. Cook Plant: Unit 2	1,060,000	Indiana & Michigan Electric Co.	1976
Midland	Midland Nuclear Power Plant: Unit 1	492,000	Consumers Power Co.	1980
Midland	Midland Nuclear Power Plant: Unit 2	818,000	Consumers Power Co.	1979
St. Clair County	Greenwood: Unit 2	1,200,000	Detroit Edison Co.	1980
St. Clair County	Greenwood: Unit 3	1,200,000	Detroit Edison Co.	1981
Quanicassee	Quanicassee: Unit 1	1,150,000	Consumers Power Co.	1981
Quanicassee	Quanicassee: Unit 2	1,150,000	Consumers Power Co.	1982
MINNESOTA				
Monticello	Monticello Nuclear Generating Plant	545,000	Northern States Power Co.	1971
Red Wing	Prairie Island Nuclear Generating Plant: Unit 1	530,000	Northern States Power Co.	1973
Red Wing	Prairie Island Nuclear Generating Plant: Unit 2	530,000	Northern States Power Co.	1974
MISSOURI				
Fulton	Callaway Plant: Unit 1	1,150,000	Union Electric Co.	1981
Fulton	Callaway Plant: Unit 2	1,150,000	Union Electric Co.	1983
MISSISSIPPI				
Port Gibson	Grand Gulf Nuclear Station: Unit 1	1,290,000	Mississippi Power & Light Co.	1979
Port Gibson	Grand Gulf Nuclear Station: Unit 2	1,290,000	Mississippi Power & Light Co.	1981
NEBRASKA				
Fort Calhoun	Ft. Calhoun Station: Unit 1	457,400	Omaha Public Power District	1973
Brownville	Cooper Nuclear Station	778,000	Nebraska Public Power District and Iowa Power and Light Co.	1974

Table 10-2 Nuclear Power Reactors in the United States (continued)

Site	Plant name	Capacity (net kw)	Utility	Commercial operation
NEW HAMPSHIRE				
Seabrook	Seabrook Nuclear Station: Unit 1	1,200,000	Public Service of N.H.	1979
Seabrook	Seabrook Nuclear Station: Unit 2	1,200,000	Public Serivce of N.H.	1981
NEW JERSEY				
Toms River	Oyster Creek Nuclear Power Plant: Unit 1	640,000	Jersey Central Power & Light Co.	1969
Forked River	Forked River Generating Station: Unit 1	1,070,000	Jersey Central Power & Light Co.	1979
Salem	Salem Nuclear Generating Station: Unit 1	1,090,000	Public Service Electric and Gas, N.J.	1975
Salem	Salem Nuclear Generating Station: Unit 2	1,115,000	Public Service Electric and Gas, N.J.	1976
Salem	Hope Creek Generating Station: Unit 1	1,067,000	Public Service Electric and Gas, N.J.	1981
Salem	Hope Creek Generating Station: Unit 2	1,067,000	Public Service Electric and Gas, N.J.	1982
Little Egg Inlet	Atlantic Generating Station: Unit 1	1,150,000	Public Service Electric and Gas, N.J.	1980
Little Egg Inlet	Atlantic Generating Station: Unit 2	1,150,000	Public Service Electric and Gas, N.J.	1981
*	–	1,150,000	Public Service Electric and Gas, N.J.	1983
*	–	1,150,000	Public Service Electric and Gas, N.J.	1984
NEW YORK				
Indian Point	Indian Point Station: Unit 1	265,000	Consolidated Edison Co.	1962
Indian Point	Indian Point Station: Unit 2	873,000	Consolidated Edison Co.	1973
Indian Point	Indian Point Station: Unit 3	965,000	Consolidated Edison Co.	1974
Scriba	Nine Mile Point Nuclear Station: Unit 1	625,000	Niagara Mohawk Power Co.	1969
Scriba	Nine Mile Point Nuclear Station: Unit 2	1,080,000	Niagara Mohawk Power Co.	1978
Ontario	R.E. Ginna Nuclear Power Plant: Unit 1	490,000	Rochester Gas & Electric Co.	1970
Brookhaven	Shoreham Nuclear Power Station	819,000	Long Island Lighting Co.	1977
Scriba	James A. Fitzpatrick Nuclear Power Plant	821,000	Power Authority of State of N.Y.	1973
Jamesport	–	1,150,000	Long Island Lighting Co.	1981
Jamesport	–	1,150,000	Long Island Lighting Co.	1983
Oswego	Sterling Nuclear: Unit 1	1,150,000	Rochester Gas & Electric Co.	1982
NORTH CAROLINA				
Southport	Brunswick Steam Electric Plant: Unit 1	821,000	Carolina Power and Light Co.	1975
Southport	Brunswick Steam Electric Plant: Unit 2	821,000	Carolina Power and Light Co.	1974
Cowans Ford Dam	Wm. B. McGuire Nuclear Station: Unit 1	1,180,000	Duke Power Co.	1976
Cowans Ford Dam	Wm. B. McGuire Nuclear Station: Unit 2	1,180,000	Duke Power Co.	1977

Table 10-2 Nuclear Power Reactors in the United States (continues)

Site	*Plant name*	*Capacity (net kw)*	*Utility*	*Commercial operation*
Bonsal	Shearon Harris Plant: Unit 1	915,000	Carolina Power & Light Co.	1978
Bonsal	Shearon Harris Plant: Unit 2	915,000	Carolina Power & Light Co.	1979
Bonsal	Shearon Harris Plant: Unit 3	915,000	Carolina Power & Light Co.	1980
Bonsal	Shearon Harris Plant: Unit 4	915,000	Carolina Power & Light Co.	1981
Davie County	Perkins Nuclear Station: Unit 1	1,280,000	Duke Power Co.	1981
Davie County	Perkins Nuclear Station: Unit 2	1,280,000	Duke Power Co.	1982
Davie County	Perkins Nuclear Station: Unit 3	1,280,000	Duke Power Co.	1982
OHIO				
Oak Harbor	Davis-Besse Nuclear Power Station: Unit 1	906,000	Toledo Edison-Cleveland El. Illum. Co.	1976
Oak Harbor	Davis-Besse Nuclear Power Station: Unit 2	906,000	Toledo Edison-Cleveland El. Illum. Co.	1981
Oak Harbor	Davis-Besse Nuclear Power Station: Unit 3	906,000	Toledo Edison-Cleveland El. Illum. Co.	1983
Perry	Perry Nuclear Power Plant: Unit 1	1,205,000	Cleveland Electric Illuminating Co.	1979
Perry	Perry Nuclear Power Plant: Unit 2	1,205,000	Cleveland Electric Illuminating Co.	1980
Moscow	Wm. H. Zimmer Nuclear Power Station: Unit 1	810,000	Cincinnati Gas & Electric Co.	1977
Moscow	Wm. H. Zimmer Nuclear Power Station: Unit 2	1,170,000	Cincinnati Gas & Electric Co.	1982
OKLAHOMA				
Inola	Black Fox Nuclear Station: Unit 1	950,000	Public Service of Oklahoma	1982
Inola	Black Fox Nuclear Station: Unit 2	950,000	Public Service of Oklahoma	1984
OREGON				
Prescott	Trojan Nuclear Plant: Unit 1	1,130,000	Portland General Electric Co.	1975
Boardman	Boardman I	1,200,000	Portland General Electric Co.	1980
PENNSYLVANIA				
Peach Bottom	Peach Bottom Atomic Power Station: Unit 1	40,000	Philadelphia Electric Co.	1967
Peach Bottom	Peach Bottom Atomic Power Station: Unit 2	1,065,000	Philadelphia Electric Co.	1974
Peach Bottom	Peach Bottom Atomic Power Station: Unit 3	1,065,000	Philadelphia Electric Co.	1974
Pottstown	Limerick Generating Station: Unit 1	1,065,000	Philadelphia Electric Co.	1979
Pottstown	Limerick Generating Station: Unit 2	1,065,000	Philadelphia Electric Co.	1980
Shippingport	Shippingport Atomic Power Station: Unit 1	90,000	Duquesne Light Co.	1957
Shippingport	Beaver Valley Power Station: Unit 1	852,000	Duquesne Light Co.-Ohio Edison Co.	1975
Shippingport	Beaver Valley Power Station: Unit 2	852,000	Duquesne Light Co.-Ohio Edison Co.	1979
Goldsboro	Three Mile Island Nuclear Station: Unit 1	819,000	Metropolitan Edison Co.	1974

Table 10-2 Nuclear Power Reactors in the United States (continued)

Site	Plant name	Capacity (net kw)	Utility	Commercial operation
Goldsboro	Three Mile Island Nuclear Station: Unit 2	905,000	Jersey Central Power & Light Co.	1976
Berwick	Susquehanna Steam Electric Station: Unit 1	1,050,000	Pennsylvania Power and Light	1979
Berwick	Susquehanna Steam Electric Station: Unit 2	1,050,000	Pennsylvania Power and Light	1981
Fuller	Fulton Generating Station: Unit 1	1,140,000	Philadelphia Electric Co.	1981
Fuller	Fulton Generating Station: Unit 2	1,140,000	Philadelphia Electric Co.	1983
SOUTH CAROLINA				
Hartsville	H. B. Robinson S.E. Plant: Unit 2	700,000	Carolina Power & Light Co.	1971
Seneca	Oconee Nuclear Station: Unit 1	886,000	Duke Power Co.	1973
Seneca	Oconee Nuclear Station: Unit 2	886,000	Duke Power Co.	1973
Seneca	Oconee Nuclear Station: Unit 3	886,000	Duke Power Co.	1974
Broad River	Virgil C. Summer Nuclear Station: Unit 1	900,000	South Carolina Electric & Gas Co.	1978
Lake Wylie	Catawba Nuclear Station: Unit 1	1,153,000	Duke Power Co.	1979
Lake Wylie	Catawba Nuclear Station: Unit 2	1,153,000	Duke Power Co.	1980
Cherokee County	Cherokee Nuclear Station: Unit 1	1,280,000	Duke Power Co.	1982
Cherokee County	Cherokee Nuclear Station: Unit 2	1,280,000	Duke Power Co.	1983
Cherokee County	Cherokee Nuclear Station: Unit 3	1,280,000	Duke Power Co.	1984
TENNESSEE				
Daisy	Sequoyah Nuclear Power Plant: Unit 1	1,140,000	Tennessee Valley Authority	1975
Daisy	Sequoyah Nuclear Power Plant: Unit 2	1,140,000	Tennessee Valley Authority	1976
Spring City	Watts Bar Nuclear Plant: Unit 1	1,169,000	Tennessee Valley Authority	1978
Spring City	Watts Bar Nuclear Plant: Unit 2	1,169,000	Tennessee Valley Authority	1978
Oak Ridge	Clinch River Breeder Reactor Plant	350,000	U.S. Government	1980
TEXAS				
Glen Rose	Commanche Peak Steam Electric Station: Unit 1	1,150,000	Texas Utilities Services Inc.	1980
Glen Rose	Commanche Peak Steam Electric Station: Unit 2	1,150,000	Texas Utilities Services Inc.	1982
Jasper	Blue Hills: Unit 1	918,000	Gulf States Utilities	1980
Wallis	Allens Creek: Unit 1	1,150,000	Houston Lighting & Power Co.	1980
Wallis	Allens Creek: Unit 2	1,150,000	Houston Lighting & Power Co.	1982
Matagorda County	South Texas Project	1,250,000	Central Power & Light Co.	1980
Matagorda County	South Texas Project	1,250,000	Central Power & Light Co.	1982
VERMONT				
Vernon	Vermont Yankee Generating Station	513,900	Vermont Yankee Nuclear Power Corp.	1972

Table 10-2 Nuclear Power Reactors in the United States (continued)

Site	*Plant name*	*Capacity (net kw)*	*Utility*	*Commercial operation*
VIRGINIA				
Gravel Neck	Surry Power Station: Unit 1	788,000	Virginia Electric & Power Co.	1972
Gravel Neck	Surry Power Station: Unit 2	788,000	Virginia Electric & Power Co.	1973
Mineral	North Anna Power Station: Unit 1	898,000	Virginia Electric & Power Co.	1975
Mineral	North Anna Power Station: Unit 2	898,000	Virginia Electric & Power Co.	1976
Mineral	North Anna Power Station: Unit 3	907,000	Virginia Electric & Power Co.	1977
Mineral	North Anna Power Station: Unit 4	907,000	Virginia Electric & Power Co.	1978
Gravel Neck	Surry Power Station: Unit 3	882,000	Virginia Electric & Power Company	1980
Gravel Neck	Surry Power Station: Unit 4	882,000	Virginia Electric & Power Company	1981
WASHINGTON				
Richland	N-Reactor/WPPSS Steam	850,000	Atomic Energy Commission	1966
Richland	WPPSS No. 1	1,206,000	Washington Public Power Supply System	1980
Richland	WPPSS No. 2	1,103,000	Washington Public Power Supply System	1977
Satsop	WPPSS No. 3	1,232,000	Washington Public Power Supply System	1981
Sedro Woolley	Skagit Nuclear Project	1,200,000	Puget Sound Power & Light	1982
WISCONSIN				
Genoa	Genoa Nuclear Generating Station	50,000	Dairyland Power Cooperative	1971
Two Creeks	Point Beach Nuclear Plant: Unit 1	497,000	Wisconsin Michigan Power Co.	1970
Two Creeks	Point Beach Nuclear Plant: Unit 2	497,000	Wisconsin Michigan Power Co.	1972
Carlton	Kewaunee Nuclear Power Plant: Unit 1	541,000	Wisconsin Michigan Power Co.	1973
*	–	900,000	Wisconsin Michigan Power Co.	1980
*	–	900,000	Wisconsin Michigan Power Co.	1982
Durand	Tyrone Energy Park: Unit 1	1,150,000	Northern States Power Co.	1982
Durand	Tyrone Energy Park: Unit 2	1,150,000	Northern States Power Co.	1983
PUERTO RICO				
Puerto De Jobas	Aguirre Nuclear Power Plant	583,000	Puerto Rico Water Resources Authority	1979
*Site not selected.				
*	–	1,228,000	Tennessee Valley Authority	1980
*	–	1,228,000	Tennessee Valley Authority	1981
*	–	1,228,000	Tennessee Valley Authority	1980
*	–	1,228,000	Tennessee Valley Authority	1981

Thermal Breeders. There are two types of breeders proposed which could use thermal neutrons, the *light water breeder reactor,* (LWBR) and the *molten salt breeder reactor* (MSBR). The advantage of the LWBR is that it uses essentially the same technology as the two commercial reactors, the PWR and BWR, which are already in operation. What is added is a blanket of thorium fuel in which the new fuel, U^{233}, is generated.

In the LWBR as in the HTGR we have already described, the breeding ratio is less than 1 so, while it stretches out the fuel life, it does not double it.

A different type of thermal breeder receiving some consideration, is the MSBR. This reactor has a core of graphite tubes (to moderate the neutrons) through which the fuel, in the form of a molten salt, is circulated. There is a blanket of thorium fluoride, also a molten salt, surrounding the core. The salts, at a temperature of about 1200°F, carry heat away from the core and steam is produced at the high temperatures needed for high thermal efficiencies. This reactor has the added advantage that the fuel and fertile material, since they are in the form of a circulating liquid, can be continuously reprocessed, and fission products and new fuel removed. It is estimated that such a reactor could be built with a breeding ratio of 1.05 and a doubling time of about 20 years.

Apparently, the AEC considers this doubling time too long; it wants new fuel more quickly. All the emphasis, therefore, has been placed on the fast neutron breeder reactors.

The LMFBR

The *liquid metal fast breeder reactor* (LMFBR) has already established a record of uniqueness, even though its first major test is still in the future. It has been the subject of a presidential message, the subject of a lawsuit and was involved in one of the most frightening of the reactor accidents.

President Nixon in his June 4, 1971 message on energy, included a committment to a 1980 target date for successful demonstration of the LMFBR. In 1973 the Federal Court of Appeals upheld the suit by the Scientists Institute for Public Information requiring the AEC to provide an "environmental impact statement" on the entire fast neutron breeder program. In October 1966 the Enrico Fermi 60.9-mw reactor at Lagoona Beach, Michigan (on Lake Erie near Detroit), the first privately owned fast neutron breeder reactor in the United States, suffered a fuel melt down that kept it out of action until 1970. The breeder reactor has thus already made its presence known. Since this reactor has been officially designated as the new model for nuclear power, it's worthy of close review.

The LMFBR will breed fissionable Pu^{239} from U^{238} by the nuclear reaction described in Appendix 5. The neutrons which trigger this reaction are fast neutrons; therefore, the LMFBR has no moderator in its core. The core will be surrounded by a blanket region composed of bundles of rods fabricated from the uranium left over at the enrichment plant: uranium depleted in U^{235}, almost pure U^{238}. Pu^{239} will be produced, therefore, in the blanket as well as in the fuel rods.

The strong interest in this reactor is generated by its breeding ratio which may be as high as 1.4 or 1.5; three new fissionable nuclei are created for each two that are burned for fuel. Doubling times of 8 to 10 years are expected. Since the LMFBR can use the already existing "enrichment leftovers" as fertile material and generate its own Pu^{239} fuel, it can be brought into commercial operation without further mining of uranium.

The case for the LMFBR rests on this efficient use of fuel. The strength of this case is shown in Table 10-3 which compares the fuel and ore requirements for coal-fired plants, light water converters and the LMFBR. The advantages of the latter are obvious.

The LMFBR is so efficient in its use of uranium, using as much as 50 to 70 percent instead of the 2 percent or so of the LWBR, that the cost of the electricity it produces is not very sensitive to the cost of that uranium. Even if uranium goes to 100 dollars per ton it would have essentially no effect on prices of electricity produced by breeder reactors, while it would cause a 75 percent increase in the cost of electric energy from the converter reactors.

An even more striking possibility is that it can operate on the traces of uranium found in ordinary granite. With its uranium content made fissionable in a breeder, a chunk of granite has more energy potential than a chunk of coal. We can burn the mountains for fuel.

Problems of the LMFBR. In spite of this glowing prospectus, there are serious problems awaiting solution before the LMFBR can come on line as a producer of electric power. The problems start with the coolant. Since the fission neutrons must be unmoderated, water cannot be used. What is needed is a material which is relatively heavy (to minimize collision energy losses) with good heat transfer properties, and a high heat capacity. The current choice is molten sodium.

The LMFBR will have the major components shown in Fig. 10-6. The reactor core will be immersed in liquid sodium which, because of its high boiling point, can be heated to 1150°F. Since this sodium will be made highly radioactive by neutron bombardment, there will be a secondary sodium coolant loop between the reactor and the steam generator which will produce steam at about 1000°F, giving an anticipated thermal efficiency of 40 percent.

In addition to the problems posed by its radioactivity, sodium, as a coolant, has other drawbacks. It is opaque, and this makes it difficult for the operator to see into the reactor during refueling and other core maintenance operations. Sodium also has troublesome chemical properties; while it is compatible and noncorrosive with stainless steel, it reacts explosively, (in other words, it burns rapidly and instantly on ex-

Table 10-3 Fuel Requirements of 1,000-mw Generating Plant (0.8 capacity)*

Fuel needs (tons)	*Type*			
	Coal-fired	*LWBR*	*HTGR*	*LMFBR*
Fuel	2,000,000	160-200†	105	1.4
Ore	2,500,000	80,000-100,000	52,500	700
Loads (railroad cars)	25,000	3,100-3,800	2,000	27

*Source: Draft Environmental Impact Statement, *Liquid Metal Fast Breeder Reactor Program*, Vol. 4 (Washington, D.C.: U. S. Atomic Energy Commission, March 1974).

†The smaller figure is obtained by recycling the Pu^{239} formed in the fuel.

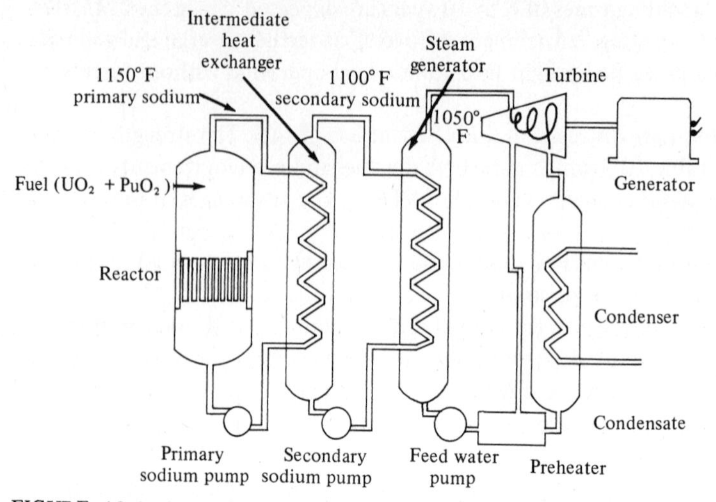

FIGURE 10-6 The LMFBR uses sodium as a coolant and has two separate sodium systems since the one in contact with the reactor core becomes highly radioactive.

posure to either air or water). Great care must be taken to insulate it from either of these materials.

While its boiling point is quite high, if it were to get hot enough to boil, the bubbles would have the opposite effect of the bubbles in the BWR reactor; now the increased numbers of fast neutrons would speed up the chain reaction (instead of slowing it down as in the BWR) and cause it to get even hotter.

At the other end of the temperature scale, at ordinary temperatures, the sodium becomes a solid which also makes maintenance difficult and poses problems for reactor start-up. The sodium is slow to begin its essential circulation and cooling. A host of minor design problems such as the need for new types of valves and pumps to handle this liquid metal await solution.

In the discussion of the control of reactors in Appendix 5 we comment on the stabilizing properties of the delayed neutrons. In the fast neutron breeders this feature is largely missing; only about 0.25 percent of the neutrons (as against 0.7 percent in U^{235} fission) are delayed in U^{233} or Pu^{239} fission. The control situation is thus much more delicate; there are not as many delayed neutrons to use for correcting mistakes. The fast neutrons add to the problem, while slow neutrons create another generation of fission events in 10^{-4} seconds, fast neutrons create the next round in 10^{-8} seconds. This is faster than even most electronic devices can respond. The LMFBR will be a tricky beast.

Meltdown at Enrico Fermi. The Enrico Fermi, sodium-cooled fast breeder reactor at Lagoona Beach, 18 miles from Detroit was the first privately owned fast breeder reactor. Its early history was one of controversy because of its new design, and its nearness to Detroit. It was run up to full power in January 1966, but shut down after a few months of operation for various troubles. In October of that year it was turned on again. There were minor problems with getting the sodium moving because of malfunctioning pumps, but it finally reached 10 percent of its maximum capacity of 200 mw at 3 PM on October 5. The operator moved the control rods out slightly further to increase the power and trouble began. Indicators reported that the core was too hot, radiation alarms rang, and an alert operator scrammed the reactor.

It took a long time to identify that several clogged cooling tubes caused the accident, and even longer (until 1970) to repair and restart it. There were nervous moments, in fact, nervous days. The reactor rods contained in addition to U^{235}, Pu^{239} created from U^{238} by the fast neutrons. Since some of the rods had melted, there was a fear that some of this potentially explosive material had collected together and might, if disturbed, get near criticality.

The accident was satisfactorily contained. No one was hurt. What is disturbing is not what happened, but what might have happened if the accident had occurred during full power operation at some later time, when much more Pu^{239} was present. The warning is, "Go slow"; fast breeders have problems not yet faced in the short history of reactor development.

Even within the community of breeder boosters, there are those who feel it unwise to put all the design effort into the LMFBR. Gas-cooled reactors, those using helium under 70 to 100 atmospheres of pressure, have backers. Because helium would absorb less neutrons than sodium, it could have a high breeding ratio and a doubling time of as low as five years. It is transparent, aiding maintenance, and, in addition, would not have the problems with bubbles which trouble sodium coolants. Its main disadvantage is that the helium coolant must be forced to circulate; a power failure would be disasterous. The molten salt breeder also has its adherents. But so far the AEC has put all its eggs in the LMFBR basket.

THE MIX OF THE BLESSING

Our description of the new entry on the list of energy convertors is now almost complete. We have described the two types of reactors. The first generation of converters which use the rare isotope U^{235} or artifically produced Pu^{239} as fuel, and the second generation of breeders, which will use the neutrons from fission to create fissionable Pu^{239} from U^{238}, or U^{233} from Th^{232}.

The proponents of nuclear energy are glowing in their predictions of the future: the converters and breeders will solve our energy problems and our air pollution problems at the same time. For the most part, we have looked only at the energy picture. Let us now quickly summarize the other advantages of nuclear reactors over fossil fuel plants, and then look behind the silver lining at some of the features that darken the cloud.

First the Good News

The major fuel advantages of nuclear reactors were summarized in Table 10-3. Nuclear reactors make accessible to us a new source of energy, uranium, at a time when there is growing concern over the adequacy of fossil fuel reserves. With successful commercial application of the breeder reactor, we will be able to count on a virtually unlimited energy source for several thousands of years.

The proponents of nuclear energy also make strong claims of greater economy of electric power production. Table 10-4 shows a comparison between a nuclear and fossil fuel plant developed by the Virginia Electric and Power Company. One sees that the higher capital (investment) costs of the nuclear plant are generally more than compensated for by the lower fuel cost.

It is predicted that the cost advantage of nuclear generated energy will increase greatly with the introduction of fast breeders. Not only do they produce fuel for themselves, and thus achieve high efficiencies of fuel use, but they are also expected to produce enough Pu^{239} to take over the fueling of the LWBRs by 1990, reducing their fuel costs dramatically. Figures 10-7*a* and 10-7*b* give some rough comparisons of the costs of electric energy (with and without breeder development), with projected costs from fossil fuel generated electric energy (shaded area). The enthusiasm for the breeder is easily understood.

The cost argument is a complex one. Opponents of nuclear energy point out that there is, for instance, no account taken of the federal subsidy of reactors in the form of low rates for fuel enrichment and waste storage, or of the huge subsidy through the Price-Anderson Act which backs up the insurance for reactors. (We will discuss this insurance problem a bit later.) There is also the question of how much of the development costs should have been assigned to the industry. Whatever the arguments, the fact is that reactor-produced electricity is competitive with fossil fuel-produced electricity. The advent of the breeder on the one hand, and the rising cost of coal and of pollution controls on coal-fired plants on the other hand, will probably make nuclear-generated electricity even more competitive in the future.

Table 10-4 Comparison of Cost of Electricity from Coal and Pressurized Water Reactor: two 940,000-kw Generating Units Operational in 1974 to 1975 near Fredericksburg, Virginia*

Costs	*Coal*	*Nuclear*
Unit investment cost of plant, dollars/kw C	202†	255
Annual capital charge rate per year i	0.13	0.13
Kilowatt-hours generated per year per kilowatt capacity k	5,256‡	5,256‡
Heat rate, million Btu/kw-hr h	0.009	0.0104
Cost of heat from fuel, cents/million Btu f	45	18
Cost of electricity, mills/kw-hr		
Plant investment, 1,000 Ci/k	5.00	6.31
Operation and maintenance	0.30	0.38
Fuel, 10 hf	4.05	1.87
Total	9.35	8.56
Breakeven cost of heat from coal, cents/million Btu	36.2	

*Source: Benedict, "Electric Power from Nuclear Fission," *Bulletin of the Atomic Scientist*, p. 11, September 1971.

†With no allowance for sulphur dioxide removal.

‡In actual system, coal plant would generate less electricity than nuclear plant.

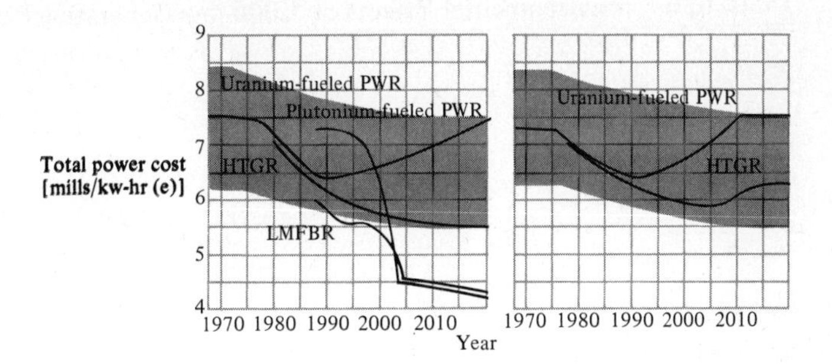

FIGURE 10-7 (a)The effect of the breeder-produced fission fuel lowers the price of electricity from the PWR reactors as well as that from the LMFBR. (b) Without the breeder the cost of electricity generated by the PWR and the gas-cooled reactor HTGR will rise, it is predicted, after an initial drop, as uranium becomes scarce. (Source: Culler and Harms, "Energy from Breeder Reactors," *Physics Today,* 35, May 1972.

Pollution

Nuclear reactors break even, at least numerically, on the pollution inventory. They will decrease air and land pollution but increase thermal and radiation pollution. The pollutant comparisons for light water reactors (LWBRs and HTGRs), the LMFBR, and coal-fired plants, all of 1,000-mw capacity, is shown in Table 10-5. The most dramatic environmental advantages of the reactors are: (1) in the category of effluents, where the reactors are assigned 1,200 to 1,500 tons per year as against the million tons of air pollutants coming from coal-fired plants; and (2) in land use, where the 300 to 400 acres of space which is needed for a coal-fired plant is compared with 70 to 140 acres for a nuclear plant and the 200 acres per year which must be stripped for the coal is compared with 13 acres of uranium strip mining for the LWBR and much less than 1 acre for the LMFBR. The cooling water needs are about even for the LMFBR and coal plants and 40 to 50 percent higher for the LWBR.

We have already seen in Chap. 7 that the effect of pollutants are not measurable by weight or volume. It is the interaction with the things man cares about that determine the extent of damage. In the case of radioactive wastes, the target for damage is man and his generations.

Radioactive Pollutants. Just as nuclear reactors bring us a new source of energy, they also bring us a new form of pollution: *radioactive pollution*. Appendix 5 provides a detailed definition of what we mean by radioactivity, and identifies the fission products as the major contributors to this form of pollution. We review this discussion briefly.

To say that a nucleus is radioactive is to say that it is unstable, that it has an excess of energy, charge and/or mass. It will get rid of this excess by ejecting an energetic particle. The most common particle from the fission-produced pollutants is the *beta particle* (β), an energetic electron. It is because these β's can penetrate deeply into matter, including human tissue, that they are feared.

The main source of the radioactive pollution from reactors is the fission reaction itself. The two pieces which remain after the nucleus is split (the fission products)

Table 10-5 Environmental Effects of 1,000-mw Generating Plant (0.8 capacity)

TYPE:	*Coal-Fired*	*LWBR*	*HTGR*	*LMFBR*
THERMAL				
Btu/sec to be dissipated				
Water	1.12×10^6	1.78×10^6	1.31×10^6	1.20×10^6
Air	0.37×10^6	0.15×10^6	0.12×10^6	0.11×10^6
Total	1.49×10^6	1.93×10^6	1.43×10^6	1.31×10^6
EFFLUENT				
Radioactivity 10^3 curies/yr	—	2-253*	2	2
Air pollution (tons/yr)				
SO_2	45,000	1,500†	1,200†	—
NO_x	26,000	900	700	—
CO	750	25	20	—
Particulates	3,500	120	95	—
HC	260	9	7	—
WASTES (10^3 ft^3/yr)				
Radioactive	—	12	10	8
Ashes	200	7†	5†	—
Limestone‡	300	10	8	—
LAND				
Acres mined	200	13	9	0.05
Plant sites (acres)	300-400	←	70-140	→

*The higher figure is for the BWR.
†The emissions charged to the reactors are computed from the electric energy used in enrichment.
‡Limestone is used in SO_2 removal. The charge made to the reactors is from enrichment.

are in a very unstable isotopic form. Generally they have an excess of neutrons in their nuclei. To get more balance, the neutrons change into protons by emitting β's.

The most important property of these β emitters is their *half-life*, the time needed for half of a given amount of a radioactive material to undergo a transformation, to *decay*, in nuclear jargon. As shown in Appendix 5, the half-life is inversely proportional to the rate at which the radioactive material decays. Thus, nuclei with short half-lives emit intense radiation, and vice versa.

There are hundreds of different isotopes formed in the various stages of fission and fission-product decay. The short-lived ones are extremely intense and, consequently, very dangerous, but a short confinement allows them to change into isotopes of longer half-life and lower radiation intensity. The very long-lived ones are correspondingly weak emitters and, therefore, not as troublesome. The most dangerous ones to man are those with intermediate half-life. Two of these latter are strontium (Sr^{90}) and cesium (Cs^{137}), which have half-lives of 28 and 30 years, respectively—about the "half-life" of a man. Therefore, it takes about 30 years to reduce these radioactive materials to half their original amount, 60 years to one quarter, 90 years to one eighth, and so on. They typify the storage problem which is faced: we must take care of these wastes for a few hundred years.

Radioactive Emissions from Reactors. The radioactive pollution from a reactor is present in all three forms: gaseous, liquid and solid. The radioactivity in the water comes, for the most part, from impurities produced by corrosion of the pipes and valves which are then made radioactive in passing through the core. Some fission products diffuse through the fuel rod cladding or escape through breaks in it. Radioactivity from the latter is reduced by the physical properties of the uranium dioxide. It is in the form of a sintered pellet, almost a ceramic, and retains most of the fission products in place, even without the cladding. The most troublesome materials are the the radioactive gases which do escape. As we have already mentioned, tritium, a radioactive isotope of hydrogen, is formed by neutron bombardment of boron in the water.

A large quantity of liquid radioactive waste is accumulated during maintenance and refueling operations. The circulating water is continually filtered to remove radioactive materials. Most of the radioactive pollutants are removed in this fashion and are then kept for a while in holding tanks to let the short-lived materials decay. The remaining water can also be held a while and is then released slowly into the cooling water of the steam turbine. It is thus mixed with the hundreds of thousand of gallons per minute of this water which is returned to the river or lake serving the reactor. The concentration of radioactive material released into the environment is regulated. We shall discuss the biological effects of this low-level radiation in a following section.

Gaseous radioactive release is only important for the BWR, in which the steam is in contact with the core and therefore carries with it some of the gaseous fission products. In a BWR the gases that pass through the turbine are stored for a short time, filtered, and released from the stack. The two pollutants which are of most concern in the gaseous waste are radioactive krypton (Kr^{85}), a *noble* gas (that is, one which is chemically inert) and tritium. Kr^{85} has a half-life of 10.7 years and tritium 12 years.

Because of these relatively long lifetimes, these materials will accumulate in the environment. It is predicted that by the year 2050 Kr^{85}, if uncontrolled, will expose man to as much additional radiation as he now receives from the natural background dose. The gaseous wastes, like the liquid ones, are carefully and continuously monitored.

An Added Burden

One can summarize the preceding by saying that while most of the radioactive residue of the fission process remains under the control of the nuclear plant (and forms the radioactive waste which we discuss in the following section), some radioactive gases are emitted from the discharge stacks, and some radioactive liquid is discharged into the cooling water source. Radioactivity is thus an environmental pollutant. In the next few paragraphs we will shape answers to such question as: How much is released? What are its effects on humans?

With these questions we approach an arena of active and acrimonious controversy which we will label "the low-level radiation controversy": How many deaths from cancer and leukemia will nuclear reactor releases cause? It is not our intent to enter that controversy. In Appendix 5 some of the background of the controversy is discussed and we show how the predictions of the number of cancer deaths and so on, are made. The task to be undertaken in this chapter is to discuss the application of radiation standards to these emissions and to try to give perspective to the added biological burden this radioactive pollutant brings to our species.

The difficulties we encounter in making this assessment are similar to those which plagued us in discussing air pollution. While we can identify levels at which lethal and near-lethal effects occur, there is great uncertainty and meager data for low-level exposure, for radiation doses at the same level as the natural radiation background to which we are all exposed. Again paralleling the air pollution case, there is also a lack of understanding of the exact cause-and-effect mechanisms by which radiation produces its variety of harmful effects at these low levels.

The AEC sets standards for the amounts of radioactive pollutants which can be released into the environment. They largely rely on the recommendations of an international body, the International Commission on Radiological Protection (ICRP) for the scientific justification of these standards. This group of radiation scientists periodically meets to consider the developing body of knowledge on radiation effects and to suggest the levels which should not be exceeded, the maximum permissible doses (MPD), for those who are occupationally exposed to radiation and for the general population. They make these recommendations from a "risk-benefit" philosophy; they set the standards as low as they can without making it impossible to use radioactive materials and sources in ways that benefit humankind. They also assume that no "threshold" for radiation damage exists, that is, that all doses of radiation, no matter how small, have a probability of doing harm.

Radiation has two very different types of effects on humans. There are *somatic effects* which are confined to the exposed individual and show themselves as cancers and leukemias (and other even less specific damages such as life-shortening), and there are *genetic effects,* which may not show up until several generations have passed. The ICRP recommendations in these two areas are summarized in Table 10-6.

The unit in which exposure is measured, the *roentgen equivalent man* (rem) is one which takes into account some of the variation of the biological effects of different types of radiation (see Appendix 5). The intent of these standards is to set the yearly exposure low enough to keep cancer induction at a minimum, and also low enough to put a limit on the accumulated radiation exposure, thus providing as much protection to the gene pool of the population as is consistent with the risk-benefit analysis.

The 30-year population exposure standard works out to an average of 170 mrem (millirems, or thousandths of a rem) per year. The AEC requires that plants under its jurisdiction operate so as to keep exposure to the general population below the levels indicated in Table 10-6; in fact, the present emissions are well below these levels by perhaps a factor of 100, but surely a factor of 10.

Much of the controversy that surrounds the AECs interpretation of the ICRP recommendations comes from the fact that radiation from the activities it is involved in is not the only radiation to which the population is exposed. A summary of the average population exposures from other sources is presented in Table A5-3. In addition to the natural background, we receive, on the average, 55 mrem per year from diagnostic and therapeutic X-rays (with much higher exposures to certain organs), 3

Table 10-6 ICRP Recommendations

Type of exposure	*Amount*
Individual exposures, occupational	5 rem/yr
Individual exposures, general population	0.5 rem/yr
Total exposure, general population	5 rem/30 yr

mrem per year from fallout from past tests of nuclear weapons, and 0.4 mrem from other occupational sources. The present added burden from nuclear power operations is estimated to be less than 1 mrem per year. This last concentration will increase as the number of nuclear reactors increase.

Karl F. Morgan, Director of the Health Physics Division of the Oak Ridge National Laboratory, has made the reasonable suggestion that the 170 mrem per year be apportioned out to the various activities that produce a general population exposure, in other words, that the nuclear power industry be permitted only a share of the 170 mrem total. His recommendations are contained in Table 10-7. He would, for instance, be willing to allow the nuclear industry a total of 67 mrem from internal and external doses and reserve 50 mrem for the medical X-rays. The values within the parentheses are indications of those he believes are attainable if reasonable care is exercised; he is particularly anxious to see the average exposure from medical X-rays reduced.

The approach Dr. Morgan suggests seems sound, it is consistent with the risk-benefit approach. If we want nuclear power, then we have to make a place for it in that part of our budget of radiation exposure which we control. The addition of radiation from this source will cause deaths and genetic damage. How many deaths and how much damage we don't and won't know: they will be concealed in the statistics of all the other similar deaths. We can and should estimate their numbers (see Appendix

Table 10-7 Suggested Levels of Permissible Genetic and/or Total Body Exposure of the United States Population to All Man-made Sources of Ionizing Radiation that Might be Adopted by the Federal Radiation Council for Application in the United States*

500 mrem/yr to any critical segment of the population
5000 mrem/30 yr as the maximum

170 mrem/yr as an average as follows	*mrem/yr*
Internal dose from nuclear power operations	50 (5)‡
Internal dose from other industrial operations	30 (3)
External dose from nuclear power operations	17 (2)
External dose from other industrial operations	15 (2)
Medical (diagnostic)	40 (5)
Medical (therapeutic)	10 (5)
Occupational§ contribution to population dose	4 (0.1)
Weapons Fallout	2
Miscellaneous (watches, television, high voltage switches, etc.)	2 (0.1)
Total	~ 170 (22)

*Source: K. Z. Morgan, "Acceptable Risk Concepts," reprinted in *Environmental Effects of Moduling Electric Power,* (Washington, D.C.: Joint Committee on Atomic Energy, USGPO, 1970), p. 129.

†This includes medical exposure but excludes natural background. It includes also exposure from natural radioactive sources such as uranium, thorium, ^{226}Ra, ^{228}Ra, ^{210}Pb, etc. that have been concentrated by man and includes exposure in uranium or thorium mining operations.

‡The values in parentheses are those which Morgan believes we can attain. Since this is a long-range extrapolation some of those values are larger and others are smaller than the present estimates of population doses from the source.

§Exclusive of the nuclear power operations.

5) for it is the general public, in the end, that must do the risk-benefit calculus, that must ask, "What are we buying with our unknown dead?"

The answer is, of course, electrical power, the same answer that justified the land wasted by strip mining, the smoke above Four Corners, the dead, injured, and diseased coal miners, and even the 1,000 people per year electrocuted in power-line accidents. It is all part of the same choice. For each new addition to our generating capacity we must ask, "What will that power cost?" and eventually, "Do we need it?"

"Hot" Leftovers

In the preceding excursion into the low-level radiation controversy, we tried to assess the effects of the small amounts of radioactive effluents released into the environment. We turn our attention now to the enormously greater amount of radioactive material that must be rigorously isolated from our environment. This is the so-called *radioactive waste.*

Most of the waste materials produced by a reactor leave it in a solid form. A small part of this is the material removed by the filtration of the water and gases. This radioactive sludge is concentrated by evaporation of the water, put in drums, mixed with concrete, and shipped to AEC approved storage sites.

The most troublesome solid wastes come, not from the reactor, but from the fuel reprocessing plants. After a year or so of operation, the fuel rods reach a point where their usefulness is so diminished that they must be replaced by fresh ones. The plant is shut down. The rods are removed and stored under water for several months to get rid of the short-lived and intensively radioactive materials. They are then placed in special containers which are heavy and strong, and shipped by truck or rail to the fuel-processing plant.

This shipment is one of the sensitive points in the system. These used-fuel elements are very radioactive, and, if spilled by a truck accident or train derailment, could contaminate a large area. It would be particularly catastrophic if such a container were to be broken and spilled into a water supply. The containers are built to stand quite a wreck and have remained unspilled by a train derailment. While the shipment of this radioactive material is not a cause for great concern at the present, although there is already criticism of carelessness in handling such materials, future needs begin to boggle the mind. AEC estimates that one 1,000-mw LWBR requires 80 fuel shipments and 65 low-level waste truck shipments per year, while for a similar breeder reactor, the respective numbers are 28 and 61. At the generation levels of the year 2000, 400,000-mw LMFBR, 500,000-mw LWBR, and 100,000-mw HTGR the total number of truck shipments per year will total about 54,000 for fuels and 58,000 for low-level wastes. By these same estimates the 1974 total was around 2,000 shipments for fuels and 1,625 for wastes.[6] One of the attractive features of the consolidation of generating and reprocessing plants in huge *nuclear parks* would be the reduction of this enormous inventory of radioactivity on our highways.

The spent fuel rods must, of course, be reprocessed; there is much valuable material in them. Most of the U^{235} is there, for as we calculated earlier, only 5 to 10

[6] Draft Environmental Impact Statement, *Liquid Metal Fast Breeder Reactor Program*, Vol. III, (Washington, D.C.: AEC, March 1974).

percent of it is actually used up by fission in a year's normal operation. The U^{238} is also worth recovering, as is the Pu^{239} generated by neutron bombardment of the U^{238}. (In the past this Pu^{239} was sold to the AEC as bomb material, but this practice has been discontinued.) Pu^{239} will continue to have a market as a reactor fuel.

Much of the fuel reprocessing is presently performed at AEC plants; there is only one privately owned and operated plant, located at West Valley, near Buffalo, New York. As nuclear plants proliferate and as the breeder reactors add greatly to the reprocessing needs (in other words, the Pu^{239} or U^{233} will have to be separated from the blanket materials) the need for more reprocessing plants will surely grow. These plants will, of course, contribute to the radioactive emissions into air and water that we have just discussed.

At the plant the solid material is dissolved by acid and the usable fuel materials are chemically separated from the hot mess of fission products. The latter are further separated chemically and some of the short-lived materials are sorted out and allowed to decay. What remains, the so-called high-level wastes, are the real problem pollutants from nuclear reactors. They contain the fairly long-lived β emitters such as the Sr^{90} and Cs^{137} which we have mentioned. They are both hot in the radioactive sense and in the physical sense, a real boiling devils' brew. With their half-lives of 10 to 100 years, this material must be stored somewhere for hundreds of years in complete isolation from the environment.

High-level waste storage is the responsibility of the AEC. There are already over 100 million gallons of it (mostly produced by the production of Pu^{239} from U^{238} for bombs) stored in huge underground tanks at the Hanford, Washington AEC site and the Savannah River plant in South Carolina.

These tanks themselves are a source of controversy. Although they are strongly constructed of stainless steel, they will not last forever. There have been several incidents of leakage, the last reported one in the summer of 1973, when one tank at Hanford leaked at a rate of 2,500 gallons a day for several weeks, losing a total of 115,000 gallons of waste before the leak was detected. So far as we know, there has not been serious contamination of the nearby Columbia River.

A Dangerous Legacy. Because of their long lifetimes, we are leaving these hot by-products of electric power generation as a legacy to our next generation and the ones beyond that. A listing of some of the stuff of this legacy and their half-lives is provided in Appendix 5, Table A5-1. Since there are several half-lives of tens of years, some agency of the government, for at least a few hundred years, must know where this material is, monitor it, and protect it. It is too dangerous to allow it to get lost and later to be dug into and built on. It must also be protected from theft or sabotage by a fanatic group.

This legacy is a rapidly growing one. We have, after all, only begun to turn to nuclear reactors for our electricity. The 25,000-mw nuclear generating capacity in 1974 is expected to be 280,000 mw in 1985, and 1.2 M mw in the year 2000.

There is already some promise of help from technology in at least reducing the volume of hot waste which we will have to contend with. Techniques of converting the liquid high-level wastes to a solid by evaporation at high temperature have been developed. With this technique 100 gallons of the liquid can be reduced to 1 cubic foot of the solid. In this form, the high-level waste output from a 1,000-mw reactor will be about 60 cubic feet per year. The "other than high-level waste" for which conventional burial is possible comes to about 2,000 cubic feet per year. Figure 10-8 shows the yearly production of these two types of waste between now and the year 2000. By that year the total amount accumulated is estimated to be 730,000 cubic feet[7] of high-level waste, and 330 million cubic feet of the less radioactive debris.

Even though conversion to a solid greatly reduces the space required, safe, dry space must still be found. It has been proposed that these "hot rocks" be stored in abandoned salt mines. Salt mines are appropriate because they are, and have been for hundreds of millions of years, dry. Salt will also melt from the physical heat of the radioactive materials, flow around and seal the waste, and will heal itself if it is cracked by earth tremors, for example.

Salt mines may turn out to be the final resting place for these radioactive leftovers. Unfortunately, the AEC chose, for its first demonstration, a mine in Lyons, Kansas which was hilariously unsuited to the job. Not only had oil exploration and other activities left many surface holes through which water could enter the mine, but it was surrounded by other cavities which previous salt mining had filled with water; in fact, it has been reported, that 180 thousand gallons of water disappeared during one operation. The Lyons project has been abandoned, but other more suitable salt formations are being studied.

Although the techniques to solidify the wastes are being developed, the AEC remains uncertain about where to put it and is now talking about aboveground storage at one of the old bomb testing stations. Whatever the final resting place, it will become a growing monument to our nuclear decision.

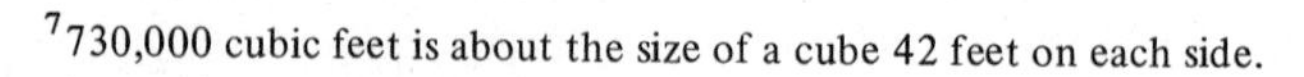
[7]730,000 cubic feet is about the size of a cube 42 feet on each side.

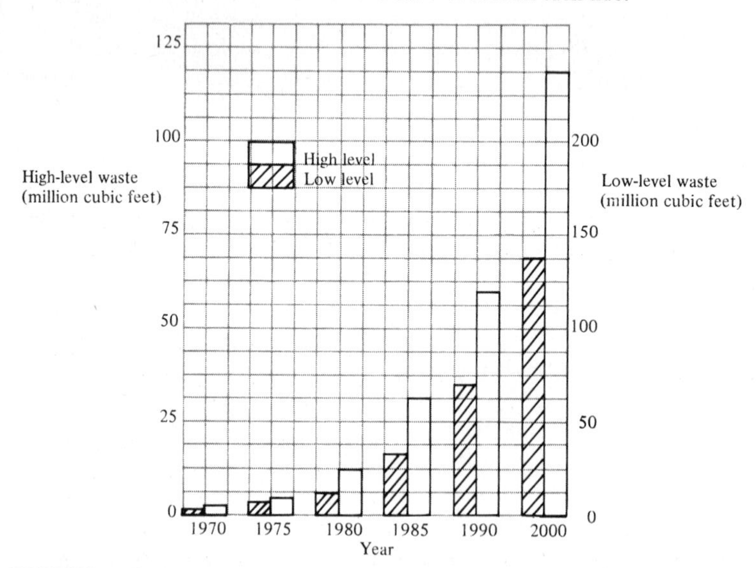

FIGURE 10-8 The increasing reliance on nuclear power will result in an expotentially growing stockpile of high-level radioactive waste. The figures shown in the figure anticipate "flash drying" of the originally liquid waste.

The Plutonium Problem

The switch to breeder reactors, which is scheduled to begin in the 1980s, will complicate the waste problem in several ways. The breeder economy is based on the production of Pu^{239}. This material is completely man-made; its half-life of about 24,000 years, while long on the scale of human history is short on the earth's scale and any which might have existed in the beginning, has long since decayed to a stable element. It has so far been produced in relatively small quantities for nuclear bomb construction. With the advent of the breeder, however, it will become much more common, with commercial production of 30 tons per year projected by 1980 and more than 100 tons a year by the year 2000.

Pu^{239} is one of the most dangerous of the radioactive elements. It is an alpha (α) emitter: it emits a short-ranged but energetic particle which can do intensive damage in the four hundredths of a millimeter, 4 to 10 times a cell dimension, which it travels in human tissue. The radiation from plutonium does little damage on the outside of the body, but inside it is a different story. The main area of concern is the lungs.

In Chap. 7 we summarized the defense mechanisms of the respiratory system and pointed out that airborne particles of a certain size can be carried into the lungs, while larger or smaller particles are excluded. It is unfortunate that Pu^{239}, in the form of plutonium dioxide, forms aerosols of that intermediate size. The danger from Pu^{239} is, therefore, that a small but intensively radioactive particle, a *hot particle,* will be deposited on sensitive lung tissue.

In the previous section we emphasized the uncertainties associated with any assessment of the damage from low-level radiation. These uncertainties are magnified in the case of Pu^{239}; there is very little experimental evidence bearing on the cancer-inducing potential of Pu^{239}, hot particles, in the lungs. In the first place, Pu^{239} has only been around for 25 years and reasonable medical statistics on plutonium exposure in the lungs go back only to about 1964. Since most cancers take 15 years, more or less, to develop, we must wait for guidance from this source.[8] In the absence of knowledge, caution is the best substitute.

With the present committment to breeder reactors gathering momentum, recognition of and preparation for eventual problems with this pollutant are imperative. The 1980s may already be too late. There are some who suggest turning to thorium-fueled breeders: U^{233} is not as potentially toxic as Pu^{239}. All the emphasis so far seems to be behind the LMFBR.

Clandestine Bomb-Making. There is a second very different but equally troublesome potential in the accumulation of Pu^{239}: nuclear bombs can be constructed from it. They can be constructed from U^{235} also, and have been. A nuclear bomb, however, needs high purity fissile material, and it is, as we have said, very difficult and expensive to separate the isotopes U^{238} and U^{235}. Only the wealthiest and most technolog-

[8]There is now some preliminary evidence that the lung cancer from smoking is caused by this kind of localization of radioactive contamination of tobacco. "Radioactivity of Tobacco Trichomes and Insoluble Cigarette Smoke Particles," Edward A. Martell, *Nature* **249**: 215-217, May 17, 1974.

ically advanced countries have been able to afford that luxury. Pu^{239}, however, is produced in a matrix of U^{238} and these two are chemically different so their separation is much less difficult. As we turn to nuclear reactors, in particular to breeders, the world's stockpile of bomb-grade Pu^{239} will grow.

A 1,000-mw LMFBR reactor produces about 1 kilogram of Pu^{239} a day or perhaps 300 kilograms (660 pounds) a year. Since it takes only 6 to 10 kilograms (13 to 22 pounds) of Pu^{239} to make a bomb, the danger of proliferation of these weapons is high.

The sensitive points in the system are the fuel-reprocessing plants and fuel transportation. It will be difficult, as the number of fuel-reprocessing plants and the amount of Pu^{239} processed grows, to assure that some is not stolen and sold "under the counter" to an ambitious and ruthless leader of some nonnuclear nation. As has been demonstrated, it is difficult to prevent truck hijackings. The price of Pu^{239} of about 10,000 dollars per kilogram adds to the temptation. The more than occasional success of undertakings by terrorist and other fanatic groups adds to our worries.

The world now exists at such a delicate balance that one can only view with despair any increase in the number of countries with nuclear weapons. It will be a great challenge to the AEC and the International Atomic Energy Agency (IAEA) to develop procedures of inspection and control to prevent the diversion of Pu^{239} from electric power production, to political power production.

Reactor Accidents

In this chapter we have emphasized the potential that the nuclear reactor has for taking the "energy load" off of our diminishing supplies of fossil fuels. We have also described the dangers that accompany the production of this energy. We have saved until the end the danger that is most catastrophic in its consequences and least likely on a probable basis: the danger of a reactor accident.

After a year of operation a nuclear reactor has in its core as much radioactive fission products as are released in 1,000 fission bomb explosions. The release of these materials over a countryside would be catastrophic. We must, however, quickly point out that a reactor is not a potential nuclear bomb. In the LWBR the U^{235} is in such a low concentration that no meltdown could conceivably cause it to collect together in a large enough and pure enough chunk to "go critical" and explode. While this statement can be made with certainty, it appears, for conventional LWBRs, it is not yet clear that one can be as certain of the safety in this respect of some of the newer LWBRs, in which some fuel rods are enriched to 80 or 90 percent (instead of 3 percent), or of the safety of the breeders with their large inventories of Pu^{239}. Even in these reactors, however, one is certain that a Hiroshima is not in the making. The danger, if it is present, is of a nuclear fizzle, small on the nuclear bomb scale but still the equivalent of a few tons of trinitrotoluene (TNT). Such an explosion, if it occurred, could spread radioactivity over an area of hundreds of square miles.

The United States reactor developers have taken great pains to avoid the possibility of an accident. Reactor safety is provided for at several different stages. It is designed into the fuel and the core, taking advantage of natural effects such as the reduction of the neutron flux by the boiling of the water moderator, and the decrease of the reactivity between fast neutrons and fissile materials as the temperature increases, the so-called *Doppler broadening.* Safety is engineered into the reactor by providing safety devices and backup safety devices, by trying to anticipate and then design safety systems for any conceivable malfunction. Safety is also constructed into the reactor; the stainless steel pressure vessel around the core will withstand a small explosion, and the entire core is surrounded by a concrete containment structure able to withstand even a small nuclear fizzle. Finally, at least to date, the reactors are required to be built at some safe distance from any large population center. This last requirement, under the pressure of energy needs, may be the first to be abandoned.

The record of the nuclear power industry is very good: there have been no deaths from accidents in commercial reactors. But there have been only about 100 reactor years of operation to date. Given the dreadful consequences of even one accident we are understandably interested in knowing what is the probability and consequence of a serious accident.

The accident most feared is a loss of *coolant,* a double break in the large pipes (24 to 28 inches in diameter with a wall thickness of 1 inch of stainless steel) which carry the cooling water to and from the reactor core. This is an unlikely event, but an earthquake, sabotage, or carelessness could cause it. (This type of accident is essentially what occurred at the Fermi plant mentioned earlier.) If this were to happen, the water would be blown out of the reactor by the steam pressure. While the loss of the moderator (and the automatic scramming) would shut off the reactor, the core, because of the high levels of radioactivity in it, would quickly heat up. Without cooling water, the following scenario is hypothesized by Ralph Lapp.

5 to 10 seconds	Core temperature jumps from 600 to 1500°F in localized sectors.
30 to 50 seconds	Temperature exceeds 2000°F; core structure begins steam reaction and energy release.
50 to 100 seconds	Core attains temperature of 3360°F.
2 minutes	Core collapse begins.
10 minutes	Melt-down debris accumulates in vessel.
1 hour	Probable melt-through of pressure vessel with possible steam explosions.
1 day	Molten mass of reactor material breaks through containment slab.
3 years	Molten material forms 100-foot glob in sand.
11 years	Cool-off of residual material shrinks to 80-foot diameter.[9]

The happening at the "1-day" mark is called the *Chinese Syndrome* since the molten mass, if it breaks through the containment slab, melts its way into the earth, China bound.

The important point made by this scenario is that the emergency cooling system, which is required of all reactors, has only a few seconds to operate: the first 5 to 10 seconds. After that the core cannot be cooled; it is too hot and beginning to melt.

[9]Ralph E. Lapp, "A Citizens Guide to Nuclear Power," *A New Republic Pamphlet,* (Washington, D.C.: *New Republic,* 1971).

This emergency system has been a source of recent controversy. There has been considerable doubt cast on present techniques by some mock-up studies and new studies and development efforts are under way.

There are many arguments and counterarguments in all aspects of reactor safety and we will not take the time to retell them. The possibility of an accident similar to the one described is certainly low, probably less than 1 in 10^4 reactor years, perhaps less than 1 in 10^6 reactor years. But the risk is there and, as we have said, the breeders will present new and, perhaps, greater risks. When we have 100 or so reactors operating in the 1980s or 500 operating by 2000, one accident in 10,000 reactor years no longer seems such good odds.

There is, in fact, one glaring inconsistency in the protestations of guaranteed safety that one receives from the reactor industry. They have not yet been able to convince insurance companies of this. The controversial study of reactor safety released in 1957 and referred to as WASH-740[9], estimated that the maximum consequences of an accident might be property damage up to 7 billion dollars and 3,000 deaths. The private insurance companies were unwilling to take on this risk, and, under the provisions of the Price-Anderson Act, the Federal Government now provides a subsidy in the form of insurance coverage up to 500 million dollars. Private companies pick up the first 75 million dollars. The AEC and the reactor industry have strongly resisted any effort to repeal this act.

The AEC has essentially rejected the WASH-740 study and report. A major new study of safety is under way, at Massachusetts Institute of Technology at this writing, and its results are expected soon. It is doubtful, however, if the odds will ever be set low enough to satisfy the most cautious. One suggestion worthy of serious consideration is to put reactors underground, to anticipate and take advantage of the Chinese Syndrome, so to speak. Estimates of the costs of such a reactor show that, while it would add millions of dollars to capital costs, it would add only mills to the cost of the electric energy produced. There is no indication as yet that the AEC or the utilities have any plans to consider this idea.

SUMMARY

In this chapter we have tried to describe the potential and the danger of the decision of turning to the heavy nuclei for our needed energy. In this turn we are making a gamble for high stakes, and we are gambling not only for ourselves but also for our future generations. In earlier chapters we have summarized some of the demand for energy that drove us to the gamble; in the chapter which follows we will look at our shrinking resource reserve which has also contributed. The next ten years will be reactor-building ones. Whether the 1980s will belong to the breeders is not yet clear, but the nuclear genie is out of the lamp. We must wait a while to find out whether he has been well-trained.

[10] "Theoretical Possibilities and Consequences of Major Accidents in Large Nuclear Power Plants," (Washington, D.C.: AEC, 1957).

11

Can We Keep the Pace?

Most of the curves of energy production and consumption that we have shown in this book have had a common characteristic; when plotted on a graph whose vertical scale is logarithmic, that is, one which ascends in powers of ten rather than linearily, they rise upward more or less as a straight line. Perhaps the most striking examples are provided by Figs. 5-2, 5-9, and 5-10 of Chap. 5, and Fig. 6-12 of Chap. 6. These curves have long sections of straight line growth.

Curves that rise as straight lines when the vertical scale is logarithmic are the exponential curves which we introduced in Chap. 5 and Appendix 1.

The exponential curve is a mark of our times. Not only is energy consumption rising roughly in this fashion, but so are television sales (and hours watched, aspirin sales, the number of books in print, and many other quantities).[1]

"NATURE ABHORS AN EXPONENTIAL"

Faced with the fact of so many exponentially increasing quantities, one tends to forget that exponential growth is not really a property of natural systems. Such growth may exist naturally for a while; it exists in a colony of bacteria put in a food-rich environment, it exists in chain-reacting uranium. But it does not continue for very long: the food runs out and the environment turns hostile for the bacteria, the control system takes over and slows down the chain reaction or, lacking control, the uranium mass blows apart.

A more familiar example of exponential growth is shown in Fig. 11-1 where the average weight of a male human is plotted on a semilog graph for the first 22 years of life. There are two periods of exponential growth (regions in which the curve is a straight line); from 0 to 6 months and from 5 to 10 years fit this pattern. We see, however, that as the boy approaches adulthood, the curve begins to turn over and flatten out; the growth has slowed down and, at least it is the hope of most of us, has become static.

The impossibility of continued exponential growth becomes even clearer in terms of the concept of a doubling time which accompanies an exponential. Nothing can keep on doubling forever and each doubling in growth doubles also the pressures from the surroundings that operate to limit growth.

[1] An interesting summary and commentary on this trend is given in Ralph Lapp, *The Logarithmic Century*, (Englewood Cliffs, New Jersey: Prentice-Hall, 1973).

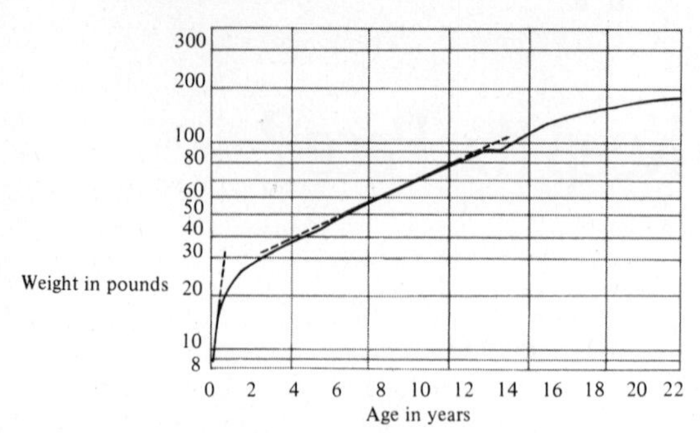

FIGURE 11-1 The weight of a (male) child shows two periods of exponential growth, from 0-6 months with a doubling time 4 months and from 5-10 years with a doubling time of about 6 years. The curve begins to level out after 18 years.

One of the most common laments about the Energy Crisis is, "Why has it happened so fast, why do we suddenly have pollution and energy shortages?" The answer is that a factor of 2 is a big factor; when we demand from the electric utility industry twice as much energy in 1970 as we demanded in 1960, it is almost impossible to respond untraumatically; the atmosphere of an industrialized valley may absorb and get rid of X tons of pollutants, but not $2X$ tons; our oil companies may be able to find X new sources of natural gas in ten years, but not $2X$, and so on. What we are seeing in these troubled 1970s are the beginnings of the environmental and other pressures which will make these rising exponentials turn over.

In the example of the growth of a colony of bacteria, the depletion of the food supply and the buildup of toxic wastes in the colony's environment are examples of pressures which cause the growth to slow. Analogous pressures which will act to slow the exponential growth of energy consumption, are the depletion of energy resources. which we will discuss in this chapter, and the growing alarm over the environmental impact of energy production and consumption, which we have already considered.

What happens to an exponential growth curve when it collides with the "facts of life"? Typically, it turns over and becomes an S-shaped curve such as is shown in Fig. 11-2 for the U.S. population (a *Sigmoid* curve, if one wants to add the sophistication of a Greek name). In Fig. 11-1 we see the beginning of that tendency at about 17 years in our average boy. Since the straight-line portion of the exponential curve signifies a constant annual percentage of growth (about 12 percent per year for the boy between the ages of 5 and 11 years), the turning over means that the percentage rate of growth becomes smaller, that the quantity being described, whatever it is, is approaching a condition of no growth, or equilibrium.

The present century, which Ralph Lapp calls the "Logarithmic Century"[2] with equal mathematical justification could have been called the "Exponential Century." The next one may be the "Sigmoid Century," for the growth must slacken. We live on a finite planet and either resource shortage or pollution surplus will soon begin to set the "limits of growth." Since growth must end, the Sigmoid ending is much to be desired; it is a smooth ending, growth slackens slowly, and approaches equilibrium.

[2] *Ibid.*

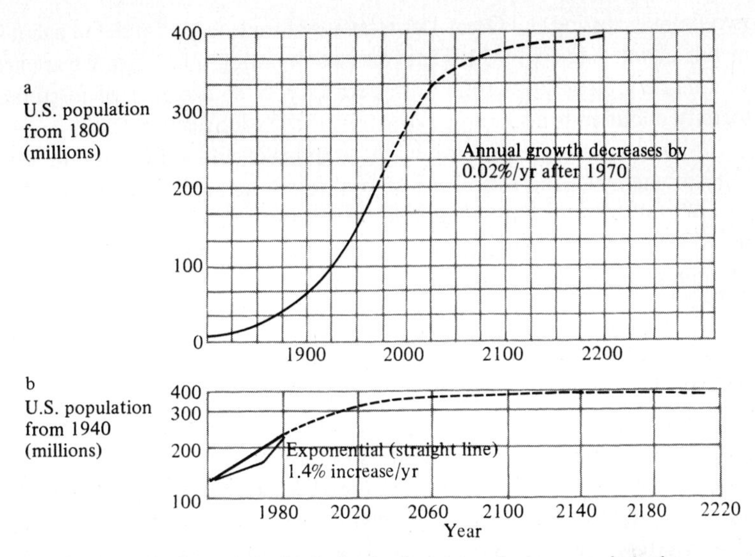

FIGURE 11-2 The U.S. population may be in the beginning stage of a turnover into the "Sigmoid" or S-shape of Fig. 11-2*a*. The actual population is shown by the solid curve; the projected showing of growth by the dotted portion. Figure 11-2*b* shows the latter part of 11-2*a* plotted on semilog paper.

There are worse ways to go. Figure 11-3 shows the now familiar linx-rabbit cycle, which is not smooth. The rabbit population grows and, stimulated by the abundance of food, the linx population expands. The growing linx population, however, reduces the rabbit population, and when the food supply gets too small, the linx population begins to decline. Then, if they have not been wiped out, the rabbits make a comeback, and the cycle continues.

It is these oscillations which we wish to avoid. There has already been a hint of them in, for instance, the energy consumption curve, which flattened out for nearly

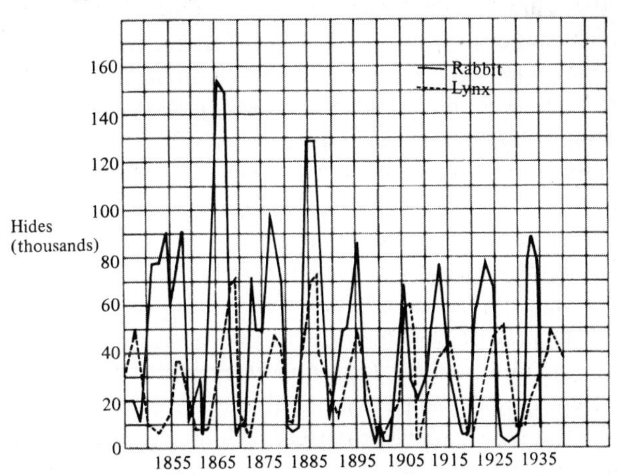

FIGURE 11-3 The variation of rabbit and linx population from data of the Hudson Bay Company. [Source: E.P. Odum, *Fundamentals of Ecology,* 3rd ed., (Philadelphia: Saunders, 1971).]

two decades during the Great Depression before it was stimulated again by the demands of war (see Fig. 5-2). That was two doubling times ago. We are again feeling a dramatic pressure, this time from a scarcity, or so it seems, of fossil fuels. Is that scarcity about to bring energy growth to the "turnover time"?

To answer that question we will have to make projections of our future consumptions of energy and to estimate our total fossil fuel resources, both tricky undertakings. We will also have to go beyond our national boundaries and become aware of the interlinking, in true ecological fashion, of our demands and supplies with those of the rest of the world.

UNITED STATES ENERGY CONSUMPTION—UP, UP, UP?

To predict how much total energy and per capita energy this country will consume in the year 2000, or even in the year 1980, calls into play the mystic art of extrapolation. We must use today's data to predict tomorrow's. There are great dangers in this (as any stock speculator knows). A look back at Fig. 11-1 will illustrate the problem. If we had extrapolated the boy's rate of growth during the first 9 months, the curve would have quickly gone off our graph. If we do the mathematics using the 4-month doubling time, we find that at 20 years we would expect him to weigh 8.8×10^{18} pounds. Even if we had extrapolated from 5 years using that doubling time, which we measure as 6.2 years, we would predict 224 pounds, an overestimate of 70 pounds.

Another example of the perils of predicting from exponential growth is provided by Schoeer in his book *Physics and its Fifth Dimension: Society*[3]. The number of Ph.D.s awarded in physics (which grew exponentially during the 1950s and 1960s), is projected and compared with the population of humans and dogs (also extrapolated) in the United States. This extrapolation predicts that by the year 2250, every person and dog in the country would be expected to have a Ph.D. in physics. In this particular case, the "environmental pressures" of the marketplace have already caused the growth in the number of physicists to slacken. We may, in that discipline, be beginning to see the unstable oscillations from shortages to surpluses of Ph.D's.

With these warnings, let us enter the projection game. There are several methods of making energy projections. The simplest is to plot the historical data (for total or per capita energy) on semilogarithmic paper and "eyeball" a curve that seems to carry out the trends. This technique depends on the accuracy of the historical data, and on the future taking its cues from the past.

In the more sophisticated methods the dependence of total energy, for instance, on other variables such as GNP, population, or prices is determined; then these variables are forecast and the final curve constructed. There are occasional examples of even more careful analysis; individual categories of energy use, such as residential heating and commercial air conditioning are projected, and, from these energy consumption can be projected into broader residential and commercial use and summed to give the total projection.

[3] D. Schroeer, *Physics and Its Fifth Dimension: Society,* (Reading, Mass.: Addison-Wesley, 1972).

A projection based on these more elaborate techniques was used back in the middle 1950s by Schurr and Netschert in *Energy in the American Economy, 1850-1975* to predict energy consumption in 1975.[4] Their extrapolations based on 1955 data were fairly good. The latest data for 1973 total energy consumption, for instance, is only about 9 percent higher than the prediction. They were, however, well off the mark in their specific predictions underestimating per capita consumption by 10 percent, oil consumption by 6 percent, natural gas by 23 percent and overestimating coal consumption by 26 percent.

With these results as a caution, we present Fig. 11-4, a compilation of predictions of total energy consumption in the United States for the years 1970 to 2000. The historical data for 1920 is provided as a reference and the various projections are shown as individual points. There are several things worth noting in Fig. 11-4. There were four projections given for 1970 which can now be checked against actual data. They are all low. These extrapolations, none for longer than 10 years, missed the actual mark by from 6.4 to 12.5 percent. This maximum underestimate amounts to 8,600 T Btu which, in terms of coal would be 430 million tons, more than three fourths the total amount of coal mined in 1970. Such comparisons warn us that it is quite possible to make rather large errors when extrapolating from historical data, and that even a 10 percent error amounts to a lot of energy. One sees why those industries dealing with nonstorable energy (electricity is the prime example) tend to overestimate consumption.

Figure 11-4 also shows the arithmetic averages of the predictions. We see that the weight of prediction rests on a continuation of the growth at a rate somewhat less than the 5.0 percent per year of the 1965-1970 period; in fact, if we give heaviest weight to predictions for 1980 and 2000, then the average prediction is that growth from the 1970 base line will take place at the 1950-1965 growth rate of 3 percent. The consensus is that exponential growth will continue for at least three more decades; only Hubbert, who made the low prediction for the year 2000 (but who also made the lowest prediction for 1970), suggests that we may see the beginning of the S-curve in the next 30 years.

[4] Schurr and Netschert, *Energy in the American Economy, 1850-1975,* (Baltimore: Johns Hopkins Press, 1960).

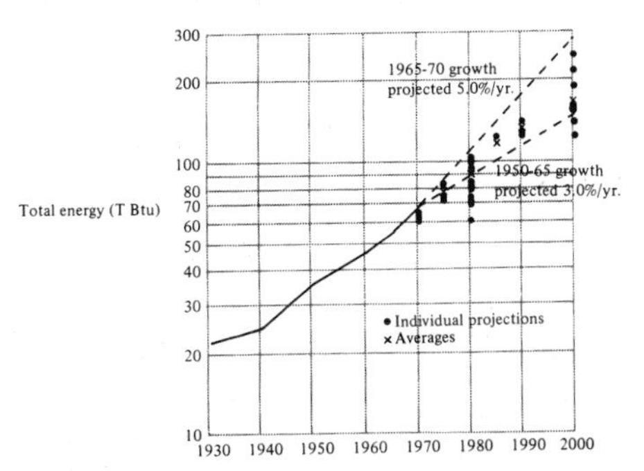

FIGURE 11-4 U.S. total energy consumption is projected to the year 2000. The individual estimates are shown as dots; the crosses are averages. [Source: Committee on Interior and Insular Affairs, U.S. Senate, *Survey of Energy Consumption Projection,* (Washington, D.C.: USGPO 74-459, 1972).]

It is of some interest to look at the assumptions which back up these projections. According to the study from which the data in Fig. 11-4 was taken,[5] the five variables and the basic assumptions concerning them were:

1 *Gross national production.* Generally, a rate of 4 percent per year was assumed.
2 *Population.* Generally, the Bureau of Census Series (II-B) projection of 1.6 percent per year was used.
3 *Price of fuels.* This assumption refers both to the price of fuel relative to other fuels and to the general price level. Most assume existing relative shares of the market will prevail in the future.
4 *Availability of fuels.* The usual assumption is that there will be no limitation on gross availability.
5 *Technology.* Most assume no revolutionary changes, but they do foresee a sizable increase in nuclear generating capacity.

Two other implicit assumptions seem to have been made: first, that there would be no business cycle swings and second, no major overseas war would occur and defense spending would continue at the present rate relative to their sectors.

We should remark on these assumptions. Many of them seem optimistic, which may be in keeping with the desire of the forecasters to overestimate rather than underestimate. The assumed 4 percent growth of the GNP seems particularily optimistic, for, since 1920 the average growth rate of the GNP has been 3.3 percent and the growth has been much less than this during the past few years (down 5.8 percent in the first quarter of 1974). We will discuss the energy/GNP ratio specifically in a later section; this has been an important indicator and there is evidence of recent significant change.

The other assumptions can be questioned also, the 1.6 percent population growth is significantly higher than the actual 1.3 percent growth of the last decade; the assumption of no change in relative prices seems to contradict the presently observable rapid rise of oil and natural gas prices relative to coal. We have just passed through several months of an oil shortage and the threat still hangs over us. The last assumption is surely the most conservative: by it we give up any hope for big changes in the next 30 years—no fusion, no fuel cells, no big changes in electric generating efficiency from magnetodynamic generators or from any of the other technologies dealt with in the next chapter.

The projections are very sensitive to small variations in these growth rates, just as the growth of a savings account is sensitive to the interest rate. The difference between a 3.5 percent and 4.0 percent per year GNP growth for a constant energy/Btu ratio is 96.7/101.7 (in Q Btu) in 1980 and 192.5/222.0 in 2000. Again, if we view these prediction differences in terms of coal to be mined or generating capacity to be built, they become large factors.

The technological change and fuel price-availability assumptions are important also. Table 11-1 illustrates some different predictions for the year 2000 which obtain from a different assumption of this type. There is almost a factor of 2 difference

[5] Committee on Interior and Insular Affairs, *op. cit.*

Table 11-1 Energy Consumption in 2000 under Alternative Technological Assumptions*

Case	Energy consumption in 2000 under assumed 3.5% GNP growth rate	
	(Q Btu)	*Average annual percentage growth rate from 1969*
"Medium-range" projection	190	3.5
Rapid improvements in electric generating efficiency (1-1½ percent annual decline in heat rate)	170	3.1
No change in electric generating efficiency (1969 heat rate prevails)	210	3.9
Single fuel (natural gas) economy; transportation sector features electric vehicles recharged by utilities	205	3.8
Single fuel (natural gas) economy; all-electric transportation sector as in preceding case; energy inputs into central power stations via hydrocarbon-air fuel cells	186.5	3.4
No purchased utility electricity; sectoral needs (including fuel-cell powered vehicles) supplied by natural gas-fired hydrocarbon-air fuel cells	133	2.3
All-electric economy, based on utility power plants fueled by conventional sources, hydro, and nuclear	241	4.3
All-electric economy, based on utility power plants, fueled by natural-gas hydrocarbon-air fuel cells; transportation sector features battery electric vehicle system recharged by utility power	211	3.8
All-electric, all-coal economy with MHD technology of centrally-operated utility power plants	190	3.5

*Source: J. Darmstadter, Appendix to *Energy, Economic Growth and the Environment*, (Baltimore: Johns Hopkins University Press, 1973).

between the fuel-cell powered future of 133 Q Btu and the all-electric economy which would need 241 Q Btu. Although neither of these extreme cases are realistic, they do indicate the range of variation which could occur.

Energy and the GNP

Forecasters have traditionally leaned heavily on predicting the rate of growth of the GNP and then using data on the ratio of total energy consumption to GNP to predict energy consumption. Since about 1950 there has been a steady trend in this latter ratio; the annual rate of energy consumption growth has been about 1 percent less than the annual rate of growth of the GNP. Stated another way, in the last half of this century there has been for an approximate 1 percent annual decrease in energy used per dollar GNP.

As Fig. 11-5 shows, this trend changed and, during the period 1965-1970 the energy/GNP ratio increased by about 10 percent. In a detailed study of the cause of this change, the National Economics Research Associates[6] conclude that three factors contributed to this increase: (1) the increase in importance of nonfuel uses of energy, which grew from 4.9 percent of the total in 1965 to 5.7 percent in 1970; (2) the leveling out of the rate of increase in the efficiency of electric power generation (see Fig. 6-12, Chap. 6); (3) the increased importance of electric heating and air conditioning, which do not add to the GNP, but are beginning to contribute importantly to energy consumption.

It would seem that these factors would continue to effect the GNP. The most recent data, however, 1971, 1972, and 1973, show a reversal of the trend and the 1973 ratio is almost identical to the 1969 ratio. The 1974 data will be interesting; on the one hand the GNP growth will be much slowed by the recession, and on the other hand energy consumption growth may have been significantly slowed by the embargo and the associated conservation efforts.

All of this is to say that the range of uncertainty displayed in Fig. 11-6 is very real. For instance, if energy use per dollar GNP were to return to its average annual 1 percent *decrease,* then a 3 percent growth in GNP would lead to a total consumption of 170 Q Btu in 2000. If energy use per dollar GNP *increases* at 1 percent annually and the GNP were to increase at the assumed 4 percent annual figure then 300 Q Btu will be needed by 2000. It is clear that careful planning and, perhaps, even control of some factors, will be needed to keep energy demand within sight of supply.

[6]National Economics Research Associates, "Energy Consumption and Gross National Product in the United States," New York, March 1971.

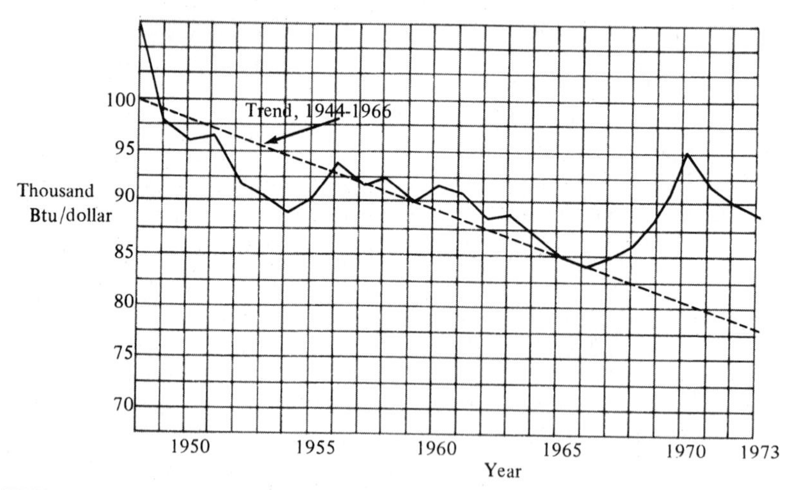

FIGURE 11-5 The ratio of energy consumed to dollar GNP values (in constant dollars) declined on the average from 1947 to 1966. From 1966 to 1970 the ratio increased; the energy cost of a dollar GNP rose. The latest data seem to show a renewed decrease. (Source: National Economics Research Associates, *op. cit.*)

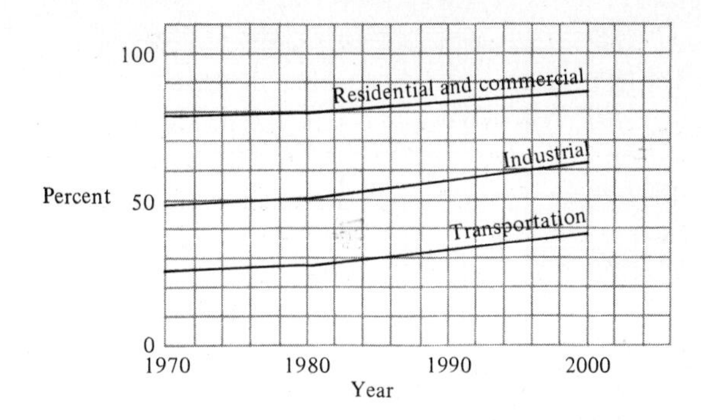

FIGURE 11-6 The energy consumption by the electric utilities and transportation is expected to increase during the next three decades at the expense of the other two. It must be remembered, however, that the electrical energy produced will be largely consumed in the "Household and Commercial" and "Industrial" categories. [Source: Ericson, *Energy and Human Welfare,* Vol. II, (New York: Macmillan, 1974).]

Growth by Economic Sector

It is of some interest to look at the prediction of growth in energy consumption in the various sectors. The uncertainties of forecasting, which have just been emphasized, will make us hesitate to place any great reliance on these figures, but the projections should be of use to those who wish to know, in more detail, just where growth is taking place.

The predictions for the share of the total energy consumed in the three sectors, residential and commercial, industrial, and transportation are shown in Fig. 11-6. Of these three, transportation is expected to be the most rapidly growing one due to an increasing number of cars per family as a result of increasing income, and to the coming of age of a large number of drivers. The inefficiency of energy use by automobiles is thus a particular calamity.

Electrical Energy Consumption

The most rapidly growing use of energy is in the generation of electricity. Most forecasters expect the use of electrical energy to grow at almost 6 percent per year until 1980 and to drop to about 5 percent from 1980 to 2000. The projected percentages of the total energy to be used in the generation of electricity during the next three decades are shown in Fig. 11-6 along with the percentages in other categories. The predicted future consumptions of electrical energy in kilowatt-hours are given in Fig. 11-7. Again we see that the predictions were already low by 1970 and that continued exponential growth is assumed.

In Chap. 6 we have discussed the causes of this rapid growth in the consumption of electrical energy. The most important causes seem to have been the steadily dropping cost of electricity, the cleanliness and convenience of its use, the increasing affluence of the consumers (leading to more appliances, air conditioning, and electric heating), and the agressive promotion of electrical energy by utilities companies, including a preferential rate structure for all-electric homes and for large industrial and commercial consumers. Projecting from this base, the quadrupling of electric energy

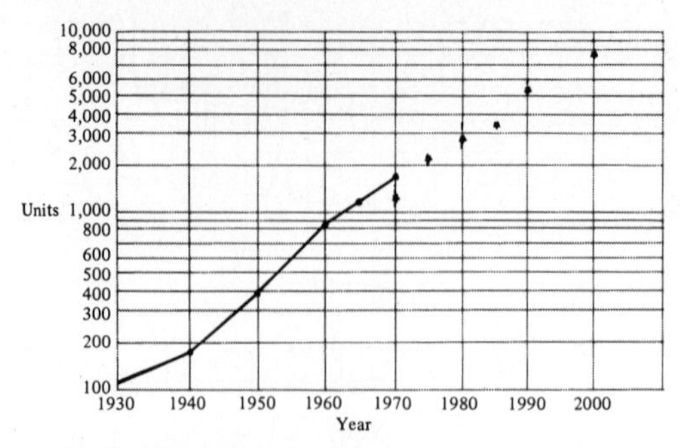

FIGURE 11-7 The projection of electrical energy consumption is for continued doubling every ten years until 2000. The dots represent individual estimates; the crosses are averages. (Source: Committee on Interior and Insular Affairs, U.S. Senate, *op. cit.*)

demand by 1990 and a near-doubling again by 2000 seems defensible.

This large relative and absolute increase in electricity generation will have the obvious effects we have discussed earlier. The amount of fossil fuel consumed, even with a major nuclear role in power production, will climb, and, with it the tonnage of ash, sulphur oxides, nitrogen oxides, carbon monoxide, and carbon dioxide to be disposed of. Since much of the increase in electric power demand will be at the expense of direct fuel sources, the average efficiency of energy use will decrease and the waste heat to be disposed of will rise from about 13.1 Q Btu in 1970 to 26.6 Q Btu in 1990 and 32.7 Q Btu in 2000.[7] The energy wasted in this fashion in the year 2000 will be about half the total consumed in 1970. Stated in other terms, the energy wasted in the generation of electricity, which was 14 percent of the total in 1965 and 19 percent in 1970, will be 19.5 percent in 1990 and 18.5 percent in 2000 if the expected improvement in efficiency takes place. The great increase in waste heat should increase considerably the pressure either to find some use for it or to replace the heat engine as motive power for electrical generators.

A Contrasting Prediction. For those who view a nearly eightfold increase in the use of electrical energy by 2000 (with its ramifications in air pollution, number of plants, miles of transmission lines, acres strip-mined, tons of radioactive waste stored, and so on) as a threatening prediction, the results of a recent study may be encouraging. Chapman, *et.al.*[8] investigated the response of electrical demand growth to four factors (which other studies had pointed to as of major importance): price, population, income, and the price of natural gas. They found the *elasticities*, that is, percentage increase (+) or decrease (−), in electrical consumption which followed the percentage changes in these indicators shown in Table 11-2. Since these changes do not take place immediately, they also show the percentage response in the first year and the number of years it will take for a 50 percent response.

One sees from Table 11-2 that an increasing price of electricity is predicted to lower consumption 1.3 to 1.7 percent (depending on consumer class) for each 1 per-

[7] These figures were obtained by converting kw-hr to Btu at 3412 Btu per kw-hr and assuming efficiencies of 30 percent in 1970, 40 percent by 1990, and 45 percent in 2000.

[8] Chapman, Tyrell, and Mount, "Electricity Demand Growth and the Energy Crisis," *Science* **178**: 703, November 1972.

Table 11-2 Elasticity in Electrical Consumption*

Consumer class	Elasticity: Electricity price	Population	Income	Gas price	First year response (percent)	Years for 50% total response
Residential	-1.3	+0.9	+0.3	+0.15	10	8
Commercial	-1.5	+1.0	+0.9	+0.15	11	7
Industrial	-1.7	+1.1	+0.5	+0.15	11	7

*Source: Chapman, Tyrrell, Mount, *op. cit.*)

cent increase, while increases in all other indicators increase the consumption. After a steady decrease (in constant dollars) from 1945 to 1970, the price of electricity increased in 1971 by 1.8 percent. The FPC predicts an increase of 19 percent between 1968 and 1990. Other estimates are as high as a 100 percent increase in the next three decades.

The six different projections from this study of the electrical demand growth between 1970 and 2000 are shown in Fig. 11-8. They correspond to the different assumptions outlined in Table 11-3. We see that the first four of these, which stem from rather reasonable assumptions about the changes of the four indicators, lead to electrical consumption projections which still agree fairly well in 1975 with the estimates in Fig. 11-7. By the year 2000, however, the projections based on this work are three to five times lower. It is only by assuming a 1.4 percent population increase, and a *decreasing* cost of electricity that a projection (curve F of Fig. 11-8) consistent with the estimates of Fig. 11-7 is obtained.

The results of this study deserve attention for two reasons. If the technique stands

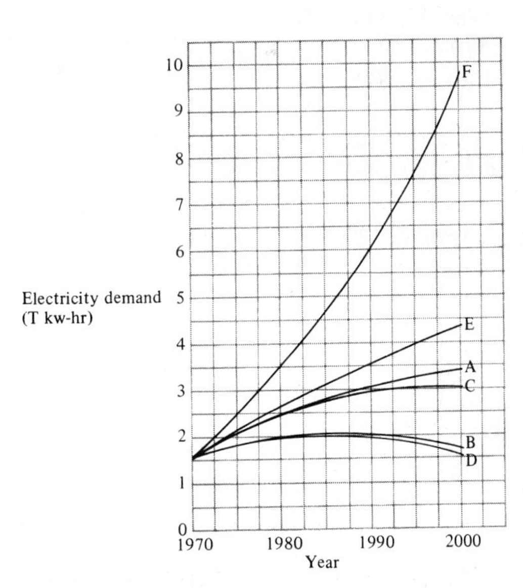

FIGURE 11-8 It is predicted that the increasing price of electricity will cause less rapid growth in electric energy consumption than expected (see Fig. 11-7). The assumptions guiding these various curves are given in Table 11-3. (Source: Chapman, Tyrrell, Mount, *op. cit.*)

Table 11-3 Electricity Demand Growth and Alternative Assumption*

Case	Population assumption	Electricity price assumption	Electricity demand (T kw-hr)			
			1975	1980	1990	2000
A	BEA	FPC	1.98	2.23	3.01	3.45
B	BEA	Double by 2000	1.88	2.07	2.11	2.01
C	ZPG 2035	FPC	1.98	2.37	2.95	3.29
D	ZPG 2035	Double by 2000	1.88	2.05	2.07	1.91
E	BEA	Constant	2.02	2.54	3.56	4.56
F	BEA	†	2.14	3.05	5.66	9.89

*Source: Chapman, Tyrrell, Mount, *op. cit.*

BEA, Bureau of Economic Analysis; FPC, Federal Power Commission; ZPG 2035, zero population growth reached in 2035. In the constant price assumption 1970 prices are maintained in each region increases annually by 3.33 percent of its 1970 value for 30 years.

†Average prices decline 24% from 1970 to 1980, and 12% each 10 years thereafter until 2000.

critical analysis, it can be applied with benefit to other areas of energy growth, and there, too, provide more realistic estimates than we now have. The predicted numbers have as much importance as the technique, however. As we pointed out in Chap. 6, predictions of electrical energy consumption have a self-fulfilling character because of the long lead times needed for the construction of generating facilities, and the short doubling time of demand growth. Generating capacity is built to specification of such prediction, especially the authoritative FPC predictions. Therefore, it is most important, if energy consumption is to be held as low as possible, that overestimation of demand be avoided. We will hope to see some discussion of, and changes in, the "official" predictions of electric energy demand which, by as early as 1985 (one doubling period from now in the traditional view) begins to differ significantly from these more careful predictions.

ENERGY SUPPLIES

In the previous section we have discussed the projections of energy demand into the next three decades. We must now examine the sources of that energy to see how much is available in various resource forms. There are really two questions to be asked: (1) How much coal, oil, gas, and nuclear energy can be produced in the next decades to meet the demands we have just projected? and, for the longer-range view, (2) What is the total amount of these resources which we can count on? We will look at question 1 first.

We use as our source for the comparisons which follow the report "U.S. Energy Outlook" by the National Petroleum Council (NPC),[9] an advisory body to the Secretary of the Interior made up, for the most part, of industrial representatives. This report focused on the period 1975-1985. The approach was the following: the various parameters which affect the domestic production of energy were identified and

[9]A Summary Report of the National Petroleum Council's Committee on U.S. Energy Outlook, *U.S. Energy Outlook*, (Washington, D.C.: National Petroleum Council, December 1972).

grouped into four cases from which supply predictions were derived. The electric utility industry's fuel needs were considered separately and the effects of changes in that fuel mix were studied. The major goal of the study was to determine the imports of gas and oil which would be needed under the various assumptions. We will briefly review these assumptions and the study results.

In order to cover the full range of possibilities, the study included the high, medium, and low demand cases shown in Table 11-4 and four different supply cases based on different assumptions about fuel production. The demand projections shown in Table 11-4 do not differ significantly from those shown in Fig. 11-4; the 1980 NPC estimates are above-average while those for 1985 span the projections in Fig. 11-6.

The NPC study puts most of its emphasis on determining supply. It identifies the main factors which influence energy production and determines their quantitative influence on supplies. These factors include prices, exploratory activities and results, tax laws, environmental opposition to power plant siting, and the amount of research and development (R & D) support to synthetic fuel production. The results of the investigation of these and other parameters were grouped into four cases.

Case 1 estimates the possible outcome from a maximum effort to develop domestic fuel sources. Case 1 assumes oil and gas drilling increases at a rate of 5.5 percent per year, and uses a high projection of oil and gas discovered per foot drilled. The nuclear power projections are based on the assumption that all new base-load generating plants ordered between now and 1985 will be nuclear. Production of coal for domestic consumption is increased at a rate of 5 percent per year. Synthetic fuels are developed and produced at the maximum rate physically possible without any restrictions due to environmental problems, economics, and the like.

Case 4, the pessimistic supply case, assumes that recent trends in United States oil- and gas-drilling activity, and the success from such efforts will continue; the siting and licensing problems with nuclear plants will continue; the incentives to develop new coal mines will not improve; and environmental constraints will continue to retard development of resources. This case results in a continued deterioration of the nation's energy supply.

Case 2 assumes a less optimistic future supply picture than Case 1. Oil and gas-drilling activity grows at a lower rate (3.5 percent per year) than in Case 1 but with the same finding rates per foot drilled. For nuclear energy, Case 2 assumes that prob-

Table 11-4 Projections of United States Total Energy Demand under Three Different Sets of Assumptions*

	Growth rate (average annual percent gain)			*Volume (Q Btu)*	
Case	*1970-1981*	*1981-1985*	*1971-1985*	*1980*	*1985*
High	4.5	4.3	4.4	105.3	130.0
Intermediate (initial appraisal)	4.2	4.0	4.2	102.6	124.9
Low	3.5	3.3	3.4	95.7	112.5

*Source: A Summary Report of the National Petroleum Council's Committee in U.S. Energy Outlook, *op. cit.*

lems in manufacture and installation lead times will be solved quickly. Coal production is increased at a rate of about 3.5 percent per year. Synthetic fuels are developed and produced at a moderate buildup rate.

Case 3 assumes that there will be improvement over Case 4 but not to the level of Case 2 in the development of indigenous energy supplies. Oil and gas *drilling* grow at the same average annual rate of 3.5 percent per year projected in Case 2, but the trends of oil and gas *finding* per foot drilled are lowered to those of Case 4, which reflect recent actual experience. The development of nuclear power proceeds at about the rate in the AECs most favorable forecast. There is no significant difference between Case 2 and Case 3 for coal and synthetics.

The potential supplies of domestic energy from all the potential sources are summarized for the years 1975, 1980, and 1985 in Table 11-5.

In terms of total supplies one can see from a comparison of Tables 11-4 and 11-5 that even the maximum supply (the most optimistic estimate) of Case 1 does not meet even the lowest demand projections in 1980 while by 1985, supply Case 1 meets the high demand of Table 11-4 and supply Case 2 comes close to meeting the low demand.

Unfortunately, the situation is more complicated than this. One cannot make a simple comparison of total supplies with total demand; the fuels are not completely interchangable. You can't run automobiles on coal or nuclear energy, for example. Since interchangability is greatest in the electric utility industry, the NPC report considered various combinations of fuels for that industry to assess the effect on the United States energy balance. The comparisons given in Table 11-6 and Fig. 11-9 assume the fuel mix projected by the FPC which was summarized in Table 6-5 of Chap. 6.

Of most interest are the amounts of energy we must import to meet the various demand levels. Since coal and nuclear energy are totally domestic sources, the imports considered were oil and gas. We can gain some idea of the range of possibilities by looking at the extreme cases: the high and low demand and the supply Cases 1 (maximum supply) and 4 (minimum supply). Figure 11-9 compares the energy import-export situation for Cases 1 and 4. The NPC message carried in Fig. 11-9 is a straightforward one. From 1975 through 1980 we will be dependent on other countries for a significant part of our energy supply. If we give support to the oil companies, however, the situation will have already improved by 1980 and by 1985 domestic production, under the optimistic conditions of supply Case 1, will equal the low demand.

The specific figures for the percentages of oil and gas imported are shown in Table 11-6. The NPC prediction is that by 1975 we will be importing 40 percent of our oil; the 1970 percentage was 26 percent. This will fall to 15 percent of the minimum demand in 1980 and to 0 percent in 1985. These assumptions are for the optimistic supply case (1). The comparable figures for the pessimistic supply case (4), are, assuming minimum demand, 61 percent (1980) and 57 percent (1985). The data in Fig. 11-9 and Table 11-6 make the case for the "dig and drill" proponents. The choice is between the optimistic Case 1 in which the energy companies are given maximum incentive for exploration, there is a maximum increase (5 percent per year) in coal production, and all new plants are nuclear, and Case 4 which is the energy

Table 11-5 Potential Domestic Energy Supply Availability (all data × 10^{12} Btu/yr)*

Year		*Initial appraisal*	*Case 1*	*Case 2*	*Case 3*	*Case 4*
1975	Oil					
	Domestic liquid production	22,789	20,735	20,630	19,754	19,502
	Shale syncrude	0	0	0	0	0
	Coal syncrude	0	0	0	0	0
	Subtotal–Oil	22,789	20,735	20,630	19,754	19,502
	Gas					
	Domestic production	20,430	24,513	24,300	22,766	22,421
	Nuclear stimulation	0	0	0	0	0
	Syngas (Coal)	0	0	0	0	0
	Subtotal–Gas	20,430	24,513	24,300	22,766	22,421
	Hydroelectric	2,840	2,990	2,990	2,990	2,990
	Geothermal	120	120	120	120	120
	Coal	16,310	16,650	15,554	15,554	15,100
	Nuclear	3,340	4,000	4,000	4,000	1,661
	Total potential supplies	65,829	69,008	67,594	65,184	61,794
1980	Oil					
	Domestic liquid production	24,323	27,758	26,456	23,789	18,112
	Shale syncrude	0	296	197	197	0
	Coal syncrude	0	175	0	0	0
	Subtotal–Oil	24,323	28,229	26,653	23,986	18,112
	Gas					
	Domestic production	18,030	26,746	25,043	21,041	17,906
	Nuclear stimulation	0	206	103	103	0
	Syngas (Coal)	190	512	329	329	165
	Subtotal–Gas	18,220	27,464	25,475	21,473	18,071
	Hydroelectric	3,033	3,240	3,240	3,240	3,240
	Geothermal	343	782	401	343	191
	Coal	19,928	21,200	18,284	18,284	17,550
	Nuclear	9,490	11,349	11,349	9,787	6,788
	Total potential supplies	75,337	92,264	85,402	77,113	63,952
1985	Oil					
	Domestic liquid production	23,405	31,689	28,477	24,346	21,426
	Shale syncrude	197	1,478	788	788	197
	Coal syncrude	0	1,489	175	175	0
	Subtotal–Oil	23,602	34,656	29,440	25,309	21,623
	Gas					
	Domestic production	14,960	31,604	27,324	21,049	15,474
	Nuclear stimulation	0	1,341	825	825	0
	Syngas (Coal)	560	2,269	1,208	1,208	494
	Subtotal–Gas	15,520	35,214	29,357	23,082	15,968
	Hydroelectric	3,118	3,320	3,320	3,320	3,320
	Geothermal	514	1,395	661	514	257
	Coal	23,150	27,100	21,388	21,388	20,300
	Nuclear	21,500	29,810	25,249	20,220	16,126
	Total potential supplies	87,404	131,495	109,415	93,833	77,594

*Source: A Summary Report of the National Petroleum Council's Committee in U.S. Energy Outlook, *op. cit.*

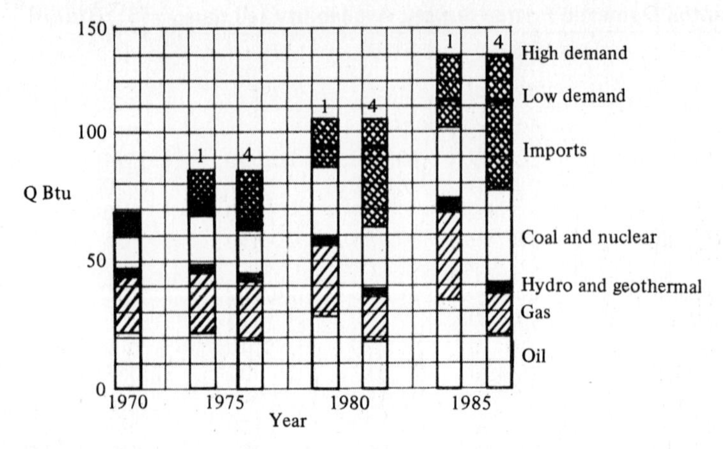

FIGURE 11-9 Large amounts of energy in the form of petroleum products (the cross-hatched areas) will need to be imported under the pessimistic supply case (4) each year until 1985 and moderate amounts even under the optimistic supply case (1).

company nightmare: no improvement in exploration and finding, environmental constraints on both resource development and power plant siting and so on. Figure 11-9 represents the choice put before us by the energy companies.

The NPC study also predicts that even in the optimistic Case 1, coal and nuclear will provide only one third of the total energy in 1985, and coal gasification that year is expected to provide only 6 percent of the total gas consumption. There is also a projected addition of natural gas from nuclear stimulation which adds 3 percent to this total. As we shall discuss later, this may be quite over optimistic. As for oil, oil from shale is estimated to comprise about 3 percent in 1985 with a similar percentage of oil from coal liquification. There is no expectation of contributions from solar energy or any other unconventional sources.

The NPC also looked ahead to the year 2000. They predict a high and low demand of 215 and 170 Q Btu which lie within the ranges shown earlier in Fig. 11-4. The potential of domestic supplies for this period are also estimated (see Table 11-7). The NPC prediction is that with sufficient incentive and with removal of some of

Table 11-6 Percentages of Oil and Gas Imported*

		Percent of total energy imported		*Percent of oil imported*		*Percent of gas imported*	
Supply		*Demand*		*Demand*		*Demand*	
case	*Year*	*(Max)*	*(Min)*	*(Max)*	*(Min)*	*(Max)*	*(Min)*
	1975	19.7		42.4		4.7	
1	1980	18.0	9.7	34.7	14.9	12.4	12.4
	1985	14.2	0.9	26.7	0	9.5	2.9
	1975	26.0		41.2		5.1	
4	1980	39.3	33.2	67.4	60.6	17.8	17.8
	1985	40.3	31.1	67.9	56.7	29.2	29.2

*Source: A Summary Report of the National Petroleum Council's Committee in U.S. Energy Outlook, *op. cit.*

Table 11-7 Domestic Energy Output Potential in the Year 2000 Based on an Intermediate Level of Supply in 1985 (Conventional Energy Sources)*

Source	*Units*	*1985*	*2000*
Oil, total domestic liquid production	Mbls/D	14	10-18
Natural gas production	Tcf/yr	27	15-25
Coal, traditional uses only	Million tons/yr	863	1,200-1,700
Hydro	B kw-hr/yr	316	340-380
Nuclear	B kw-hr/yr	2,463	7,500-9,500
Oil, total domestic liquid production	Q Btu/yr	29	21-37
Natural gas production	Q Btu/yr	28	15-26
Coal, traditional uses only	Q Btu/yr	21	30-42
Hydro	Q Btu/yr	3	4
Nuclear	Q Btu/yr	25	61-102
Total	Q Btu/yr	106	131-211

*Source: A Summary Report of the National Petroleum Council's Committee in U.S. Energy Outlook, *op. cit.*

the limiting environmental restrictions, the nation could be self-sufficient by the end of this century. The fulfillment of this prediction will depend, to an extent, on the development of new technology which we will discuss in Chap. 12. As we begin to look beyond this century, however, we begin to question the sufficiency of our resources of fossil fuels and uranium. We will turn now to the estimation of that treasure.

How Much Is Left?

The uncertainties we underlined in discussing projections of energy demand have their counterparts when we attempt to assess resources. It is hard to know how much coal, oil, gas, or uranium is really there until you actually dig or pump it out. It is also true that most past estimates of resources have underestimated the totals and, because of this, many people hear present day warnings as just another cry of "Wolf!" to be ignored.

There are some facts, however, that are indisputable and these make a good starting place. The first and most important one is that this planet is finite and therefore has finite resources. We also know that only a finite amount of carbon has been laid down in fossil form and, from some intelligent conjectures about the biology and geology of the past epochs, this can be estimated. The most recent calculation of this number is 68×10^{20} grams.[10] This will serve as an upper limit on our fuel supplies. It should be pointed out, in passing, that we could not safely burn all these carbon atoms. We have in our atmosphere only about 2×10^{20} grams of oxygen; burning only 6 percent of the total carbon would thus bind all our oxygen into carbon dioxide.

From this upper limit on total fossil carbon, we look to its estimated distribution in the form of coal, oil, gas, and oil shale. These estimates are summarized and

[10] W. W. Rubey, "Geological History of Sea Water," *Bulletin of the Geological Society of America,* 1951.

compared with the previously mentioned upper limit in Table 11-8. We notice that much of the carbon is locked into the oil shales, a resource just now coming into the range of our technology.

Two columns are shown in Table 11-8: "total carbon" and "ultimately recoverable resources." This distinction is an important one to make. Much of the confusion over resource estimates has come from the use of various terms: *reserves, resource base,* and *ultimately recoverable resources.* We will follow Williams[11] in choosing the latter for our comparisons.

The term reserves refers to resources whose location is definitely known and which can be (and, in most cases, will be) extracted with current techniques. The resource base, on the other hand, is the total amount of a resource which is estimated to be in a geographical area; no consideration is given to the cost or possibility of extracting all of it. It is unlikely that all of a resource will ever be discovered and even more unlikely that, if found, it will all be extracted. At present, for instance, we only extract 50 percent of the coal in an underground mine and even less (about 30 percent) of the oil in a given producing area.

The ultimately recoverable resources is an estimate of that part of the resource base that we can reasonably expect to recover at some future time. It lies somewhere between the two extremes and, like them, is subject to change as our techniques and geological knowledge improve. We see from Table 11-8 that the present best estimate of ultimately recoverable resources of carbon is 8.2×10^{18} grams, one thousandth, more or less, of the estimated total.

Burning this amount of carbon would only deplete the atmospheric oxygen by 2 percent. It would, however, greatly increase the carbon dioxide content, by a factor of 12. Although over the centuries this amount would be reduced by absorption

[11] R. Williams, "Fossil-Fuel Resources," Chap. 2, Vol. II of Commoner, Boksenbaum, and Corr, *Energy and Human Welfare,* (Riverside, N.J.: Macmillan, 1974).

Table 11-8 Estimates of World Total Fossil Fuel Resources

Source	Total (g)	Ultimately recoverable resources (g)
Fossilized carbon formed photosynthetically*	68×10^{20}	$1.7 \times$
Shale containing more than 5% organic matter†	7×10^{20}	1.7×10^{18}
Coal‡	0.12×10^{20}	6×10^{18}
Petroleum §	0.01×10^{20}	0.26×10^{18}
Natural gas §	0.005×10^{20}	0.25×10^{18}
		8.2×10^{18}

*Rubey, *op. cit.*

†D. C. Duncan and V. E. Swanson, *Organic-Rich Shale of the United States and World Areas,* U.S. Geological Survey Circular, No. 523, 1965.

‡P. Averitt, "Coal Resources of the United States," *U.S. Geological Survey Bulletin,* **1275**, 1969.

§T. A. Hendricks, *Resources of Oil, Gas, and Natural Gas Liquids in the U.S. and World,* U.S. Geological Survey Circular, No. 572, 1965.

in the oceans, the anticipated effects of burning even a small amount of this total carbon cause the worries discussed in Chap. 8.

United States Fossil Fuel Resources

From a nationalistic viewpoint, our major resource concern is focused on this country. What amounts of coal or oil, for example, can we draw on during the next two or three energy-hungry decades and when will they run out? The United States has been rather thoroughly explored and mapped. In spite of this, however, it is estimated that half of the coal and oil is yet to be actually discovered. We begin the inventory of our fuel supply with coal.

Coal. The job of estimating coal resources is the easier one; coal beds are usually found where they are expected to be from geolcgical information. From an outcropping one can fairly accurately judge the extent of the whole field. Most of our information concerns coal no more than 3,000 feet below the surface; the present depth for profitable mining is 1,000 feet.

The most authoritative estimate of United States coal resources is that which Averitt made for the United States Geological Survey in 1969.[12] His estimate for coal of all ranks (that is, including the low Btu content coals such as lignite) is summarized in Table 11-9. The distribution of this coal by geographical region and by mining technology (underground or surface) was summarized in Fig. 9-12 (see Chap. 9). In Table 11-10, the recoverable reserves available either to strip mining or underground mining are estimated, compared with 1970 production, and their expected lifetimes at different rates of growth in production are given. We see from an analysis of this data that the reserves consist of 104.6 billion tons of underground coal (at a 50 percent recovery rate) and 45 billion tons of coal which can be strip-mined with present techniques, a total of about 150 billion tons. Most of the underground coal is in the Pennsylvania-West Virginia and Illinois-Indiana region and most of the coal to be strip-mined is in Montana-Wyoming and New Mexico. The lifetimes calculated for the different growth rates are to be compared with the present growth rate of coal production of about 3 percent and the maximum possible growth rate used in the NPC calculations which was 5 percent. Since underground coal production grew at a rate of 8 percent per year in the early 1940s and surface mining presently is grow-

[12] Averitt, *op. cit.*

Table 11-9 Total United States Coal Resources*

Type	*Amount (T tons)*
Mapped and Explored: 0-3,000-ft overburden	1.56
Probable additional resource in unmapped and unexplored areas: 0-3,000-ft depth	1.31
3,000-6,000-ft depth	0.34
Total	3.21

*Source: Averitt, *op. cit.*

ing at about 6 percent, it seems that the nation could produce more coal than these projections assume over the next 15 years.

Even the tremendous coal resources can't stand continued growth in consumption. At the present level of coal production, the reserves of 150 billion tons would last 479 years (see Table 11-10). If the 1973 consumption, 569.3 million tons grows at 5 percent per year; however, the 1,600 billion tons (50 percent recovery of the 3,200 billion ton resource base) would last only 97 years. This short lifetime for the tremendous resource of coal reminds us again of the power of the exponential. Continued growth at 5 percent means a doubling time of 14 years. Thus, by 2070 (93 years from 1973) we would be using almost 7 times as much coal as we did in 1973.

Over the short range of the next few decades coal production must rise. There is still plenty of coal. The productivity of the mines will not be constrained by resource availability. The real constraints are imposed by investment requirements, manpower deployment, environmental restrictions, and so on.

The Western coal is especially attractive, as mentioned earlier, because it can be surface mined. It is here that huge mine-mouth generating plants and coal gasification and liquifaction plants are expected to be constructed. There is plenty of coal for them, 26 billion tons of reserves available for strip mining in that region alone. If we estimate productivity at 10,000 tons per acre, the mining of this coal will leave 2.6 million strip-mined acres. As we have stated earlier, whether the arid western land can ever be reclaimed after this assault is a matter of concern. We can, therefore, summarize the picture of coal reserves in the following way. There is still a great abundance of coal in this country which can be produced with present technology. The technology of the future (a higher percentage recovery, for instance) can expand the estimates further. For the next century at least, the use of coal as an energy source, will not be resource-limited, but will be limited by economic and environmental considerations.

Oil. The uncertainty of oil reserves is much larger than that of coal. It is not as well-understood geologically. Oil migrates and the same kind of rocks which are oil-rich in one area may be dry in another. When the speculation is confined to the United States, however, the extrapolation from the known to the unknown can be made with some confidence. Nonetheless, the estimates of "ultimately discoverable petroleum liquids" made by recognized experts, differ by a factor of 3, from the 594 billion barrel estimate by Hubbert to the 1,895 billion barrels of the United States Geologic Survey (see Table 11-11).

Table 11-10 Coal "Reserves" (Surface and Underground) Compared to Production Statistics*

Source	Recoverable reserves (billion tons)	1970 production (million tons)	Life of reserves at annual rate of production growth (years) 0%	3%	5%
Underground	104.6	338.8	309	80	58
Surface	45.0	264.1	170	61	46
Total	149.6	602.9			

*Source: A Summary Report of the National Petroleum Council's Committee in U.S. Energy Outlook, *op. cit.*

Table 11-11 Estimates of Ultimately Discoverable Petroleum Liquids Originally in Place*

	Billion barrels				
Area	*1972 USGS*	*1969 Hubbert*	*1959 Weeks*	*1970 Moore*	*1968 Elliott and Linden*
Lower 48 states	1,519	516	———	not estimated	———
Alaska	376	78			
Total United States	1,895	594	1,315	670	1,286

*P. K. Theobald, S. P. Schweinfurth, and D. C. Duncan, *Energy Resources of the United States,* U.S. Geological Survey Circular No. 650, July 1972.

Any of the figures in Table 11-11 represents a large amount of oil; the amount produced to date (1970) is 93 billion barrels. The somewhat conservative NPC summary of oil-in-place resources is shown in Table 11-12. The total discoverable oil is put at 810 billion barrels. Half of this, 425 billion barrels, has already been discovered. In Table 11-12, the geographical location of this oil is also indicated. We see that the NPC experts expect to find twice as much additional oil in Alaska as has already been found there, and also that 160 billion barrels, almost half the discoverable oil, is expected to be found in offshore areas including the Atlantic Ocean area which, so far, has had practically no exploratory drilling.

The 810 billion barrels of oil-in-place indicated in Table 11-12 is not all available for consumption. Just as the case in underground coal mining, much of this oil is not recoverable by present techniques. The actual recovery efficiency has risen from about 26 percent in 1955 to a present 31 percent. It is projected to rise to 36 or 37 percent by 1985. Thus, 2 barrels of oil remain underground for each barrel removed, 186 billion barrels for the 93 billion barrels produced to date.

Assuming an average recovery rate of 35 percent during the next decade or so, we can expect to produce only 284 billion barrels of those 810 billion barrels of "oil-in-place." The actual "reserves"–the oil identified and considered economically recoverable–is estimated at 52 billion barrels (see Table 11-15). For 1973 we produced 3.36 billion barrels of oil. If that rate of production continues, the "reserves" will last only 15.5 years. We did not, of course, produce all the oil we needed in 1973; our total consumption including petrochemical raw materials and other nonfuel uses was 6.30 billion barrels. If we were to provide all that oil from these domes tic reserves, they would last only 8 years.

These lifetime calculations assume constant consumption at the 1973 level and are based on the "reserves." A different perspective is given if we base our calculations on growing consumption and on the total estimated resources (the "resource base" of Table 11-15). From 1960 to 1970 United States oil consumption grew at about 4.5 percent per year. If the 1973 consumption of 6.30 billion barrels per year were to grow at 4.5 percent per year, our domestic oil resources, 284 billion barrels would last 17.5 years.

These two computations show some of the uncertainty in making predictions of resource lifetimes. The results depend in an important way on whether "reserves"

Table 11-12 Oil-in-Place Resources*

Region	Billion barrels: Ultimate discoverable oil-in-place	Billion barrels: Oil-in-place discovered to 1/1/71	Remaining discoverable oil-in-place: Billion barrels	Remaining discoverable oil-in-place: percent of ultimate
Lower 48 states–onshore				
2 Pacific Coast	101.9	80.0	21.9	21.5
3 Western Rocky Mountains	43.6	5.8	37.8	86.7
4 Eastern Rocky Mountains	52.4	23.9	28.5	54.3
5 West Texas area	151.6	106.4	45.2	29.8
6 Western Gulf Coast basin	109.0	79.7	29.3	26.9
7 Midcontinent	63.0	58.4	4.6	7.3
8-10 Michigan, Eastern Interior, and Appalachians	36.5	30.5	6.0	16.4
11 Atlantic Coast	3.8	0.2	3.6	94.7
Total	561.8	384.9	176.9	31.5
Offshore and South Alaska				
1 South Alaska including offshore	26.0	2.9	23.1	88.8
2A Pacific Ocean	49.6	1.9	47.7	96.2
6A Gulf of Mexico	38.6	11.5	27.1	70.0
11A Atlantic Ocean	14.4	0.0	14.4	100.0
Total	128.6	16.3	112.3	87.3
Total United States (except North Slope)	690.4	401.2	289.2	41.9
Alaskan North Slope				
Onshore	72.1	24.0	48.1	66.7
Offshore	47.9	0.0	47.9	100.0
Total	120.0	24.0	96.0	80.0
Total United States	810.4	425.2	385.2	47.5

*Source: A Summary Report of the National Petroleum Council's Committee in U.S. Energy Outlook, *op. cit.*

or the "resource base" are used and on assumptions as to the growth of consumption.

No matter what the assumptions, however, we see that the oil resource picture is thus quite different from the one we saw for coal. We are just beginning to make inroads into the vast resources of the solid fuel while we can almost see the end to the liquid. We must emphasize again, however, that the actual amount of oil we will produce each year from this country's still-rich fields will not be resource-limited in the next decade or two, but will be determined by economic and environmental constraints.

It is of interest to note in these days of headlined oil shortages and threatened gas rationing that the amount of drilling for oil (in millions of feet per year, for instance) undertaken by the oil companies has decreased each year since 1957. With the exception of the large North Slope discovery in Alaska, the ratio of reserves to production has been steadily dropping. For whatever reasons—lack of investment money, increasing difficulty of new drilling, or the like—the inadequacy of domestic supplies has arisen in part, at least, because the oil companies have not been looking for oil with sufficient intensity.

To reach the optimistic production rates of the NPC Case 1 (11.75 million barrels per day in 1980 and 13.54 million barrels per day in 1985), the oil companies will demand favorable leasing and tax arrangements and insist on completion of the Alaskan pipeline and on rights to explore and drill off the Atlantic Coast. The oil is there, at least for our generation, we must decide what price we are willing to pay for it.

Natural Gas. The resource picture for natural gas is similar to that for oil; they were formed in the same general geological processes and are often found together. In discussing natural gas resources, a distinction is made between *associated-dissolved gas* and *nonassociated* gas. The former is found with oil, and its total quantity can be estimated by projecting the historic ratio of gas to oil. The latter is found alone. Estimates for nonassociated gas are much more difficult to make. The amount found per foot of drilled well, for instance, fluctuated from 140 million cubic feet to 408 million cubic feet between 1955 and 1970. The NPC estimate of recoverable gas supplies is shown in Table 11-13. We see that the nonassociated gas

Table 11-13 Recoverably Gas Supply

Region	*Ultimate discoverable gas* (Trillion cubic feet)	*Gas discovered to 1/1/71* (Trillion cubic feet)	*Remaining discoverable* TCF	*Remaining discoverable* Percent of ultimate
		Nonassociated		
Lower 48 States–Onshore				
2 Pacific Coast	25.7	8.1	17.6	68.5
3 Western Rocky Mountains	50.1	17.9	32.2	64.3
4 Eastern Rocky Mountains	51.6	10.0	41.6	80.6
5 West Texas area	101.5	27.2	74.3	73.2
6 Western Gulf Coast basin	397.9	211.7	186.2	46.8
7 Midcontinent	223.3	104.8	118.5	53.1
8-9 Michigan, Eastern Interior	12.5	0.4	12.1	96.8
10 Appalachians	95.9	33.0	62.9	65.6
11 Atlantic Coast	4.6	0.01	4.6	99.8
Total	963.1	413.1	550.0	57.1
Lower 48 States–Offshore				
2A Pacific Ocean	3.8	0.5	3.3	86.8
6A Gulf of Mexico	201.8	45.4	156.4	77.5
11A Atlantic Ocean	54.5	–	54.5	100.0
Total	260.1	45.9	214.2	82.4
Total United States (except Alaska)	1,223.2	459.0	764.2	62.5
Alaska	277.4	5.1	272.3	98.2
Total United States	1,500.6	464.1	1,036.5	69.1
		Associated-Dissolved		
Total United States	356.7	215.2	141.5	39.7
		Nonassociated and Associated-Dissolved		
Total United States	1,857.3	679.3	1,178.0	63.4

*Source: A Summary Report of the National Petroleum Council's Committee in U.S. Energy Outlook, *op. cit.*

makes up 81 percent of the total. According to industry estimates, 60 percent or so of the remaining discoverable supply of 1,178 trillion cubic feet is in operationally difficult areas: approximately 14 percent is below 15,000 feet, 20 percent is offshore, and 78 percent is in Alaska. As in the case of oil, there is a fairly large variation between estimates of the total gas supply made in different studies. Table 11-14 gives a summary of other estimates to be compared with the 1,857 trillion cubic feet estimate made by the NPC.

Since gas is more mobile than oil, its recovery efficiency is higher, about 80 percent as opposed to 30 percent. Therefore, we can expect to obtain about 1,500 trillion cubic feet of the 1,857 trillion cubic feet potential supply. As was the case with oil, the lifetime we predict for this resource depends on the assumptions. The consumption of natural gas in 1973 was 22.85 trillion cubic feet. At this rate the reserves of 290 trillion cubic feet would last 12.5 years (see Table 11-15). If we look instead at the 1,500 trillion cubic feet, at the growth in consumption of 4.5 percent characteristic of the 1960-1970 period, this resource would last 22.5 years.

Nuclear-Explosive Stimulation. The AEC Ploughshare Program has long held out the hope that additional gas can be produced by exploding nuclear devices deep underground to break up tight reservoirs, regions of porous rock which contain gas, but in which the gas does not flow, and therefore cannot be pumped. It is suggested that a nuclear explosion could crush a large volume of this rock and release the gas from the pores. It is estimated that there are 90 trillion cubic feet of gas in the 250,000 acres of Rocky Mountain basins already under lease and that the total resources may be much larger; a Bureau of Mines estimate for those basins gives 300 trillion cubic feet as the potential.

There have been two tests of this procedure so far, "Gasbuggy" in northwest New Mexico and the Rulison experiment in Colorado. The results have had a mixed interpretation. Gas has been released, but it is radioactive. Not only is the radioactive gas, Kr^{85}, produced by the fission explosion, but tritium, radioactive hydrogen, can become part of the methane gas molecule. The AEC is now working to develop a low tritium yielding explosive in hopes of reducing the radioactivity problem.

The seismic shock of the explosion has also been troublesome. In the Project Rulison experiment in which a 40-kiloton explosive was set off 8,440 feet underground, the AEC has paid off 260 claims for damages (cracked cinder block, dislodged brick, damaged plaster, to name some) totalling 72,400 dollars. It is difficult

Table 11-14 Estimates of Nonassociated and Associated-Dissolved Gas*

	TCF					
Area	*1970 PGC*	*1972 USGS*	*1969 Hubbert*	*1959 Weeks*	*1970 Moore*	*1968 Elliott and Linden*
Lower 48 states	1,877	3,556	1,312	not estimated	not estimated	not estimated
Alaska	447	862	188			
Total United States	2,324	4,418	1,500	1,250	1,934	2,175

*Theobald, Schweinfurth, and Duncan, *op. cit.*

to imagine that local residents will tolerate the present plans which call for many such explosions in that area. It may be possible to use conventional explosives instead of nuclear. There is, however, a further complication at present. The Colorado basin targeted for explosive stimulation is also an area of rich potential for oil shale exploration. These two types of resource exploitation are probably not compatible. It is quite likely that the seismic shock of the explosion would fracture the pillars of any existing underground mines and prevent pillar construction in new ones. Even the deep open pit mines would be damaged. At best, the future of this controversial means of gas production is still uncertain.

Oil Shales. In Table 11-7 we saw that a major portion of the world's fossil carbon is locked into *oil shales.* These shales are the rocky remains of the bottoms of the great lakes and seas which covered much of this country and the world millions of years ago. Organic material which settled into the bottom muds and sands has been converted, by the action of heat and pressure, into the large organic molecules that characterize petroleum.

Almost all shale has some organic constituent but most of it is not of commercial interest. There are, however, many large shale deposits so rich in the petroleum compounds that commercial recovery is, or is becoming, feasible. The recovery technique is simple: the shale is heated in a closed vessel or a retort, and the petroleum liquids and gases are driven from the rock and collected. The heat for this can be provided by burning recycled low Btu gas obtained from the shale itself. This retorting technique predates conventional oil production and was used to produce oil for illumination in the nineteenth century.

The richness of oil shales is designated by the gallons of petroleum liquid which can be distilled from a ton of rock. The richest deposits are in the Soviet Union in the prewar county of Estonia where deposits average as high as 1.5 to 2 barrels per ton, or 60 to 80 gallons per ton. Shale of this richness can be burned as is. The richest United States deposits are the remains of three large fresh-water lakes that existed in the Utah-Wyoming-Colorado area 40 to 50 million years ago during the Eocene Era. The location of these lakes and the deposits that remain from them are shown in Fig. 11-10. Although the existence of these fuel resources has been known for some time (the Ute Indians burned shales in their campfires) it is only in the past few years that real commercial interest has developed.

The western oil shale shown in Fig. 11-10 underlies 25,000 square miles (16 million acres) of the Green River Formation, as this area is known; 17,000 square miles or 11 million acres of this shale is potentially valuable.

It is estimated that the oil content of these Green River shale deposits, of at least 10 feet thickness and containing at least 25 gallons per ton, total 600 billion barrels of oil, an amount comparable to the 810 billion barrels of oil-in-place of Table 11-12. An additional 1,200 billion barrels could be recovered, it is estimated, from shale deposits of the same minimum thickness with yields of 15 to 20 gallons per ton. The Department of the Interior, in its massive study of the potential and environmental impact of oil shale development,[13] set as its most optimistic projection the beginning

[13] United States Department of the Interior, *Final Environmental Statement for the Phototype Oil Shale Leasing Program* (6 vols.)(Washington, D.C.: USGPO, 2400-00786, 1973).

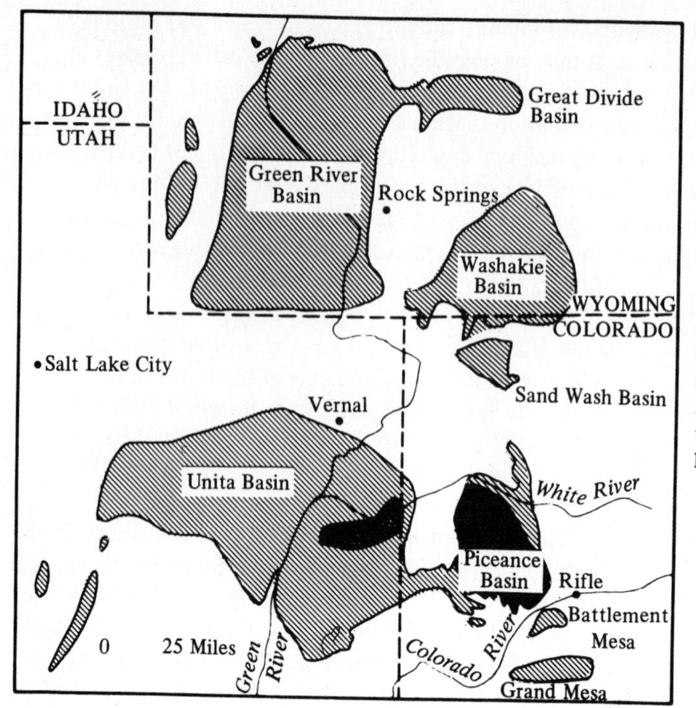

FIGURE 11-10 Large oil shale resources are concentrated in the prehistoric lake beds of these four basins in Wyoming, Colorado, and Utah. (Source: A Summary Report of the National Petroleum Council's Committee in U.S. Energy Outlook, *op. cit.*)

of commercial production in 1975 at a rate of 50,000 barrels per day (to be compared with a total domestic oil production of 9.6 to 10.2 million barrels per day) and sees this rising to 300,000 barrels per day in 1980 as "second generation" technology is established. By 1985 a cumulative capacity of 1 million barrels per day might be established and contribute importantly to the 15.5 million barrels per day total domestic oil production of the "optimistic" Case 1 projection of the NPC.

The environment will not give up the oil from the shale free-of-charge. The 1 million barrels per day production rate will require the processing of about 1.5 million tons of rock a day and the disposal of 1.2 million tons of spent shale. The mining will be both surface and underground and the 1985 production rate will involve 75,000 to 80,000 acres of land, about 1 percent of the total oil shale area. Waste will be disposed of by dumping in, for instance, dry canyons, and, after a sufficient number of years, by back-filling in the mines. Land recovery and revegetating in these arid areas will be difficult; wildlife habitat will be destroyed and recreational value diminished. The population influx would double the present one, creating both opportunities and problems. Water demand will increase, but much of the water needed for the actual processing operation (mostly for carrying the wastes in slurry form to the dumping area) is expected to be available from the mine site itself, as the mine is deepened below the groundwater level.

Tar Sands. Another source of solid oil is the *tar sands,* oil-impregnated sands, sandstone, and limestone. These asphalt-like deposits can also be recovered by mining the rock and heating it. The largest United States tar sands are in Utah and are estimated to hold the equivalent of 17.7 to 27.6 billion barrels of oil. The Department of the Interior projects an industry capable of producing 1 million barrels per day for 50 years as supportable from the Utah sands, but considers it unlikely that such an industry will develop before 1985. The environmental impact of this industry (strip mining, waste disposal, water demand) will be similar to that of the oil shale recovery.

The largest known tar sands deposits are in Canada, at Athabasia in the province of Alberta. Upwards of 174 billion barrels are estimated as being economically recoverable from this deposit, and commercial exploration is already producing 45,000 barrels per day from 100,000 tons of sand strip-mined per day. By 1985, this source may add as much as 1.25 million barrels per day to North American production.

Total Energy from Fossil Fuels

In Table 11-15 we summarize the energy available from the United States fossil fuel resources. We show the three categories, the resource base, the ultimately recoverable resources which take into account recovery percentages and economic factors, and the actual reserves, which are, of course, the smallest figures. We also show these data visually in Fig. 11-11. We see that the total resource of coal dwarfs the others and is followed by oil shale. For comparison, the total United States energy demand in 1970 was 69 Q Btu, the average projection for the year 2000 is 175 Q Btu (see Fig. 11-4). Using Fig. 11-4 as a guide and working from the average prediction, the

Table 11-15 Energy of Fossil Fuel Resources in the United States*

Resource	*Resource base*	*Ultimately recoverable*	*Reserves*
Coal			
(billion tons)	3,200	590§	150
(Q Btu)	83,500	15,200	3,900
Oil			
(billion barrels)	810	149‡	52§
(Q Btu)	4,700	870	302
Natural gas			
(TCF)	1,857	543‡	290§
(Q Btu)	1,880	560	300
Oil shale			
(billion barrels)	1,800†		
(Q Btu)	10,400		
Tar sands			
(billion barrels)	30		
(Q Btu)	174		

*Estimates are from the National Petroleum Council except where noted. Conversion to Btu are 5.8 M Btu per barrel of oil, 1030 Btu per TCF for gas and 26 M Btu per ton for coal.

†Includes all shale with greater than 15 barrels per ton.

‡NPC data for "discovered" oil and gas with 35 percent recovery assumed for oil and 80 percent recovery assumed for gas.

§Ford Foundation Energy Policy Project, "Exploring Energy Sources," (Washington, D.C.: Ford Foundation, 1974).

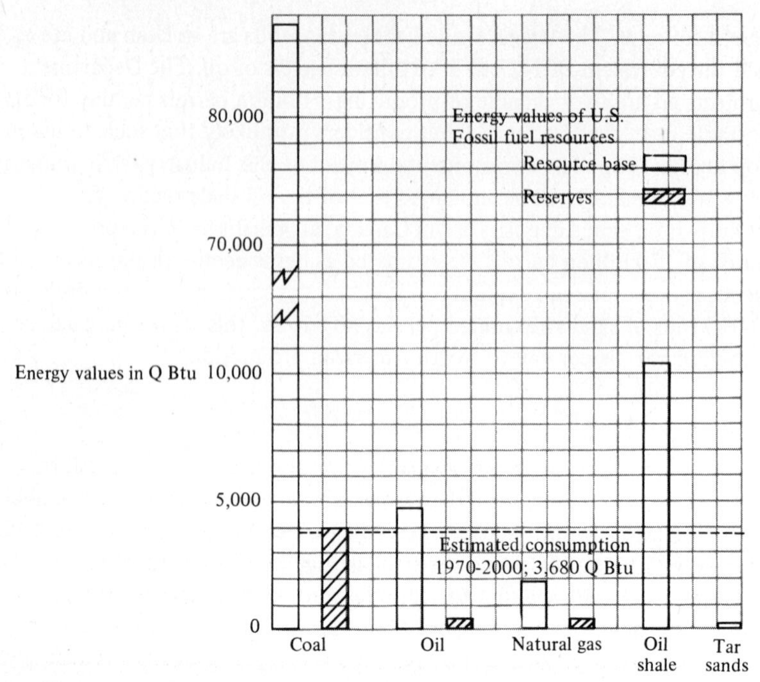

FIGURE 11-11 Energy values of U.S. fossil fuel resources. The total resources of coal are impressively large but none of the "economically recoverable" resources could alone supply the energy needs of the U.S. for the next 30 years. Data for this figure are given in Table 11-15.

cumulative energy consumption for the period 1970-2000 is 3,680 Q Btu. Displaying this total in Fig. 11-11 reminds us that, as far as the fossil fules are concerned. we must look more and more to coal.

Nuclear Energy Resources

If the energy needs of this country were to be satisfied only from the fossil fuels, the future we would face would be even more disturbing than it is. As we have shown in previous sections, the easy fuels, natural gas and oil, are already on the wane. To keep the heat engines turning and the buildings warm by continuing to draw on this stored solar energy, we will have to turn to the more difficult-to-manage fuels: to take oil from the shales and tar sands, take the sulphur out of coal and, ultimately, liquify and gasify it. We face a steady rise in the prices of these fuels.

It is the pressure of the demand for cheap fuel, especially electricity, and for more and more of it, that is causing us to accept the risk of playing with the nuclear fires. We are committed, for the remainder of this century, to turning more and more of our generators with heat from the controlled chain reactions. Thus, it is now of critical importance to look at the extent of these new resources.

In comparison to carbon, uranium is rare, it occurs in only 2 parts per million, on the average, in the earth's crust. It is, fortunately, concentrated much more than this in certain rocks. In granite, for instance, the basic rock of the earth's crust, there are 4 parts per million, on the average; in certain copper ores there are 6 parts per million, and the Conway Granites of New Hampshire contain 12 parts per million. These concentrations range upwards to the 2,000 or 3,000 parts per million of the ores which are now being mined (see Table 11-16). The geographical occurrence of uranium ore is shown in Fig. 11-12.

The data in Table 11-16 refer to natural uranium which is largely U^{238}. Only 0.7 percent of this is U^{235}, the major fuel for the present day light water reactors. The enormous amounts of energy locked in the U^{235} nucleus makes it feasible, however, to search out these small amounts. A single gram of U^{235} is equivalent, if all its fission energy is released, to 3 tons of coal. The various ores, whose uranium concentration we show, are equivalent in energy content to the amounts of coal shown in the second column. We give some idea of the costs of these ores in the third column. These dollars per ton are based on the cost of producing U_3O_8 from the raw ore.

While some of the more dilute sources of uranium, the granites, shales, and phosphate rocks, may become fuels for the breeder reactor in the next century, the only ores now in commercial use are those with the largest concentrations and costs of 8 dollars per pound or less. About 80 percent of the uranium *reserves* (known deposits, recoverable by present technology) in the 5 to 10 dollars per pound category are in New Mexico and Wyoming, split 50-50 between the two states. These reserves, ac-

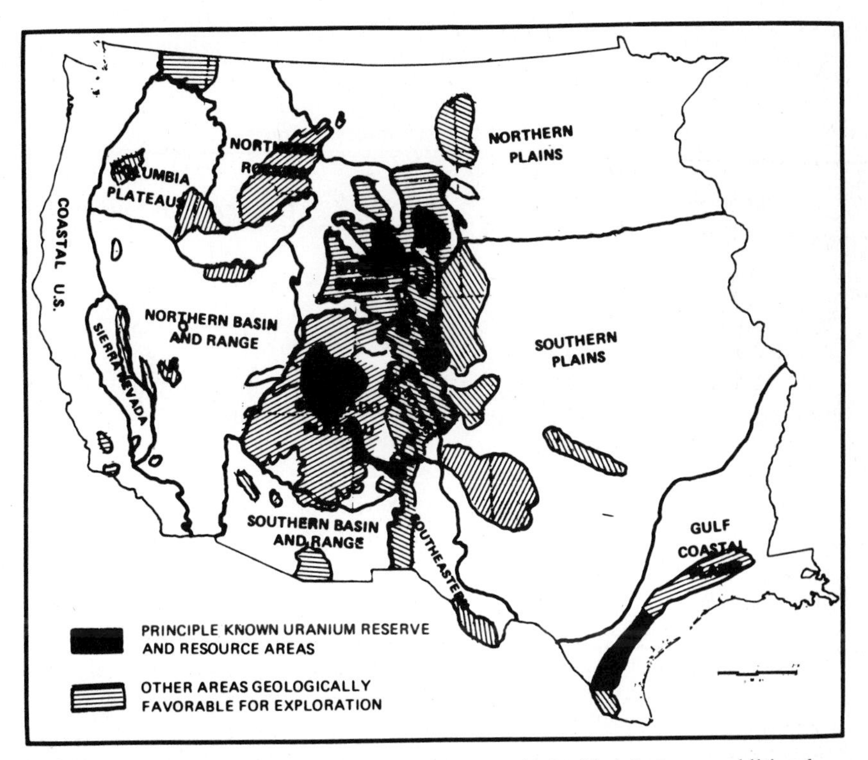

FIGURE 11-12 The major proven reserves of uranium are in the West, but some additional discoveries may be made in the Gulf coastal states. (Source: A Summary Report of the National Petroleum Council's Committee in U.S. Energy Outlook, *op. cit.*)

Table 11-16 Costs, Energy Value, and Concentration of U^{238} from Different Sources*

Source	U^{238} concentration (ppm)	Coal equivalent tons coal per ton of rock	Cost/ton of U_3O_8
Conway Granite (N.H.)	12	2	>$100
Chattanooga Shale (Tenn.)	70	13	~$ 70
Phosphate Rock (Florida, Montana, Idaho, Wyoming, Utah)	100-200	19-38	no estimate
Present ore (New Mexico and Wyoming)	2,000-3,200	380-610	$5-$10

*Source: U.S. Atomic Energy Commission.

cording to the last AEC estimate, amount to some 300,000 tons, enough ore to fuel 150 of the 1,000-mw LWR plants for 10 years (see Table 10-3, Chap. 10).

A more realistic comparison with fossil fuel resources must use the broader category, ultimately recoverable resources, which will include extensions of presently defined ore beds and estimates of the content of ores in regions which should, according to geological evidence, contain similar rich ores. A summary of the AEC resource estimate of uranium ore available in different price categories is provided by Table 11-17. The category, "by-product," is an interesting one. It is estimated that at least 1,000 tons per year of under 10 dollars per pound uranium could be recovered as a

Table 11-17 Estimates of United States Uranium Resources*

Source	Amount (thousands of tons of U_3O_8)
$5-$10/lb	
Reasonably assured	
Conventional	300
By-product	90
Estimated additional	680
Total	1,070
$10-$15/lb	
Reasonably assured	
Conventional	140
By-product	20
Estimated additional	
Conventional	300
By-product	50
Total	510
$10-$30/lb	
Reasonably assured	
Conventional	280
By-product	20
Estimated additional	1,000
Total	1,300

*Source: U.S. Atomic Energy Commission.

by-product from copper refining. The AEC gives an expectation of 90,000 tons from this process. In the next price range there is an additional 85,000 tons of uranium available as a by-product of the process by which phosphoric acid is made from phosphorous rock. We can add to the totals shown in Table 11-17, 50,000 tons of U_3O_8 which the AEC is expected to release for commercial use from its own stockpiles.

The only meaningful way to judge these amounts is to compare them to electric energy needs. From the data in Table 10-3 we find that a 1,000-mw LWBR operating at 80 percent capacity uses 200 tons of U_3O_8 in a year. During that year (at 80 percent capacity) it will generate 7 billion kw-hr of electric energy. The conversion of U_3O_8 to electrical energy is thus at a rate of 35 million kw-hr per ton. In 1970 the total consumption of electrical energy amounted to 1.64 trillion kw-hr. If all that electrical energy had been derived from U_3O_8, it would have required 47,000 tons of this ore. If the consumption of electricity grows at its 1970 rate of 6.5 per year, the 390,000 tons of "reasonably assured" U_3O_8 at under 10 dollars per pound of the AEC estimate will last only about 7 years (see Table 11-17). The total of "reasonably assured" and "estimated addition"–1.07 million tons available at this price–will only last 14 years. Even though nuclear fuel is not expected to be supplying all the electricity even by the year 2000, these calculations drive home the point that present resources are quite limited. It is from this sort of comparison that the pressure to build the breeder reactors derives. Using the U^{238} as well as the U^{235} multiplies the resource base and the years of adequate supplies by a factor of 50.

Thorium. The gas-cooled breeder reactors can also use thorium, Th^{232}. There has not been an intensive search for Th^{232} so far. Interest in it should develop in the mid-1980s if the proposed breeder development goes according to schedule. The latest AEC estimate (1967) is that there are 100,000 tons "reasonably assured" at under 10 dollars per pound and an "estimated additional" of 300,000 tons for a total of 400,000 tons. There are also significant totals at higher prices, 200,000 at 10 to 30 dollars per pound and at least 10,000,000 tons at 30 to 50 dollars per pound.

It is clear that present resources of U^{238} and Th^{232} assure, through the breeder reactor, ample electrical energy for years to come. Given the possible hazard of the breeder, however, it is of interest to ask a different question: Can we survive until 2000 without the breeder? In other words, do we have enough U^{235} to provide the expansion predicted in nuclear generation of electricity?

We will use as the target generating capacity: 6,500 mw in 1970, 140,000 mw in 1980, 475,000 mw in 1990 and 1 million mw in 2000–capacities which are in line with present forecasts. As we have seen earlier, a 1,000-mw LWR uses 200 tons of U_3O_8 per year when it is operated at 80 percent capacity. The amounts of U_3O_8 needed each year over the next three decades are shown in Fig. 11-13. The total fuel needs for the period 1970-2000 can be computed from Fig. 11-13 to be 2.4 million tons. From Table 11-17 we see that our resources in the under 30 dollars per ton price range can provide for somewhat more than that.

Since we would be using more expensive uranium ore than was used to estimate costs in, for instance, Table 10-4 of Chap. 10, the cost of electricity would rise. It is estimated that this cost increases 0.06 mill per kw-hr for each dollar per pound price increase of U_3O_8. Taking the cost of the electricity generated from the 1 million tons

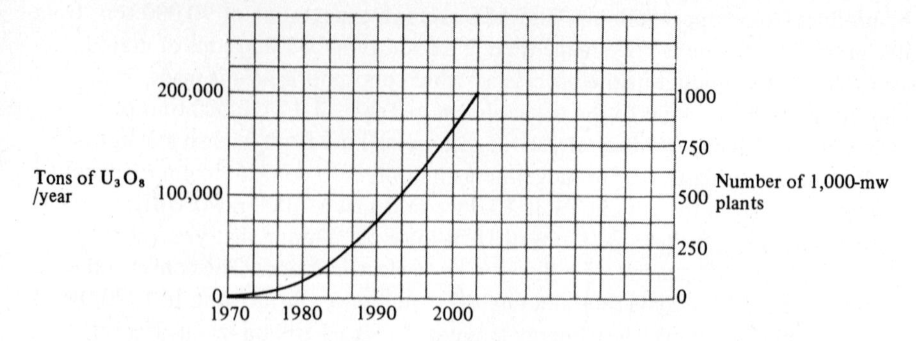

FIGURE 11-13 The need for U_3O_8 will grow rapidly. If all the projected generating capacity in the next 30 years depends on light water reactors the cumulative demand (the area under this curve from 1970 to 2000) will be for 2.4 million tons.

of U_3O_8 available at under 10 dollars per pound as 8.6 mills (in keeping with Table 10-4) and the additional 1.4 million tons at an average price of 20 dollars per pound, the cost of electricity generated from this more expensive U_3O_8 will only rise by 0.6 mill to a little more than 9 mills, an increase in the total price of less than 10 percent. With the projected increases in coal costs, nuclear fuel would, it would appear, remain competitive.

What this rough calculation suggests is the following. We can get electric energy during the decade 1990-2000 at a somewhat cheaper rate if we successfully undertake a crash program of breeder development. The real savings will come in the first two decades of the twenty-first century when the LWRs would begin to use even more expensive U_3O_8. It would be possible, however, with our reasonably priced U_3O_8 resources, to exist comfortably without the breeder until 2000 when fusion reactors, solar power, or some other source could begin to take over. This is an option which, in the present climate of opinion, is being successfully concealed. It should be more openly examined.

THE WORLD PICTURE

Demand

We have already looked at the historical record of growth in world energy consumption (see Figs. 5-23, 5-24, and 5-25 of Chap. 5). In a comparison with the United States record several things stood out: The United States has, by far, the highest per capita rate of consumption, among the major countries, 20 percent higher than the next country, Canada. (Actually, the tiny, oil-rich country of Kuwait has a per capita consumption even higher than the United States.) Our country, therefore, accounts for the consumption of 35 percent of the world's total energy with only 6.4 percent of its population. But, in a sense the United States has led the world into the energy-intensive age and therefore has reached the potentially retarding barriers earlier. As we have seen, several other countries now have energy consumption growth rates higher than the United States; in fact, the United States share of the total energy, as disportionate as it is, has been falling; it was 48 percent in 1950, fell to 39 percent in 1960, and this relative decline is expected to continue.

The extrapolation of world energy demand, therefore, is based on higher rates of per capita energy growth and higher rates of population growth than were used in the United States projections. The difficulties of projection which we warned of in connection with the estimation of future United States demand are compounded when the world's needs are under consideration. The number of political, technological, and other uncertainties are multiplied, and the data and reference values are less reliable.

It is still useful, however, as a means of giving perspective to energy resource estimations, to make some kind of projection of world demand. The one shown in Fig. 11-14 is due to Darmstadter.[14] It is based on a population growth of 2 percent per year (slowing a bit toward the end of the century) and energy consumption growth of 5 percent during the next decade and 4.5 percent afterwards. This is lower than the 5.5 percent growth during the 1960s and higher by 1.5 percent than the projected United States growth.

Figure 11-14 also shows the projected values of world per capita growth and compares them with the average projections of United States per capita consumption. It is easy to see that the possibility of equality is still a long time off, even if the United States growth drops to zero.

It is also of interest to look at the anticipated fuel mix for the next three decades. As Fig. 11-15 shows, the pattern is again a time-delayed reproduction of the United States experience: coal declining, oil and gas rising into the 1980s and then declining.

The forecast of world electrical energy consumption is shown in Fig. 11-16. Both the world total and the world per capita consumption are predicted to grow more rapidly than the United States consumption; in addition, world per capita consumption of electricity, on the average, will reach the United States 1945 level in 1980, and, by 2000 almost the United States 1960 level.

[14] Darmstadter, *op. cit.*

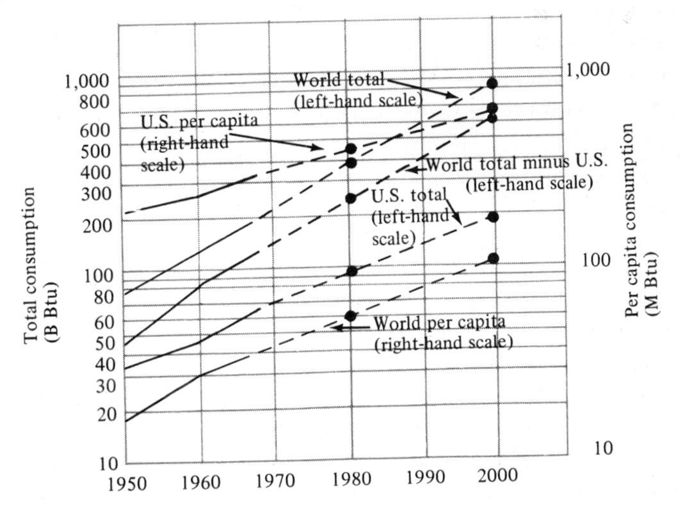

FIGURE 11-14 Projections of worldwide total and per capita energy consumption. The world total energy consumption and per capita energy consumption are both growing more rapidly than the U.S. curves. (Source: Darmstadter, *op. cit.*)

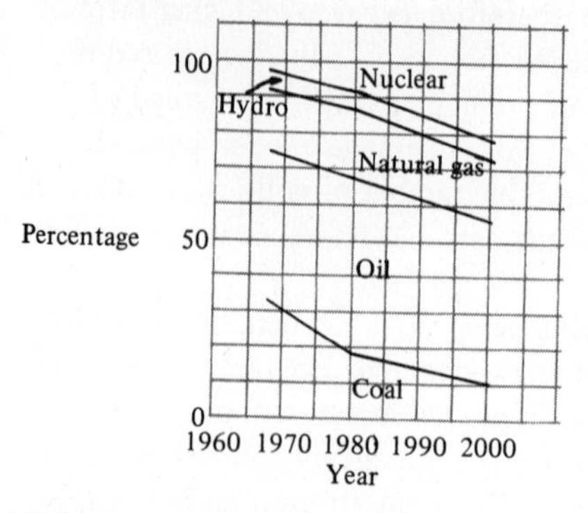

FIGURE 11-15 In the world, as in the U.S., the percentage importance of coal is expected to decline while oil and gas increase in importance until 1980 and then remain about constant. Nuclear energy is expected to provide 20 percent of the world's energy by the year 2000.

A large amount of this new electrical generating capacity is expected to be nuclear; the nuclear contribution to the world generating capacity which was 1.4 percent in 1968 (as compared with 0.9 percent in the United States), is predicted to grow to 12 percent by 1980 and 40 percent by 2000. This is a risky prediction; nuclear generating plants are expensive, and the developing countries, in particular, will be dependent on imports. If they are not to lag far behind in this development, they will have to be given massive financial as well as technilogical assistance.

The world's resources of uranium also enter into our evaluation of this projection. The largest "free-world" reserves of uranium outside the United States lie in Canada and South Africa. The estimated extent of these resources in the various price ranges are given in Table 11-18. We see that the free-world resources are of the same size, according to present estimates, as those in the United States.

The anticipated large increase in demand for uranium during the next decade may have two different effects. It will stimulate exploration and production and may lead to significant increases in these resource estimates. It may also drive the price of uranium up, as has occurred in Europe and Japan (where the fossil fuels are already imported and expensive) when they began to compete in the market for the nuclear fuels.

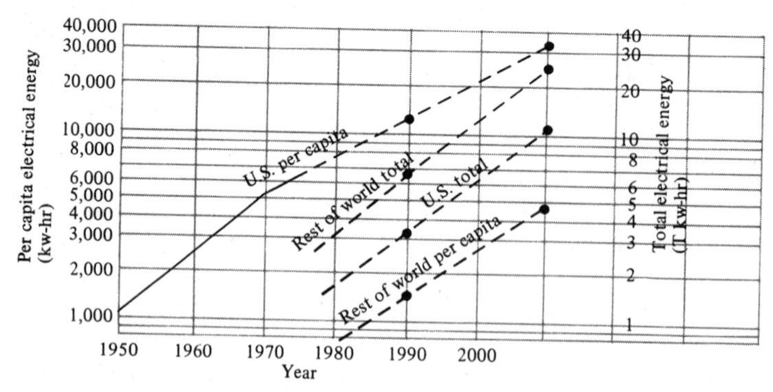

FIGURE 11-16 Consumption of electrical energy. The world is following the U.S. pattern and quickly converting to electrical energy, but the U.S. has a 40-year head start. (Source Darmstadter, *op. cit.*)

Table 11-18 Free World Uranium Resources

Source	Amount (thousands of tons of U_3O_8)
$5-$10/lb	
Reasonably assured	629
Estimated additional	294
Total	923
$10-$30/lb	
Reasonably assured	1,078
Estimated additional	960
Total	2,038

The main interest in the projections of Figs. 11-14 and 11-15 is the estimation they allow of the cumulative energy needs of the world during the next three decades and the specific predictions of the consumption of coal, oil, and gas. From Fig. 11-14 we can estimate that the total energy consumed by the world, from 1968 to 2000 will be 14,300 Q Btu.

To find the estimated amounts of coal, oil, and natural gas to be consumed we need to use the projected totals in Fig. 11-14 and the percentages in Fig. 11-15. This computation sets world needs during the next 30 years at 132 billion tons of coal, 1,100 billion barrels of oil, and 2,680 trillion cubic feet of natural gas. These numbers, for all their inherent uncertainty, give us some perspective to apply to the next question: How much fossil fuel remains in the world?

The World Picture—Supply

As was the case with United States fossil fuel resources, coal is the major world energy resource. It is also to be expected that the estimates of coal resources are more accurate than those of oil or natural gas because coal occurs, most of the time, where geological mapping says it should. Figure 11-17 shows the global geographic distributions of known coal beds.

The petroleum and natural gas picture is not so definite. The areas which are difficult to explore, have not been explored. New discoveries such as that in the North Sea a few years ago, and the more recent Alaskan discovery, have added significantly to the totals. One can say, however, that oil will be found only in the old sea beds, the sedimentary basins, and these are at least roughly mapped. Their worldwide distribution is shown in Fig. 11-18 and the discovered oil regions are marked.

Quantitative estimates of coal, oil, and natural gas, as well as oil shale and tar sands are displayed in the last two figures. Figure 11-19 shows the recoverable resources by region. These data, which are from Williams,[15] use the same assumptions as were employed in the United States estimates. Coal is assumed to be 50 percent recoverable and only seams of thickness greater than 14 inches and depth less than 6,000 feet are considered. The recoverable petroleum is taken to be 40 percent of the oil-in-place, and gas, 80 percent of the total potential for natural gas. The oil

[15] Williams, *op. cit.*

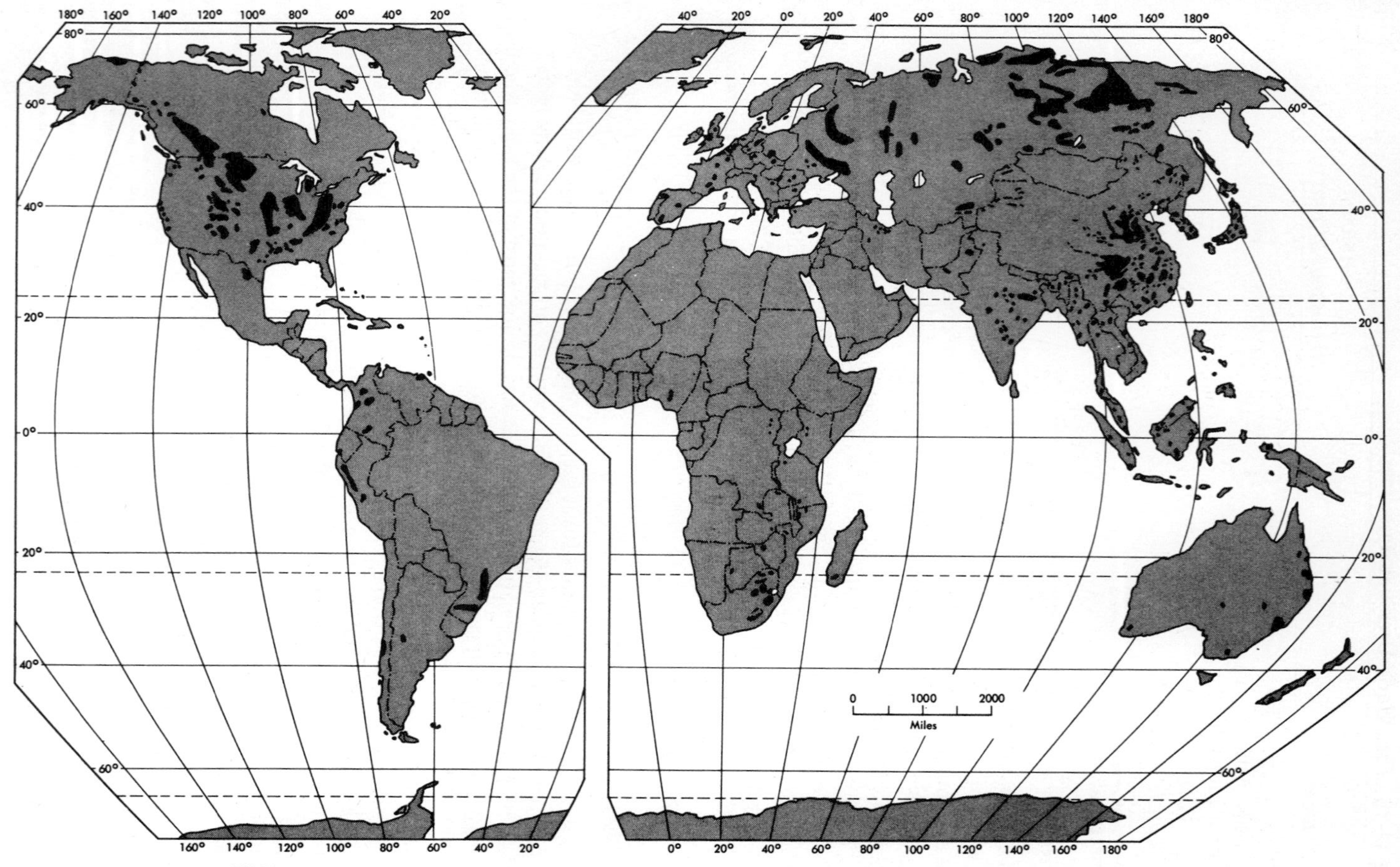

FIGURE 11-17 The importance of the coal resources of the U.S., the Soviet Union, and China are clearly shown on the map. (Source: B.J. Skinner, *Earth Resources*, Prentice-Hall, N.J.)

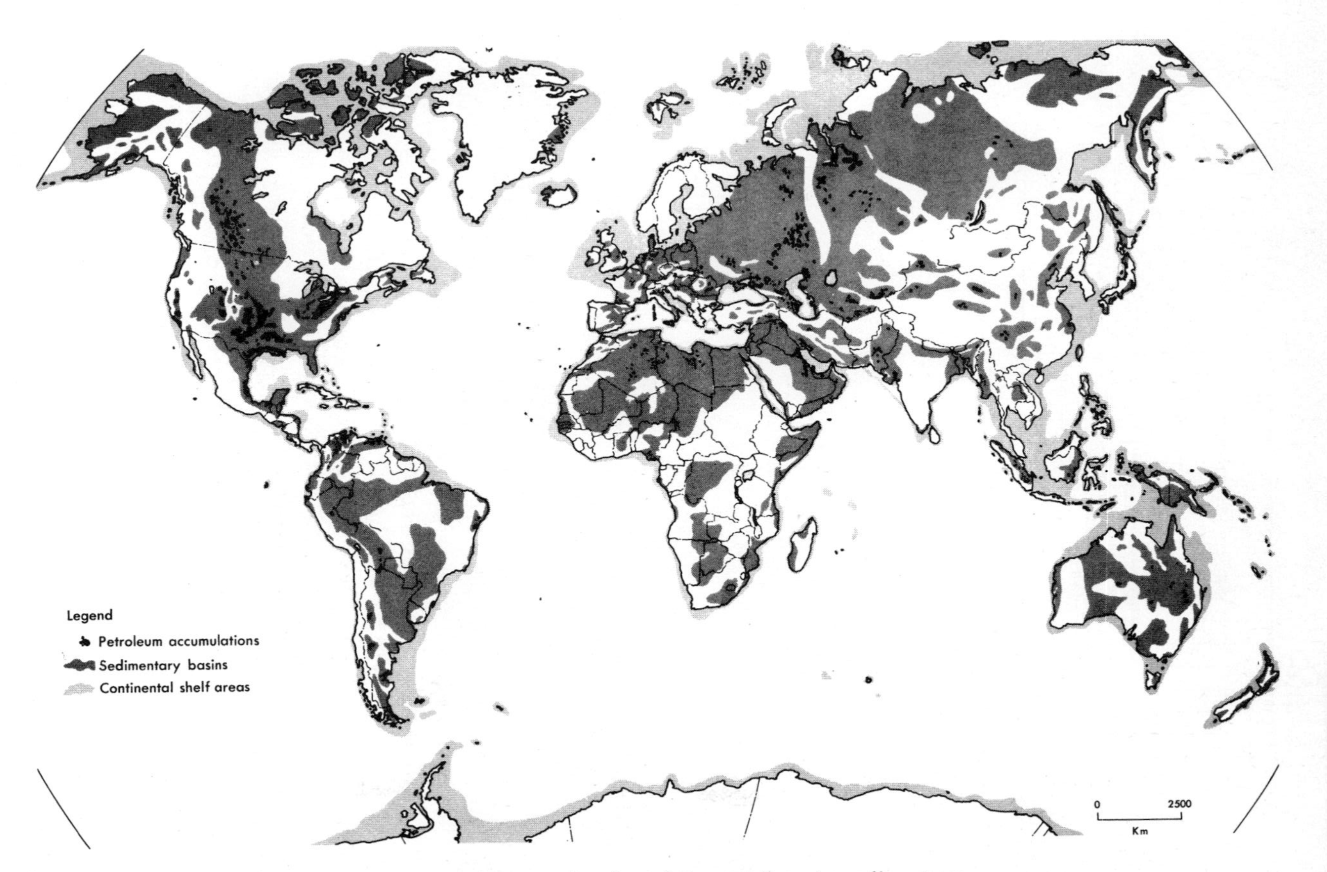

FIGURE 11-18 Oil has been found so far in only a few of the world's major sedimentary basins. (Source: Skinner, *op. cit.*)

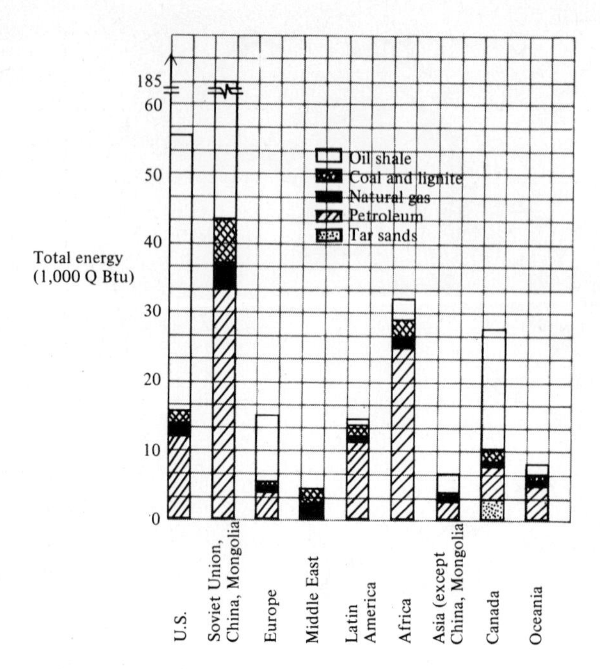

FIGURE 11-19 The earth's fossil fuel resources are not evenly distributed throughout the globe. (Source: Williams, *op. cit.*)

shale estimate is for shale with potential yield greater than 25 gallons per ton. (This resource estimate would be much larger if lower grade shale were included.)

Figure 11-19 shows the inequalities of resource energy distribution. The United States and the Soviet Union are superpowers in energy as well. Europe, Asia, and Oceania are without deep resources. The Middle East has all its energy eggs in one basket: petroleum. We also see that the Third World, at least Africa and Latin America, has a strong future when oil shale development becomes feasible.

These data are summarized in Fig. 11-20. Also shown for reference, is the total (14,300 Q Btu) of projected energy consumption for 1968 to 2000 and, in the appropriate bars, the projected cumulative consumption of coal oil and gas that we computed earlier. Again, we find comfortable reserves only in coal.

SUMMARY

In this chapter we have examined the Energy Crisis from the narrow viewpoint of resource adequacy. We have looked at demand, production capability, and total resources. What should have come through from the data and discussion is, that even with the soaring demands which are projected, our activities during the last thirty years of this century will not be resource limited. There is plenty of potential energy for the United States and for the world. Due to short-term lack of planning, unforeseen political events, and perhaps the search for higher profits, the production of energy in this country, long accustomed to energy self-sufficiency, will fall below

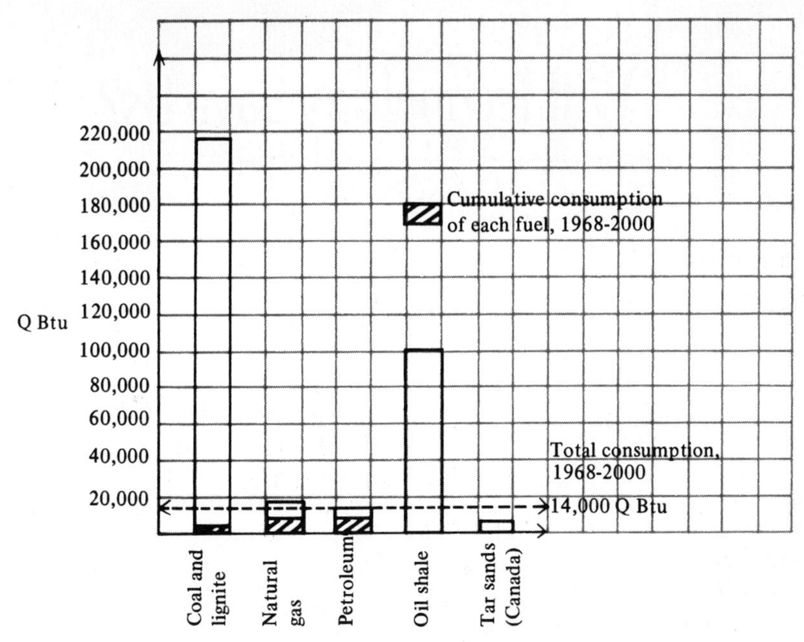

FIGURE 11-20 The coal resources of the earth are the largest remaining sources of fossil fuel energy, followed by the oil shales. The anticipated consumption of the various fossil fuels are shown by the cross-hatched regions, and the resources are compared with the projected cumulative consumption during the next three decades. (Source: Williams, *op. cit.*)

demand at least until 1980 or perhaps even 1985. It takes five to ten years to reverse trends in energy supply and demand, and the necessary measures were not taken in the late 1960s.

The longer view is somewhat rosier as far as resources are concerned; but, as we begin that accounting we must remember the lessons in Chaps. 7, 8, and 9 as well as the warnings in Chap. 10. Coal, the most abundant fuel, is the most troublesome environmentally. We must be prepared either to pay the price to neutralize its hazards, or to live with them. Uranium, the "clean fuel," has in its use the small probability of disaster and long-term damage. As we burn the last of our easy fuels, we must use some of the energy released to prepare adequately for dependence on the hard ones. We must also use some of this energy to make the technological breakthroughs which will fuel the twenty-first century.

12

Will Technology Save Us?

The great forward steps of civilization are at least connected in part to breakthroughs on the energy front. The discovery of fire gave primitive man security and comfort on the ground; the domestication of animals added their greater muscle capacity to his. Later on, the waterwheel opened up a new source of energy to exploitation, greatly increasing the power available to his tasks. Then, in the nineteenth century the industrial revolution was fueled by coal. Each of these breakthroughs combined advances in technology with discovery of new sources of energy. The technology of fire put wood alongside food as a source of energy. The invention of the waterwheel plugged us into a small additional share of the huge reservoir of solar energy. The discovery that coal would burn (and then oil and natural gas) would not have carried us far without the invention of the steam engine, electric generators and motors, the internal combustion engine, and so on.

In the previous chapter we have looked at the soaring rate of energy consumption and at the finiteness of our energy resources. There was good news and bad news there. In the short-term perspective there is adequate energy for the United States and the world, but the operational difficulties of extracting it and providing it in a form which matches our needs have led us into shortages. In the long-term these supplies of energy (fossil or fissionable fuels) will run out. What we did not see in Chap. 11 was the burning fuse of the next explosive breakthrough. We saw only the struggle to maintain our present level of existence.

THE PROMISE OF NEW TECHNOLOGY

In this and the following chapter we will try to identify areas in which that fuse could be lit. Again, it will no doubt require invention and discovery; a technological invention or an advance in science that will add yet another source of energy to those already domesticated by man. We will first examine some of the hoped-for ways in which technological advances may help us in the interim problem of carrying our present momentum into that future. These advances, which are not too far in the future, should help us get more energy and new kinds of energy out of old fuels. We will discuss some of the new sources of energy that can be made available within the next five years. Then, in Chap. 13 we will look at the "big ifs," solar power and fusion, which may give us the energy answer for the next century.

Clean, Convenient Energy from Coal

The message of Chap. 11 is clear: most of the forecasted increases in energy consumption over the next three decades will be provided for by a mixture of the old and the new, by a resurgence in the importance of coal, the first of the fossil fuels to be used, and by a dramatic growth in nuclear energy. Unfortunately, both these energy forms are limited in their application. Coal is a solid and its energy can be released only by burning it in complicated stationary conversion systems. Nuclear reactors suffer from similar drawbacks. Neither of them can service that part of the energy demand that is for piped energy, for liquid and gaseous fuels of high energy density. Only these fuels give the kind of quick-starting, controllable, and clean fires that are needed for mobile heat engines and many other specialized heating chores. Nuclear energy, converted to electricity, will fill some of the heating gap. Until economical storage of electrical energy is provided for, however, it will not power our cars. There is, therefore, a determined research and developmental effort under way to obtain liquid and gaseous fuels from coal.

Pipeline Gas from Coal. Techniques of producing a gaseous fuel from coal have been available for many years; the use of coal gas predates the use of natural gas. In the production of coke for steelmaking early in the eighteenth century the volatile gas driven from the coal in coke production was used to provide the heat for the process. As is usually the case, the existence of the new product created new markets and by 1800 gas for illumination was being produced deliberately by heating coal to produce *town* or *water* gas, a low Btu fuel (about 300 Btu per foot^3 as against 1000 Btu per foot^3 for natural gas). With the advent of cheap natural gas, interest in this process waned and has only recently been renewed by the shortage of the natural product.

The basic chemistry of the production of pipeline quality gas from coal is to bring the carbon from coal together with steam to form methane, the major constituent of natural gas. A flow chart for one of the processes now being tested is shown in Fig. 12-1. The pulverized coal is first heated to drive off the volatile components which can then be burned to provide heat; it is then mixed with organic solvents to form a slurry and pumped to the *hydrogasifier.* It is in this unit that the main conversion takes place.

Coal, steam, and hydrogen-rich gas at very high temperature and pressure (1500°C

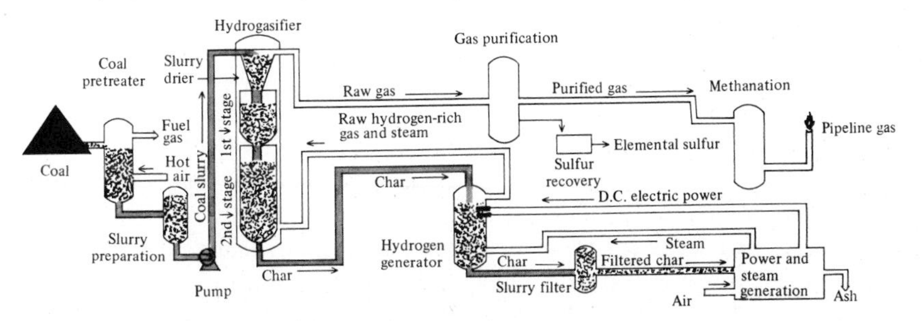

FIGURE 12-1 Flow chart for the Hygas process of coal gasification. (Courtesy of Institute of Gas Technology, Chicago, Ill.)

and 20 to 70 atmospheres) are brought together in the gasifier to form a gas composed mostly of carbon monoxide and hydrogen with water and sulphur compounds as impurities. These impurities are removed by the purifier. One of the important advantages of this and similar techniques is that the sulphur is in the form of hydrogen sulphide (H_2S) which can be easily removed and converted to elemental sulphur. The sale of this sulphur can add to the return from the process and thus lower the price of the gas, or it can be stockpiled.

After passing the purifier the gas then undergoes what is called *catalytic methanation,* in which the reaction $3\ H_2 + CO \rightarrow CH_4 + H_2O$ takes place in the presence of a suitable catalysis. The output pipeline gas, synthetic natural gas (SNG), is then 95 to 98 percent methane with a Btu value of about a 1000—essentially identical to natural gas.

We can see that the *Hygas process* (developed by the American Gas Association and the Institute of Gas Technology, Chicago, Illinois) whose flow chart is shown in Fig. 12-1 uses coal in several ways. The *char* (partially gasified coal) is burned to provide steam both for the gasification process and to produce electric power. It is also the char at the earlier stage, the *hydrogen generator,* which reacts with steam in the reaction $C + H_2O \rightarrow CO + H_2$ to produce the hydrogen-rich gas for the gasifier. Since these processes all occur at high pressures, the output is at a pressure suitable for pipeline transport, an important consideration.

The Hygas process is one of four processes in various stages of development in the United States. The only commercial coal gasification system presently available is the Lurgi process developed in Germany. The first commercial facility in this country, a 500 million dollar, 250 million cubic feet per day plant to be built by the El Paso Natural Gas Company in northwestern New Mexico, will use the Lurgi process. The speed with which the American processes can be brought to a competitive stage will depend on the amount of money available. Only about 40 million dollars has been spent by the Office of Coal Research (OCR) on coal gasification during the past 11 years. (The OCR is, in fact, the major source of support.) This amount is about one tenth the yearly funding for fission research and development and increased funding is desperately needed.

At the present, SNG is more expensive than natural gas (see Table 12-1). The expected tripling of the cost of natural gas, the high cost of liquified natural gas, and the scarcity of low-sulphur fuel oil make it only a matter of time before it will be able to compete.

With funding increased (the OCR expects to provide about 125 million dollars for coal gasification in the next three years) commercial operation might begin on a large scale in the early 1980s and add significant amounts of gas. The NPC Case 1 estimate (see Chap. 11) was 2.5 trillion cubic feet per year, 8 percent of the total gaseous fuel supply by 1985. Backed by the huge reserves of coal, this could be an important source of gas for decades.

There will, of course, be environmental consequences. For economic reasons the plants will probably operate on strip-mined coal. They will be only about 65 percent efficient, so 35 percent of the coal's energy will be lost, mostly as waste heat. It is probable that cooling towers will be built to discharge this heat to the atmosphere.

The fact that the sulphur and particulate matter are removed before combustion

Table 12-1 Cost of Various Clean Fuels*

Fuel	Average selling price (cents/M Btu)
Natural gas—at wellhead†	
Texas	16.1
Louisiana	20.2
Alaska	23.8
U.S. average	18.2
Natural gas—at New York City gate	35 to 45
Liquid natural gas—East coast	80 to 100
Low-sulfur fuel oil—East coast	65 to 75
SNG from light petroleum fractions	110 to 120
SNG from coal—Lurgi	105 to 115
SNG from coal—U.S. processes‡	70 to 95

*Source: Hammond, Metz, and Maugh, *Energy and the Future,* (Washington, D.C.: AAAS, 1973), p. 12, Table 2.

†January 1972.

‡Assuming use of strip-mined western coals.

will be an environmental plus, but this also raises the question of solid waste disposal. Pipeline gas must be essentially free of hydrogen sulphide or carbon dioxide since these gases form acids which will corrode the pipelines when moisture is present. Some of the secondary removal processes will produce concentrations of hydrogen sulphide mixed with carbon dioxide and it will be uneconomical to remove the hydrogen sulphide. The temptation will be to vent this in the air. It is a temptation that must be anticipated and frustrated.

Coal gasification can make a healthy contribution to the energy situation. It will take firm regulations to assure that it does not make unhealthy contributions elsewhere.

Oil from Coal. It is also possible to make a liquid fuel from coal. This is accomplished chemically by *hydrogenating* the coal, by adding hydrogen to the carbon to create petroleum-like molecules. This process was successfully used in Germany during World War II to produce oil and gasoline.

Early emphasis in this country was on the production of gasoline from coal and some 100 million dollars was spent on a plant at Louisana, Missouri before it was shut down in 1953 with the conclusion that competitive costs could not be reached. Here again, one is impressed with the contrasting federal policy toward nuclear energy and coal. If that effort could have been continued, even if unprofitable, who can doubt that coal today would carry a much greater share of the energy burden.

The problem of producing gasoline is that the gasoline molecule needs lots of hydrogen and this hydrogen comes from an expensive source, usually natural gas. The present efforts are directed toward developing a heavy, low-sulphur fuel for the electric utilities.

We will describe two of the several approaches. The first of these, *solvent refined coal* (SRC), produces a solid much like solid tar or *pitch.* Since SRC melts at about 180°C it can be heated and shipped in tank cars, or, perhaps, even pipelines, and can be preheated and used as a liquid fuel.

In the SRC process, raw coal is mixed with an oil (which can itself be derived from the coal) at high temperature and pressure (1000 pounds per inch^2 and 450°C). Under these conditions almost all (90 percent) of the carbon in the coal is dissolved and the liquid can be filtered to remove the unburnable ash and most of the sulphur. Some hydrogen is added before filtering to cause the formation of lighter molecules and to react with sulphur to form hydrogen sulphide, which can then be filtered out. The cleaned liquid is now heated to evaporate the solvent (and some light oils and gases from the coal) and the hard, brittle, and high energy-content residue, the SRC, can be shipped. The solvent and the hydrogen from hydrogen sulphide are recycled. The overall efficiency is 67 percent, that is, 67 percent of the energy content of the coal plus the hydrogen-containing natural gas is retained by the SRC and the distilled oils.

The main advantage of SRC is that it produces a uniform, energy-rich, low-sulphur fuel from any kind of coal. The energy value of SRC is 32 M Btu per ton as against 20 to 26 M Btu per ton for coals. The ash level may be as low as 1 percent and the sulphur content 1 percent or less—not low enough to meet some of the move stringent air quality standards. The SRC process will probably find its most competitive application to the high ash-content, low-sulphur, and low-Btu Western coals. A 50 ton per day pilot plant supported by the OCR is nearing completion at Tacoma, Washington; a similar plant with private funding is planned for Alabama.

A second process, the *hydrosulphurization* coal-to-oil process of the Bureau of Mines is at an earlier stage of development, but shows promise. In the Bureau of Mines process, raw coal and a recycled portion of its product oil are carried along above an appropriate catalysis by a rapid, turbulent flow of hydrogen. This is similar to the fluidized bed techniques we will refer to later, in that the coal is kept moving and from clogging the catalysis bed by the turbulence of the hydrogen gas. The turbulence also increases the mixing which is necessary for hydrogenation and for removal of sulphur as hydrogen sulphide. The latter is the major goal and use of expensive hydrogen in hydrogenation is kept at a minimum.

The Bureau of Mines pilot plant, capable of 5 to 10 pounds per hour operation, successfully converted a Kentucky coal having 4.6 percent sulphur and 16.0 percent ash into a fuel oil with 0.19 percent sulphur and 1 percent ash. Plans called for the design and construction of a 5 tons per day pilot plant to begin in 1973.

Generally, liquifaction of coal lags well behind gasification in its development and, unless large amounts of money are made available for research and development, it will not make a significant contribution to the energy problem for the next two decades. The NPC optimistic Case 1 forecast is for about 680,000 barrels per day in 1985 (out of a total domestic oil production of 16.89 million barrels and the pessimistic Case 4 assumption is for no coal-derived oil in 1985).

The synthetic oils are on the verge of being competitive with residual fuel oil; the estimated costs of SRC and the liquid fuel just described are in the range of 45 to 70 cents per M Btu and therefore should compete favorably with low-sulphur fuel oil (see Table 12-1). Their environmental effects will be generally similar to those we described for coal gasification.

Coalplexes in Our Future? Some of the coal engineers are looking to a future in which coal-conversion plants will serve several different purposes and produce different products. One of the most articulate and farsighted spokesmen for this view is Arthur Squires of the City College of New York.[1] He proposes a coal treatment complex, a *coalplex,* that has as its main goals sulphur and ash removal before combustion, in other words, clean fuel, and much higher efficiency of thermal to mechanical energy conversion—cheap power. The important ingredients in the accomplishment of these goals are a new boiler design, the *fluidized bed,* to replace the present pulverized fuel (PF) boilers, and the addition of a gas turbine *topping cycle.*

In the PF firing, powdered coal is injected by an air blast and burned, the ashes settling through a grate. Present day furnaces of this type are huge; the 1,000-mw Ravenswood Station of the New York's Con Edison has two firing chambers, each 34 by 64 feet and 138 feet high.

When the concern over the fine particulate material emitted began to grow with the growing size and number of these furnaces, it was found that one could add on an electrostatic precipitator and remove these particles. These are also huge: the one at the Ravenswood Station is larger than the furnace just described. Its dust-removal efficiency is better than 99 percent.

In the late 1960s and early 1970s, however, concern has been growing over other pollutants from coal-burning, in particular, over the sulphur and oxygen compounds which we have been calling the sulphur oxides. The PF furnaces have begun to run into serious trouble with regard to these pollutants. The precipators do nothing to the sulphur oxides. Furthermore, if the problem is attacked at the source by burning low-sulphur coal, it is found that, because of different electrical properties of this coal ash, the precipitators no longer work properly.

In the fluidized-bed concept, the coal is buoyed up, or floated, by a rapid flow of high pressure air from below. The coal, crushed to particle sizes of ¾ inch or less, is burned in a shallow combustion zone near the air inlet at the bottom. At the combustion temperatures, which are near 2000°F, ash has the fortunate property of being self-adhering: ash sticks to ash and not to coal. Thus, it agglomerates and ash particles drop out through the grate.

The hot combustion gases rise up through the coal above, driving off the volatile gases which become a fuel gas for the turbine. The now heavier coal particles then sink and are burned in the combustion zone.

The fluidized bed has several advantages over the PF boiler. Two of them are of interest here. It can be operated at high pressures with a resulting great reduction in size. The smaller boiler should be cheaper to build and produces no dust, so that an electrostatic precipitator is unnecessary. It is also possible to remove the sulphur from the coal in such a boiler, either by mixing ground limestone into the bed and later removing the sulfur from it, or by creating hydrogen sulfide with the addition of steam and removing the sulfur from the hydrogen sulfide. Several of these processes are under intensive investigation.

The second major component of the coalplex is the gas turbine topping cycle. As we discussed in Chap. 5, the efficiency of a thermal to mechanical energy conversion depends on $1 - T_{out}/T_{in}$. Thus, it is necessary for high efficiencies to have high input temperature, T_{in} and low exhaust temperatures, T_{out}. The steam turbine and boiler systems, however, are now operating at about as high temperatures (near-

[1] A. Squires, "Clean Power from Dirty Fuels," *Scientific American,* **227**: 26-35, October 1972.

ly 1000°F) and pressures as they are capable of. Gas turbines, developed for aircraft jet engines, however, can operate at much higher temperatures. Squires anticipates that turbines operable at 2400°F will be available for stationery use by 1980.

Wedding a gas turbine to a steam system makes a complementary combination. The gas turbine can stand high input temperatures but, since it exhausts gas at relatively high temperature, its efficiency is low, around 25 percent. The steam turbine can't take as high input temperatures but achieves relatively low output temperatures. If the hot exhaust gas from the gas turbine is used to create steam for the steam turbine, overall efficiences of about 50 percent may be attained with a consequent lowering of waste heat and thermal pollution. Such a *combined-cycle* plant, burning a sulphur and ash free fuel (and also discharging much fewer nitrogen oxides because the fuel gas is oxygen-difficient) might be clean enough to be located in a city where its discharge heat could be used.

This combined-cycle electrical generating plant would be the heart of the coalplex. Variations on it are possible; in the topping cycle the gas turbine might be replaced by the magnetodynamic generator which we will discuss in the following section. It is also possible to add as a *bottoming cycle* a turbine run by the expansion of a fluid with a boiling point lower than water and thus extract energy from the exhausted steam. Efficiencies as high as 60 percent are not inconceivable.

The overall coalplex operation is shown in the flow diagram in Fig. 12-2. In the main chamber a fuel gas is distilled from the coal; part of it is combusted to provide heat and part goes to run the gas turbine. The desulphurization also takes place at that point. The heat produced in this chamber creates steam for a steam power plant which also receives a contribution from the gas turbine exhaust. The products in addition to electric power are sulphur and coke, or char, which, with its high Btu value and low-sulphur content is worth shipping to other, distant, power plants.

For most economical operation these large coalplexes would be located at the mine. One of the scale being described here might process 13 million tons of coal a year, produce 1,000 mw of electric power itself and provide sulphur-free fuel for

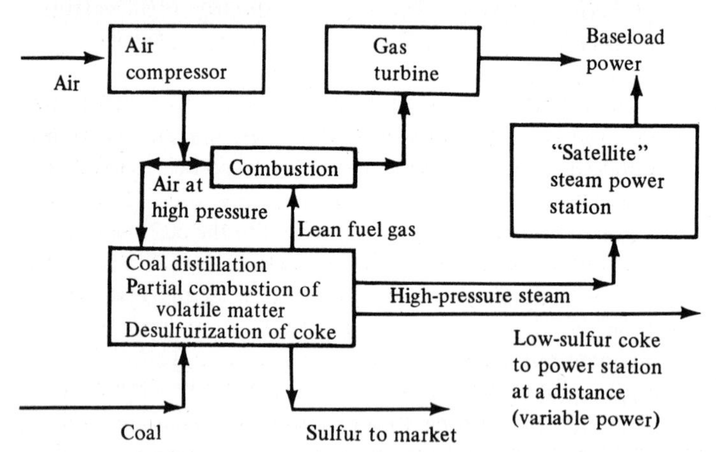

FIGURE 12-2 Flow diagram of the processes occuring in a coalplex. Electric power, low-sulphur fuel, and sulphur are produced. (Source: *Power Generation and Environmental Change*, MIT Press, Mass.)

another 5,000 mw as well as 400,000 tons of sulphur. Squires further projects as evolutionary steps:

> *1* The first Coalplexes would be justified simply for their economy in dealing with sulphur.
>
> *2* Later, modifications would "cream off" limited amounts of pipeline gas and liquids from volatile matter. Simplicities in the processing of volatile matter would result from opportunities to throw off high-level waste heat to steam for power.
>
> *3* As time passed, further modifications would expand production of gas or liquid.
>
> Ultimately, in an economy powered principally by breeder reactors, a Coalplex would evolve for which power might be a relatively minor by-product, and fixed carbon would be shipped mainly for metallurgical or electrochemical use.[2]

In many ways this is an attractive scenario. It offers hope of improving the energy situation as well as abating air and thermal pollution. It is built on coal, however, and as we have seen in Chap. 9, coal mining itself poses environmental hazards which will have to be dealt with, if the overall environmental impact is to be reduced.

MHD

A competitive topping cycle to the gas turbine is the *magnetohydrodynamic* (MHD) *generator.* This device operates on the generating principle described in Appendix 4; if a conductor is moved at right angles to a magnetic field, an electric current is generated in it. In the conventional generator a conducting wire is moved by the mechanical effort of a turbine through the magnetic field and the generated current flows in the wire. In the MHD generator a hot gas is carried by its own momentum through the magnetic field and the current is generated in the gas, electrons moving in one direction and positive ions in the other.

A sketch of a MHD generator is shown in Fig. 12-3. The gaseous fuel, which could be produced by a fluidized bed of coal as we have described earlier, enters and burns in the combustion chamber reaching high temperature and pressure. Since the gas is the moving conductor, it must be partially ionized; that is, some of the electrons must be separated from the atoms. This occurs naturally only at very high temperatures, 10,000°F or more. Lacking metals and electric insulators which can withstand such temperatures, it is necessary to *seed* the gas with a material which can be ionized at much lower temperatures. Alkali metals such as potassium are used in this role. Seeded gas can be made adequately conductive at temperatures of 2000 to 3000°F.

As is discussed in Appendix 4, the force on a charged particle q moving with velocity v in a magnetic field of strength B is given by the product qvB. Thus, to get a large force on the electrons in the conducting gas, and, consequently, a large voltage, it is important to have high velocity (which means large kinetic energy and,

[2] Squires, *op. cit.*

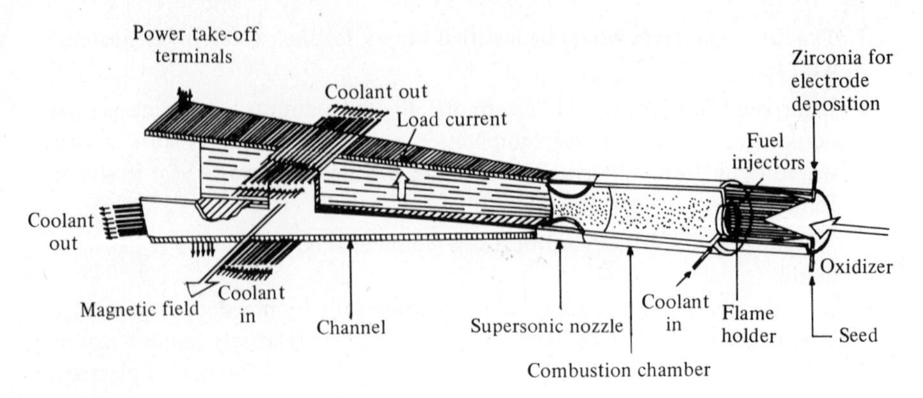

FIGURE 12-3 Sketch of a MHD generator. The gaseous fuel enters from the left and the power is generated across the power take-off terminals. (Courtesy of AVCO Corp.)

therefore, high temperature) and a large magnetic field. To get the latter, the power engineers are looking toward superconducting magnets, magnets whose current-carrying wires are at liquid helium temperatures. Such wires, as explained in Chap. 6, have no resistance and can carry the huge currents needed for the large magnetic fields.

The current that is generated in the moving gas, the load current in Fig. 12-3, is at right angles to the motion of the gas and to the magnetic field. It is collected by the electrodes, the "power take-off terminals" in the drawing. The generated power is dc and must be converted to ac for general use.

Acting alone, the MHD generator is not impressively efficient, in the range of 20 to 30 percent. As a topping cycle, however, it can utilize the high temperature energy from combustion and, its exhaust gases, still very hot, can produce steam for a steam turbine. Overall efficiencies may be as high as 50 to 60 percent.

Since the MHD generator requires little cooling water, it promises reduced thermal pollution of rivers as well as more efficient use of fossil fuel reserves. The MHD system also offers significant advantages in air pollution reduction. The seed material, for economic reasons, must be reclaimed from the exhaust gas. Sulphur reacts with the alkali seed particles forming potassium sulphate, for example, which is removed by cloth traps or electrostatic precipitators. Nitrogen oxide could be a problem since very high temperatures are reached; however, it appears that this pollution is also controllable through the composition and temperature of the combustion.

Several countries are farther along than the United States in their MHD development efforts. England, France, Germany, Japan, Poland and the Soviet Union all have work under way. The most ambitious effort is in the Soviet Union where a 75-mw unit fueled by natural gas is being tested.

There are several major problems which must be solved before MHD can be put into routine commercial operation. Corrosion of the duct and the electrodes has seriously reduced operating lifetimes so far. Better refractory metals are needed if the corrosive coal fuel gases are to be used. The use of superconducting magnets on a device operating at such a high temperature is also a problem awaiting solution.

Research on this device will be accelerated. When its problems are solved, the low maintenance expected from its lack of moving parts and the absence of pollution (thermal or air) should make it a useful addition to the roster of electric power generators.

Other Exotic Heat Engines

There are several other techniques which have not as yet progressed much beyond the laboratory stage of testing but which may have some special application. All of these are, like the MHD generator, designed to convert thermal energy directly into electric energy. We will review their principles and present status briefly.

Electrogasdynamics. In the electrogasdynamic (EGD) generator, a hot gas expands through a nozzle similar to the one employed in MHD, is ionized and then the kinetic energy of the gas carries it "upstream" against an electric field. In a sense, the motion of the electrons of the gas against the field "lifts" them and converts their thermal kinetic energy to electric potential energy. The electrons are captured by a collecting electrode and a load current can then be obtained by letting them flow through an external circuit back to the positive electrode. EGD enthusiasts believe that large scale devices of this type will be competitive with steam generators in efficiency and will be cheaper to build. They will make no demands for cooling water and would be, in theory at least, adaptable to any source of heat. One experiment has progressed to the pilot model stage but is not yet considered practical.

Thermionic Generators. These devices for converting thermal energy to electric energy operate on the same principle as the filaments of the "vacuum tubes" of early radio days. If a metal becomes hot enough it will "boil" electrons out of its surface; in other words, the electrons in the metal obtain enough thermal kinetic energy to overcome the forces which bind them into the metal. Temperatures of 2500 to 3000°F are needed for efficient operation. If this hot *emitter* is enclosed by a cylindrical collector, a negative charge will build up on the collector. The emitted electrons travel to the collector against the repelling force of this negative charge and the electric potential energy this represents can be used when the electrons flow back to the emitter through an external circuit. This too, is a heat engine and the efficiency depends on the difference between emitter and collector temperature. Thus, the requirement of close geometry to maximize electron collection is opposed by a need to reduce heat transfer between the two. Efficiences of 15 to 25 percent have been achieved in laboratory experiments and there is some interest in using these as topping devices also. The major United States interest is in reactor-powered devices or smaller ones heated by radioactive isotopes which can be used in the space program. Several German industries are entered on a six-year, 50 million dollar program to develop a reactor-fueled thermonic generator for use in a television satellite.

Thermoelectric Generators. Specialized power-generating devices have been built which use the *Seebeck effect,* the generation of a potential difference (a voltage) created by a temperature difference between the junctions of two conductors made of different materials. This effect is used in a sensitive temperature-measuring device called the *thermocouple.* It is also a heat engine and its efficiency depends on the temperature difference. Large differences are obviously difficult to maintain between the joined conductors. Bell Laboratories is experimenting with a generator producing 160 watts at 2 percent efficiency from solar heat for use as a power source in remote locations. An interesting feature of this effect is that it can be turned around. When current is driven through a junction a temperature difference known as the *Peltier effect* is established. Thermoelectric generators can thus be used for heating or cooling. These devices, however interesting, show no short-range promise for our massive energy needs.

Around the Thermal Bottleneck: Fuel Cells

All of the energy-converting devices discussed in this chapter suffer from the same frustration: the energy has to go through the thermal bottleneck described in Chap. 4, thus putting a lid on the efficiency of the energy conversion. They all have their advantages and disadvantages, but in the end their efficiency is governed by $1 - T_{out}/T_{in}$. For efficiences greater than 50 to 60 percent, we must find ways to convert chemical potential energy, for instance, directly to electrical energy in order to detour this bottleneck. A most attractive device for accomplishing this is the *fuel cell.*

The fuel cell operates, in a sense, as a slowed-down combustion reaction. In combustion, electrons are removed from the fuel molecules (oxidizing them, in chemical terms) and are attached to the oxidizer molecules (reducing them). A simple example which has application in fuel cells is the burning of hydrogen

$$H_2 + O \longrightarrow H_2^{++}O^{--} + \text{energy}$$

Hydrogen fuel molecules are oxidized (electrons are removed) and are attached to the oxidizer molecules which, in this case, are oxygen. When hydrogen burns in air or oxygen, this reaction takes place very fast; the fuel and oxidizer molecules are mixed together and electrons pass directly from fuel to oxidizer. The energy, which in the burning of hydrogen amounts to about 4 Calories per kilogram, is released as heat. In the fuel cell the fuel molecules are separated from the oxidizer molecules and the electrons are transferred by an external circuit, thus allowing them to perform work. The total energy released is, of course, the same.

Although fuel cells can work, in theory, on any oxidizable (burnable) fuel, the hydrogen-oxygen cell (H-O) was the first studied and is still the most important one in present plans. Its operation, shown schematically in Fig. 12-4, provides a model for other types.

The fuel-cell structure is similar to that of a battery. It has an anode and a cathode; hydrogen enters and diffuses through the anode (typically containing a platinum catalyst) and the reaction $H_2 \rightarrow 2H^+ + 2e^-$ takes place. Electrons are stripped off the hydrogen and remain on the anode which thus builds up a negative electric charge. At the cathode, an electrode containing a platinum or gold catalyst, OH^- ions are created by oxygen, water, and the electrons flowing through the external circuit from anode to cathode. The OH^- ions migrate through the electrolyte (whose

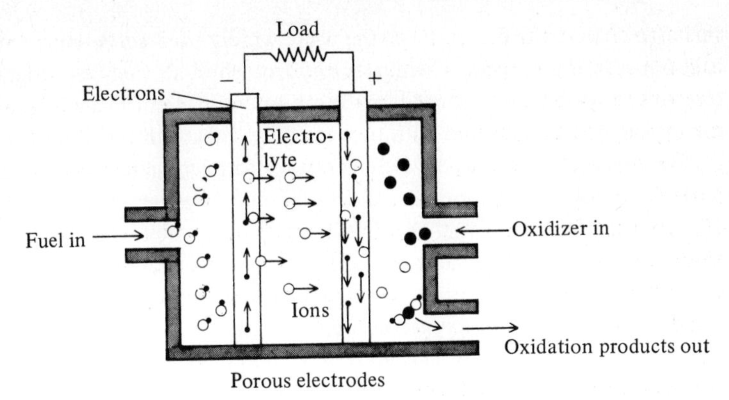

FIGURE 12-4 Schematic diagram of a fuel cell. Electrons are removed from the fuel atoms at the anode on the left and are picked up by the oxidizer atoms on the right, after travelling through the external "load" circuit.

purpose is to provide a conduction path for them) and join with the H^+ ions to produce water, the sole residue of the H_2-O_2 fuel cell.

A single H_2-O_2 fuel cell generates 100 to 200 milliamperes (thousandths of an ampere) of direct current per square centimeter of electrode surface at a voltage of about 1 volt, producing, therefore, 0.1 or 0.2 watts of power per square centimeter. Power levels to 0.6 watts/per square centimeter are obtained at somewhat lower voltages. A single fuel-cell of this power capability, however, is very small and many cells can be connected in sequence to build up voltages of 100 to 1,000 volts and powers in excess of 100 mw. Their size is one of their significant advantages; a 1-mw unit, it is estimated, would occupy a volume of an 8-foot cube.

The fuel cell, like the battery or electric generator, can have very high efficiencies. H_2-O_2 cell efficiencies of 70 to 80 percent are obtained in the high performance alkaline electrolyte cells. Components for removing waste heat and the water absorb energy resulting in H_2-O_2 fuel cell power-plant efficiencies of 65 to 75 percent.

The most glamorous applications of H_2-O_2 fuel cells has been in the Gemini and Apollo space programs where low system weight is necessary. At the height of the space program in 1964 to 1965, 16 million dollars was spent on fuel cell research. Funds for space research dropped in the late 1960s. Recently, development has been started by Pratt and Whitney Aircraft on a 12-kw H_2-O_2 power plant for the reusable Space Shuttle and on 20 to 30-kw power plants for Naval undersea use. H_2-O_2 fuel cells are also being investigated as part of an advanced energy storage system in which excess off-peak electricity could be used to electrolyze water into hydrogen and oxygen to then be used in a fuel cell when demand increased.

Except for the energy storage potential, most terrestrial applications of fuel cells must obtain the hydrogen they consume from readily available fuels such as natural gas, petroleum distillates, methanol ("wood alcohol") or liquid and gaseous fuels from coal. (We will discuss other possible sources of hydrogen later in this chapter.) If methane is the fuel, then energy is needed to separate the hydrogen from the methane, thus reducing the efficiency to perhaps 40 or 50 percent. In addition to the reformer, as the hydrogen producing device is called, a fuel cell needs an inverter to change the dc power to the more widely used ac. A small amount of energy is also lost in the process. The current hydrocarbon-air fuel-cell programs use an acid electrolyte system because acids are more tolerant than the alkaline electrolytes to carbon-bearing gases found with processed hydrocarbon fuels.

The advantages of the fuel cell over other forms of generation are numerous. Even with the penalty of hydrogen production from methane, its efficiency is competitive with the best fossil fuel or nuclear plants. It is quiet when in operation since there are no moving parts except, perhaps, fans and pumps. It poses no pollution threat: the waste products are water and carbon dioxide. It does release some heat when the hydrogen and oxygen combine to make water, but this heat is released to the atmosphere so that there is no need of cooling water; in fact, the cleanliness of fuel cells makes it possible to consider using them at a power consumption site, an office or apartment building, or private residence where this heat could even be used to heat space or water as well as to provide for air conditioning, dehumidification, and waste disposal.

The versatility of the fuel cell is important. Cells can be stacked in series to provide varying amounts of power and their efficiency is largely independent of the amount of power they deliver. Figure 12-5 gives a comparison of the efficiency of hydrocarbon-air fuel-cells at different loads with other generating systems. The fuel cell is clearly the choice over the entire range.

There are two different general approaches under way exemplified by two projects: one at Westinghouse sponsored by the OCR to develop a large-capacity, central power station fuel cell to run on coal, and a second project carried out by the Pratt and Whitney Aircraft Division of United Aircraft Corporation. The Pratt and Whitmey efforts are the more advanced and, in many ways, the more interesting. The first of two Pratt and Whitney programs is sponsored by a group of 30 natural-gas companies through a nonprofit Team to Advance Research for Gas Energy Transformation (TARGET) and Pratt and Whitney Aircraft. A total of about 46 million contract dollars had been spent on the TARGET effort by the end of 1973. The goal of the TARGET program is to develop small-capacity fuel cells for on-site conversion of natural gas to electricity in offices, apartment build-

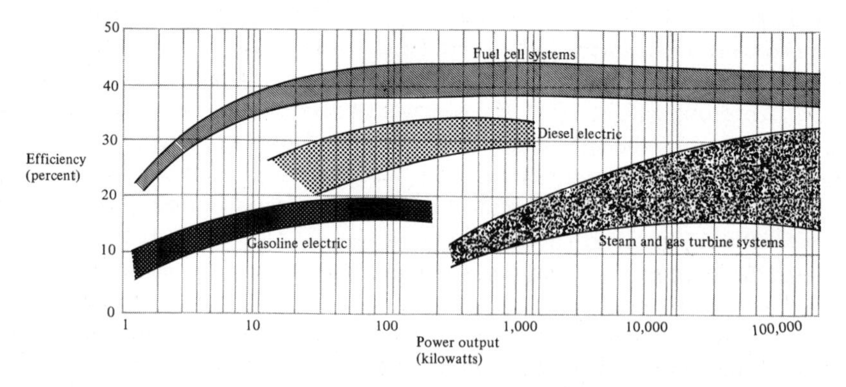

FIGURE 12-5 The efficiency of various electric generators as a function of power output. (Courtesy of American Association for the Advancement of Science, Washington, D.C.)

ings, small businesses, and residences. Hydrogen and carbon dioxide will be formed from the methane in natural gas by decomposing it with steam. Oxygen will be furnished by air. These substitutes will lower the efficiency to perhaps 40 percent but the cells will remain pollution-free. The nitrogen in the air will not reach high enough temperatures to form nitrogen oxides; it will be discharged as nitrogen.

Testing the concept of on-site fuel cells has been accomplished by TARGET. About 65 fuel-cell power plants, each having a power capacity of 12.5 kw, were tested at 35 locations in homes, businesses, and the like. Costs estimated by Pratt and Whitney Aircraft in 1972 were 350 to 450 dollars per kw (as contrasted to about 250 dollars for fossil fuel plants), but the research team expects these costs to be lowered by 1976 or 1977. The advantage of on-site conversion to electricity are made clear by reference to the data in Fig. 6-3, Chap. 6. A given quantity of energy is some 3 to 5 times cheaper to transport as a gas by pipeline than is electrical energy in a transmission line. Even though the thermal efficiency of a modern power plant (about 40 percent) may be as high as that of one of these on-site fuel cells, the loss in transmission (10 to 15 percent) and the necessity of adding less efficient auxiliary generators at peak demand, results in the fuel cell producing 10 to 25 percent more energy per unit of fuel. TARGET has, therefore, the ambitious aim of substituting natural gas for electricity in most small-user situations. Even with the growing natural-gas shortage, they view the switching of this premium fuel from industrial to residential and commercial users, and the substitution of SNG as viable future options. High system efficiency and practically zero pollution makes this an exciting possibility. It is anticipated that on-site fuel cells will be in service by the end of this decade. There may be a fuel cell in your future.

The use of modular fuel cell units of the multimegawatt power capacity to supplement utility systems in rapidly growing areas is also attractive. They can be located close to the load and can be set up quickly, have no environmental impact, and can be taken down when the need disappears.

This concept is sponsored by a group of investor owned electric utilities and Pratt and Whitney Aircraft. The goal of this 42 million dollar program is to develop a 26 mw hydrocarbon-air fuel-cell generator for use within existing electric systems in locations such as substations and load centers. The electric utilities are attracted to the fuel cell because of its environmental qualities, its high system efficiency (particularly at part load), elimination of transmission losses, and the capital cost of the lines, ease of siting, flexible operating capability, and modular construction. Pratt and Whitney Aircraft also hope to have fuel cells in utility service by the end of this decade.

The OCR has invested almost 4 million dollars over the past seven or eight years in search for a large, high-temperature fuel cell of several hundred-mw capacity to be run on coal. Although progress in this direction has not advanced far, there are some interesting possibilities. At very high temperatures (about 1000° C) the oxidization reaction does not need to be catalyzed by expensive platinum, but proceeds spontaneously. This advantage is, at present, more than offset by difficulties caused by the high temperatures. Commercialization of these large cells, which, when used along with coal-gasification plants, might achieve overall efficiencies of 60 percent, is not expected in the next decade.

Biochemical Fuel Cells. A most interesting idea that has attracted some research attention is the biochemical fuel cell in which microorganisms carry out the chemical reaction at one or both electrodes. Bacteria may act on organic matter, for instance, and release methane or, even hydrogen, as fuel. The efficiency of these cells will be low for the organism uses 50 percent or so of the energy to support its own metabolism. But the raw fuels could be any organic waste such as sewage. It may be that some fuel cell such as this will provide power during extended manned space voyages.

Nuclear Batteries

The fuel cell and the solar cell (which we shall discuss in Chap. 13) have been important power sources on the space vehicles. There is another device, however, that has also been important in our exploration effort and which may have some application on this planet, the *nuclear batteries* developed under the AEC Systems for Nuclear Auxilary Power (SNAP) program. In most of these SNAP units the kinetic energy of the emitted radioactive particles is converted to heat and the heat generates electricity by the thermoelectric process, one of the exotic heat engines described earlier. These SNAP units, also called radioisotope thermoelectric generators (RTG) use long-lived isotopes and produce power from a few milliwatts to 100 watts for missions on the surface of the moon or underneath the opaque atmosphere of Venus, for instance, where solar cells won't work. On the first lunar mission, Apollo 11, a radioactive heater using Pu^{238} (half-life 90 years) was used to keep one of the instrument packages warm on the lunar surface. The second set of experiments, Apollo 12, utilized a 60-watt RTG powered by the same radioisotope. These devices also provide power on weather and military observation satellites. The Pioneer series, which is exploring Jupiter, has four 30-watt RTGs aboard.

Another successful nuclear battery is a direct-conversion device which collects the radioactive particles. Radioactive particles are charged; if they are captured by a collector a charge is built up. As further particles use their kinetic energy to climb the "electric hill" to that charge, they convert this kinetic energy, derived from nuclear potential energy, to electric potential energy. One of these batteries available commercially uses Kr^{85}, a radioactive product of fission. The Kr^{85} gas is confined in a thin glass sphere. The high energy electrons (β particles) emitted from the Kr^{85} penetrate the glass wall and are collected on a concentric metal sphere. Voltages as high as 10,000 volts have been attained, but nuclear batteries are usually operated at 1,000 volts and produce currents of 10^{-9} ampere. Kr^{85} has a half-life of 10.4 years so that these batteries are useful for maintaining delicate equipment far from other sources of power.

From what has been described so far in this chapter we see that technology is advancing—much too slowly, it seems, but advancing—and holds out the twin possibilities of improving the efficiency of our heat engines and developing direct electric power generators in which the energy detours the thermal bottleneck. These new technologies offer us more Btu from old energy sources and, as we have seen, some of these new Btu may be obtainable within the next five years.

Ultimately, we must turn away from the scarce fuels to more abundant ones. In the following chapter we will consider the continuous sources of energy which we have so far neglected. But before we look at these potential new sources of energy, we will describe another way in which technology may help us by providing a new intermediate form of energy, hydrogen, which offers significant advantages in several troublesome areas including energy storage and pollution.

HYDROGEN–THE NEXT PREMIUM FUEL?

One of the major trends we have seen developing in the preceding chapters is the emergence of electrical energy as the important intermediate energy. There are only two major drawbacks to this "all-American energy": it can't be stored, and thus, is difficult to use to run automobiles, and it is inefficient and expensive to transmit.

The storage problem is the most serious, and the invention of an economical, high-density electrical storage technique would, of course, make the electrical automobile more feasible. It would also reduce the cost of electricity; this cost is increased now by the demand fluctuations which result in less efficient use of base-load generators. Overall, and on the average, only about 50 percent of the United States generating capacity is used. To meet the demand fluctuations, utilities have relied on inefficient but quick-starting gas turbines and diesel engines. A much more efficient system would result if the large plants—nuclear or fossil—could run at full capacity and store their off-peak power. We will see the same need anticipated if electric power is to be obtained from the erratic continuous sources, (wind and solar, for example) or if thermal energy for home-heating is to be obtained from the sun.

Our present means of storing energy are not economically satisfactory. The favored large-scale system is pumped storage discussed in Chap. 6; for small demand we use batteries.

Storing Electrical Energy

One line of research which is being vigorously pushed is aimed at the production of cheaper and more efficient batteries: batteries with higher energy densities than the present lead plate automobile batteries, and which can be recycled (charged and re-charged) many times. Much attention has been given in recent years to batteries with solid electrolytes and liquid electrodes (the reverse of the common structure). The current choice of batteries of this type uses sodium and sulphur (in liquid forms) as electrodes and a solid sodium aluminate as an electrolyte and separator. These are abundant, low-cost materials and, in theory, could store 600 watt-hr per pound as compared to 10 watt-hr per pound in lead batteries. Test models have iasted 2,000 to 3,000 hours and been recycled 200 to 300 times. As the cost of these batteries is reduced, it is conceivable that they will begin to compete with pumped storage. They already are competitive on space considerations. A 60-mw pumped storage plant with a 20-foot dam and a *head* (the water elevation above the turbine) of 200 feet requires 15 acres of land. A sodium sulphur storage system of batteries stacked in a 20-foot high bay would require 0.1 acre.

Hydrogen for Storage and Transmission

The need to store energy is one of the reasons there is growing interest in the use of hydrogen as a secondary fuel. It has much to recommend it. It is abundantly available since it can be obtained from water; it is a gas and, thus, easily shipped. Environmentally, it is an almost ideal fuel since the combustion product is water.

Hydrogen offers advantages not only for energy storage, but for energy transmission, at least when compared to electricity. As we have seen in Chap. 6, a given amount of energy can be transported 3 to 5 times more cheaply as gas than as electricity. Hydrogen shares in this advantage. It will flow faster then natural gas through a pipeline, but carries less energy per cubic foot: 325 Btu versus the 1000 Btu per foot^3 of natural gas at the same pressure. Even with the difficulties of pumping hydrogen, its lower Btu value, the necessary tighter piping to prevent leaks, and its greater danger of explosion, estimates are that hydrogen could be transported for perhaps 1 dollar per M Btu as against $2.20 per M Btu for electricity.

Hydrogen as a fuel can also compete with electricity in versatility. It can be burned for heat, is useful as a reducing gas in metallurgy, can be used in coal gasification and can, as we have seen, be converted to electricity via a fuel cell. These advantages, coupled with the enforced decrease in our direct use of fossil fuels, make it worthwhile to explore the means and costs of producing hydrogen and examining some of the technical details of its use as a fuel.

Hydrogen Production

The most abundant source of hydrogen is water. Hydrogen can be (and is, presently) produced from hydrocarbons (oil and natural gas) but their growing scarcity does not recommend them as sources. We will, therefore, concentrate on ways to dissociate water.

There are several methods to bring about the separation of hydrogen from oxygen: electrolysis, photolysis, bioconversion, and thermal decomposition. *Bioconversion* is the catalytic action of the enzyme hydrogenose from algae, in cooperation with photosynthetic processes. *Photolysis* is the dissociation of water under bombardment by ultraviolet radiation, which, in one scheme, could be produced by a fusion reactor. *Thermal dissociation* occurs at temperatures of 2500°C and above and might be possible in a very hot reactor or at elevated temperatures obtained from concentrated solar radiation. The only practical scheme at this time is electrolysis.

Electrolysis, which has been discussed previously, is accomplished by forcing an electric current through water and breaking the bond between hydrogen and the oxygen molecule. This requires energy; commercial electrolysis plants operate at efficiencies of 60 to 70 percent. It has been pointed out, however, that since some of this energy could be obtained as heat from the generating plant which produced the electricity, electrolysis could proceed, as far as electrical energy is concerned, at an efficiency approaching 100 percent. The inefficiency of producing the electricity reduces this to 30 to 40 percent efficiency overall.

The cost of hydrogen is thus closely tied to the cost of electricity. One estimate is that hydrogen will cost $1.50 to $2.50 per M Btu if electricity costs 4 to 7 cents per kw-hr.[3] Natural gas currently sells for 50 cents to 1 dollar per M Btu. Hydrogen is thus on the threshold of competitiveness and the cost of natural gas, dictated by its scarcity, has been rising rapidly.

Hydrogen As a Fuel

If increased production of electricity by nuclear reactors or by the devices of the more distant future enable it to remain economical as the price of fossil fuel continues to climb, we may soon reach the crossover point for hydrogen. It is useful, therefore, to examine in more detail the ways in which it can be used.

One such future is visually summarized by Fig. 12-6. The hydrogen (here, assumed to be produced by electrolysis) can be burned to produce the thermal energy which is a major end use of most of the sectors of our economy. It could supply heat for homes and for industry, and, by the absorption-desorption process we discussed earlier, provide air conditioning. It may even be economically feasible to burn hydrogen in a steam-generating plant. The hydrogen could be produced with cheaper offpeak power and burned to produce power to supply peak demands. It might also, in a more futuristic scheme, be produced by a solar sea plant and piped or shipped to a combined gas turbine steam-generating plant on shore (to be discussed in Chap. 13). The low environmental impact would allow placement in a community where its waste heat could be used.

The second attractive use of hydrogen would be to produce electricity by fuel-

[3]Gregory and Long, "The Hydrogen Economy," in Bockris, Ed., *Electrochemistry of Cleaner Environments,* (New York: Plenum, 1972).

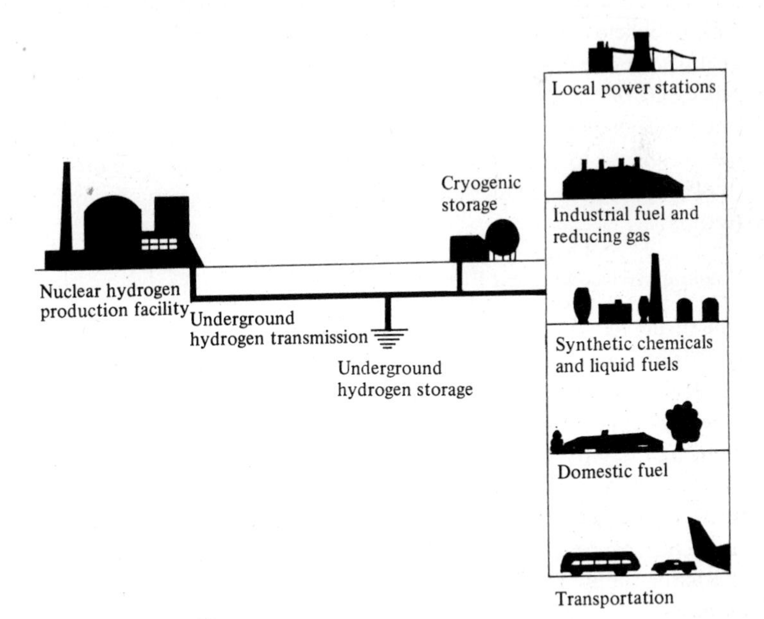

FIGURE 12-6 A schematic representation of some of the storage and end-use options available if hydrogen becomes available as an intermediate energy form. (Source: Hammond, *op. cit.*)

cell conversion. As we have said earlier, location at the point of need would reduce the loss of energy in transmission and allow utilization of the exhaust heat.

Hydrogen is also being considered as a practical fuel for that most difficult sector, transportation. Electric automobiles have been powered by hydrogen fuel cells. They can also use hydrogen directly in the gas tank. Gaseous hydrogen occupies too much space; liquified hydrogen, while it has almost three times as much energy per pound as gasoline, also occupies more room and needs thermal insulation. One way around this may be to make use of hydrogen's proclivity for penetrating metals and becoming bound into the metallic structure as metal hydrides. For some metals, the diffusion of hydrogen is so thorough that they hold, in a given volume, about as much hydrogen as would be in the same volume of liquid hydrogen. The gas tank could be filled with this metallic hydride (titanium is an appropriate metal) and the hydrogen driven off by heat as needed. The search is on for less expensive metals to serve this purpose.

The ease of penetration of metal is, at the same time, one of the drawbacks of hydrogen. In a hydrogen pipeline or container, the penetration can cause the metal to become brittle and break. For similar reasons, hydrogen leaks readily and much tighter seals will be needed if it is to be successfully transported by pipelines.

One of the benefits of hydrogen as a fuel in transportation (liquid hydrogen shows promise as an airplane fuel) is its low pollution contribution. It does not produce hydrocarbons or carbon monoxide but, since it heats air, it will produce some nitrogen oxide. However, the production of nitrogen oxides will be a tenth or less than with gasoline.

Other uses will, no doubt, follow if hydrogen becomes available. Coal gasification and liquification proceed much more efficiently with plentiful hydrogen. These processes could also use much of the oxygen which also will be produced in mammoth quantities by electrolysis. Hydrogen is also needed in petroleum refining, and the processing of oil shale will increase this demand. It is also possible to make fuel (methanol or gasoline, for instance) out of hydrogen and atmospheric carbon dioxide. Abundant hydrogen could also contribute importantly to processing iron ore (replacing coke as a reducing agent) and to the production of ammonia (NH_3) for fertilizer.

Possible Hydrogenerated Futures

One of the most thorough studies of the optional futures which large scale production of hydrogen might allow was carried out by Winsche, *et.al.*[4] They have evaluated the economic and environmental effects of using hydrogen as an intermediate fuel to supply the energy needs, including automotive fuel, of a typical household and compared it with all-electric supply systems and systems that are based on other combinations of energy supplies. The results of this study are summarized in Table 12-2. Systems 1, 6, and 7 are reference systems. System 1 is the conventional all-electric home plus gasoline for the car. In System 2 the home is all-electric but the automobile is run on hydrogen, part of which (37 percent) is provided by off-peak electrici-

[4]Winsche, Hoffman, Salzano, "Hydrogen: Its Future Role in the Nation's Energy Economy," *Science,* **180**: 1325, June 19, 1973.

Table 12-2 Annual Energy Costs per Household for Various Alternative Energy Systems*

System	*Annual cost per household* — *Aboveground electric transmission (dollars)*	*Underground electric transmission (dollars)*
Coal-electric and gasoline	764	1239
Coal conversion to methane	515	642
Coal conversion to hydrogen	486	613
Hydrogen derived from nuclear energy and coal	815	1290
Nuclear energy conversion to hydrogen	928	1055
Nuclear-electric and gasoline	822	1297
Nuclear–all electric	815	1504

*Source: Hammond, Metz, and Maugh, *Energy and the Future,* (Washington, D.C.: AAAS, 1973), p. 122, Table 14.
Consumption per household is the same in all cases, and is distributed in the following manner: space heating, 75 M Btu/year; air conditioning, 25 M Btu/year; water heating and cooking, 15 M Btu/year; miscellaneous electric, 25 M Btu/year; automobile, 25 M Btu/year.

ty; the remainder comes from coal. System 3 is all hydrogen, formed by electrolysis and piped to the local area. Some of this is reconverted to electricity in a fuel cell. System 4 is similar to 3 except that the hydrogen comes from coal and, in System 5 methane is generated from coal and such hydrogen as is needed is formed from it for fuel-cell conversion to electricity. System 6 is an all-electric (including automobiles) system. System 7 is all coal. It is clear that systems based on coal are the most economical, but that systems based on electrolyzed hydrogen and fuel-cell reconversion are competitive with conventional systems.

A hydrogen-fueled future is thus technologically possible and, it appears at first glance, economically possible. It would be a less polluted future also. Replacing gasoline by hydrogen eliminates the hydrocarbons, carbon monoxide, and particulates, and significantly lowers the nitrogen oxides. Replacing coal-fired generation of electricity by nuclear generation of electricity, the production of hydrogen by electrolysis and then, fuel-cell regeneration of electricity at the point of use cuts out sulphur oxides and particulates and allows underground gas pipes for transmission of energy.

A final note on these possible futures concerns the amounts of hydrogen they will demand. As a perspective, current United States consumption is 2.28 trillion cubic feet per year and world consumption about 6 trillion cubic feet. If hydrogen replaces natural gas, the United States consumption would rise to 60 trillion cubic feet. At current electrolyzer efficiencies, more than three times the current electric generating capacity would be needed for this output. If the more ambitious replacement of all fossil fuels by hydrogen were the goal, then 295 trillion cubic feet of this new fuel would be needed by the year 2000, requiring 15 times the 1970 generating capacity.

These figures make it seem unlikely that an all-hydrogen economy will exist by

2000. What is necessary, however, is that the basic research and development begin now so that hydrogen can be at least undergoing large-scale experimental usage by that time. We must begin to find out by then how much help this clean, light fuel can give us as we seek better ways to store and transport energy.

SUMMARY

In this chapter we have outlined some of the technological developments in progress which will, we hope, help us to get through the next 30 years of energy-environmental difficulty.

A major part of the research and development effort is going into getting clean, cheap energy from coal. We described three different techniques: the first, coal gasification, is nearest commercial realization and, in the most optimistic of the NPC estimates, could be providing for 8 percent of our gaseous fuel needs by 1985. Liquid fuel (gasoline or oil from coal) is more difficult to bring into commercial production as it requires plentiful, inexpensive hydrogen. The first new fuel from the liquification experiments will be a tar-like fuel, solvent refined coal, of high Btu value and low sulphur content.

An exciting possibility proposed by Arthur Squires is the coalplex, a new type of boiler (the fluidized bed) which produces a power gas usable in a gas turbine topping cycle and has a potential efficiency of greater than 50 percent. All of these new coal utilization techniques provide for removal of the sulphur before burning, a much more practical method than the present emission control devices.

We also considered the magnetohydrodynamic (MHD) generator which is being developed as another topping device and converts heat energy directly to electrical energy. This, and the other exotic heat engines are somewhat further in the future, but all offer the hope of cleaner fuel and higher efficiencies of energy conversion.

Perhaps the most exciting device that is being developed is the fuel cell, in which the burning of hydrogen and oxygen to produce water is slowed down and the energy is released as electric kinetic energy rather than heat. Some 60 of these flexible, efficient, and clean energy-conversion devices are now being tested in homes, offices, stores, and apartments. Their major commercial advantage comes from the substitution of pipeline gas for transmitted electric energy.

In the final section we looked at the advantages, disadvantages, and practicality of producing and using hydrogen as a fuel. It is a clean fuel; the combustion product is water, and, since it is a gas, it can be shipped by pipe (although the pipes will need major improvements in order to hold the tiny hydrogen molecules). It can be used in may ways: as a fuel for transportation, as a source of electricity through fuel-cell conversion, and as a reducing agent in metallurgy and coal liquification. It also offers a way to store electric energy; cheaper off-peak power, or electric power produced by windmills or the solar sea plants can be used to electrolyze water and the hydrogen stored until it is needed.

These technologies should help us lessen the effects of the intermediate-range Energy Crisis and get through the next 30 years. After that time, we will have to look to the continuous sources of energy which will be discussed in the next chapter.

13

Energy from Sun and Sea: The Continuous Sources

In Chap. 12 we reviewed the technological struggle to postpone the inevitable. The clever ways by which we can obtain more energy from coal, shale oil, and so on will stretch our fossil fuels by decades, at least. They will not, however, change the fact of finiteness; the fact that we have almost reached the limits of these depletable sources. Set against our enormous demand, even the great legacy of coal, left for us from the rich vegetation of the Carboniferous Age, vanishes in a puff of time if we deal in centuries rather than years. One lesson we are learning from the Exponential Century is that unlimited growth is incompatible with the finite resources of Spaceship Earth.

But Spaceship Earth is not really on her own; she has a vast and continuous source of energy from the Mother Ship: Sun. In Chap. 3 we examined the various ways this solar energy appears, in the gravitation potential energy of lifted water and the chemical energy of plants, for example. This is a continuous source of energy rather than a finite or depletable one. It is time to examine these various sources of continuous energy, determine their relative magnitudes and look at the progress, or lack of it, in turning them to our needs.

THE SOURCES OF CONTINUOUS POWER

Since we will now be describing energy that is steadily renewed on the earth, we will deal with power, the time rate of energy consumption or production, rather than with total energy. In Table 13-1 we compare the power available from all the continuous sources which appear exploitable for our purposes. Most of these fluxes of continuous power are distributed, evenly or unevenly, over our land area; we display them, therefore, in terms of kilowatts per acre. For perspective we also show the level of total power consumption, (about 2 M mw in 1971) averaged over the 1930 million acres of the 48 contiguous states.

All but two of these sources of power ultimately come from the sun. Only tides, whose source is the rotational kinetic energy of the earth-moon system, and geothermal power, whose source is radioactive decay in the molten interior of the earth, are nonsolar. We have labelled these as "continuous sources," but, of course, "forever" is meaningless, and their lifetimes, while long compared to ours and our species' history, are finite. Solar sources will deplete over the next few billion years, although

Table 13-1 Power Per Acre on Continuous United States

Source	*Kilowatts per acre*
Sunlight at top of clouds (day-night average)	1,420
Sunlight absorbed at ground level (averaged)	810
All the winds (below 250 ft)	~ 5.3
All the tides	~ 0.02
Natural geothermal (steam and hot water)	~ 0.25
All photosynthesis	~ 0.32
All hydropower	~ 0.12
1971 U.S. energy consumption, average per acre	1.1

less than the two nonsolar sources, but in our myopic view of time they are all inexhaustible.

The size of our power needs measured against these sources is impressive. It is already bigger than most of them. Even if we burned all the new growth each year and generated electricity with all our runoff water, we could not satisfy our demand. Covering the country with tall windmills could just meet the present demand. The tides and geothermal steam and hot water can make only fractional contributions to these averaged needs. Only by pointing power receptors at the sun can we find abundance.

These comparisons do not mean that we can afford to neglect all these other sources of power. Some of them can be important locally, and all of them, since they are free for the taking, should be fitted into the total picture where they can be. We will, therefore, briefly consider each of them, seek to make estimates of the level of power each might realistically provide, describe the technology available or needed, and look at the environmental costs of the addition of their kilowatts to those we are presently consuming.

"SECONDHAND," RENEWABLE RESOURCES

We will look at the smaller sources before we consider the two fusion and solar power which offer us the hope of large returns. We call these smaller sources "secondhand" because each of them is converted to the form in which we are interested; wind and water power from solar energy, geothermal from nuclear energy of the earth's radioactive ores, and tidal energy from the spinning energy of the earth-moon system. We will look at them in a descending order of potential importance.

Water Power

We have already discussed hydropower in some detail in connection with the generation of electricity (see Chap. 6). Water power presently produces a significant amount of electric power; it accounts for 56,000 mw of generating capacity out of a total of 340,000, or 16 percent. From the stream flow records one can calculate that the total power potential of this form of solar energy in the 48 contiguous states is about 150,000 mw. We are, therefore, already using about a third of our total potential.

Of the remainder, a significant amount of potential capacity is protected by legislation. The Colorado River Basin Project Act excludes a potential of 3,500 mw between the Glen Canyon and Hoover Dam projects; the Wild and Scenic Rivers Act excludes about 9,000 mw of potential on the 37 river stretches included in that Act. It is thus clear that direct hydropower cannot be expanded to provide much additional power. Pumped-storage plants will, it is projected, be built rather extensively, but these store energy rather than produce it.

The world picture is somewhat different. As shown in Table 13-2, there is still a lot of undeveloped water power capacity in the world. Much of it is to be found in Africa and South America whose coal resources, as we have seen, are not very large. With a period of intensive development, hydropower could make a large percentage contribution to the electric power of the world. We must, however, guide our enthusiasm for such development by the recognition of the environmental impact summarized in Chap. 8.

Geothermal Power

One of the oddities in the history of energy is that we have dug and pumped large quantities of fossil fuel from the earth's crust and burned it to produce steam and, yet, have largely ignored the steam itself which is there in the crust. For centuries, the existence of hot springs and steam geysers have been a cause of wonder and provided for recreation and medication. It is only within this century, in 1904, that geothermal steam, from the large Lardello geothermal field in Italy, began to turn turbines and generators. It was not until 1960 that the first United States geothermal plant was commissioned: a 12,500-kw generator at the Geysers field in California, the world's largest known field. The rising fuel prices, threatened shortages, and booming electrical demand of the past decade have awakened interest in this "free" steam. With the passage of the Geothermal Steam Act in December 1970 authorizing exploration, leasing, and operations on public lands, a minor "steam rush" began.

Electric power capacity dependent on geothermal energy in 1972 totaled 192 mw in this country, all from the Pacific Gas and Electric Company's six turbines at the

Table 13-2 World Water-Power Capacity*

Region	*Potential (10^3 mw)*	*Percent of total*	*Development (10^3 mw)*	*Percent developed*
North America	313	11	59	19
South America	577	20	5	
Western Europe	158	6	47	30
Africa	780	27	2	
Middle East	21	1	–	
Southeast Asia	455	16	2	
Far East	42	1	19	
Australasia	45	2	2	
Soviet Union, China, and satellites	466	16	16	3
Total	2,857	100	152	

*Source: M. K. Hubbert, "Energy Resources," in *Resources and Man,* (San Francisco: W. H. Freeman, 1969), p. 209, Table 8.7.

Geysers. By the summer of 1974, 11 units are expected to be producing 502 mw, making it the world's largest geothermal operation. Even then, it will contribute less than a percent of the expected total megawatt-generating capacity of all types in that year. Geothermal energy, while small, is a rapidly growing source, worth our brief attention and some conjectures on future growth and total potential.

Geothermal energy derives from the molten core of the earth. While most of this heat is at depths beyond present drilling capability, there are many areas where geologic forces push molten rock, *magma,* near the surface (see Fig. 13-1). This magma dome heats the rocks above it to temperatures as high as 700°F. If these rocks are porous or cracked, and contain water, and if they are capped off by impermeable rocks, reservoirs of hot, high pressure steam may develop. These are the *dry steam* fields, whose steam can be channeled, with little treatment, directly into steam turbines. They are the only types of geothermal fields used in the United States at present.

The major attraction of geothermal steam is, of course, its low cost. Pacific Gas and Electric buys its steam from the Union Oil Company which operates the wells. The cost of steam alone is 2.4 mills per kw-hr and the total cost of steam-produced electricity is about 5.25 mills per kw-hr, well below the 6 or 7 mills per kw-hr of conventional electric generation.

There are several other types of geothermal fields of potential interest: *wet steam* fields, in which steam must be separated from boiling water, and the fields of much larger potential size in which the water is hot but below the point of boiling. There are, in addition, an unknown but probably large number of *dry geothermal reservoirs* in which the rocks are hot but are not porous or fractured enough or, for some other reason do not hold water.

The technology of exploitation of the dry steam fields is relatively simple. To be used, the steam reservoir must be hot enough (350 to 400°F) to keep the steam dry and superheated, and shallow enough (3 km or less) to be easily reached. A well is then put down to the reservoir, and the steam is fed directly to the turbine with only minor filtering to remove abrasive particles which might damage the turbine blades.

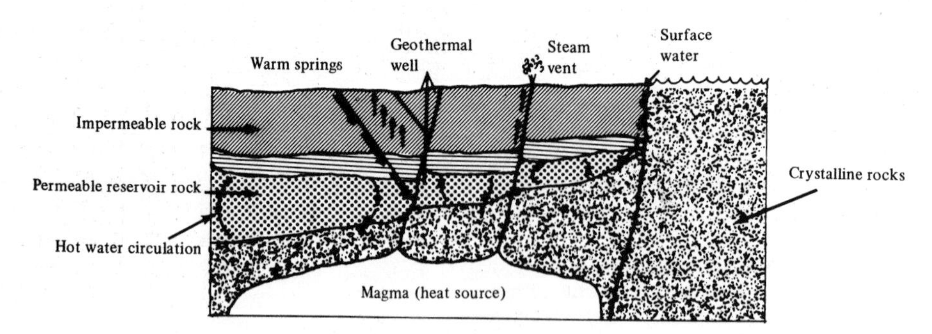

FIGURE 13-1 The magma, molten rock pushes up near the surface and heats the water-impregnated reservoir. If the reservoir is capped with impermeable rock, high-pressure steam will be produced. [Source: Draft Environmental Statement, Liquid Metal Fast Breeder Program, Vol. IV, (Washington, D.C.: USAEC, March 1974).]

Since the temperature of geothermal steam is considerably lower than the 1000°F steam in modern fossil fuel plants, the efficiencies of these plants are low, perhaps as low as 20 percent, and with the low pressures, small turbines must be used. The units now being installed in the Geysers are 50 to 60 mw each in generating capacity. The plants are spaced 1 or 2 miles apart and fed by steam lines from several wells.

The environmental impact of these dry steam wells is minimal. They are ugly and noisy and there is some minor emission of odorous gases. They do not, as presently run, use external cooling water sources; the condensers are cooled by water from the liquified steam circulated in cooling towers. The extra water, about 20 percent, is reinjected into older wells where it may help prevent subsidence.

The environmental impact of other types of geothermal operation such as wet steam wells may be more noticeable. In such wells, much larger amounts of water will have to be dealt with. The water itself may be heavily contaminated with minerals and the withdrawal of such large amounts of water from the ground will probably cause subsidence. There has been some subsidence in the Wanakei wet steam field in New Zealand, where 70 million tons of water per year are removed. It is expected that this problem can be controlled by reinjected water.

Geothermal exploration is in a primitive early state. Most exploration so far has consisted of searching for hot springs and steam vents, a stage not unlike the early oil prospecting for oil seepage. Because of this, estimates of total resources are apt to be quite divergent. Muffler and White[1] for instance, estimate the identified, recoverable geothermal resources of the United States as 60 Q Btu at the well head, or, applying a 15 percent recovery efficiency, as 10 Q Btu. This estimate is for the dry steam reservoir of the Geysers, as well as several other high-temperature hot water reservoirs. R. W. Rex[2] on the other hand, estimates that there is twice this capability in the Geysers area alone (enough steam to produce 1,000 mw of power for 100 years), and that the deeper known reservoirs could provide 30 times as much or 600 Q Btu. This is to be compared with, for instance, the 3,000 Q Btu of known coal reserves. The estimates for undiscovered recoverable reserves (those recoverable at cost competitive with other forms of electric power generation) range from 60 to 120 Q Btu from White upwards to the Rex estimate of greater than 20,000 Q Btu.

These estimates are for dry steam wells. If the technology is developed to utilize the high-temperature hot water or wet steam reservoirs the resource becomes considerably larger; 4,000 Q Btu is the low estimate. Rex believes that there are about 50,000 Q Btu of this type of geothermal reserves in the Western United States. This amount of energy would allow the generation of 2.4 M mw for a century. The present United States generating capacity (1972) by comparision is about 400,000 mw.

Beyond these high-temperature hot water reserves are a variety of other possible sources also awaiting technological breathroughs: intermediate-temperature hot water reservoirs (200 to 350°F), hot rocks, and reservoirs deeper than present drilling capability (greater than 10 km). The estimates of these potentials range from 40,000 to 14.5 million Q Btu or 72.6 M mw centuries of generated electric energy.

[1] L.J.P. Muffler and D.E. White, "Geothermal Energy," *The Science Teacher,* **39**: 3, 1972.

[2] R.W. Rex, submitted to "Geothermal Energy Resources and Research," Hearings Before the Committee on Interior and Insular Affairs, Serial No. 92-31, June 1972.

The realization of any of the energy in other than dry steam wells wait on improvements in technology. The hot-water wells, in particular, will be tapped most efficiently by heat exchange between the water and a low boiling-point liquid such as isobutane or freon which can be vaporized to drive a gas turbine. The Soviet Union is reported to be using this heat exchange technique with water at 108°F in Kamchatka. Such a plant using isobutane has been designed for the United States but money for pilot plant construction is not yet available.

In addition to these low-temperature cycles (which could have applications in regular steam generation), other technological barriers awaiting penertration are means of recharging how sources with new water, techniques of artificial stimulation (explosive fracturing, for instance), and deep, high-temperature drilling techniques.

Table 13-3 presents a projection (from an official of one of the companies developing geothermal power) of the role of electrical generating capacity from this source through the end of this century. Results are shown in three columns: from a moderate research and development program, from an "intensive and accelerated" research and development program, and, for comparison, the FPC projections of generating capacity. It would again appear that determination and planning could result in a dramatic increase in the importance of this domestic and fairly clean energy source.

Beyond pure electric power the possibilities of geothermal energy broaden in a very interesting way. In many well-situated areas such as Klamath Falls, Oregon, and Boise, Idaho in this country as well as in the Soviet Union, Hungary, Japan, Iceland, and New Zealand, the hot water is also used for home heating, industrial processes (a paper and pulp company in New Zealand) and even air conditioning (working on the same general principle as the old gas flame refrigerator).

The other output from these wells, rivaling heat in importance, is water. The next big geothermal project in the United States is planned for the Imperial Valley. The Bureau of Land Reclamation and the Office of Saline Water have begun a seven-year, 16 million dollar project to determine the feasibility of using the extensive hot brines under that valley to produce fresh water for the Colorado River. Under the

Table 13-3 Potential Geothermal Generating Capacity*

Results from a moderate research and development program		*Results from an intensive and accelerated research and development program*	*Total U.S. generating capacity‡*
Year	*(mw)*	*(mw)*	*(mw)*
1972	192		367,396
1975	1,500	750†	
1980	10,500	36,000	665,000
1985	19,000	132,000	
1990	35,000	242,000	1,260,000
2000	75,000	395,000	

*Source: "Geothermal Energy Resources and Research," Hearings, Committee on Interior and Insular Affairs, Serial No. 92-31, June 1972, p. 245.

†This estimate for 1975 is smaller than that shown in the first column because current research and development does not appear likely to be able to deliver 1,500 mw on line by 1975.

‡The 1970 National Power Survey, Part I, Federal Power Commission (Washington, D. C., December 1971).

Colorado River Basin Act, the Department of the Interior is asked to provide 2.5 million acre-feet of water per year for the Colorado River. It is being proposed that a multipurpose geothermal development in the Imperial Valley could produce power, water, and minerals. To produce clean water, the hot brine would be pumped into a succession of "stills" at less than atmospheric pressure, the water would boil, and the steam would be piped off and allowed to liquify. The material that remains in the end is rich in many types of chemicals leached from the rocks and can, perhaps, be sold.

Withdrawing water at the rates being considered for the Imperial Valley brings on the danger of subsidence. It is proposed that water be reinjected; either irrigation runoff water or ocean water pumped inland. If this works, the Imperial Valley will, in fact, be a huge *geothermal desalination plant.* Over a period of, perhaps, 100 years hot water will be withdrawn, distilled with its own heat, replaced by salt water (or contaminated irrigation runoff water); this water will, in turn, be heated by the magma, drawn off, purified, and so on. The projected cost is much less than any presently operating desalination plant.

World Geothermal Resources. Geothermal energy can make modest but important additions to United States energy production. Its world value is difficult to estimate but potentially huge. White[3] has estimated the total heat content in the earth's crust to a depth of 10 km to be 10^{24} Btu, equivalent to 2,000 times the estimated world coal reserves. While we can't think of this heat as a resource of the same nature as coal, there are promising geothermal hot spots all over the world. They occur in the volcanically active, earthquake-prone areas of the earth's crust: down the Western coast of the Americas from Alaska to Chile, down the Rift Valley in Africa and extending up through Turkey, and in the Far East along the Circle of Fire that borders the Pacific Ocean. Individual countries are unusually well blessed; in Turkey two thirds of the country is believed to have geothermal potential, and surveys carried out by the government of Ethiopia and the United Nations (by aerial scanning with an infrared camera) suggest that one Ethiopian field has the potential to produce enough electricity to satisfy all the present needs of Africa. There is also a potentially large geothermal development effort under way in Mexico. There has been a 3.5-mw pilot plant in operation since 1959 at Hidalgo near the Gulf Coast, and two 37.5-mw units are under construction near the Cerro Prieto volcano.

The heat is there, underlying much of the world. Imagination and money are needed to bring it to the surface in the service of man. This heat could substitute for that we obtain now by burning those precious carbon-hydrogen-oxygen molecules of oil, natural gas, and coal which are of so much greater value as chemical raw materials for the synthetic wonders of the present and future.

Energy from Waste

As a secondhand energy source, organic waste could make significant contribution to our energy needs, and, at the same time, lighten an environmental burden. There are

[3]D.E. White, Geothermal Energy, USGS Cir. 519, 1965.

several feasible conversions of the chemical energy stored in solid and liquid waste; much of it could be burned to produce heat for power-plant steam generation, there are also well-developed techniques for producing methane gas and fuel oil from this same source.

The total potential Btu content from organic solid waste is considerable (see Table 13-4). If converted to oil, for instance, the waste could provide 1.1 billion barrels per year, about 20 percent of our current oil consumption. As is emphasized by this table, however, much of this waste material occurs in small amounts at many sites and is not readily collectable. The estimated amount that occurs at large operations, big cattle feedlots, municipal collections, large sawmills, and the like is, however, sufficient to produce 170 million barrels per year of oil which is 12 percent of the amount of crude oil imported in 1971.

There are several pilot plants in operation to test the various methods to produce power and dispose of waste. In the pyrolysis process (burning in the absence of oxygen) developed by the Garrett Research and Development Company of LaVerne, California, organic waste is shredded, dried, and the inorganic materials such as glass and metal are removed for recycling. The material is then reshredded and heated to 500°C in an oxygen-free atmosphere. Each ton of refuse produces 1 barrel of oil in addition to burnable char and some low Btu gas; these latter products provide the heat for the process. Garrett has received a contract to build a 200 ton per day plant to handle all the solid waste from the California cities of Escondido and San Marcos. They will sell the fuel oil to the San Diego Gas and Electric Company and project operating costs to be 8 dollars per ton, less than the costs per ton of disposing of the refuse in that area. They estimate that a 2000 ton per day plant could operate at 5 dollars per ton and handle the output from a city of 500,000.

There are other less sophisticated experiments with burning waste for power which also show promise. St. Louis now generates 10 percent of its electricity by burning solid waste. The Bureau of Mines, in a pilot plant at their Pittsburg Energy Research Center, is successfully extracting oil from the cellulose carbon in waste, the major component of solid waste because of the newspaper content. In this proc-

Table 13-4 Amounts of Dry, Ash-Free Organic Solid Wastes Produced in the United States in 1971*

Source	*Wastes generated (million tons)*	*Readily collectable (million tons)*
Manure	200	26.0
Urban refuse	129	71.0
Logging and wood manufacturing residues	55	5.0
Agricultural crops and food wastes	390	22.6
Industrial wastes	44	5.2
Municipal sewage solids	12	1.5
Miscellaneous	50	5.0
Total	880	136.3
Net oil potential (10^6 barrels)	1,098	170
Net methane potential (10^9 cubic feet)	8.8	1.36

*Source: L.L. Anderson, Bureau of Mines.

ess, waste is heated at high pressure and temperature in an atmosphere of carbon monoxide and steam. The technique is similar to coal liquifaction described earlier. Some 99 percent of the carbon content can be converted to fuel oil and, even after recycling some of this oil, 1.25 barrels per ton of dry waste are obtained. The sale of oil and the reduction in volume of waste to be disposed of makes this a commercially viable operation.

In many ways the simplest way to obtain energy from organic waste is to produce methane from it; in fact, this occurs naturally, and many sewage disposal plants trap this gas and use it for heating. Methane is produced by *anaerobic* (oxygen-free) digestion of organic materials by micro-organisms. This *fermentation* process could, it is estimated, produce 10,000 cubic feet of methane or 10^5 Btu of energy from a ton of solid waste. The estimated potential of collectible waste (from Table 13-4) is 1.36 trillion cubic feet, 6 percent of the 1971 natural gas consumption.

A commercial operation using this process would have to sort out the inorganic materials, shred the refuse, and dispose of the sludge. The latter can be used as land fill or as fertilizer, or be converted to oil by the processes described. An advantage of this system is that sewage has already a large component of the water which is needed as a growth medium. Methane as a fuel is, or course, very desirable and, by existing techniques, can be easily cleaned of potential pollutants.

The maturation of some of these processes can have great benefits, supplementing our energy sources with clean fuel and at the same time, reducing the amount of solid waste that is literally piling up around us.

Energy from the Winds and Tides

We will now discuss two secondhand renewable sources of quite different origin. The winds carry kinetic energy generated by the solar heat engine; the tides gain their energy from that originally stored as rotational energy of the moon about the earth. Although both of these sources have been used, neither is expected to play a big future role.

As we saw in Table 13-1, the wind is the larger potential source. It suffers the same disadvantage as solar energy in general: it is erratic and diffuse. It has been harnessed to man's activities successfully on a small scale, however, and there have been some interesting recent experiments. The most ambitious United States effort was a 1.25 mw wind-powered generator installed on "Grandpa's Knob" in Vermont. Unfortunately, it suffered from the World War II materials shortage and one of its two counter-rotating blades, of inferior construction, broke off and was never repaired.

Some idea of the potential of wind power can be gained from the following data: In Oklahoma the average wind power is 18.5 watts per foot^2, which is equivalent to the average solar energy per foot^2 in that state. William Heronemus of Amherst University, a long-time advocate of wind power, has calculated that 300,000 wind turbines placed in a belt across the Midwest from Texas to North Dakota could produce 0.19 M mw of power, equivalent to about one half our present generating capacity. The prospect of building 15,000 towers across our Midwest, each 850 feet high and carrying 20 turbines, lessens the appeal of this idea.

A different use of wind power is being tried out on the Virgin Islands. Fresh water is to be produced by collecting moisture from the air in a seawater-cooled condenser. The seawater is to be pumped from the ocean by wind power. For wind to be a realistic contributer, however, the mechanical problems of big turbine construction, as well as the problems of energy storage, will have to be solved. The electrolysis of water to hydrogen in offshore wind generators has been proposed as one possible answer to the latter. It may be that the present crunch will revive interest in the use of this clean free energy which is carried on the wind.

Tidal Energy. The tides are not likely to help us since there is just not enough energy there. The high tides of Maine have been sporatically considered as a potential source, but design calculations have not justified installation of the necessary equipment. There are two operating tidal plants producing electricity now in other parts of the world: a 250-mw plant on the Rance River Estuary on the Brittany Coast of France and another reported at Kislaya-Guba in the Soviet Union.

"MAINLINING" ENERGY: POWER FROM THE SUN

A glance at Table 13-1 is all that is needed to identify the continuous source with the largest potential. The average input power per acre from the sun absorbed at ground level is some 700 times larger than the per acre average of power consumed per day in the United States. Most of our uses of solar energy so far have been of the secondhand type just described: water and wind power and, of course, photosynthesis. Our needs, however, have now grown so large that we are looking beyond this secondhand power to the source itself. The search is on for ways to convert direct solar energy to our uses, to "mainline" it, so to speak.

This search is not a new one. Through the ages, man has been impressed by the power of the sun and even occasionally turned it to his uses. Archimedes, according to the legend, used focusing mirrors to set fire to invading Persian ships and small, solar-heated steam engines had some success a century ago. But techniques to use solar energy as a reliable and large-scale source of power have not been developed for obvious reasons. Solar energy is dilute so that a large capacity requires a large collection area. To obtain 1,000 mw of power, for instance, requires, according to Table 13-1, about 1,250 acres or about 2 square miles of surface area at 100 percent collection efficiency. At lower collection efficiencies the areas are correspondingly larger: 20 square miles at 10 percent efficiency.

It is also an erratic, rather than a reliable, source, fluctuating on the predictable day-night cycle and unpredictably with the cloud cover. We are, therefore, faced not only with the need for a large collection area, but also with the problem of storing the energy once it is collected.

These handicaps, the need for large area collection and for energy storage, have served so far to keep experimentation with this energy at the hobby level. Our newly realized energy shortage has, however, caused a new flow of federal research support into solar energy research and several methods for domesticating the sun are now being actively investigated. Work is proceeding along four different lines: (1) solar home-heating and cooling; (2) solar cells for direct conversion to electricity; (3) large central power plants for the production of electricity; and (4) improved photosynthesis. We will briefly discuss these techniques in their order of feasibility.

Solar Homes

The most simple energy conversion for the incoming solar radiation is into heat energy. There is much to recommend emphasis on direct use of solar energy in the home. Some 20 percent of the total energy bill goes into heating and cooling residential and commercial buildings. Furthermore, the conversion to heat energy is a highly efficient one. It follows the arrow of the Second Law of Thermodynamics. Collectors capable of converting 82 percent of the incident radiation to heat have been demonstrated.

The techniques of heating and cooling with solar energy are quite simple. The roof provides ample collector area. The solar incidence of Table 13-1 is equivalent to 1400 Btu per foot^2 per day. A typical residence of 1,600 feet^2, holding a family of four, will, therefore, receive, on the average, 800 M Btu of solar energy in a year, some 6 times the average 130 M Btu required for space heating. These averages, of course, do not display the difficulties: the major fraction of this energy arrives during the summer months when heat is not needed. Even with this disclaimer in mind, however, there are large areas of the United States which do receive enough steady solar power to take care of their total need for space and water heating. These areas are shown as the zone of "maximum feasibility" in Fig. 13-2. The second zone, roughly covering the middle latitudes of the United States, is one of "engineering feasibility"; some auxiliary heating would have to be provided. For the zone of "minimum feasibility," solar heat would be the auxiliary heat source, supplementing conventional systems in the spring and fall.

Heating is accomplished by *flat plate collectors,* essentially, a blackened plate covered by one or several transparent insulating layers (see Fig. 13-3). The transparent layers prevent cooling of the blackened absorbant plate by air currents. They can be made selectively transparent, admitting the incident short wavelength radiation but reflecting (back to the absorber) the longer wavelength infrared *heat radiation.* It is also possible to select absorber materials which are poor infrared emitters and further increase their heat-holding abilities. Water pipes are then put in thermal contact with the back of the plate and carry the heat energy into some kind of a storage reservoir.

Several storage techniques have been tested. Hot-water storage is sufficient for a few days as are heated rocks or pebbles. More sophisticated techniques now being

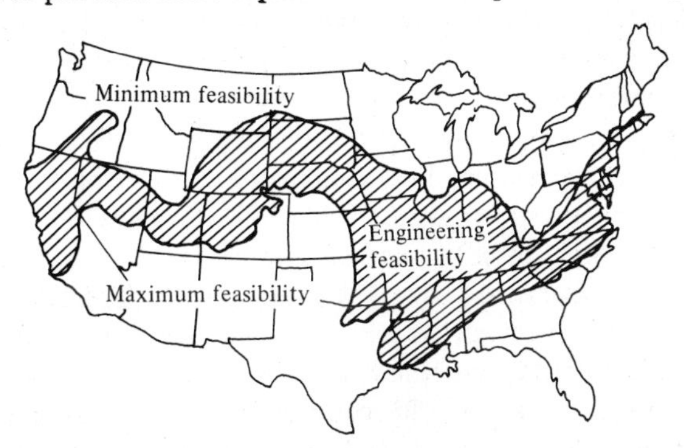

FIGURE 13-2 The various zones of feasibility for solar space heating in the U.S. [Source: "Solar Energy Research," Staff Report of the Committee on Science and Astronautics, U. S. House of Representatives, (Washington, D.C.: USGPO, December 1972).]

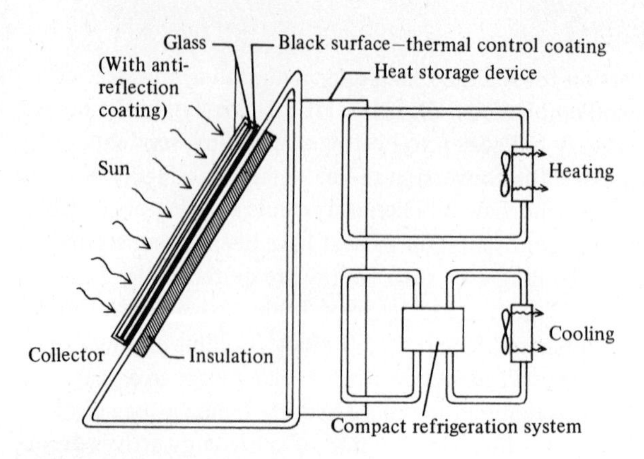

FIGURE 13-3 The most simple solar collecting device would use a blackened flat plate with a transparent, antireflection coating. Circulating water removes the heat for storage or use. (Source: "Solar Energy Research," *op. cit.*)

developed use salts that are melted, absorbing energy, during the collection period and then solidify, releasing energy when it is needed.

Such a system could carry the heating load by itself in the zone of maximum feasibility (see Fig. 13-2). In the zone of engineering feasibility a backup heating system would be needed for a succession of cold cloudy days. If these two systems are combined (if an electrical backup system is provided, for instance), the storage system can be recharged when necessary by electrical heaters during the nightime using off-peak power at correspondingly low rates.

The technology to accomplish what has been described already exists; in fact, solar-heated houses have been operating in various parts of this country for many years. In one which has been in operation for 20 years in Boston solar energy has consistently supplied about half of the total heat needs. They have already demonstrated cost competitiveness. Table 13-5 shows the results of the most recent analysis of the relevant data and compares the cost of solar heat (dollars per M Btu) with electric, gas, and oil heating. In this analysis the comparison takes into account not only costs of fuel, but also the capital requirements and operating costs, and assumes a 20-year life cycle for the building.

It is clear from even this data that solar space and water heating is the most economical process in some parts of the country. (It appears to be competitive with gas and oil in Santa Maria, California.) In most parts of the country it is more economical than electric heating. In the light of this knowledge, one wonders why electric home-heating has been enjoying the boom we described in Chap. 6 (a 470 percent increase from 1960 to 1968) while solar heating is still a novelty.

The answer to that question underlines the real barriers to wide utilization of solar heat. These barriers are not technological, but economical and practical. The dominant consideration in the purchase of a home seems to be the purchase price: the capital investment. Solar heating is add-on heating since it has to be at least backed up by a conventional heating capability. But even in the zone where solar heating alone is enough, the initial cost of the solar equipment is larger than the cost of conventional facilities. Solar heating systems are capital intensive. This means that

Table 13-5 Costs of Space Heating (1970 Prices) In Dollars Per Million Btu Useful Delivery*

Location	*Optimized solar heating cost in 25,000 Btu/degree-day house, capital charges @ 6%, 20 years* Collector @ $2/ft²	Collector @ $4/ft²	*Electric heating, usage 30,000 kw-hr/year*	*Fuel heating fuel cost Only* Gas	Oil
Santa Maria	1.10	1.59	4.28†	1.52	1.91
Albuquerque	1.60	2.32	4.63	0.95	2.44
Phoenix	2.05	3.09	5.07	0.85	1.89
Omaha	2.45	2.98	3.25§	1.12	1.56
Boston	2.50	3.02	5.25	1.85	2.08
Charleston	2.55	3.56	4.22	1.03	1.83
Seattle-Tacoma	2.60	3.82	2.29‡,§	1.96	2.36
Miami	4.05	4.64	4.87	3.01	2.04

*Source: NSF/NASA Solar Energy Panel, "Solar Energy as a National Energy Resource," (Washington, D.C.: National Science Foundation, 1972).

†Electric power costs are for Santa Barbara; electric power data for Santa Maria were not available.

‡Electric power costs are for Seattle.

§Publicly owned utility.

Solar heat costs are from optimal design systems yielding least cost heat.

the mortgage company gets into the act, that the house may be harder to sell, and so on. Electric heating, on the other hand, is, as we have mentioned, initially economic; it requires no complex equipment, no furnace ducts or pumps. It is convenient to operate and maintain, a consideration of primary concern to the purchaser. With the present mobility of Americans, we are not accustomed to considering 20- or even 10-year life cycles of houses and this set of mind is present not only in buyers, but in builders, insurers, and all the others involved in the housing business. It is only on such a basis, however, that the savings from free fuel will become impressive.

These savings are now more impressive than what is shown in Table 13-5. Even in the instance since 1970, the overall energy picture has changed drastically and greatly enhanced the competive position of solar energy. The $1.85 per M Btu for natural gas which Table 13-5 shows for Boston is now 3 dollars; other fuels, even electrical energy, are rising in a similar fashion. The clear message of all our energy studies is that this increase will continue. It would seem that new efforts to stimulate mass production of solar heating units and to change building practice codes, insurance and mortgage regulations, and the like can now be justified on an economic as well as an energy ethic.

Additional economies accrue if solar energy is used to provide cooling as well as heating. The technological feasibility of this has also been demonstrated. The technique is similar to that one used in the old Servel gas refrigerator in which a gas flame provided the energy for cooling.

A simple version of this is shown in Fig. 13-4. Two bulbs, A and B, are connected by a tube. Bulb A contains a strong solution of ammonia (which is very soluable in

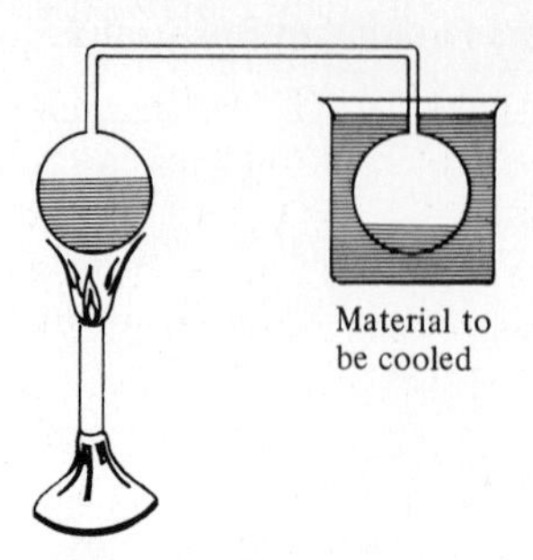

FIGURE 13-4 The absorption cooling process. Ammonia vapor driven out of solution in A by heat is condensed in B. If A is now cooled, the vapor will be reabsorbed in A and take heat energy from the medium surrounding B.

water). When A is heated (by a flame or solar heat) ammonia vapor is driven off, goes into B, absorbs heat, and condenses into a liquid. If the heat is now removed and A is cooled the gas flows back to it, removing heat energy from B. An operating system of course, accomplishes this in a cyclic fashion.

In the refrigerator, ammonia is chosen because of its low boiling point (-40°C) and because its heat of vaporization (the amount of heat energy per gram needed to cause it to boil) is 300 Calories per kg—only water's 540 Calories per kg is larger. In air conditioning, where smaller temperature differences are encountered, other liquids with higher boiling points could be used. Air conditioning application of solar energy suffers from the same disadvantages as other solar forms. For ordinary use, for instance, the solar collector has to be of about the same surface area as the square footage to be cooled. For actual refrigeration a focusing collector may be needed to provide high enough temperatures. All in all, however, the expert opinion is that the technology for cooling is at least as ready for large-scale application as is that for heating. It is also clear that a combined heating-cooling system will have economic advantages.

Given the ready status of technology on the heating-cooling front and the lack of action to date, it is encouraging to see even small movements in the right direction. There has been increased support from Congress and discussion of a law which provides for the construction of 5,000 or so demonstration solar homes in this country. This project, if it is carried through, should lead to the breakthroughs on the architectual, building, regulatory, financial, and other fronts that are needed. Within five years we may be able to judge what portion of the energy we now spend on home comfort can be taken from the sun.

Solar Cells

The most glamorous conversion techniques use solar cells to convert radiant energy directly into electrical energy. The energy conversion is accomplished by the photovoltaic process: solar radiation on striking certain materials is absorbed and causes a separation of the electrons from the atoms. The migration of these electrons in one direction and the positively charged ions in the other produces a small potential difference, typically about 0.5 volts. An array of these cells can produce a useful electric power capacity.

The present handicaps are many; the cells must be made from very high purity materials. Single crystals of silicon are a preferred material. While silicon is the most abundant material in the earth's crust, it is not easily refined (and refining uses energy) and shaped for use in solar cells. With present technology the cells are produced almost one-by-one in what is essentially a cottage industry. They cannot yet be mass-produced. Although efficiencies of conversion have increased from 9 percent in 1965 to 11 percent in 1971, and energy conversion costs in dollars per watt have decreased by nearly a factor of 10, the cost of the energy they produce is still beyond any reasonable competitive range. The present costs of the best solar cells for space craft usage are 300 to 1,300 dollars per watt. This has to be compared with 200 to 300 dollars per kw in fossil fuel and nuclear reactors, and 3,000 to 13,000 mils per kw-hr (10-year operation in full sunlight) compared to 7 to 8 mils per kw-hr from conventional sources. It is clear that reductions of several hundredfold in solar cell costs will be needed to allow their serious consideration for any use other than in space or very remote terrestrial locations.

There are many who believe these reductions in costs can be achieved, and advance planning for their use continues. Two very different but illustrative proposals are: (1) for their incorporation into the roof of a residence, now being implemented at the University of Delaware; and (2) for putting a large array into a synchronous orbit to collect and convert solar energy and beam it by microwave back to earth.

There are several interesting features of the Delaware house. The roof is covered with transparent panels which shield the solar cell collectors. The panels also collect heat which is stored in a frozen salt reservoir of the type we have described. A heat pump (see Chap. 6) transfers heat from the reservoir to the house or pumps heat from the house to cool it. The solar cell generators are also connected to utility lines in a two-way arrangement. When they are generating more power than the house needs, it is sent through the utility lines and credit is received. When the solar cells cannot produce enough power, the flow is reversed. Since the solar cells generate power during the daylight peak-load period, the system works to the utility company's advantage in most instances. The test of this interconnected system will be watched with interest.

At the other extreme is the proposal by Peter Glaser of Arthur D. Little, Inc. to put a large array of solar cells into a synchronous orbit about the earth and to transmit the generated electric power back to earth as microwave radiation (see Fig. 13-5). The advantages of the satellite solar array come from its nearly constant exposure to full sunlight. By intercepting the sunlight before it is reflected by the atmosphere, by removing the day-night cycle and avoiding clouds, it could be expected to receive at least 6 and maybe as much as 15 times more energy per square foot than will a ground location.

Even with this higher level of input power, the size of the array is impressive if it is to generate a significant amount of power. A 5,000-mw generator would need 20 square miles of solar cells and these arrays would have to be steered to face the sun. The collecting surfaces in Fig. 13-5 would each have to be about 3 miles wide and 7 miles long; the microwave antenna which beams the power to earth would be a mile on a side, and the receiving microwave array on earth would be 6 miles on a side.

FIGURE 13-5 Artist's conception of a solar satellite power station (SSPS). With a collecting array of solar cells 4.3 kilometers by 11.73 kilometers (2.7 by 7.3 miles) the SSPS could deliver 5000 mw of power to earth. [Source: Peter E. Glaser, *Solar Satellite Power Station,* (Cambridge, Mass.: Little)

It is clear that the present solar cells used in spacecraft and costing 3,000 to 13,000 dollars per square foot cannot be used in such an array; 20 square miles of these solar cells would cost 30 trillion dollars, 30 times the United States GNP. If the costs can, by technological breakthroughs and mass production efficiencies, be lowered by a factor of 100 or so, if the beaming and receiving of such large amounts of power can be accomplished efficiently, and the transportation provided at reasonable cost, these satellite stations might become feasible. They are now some decades at least in our future.

Improved Photosynthesis

A third approach to increased use of solar energy is to improve on nature's method, photosynthesis. There are three ways to accomplish this. The first would be to search for improved plant species which would maximize the production of *biomass,* organic solids such as wood. These could then be either processed into fuels and burned, or "digested" by bacteria to produce methane: the same process which we considered earlier for converting organic wastes to fuel gas. Some combination in which organic wastes (sewage, agricultural wastes, wood pulp) are used as nutrients to algae, water hyacinth, or some similar fast-growing plant is attractive. There are environmentally related problems in this which stem from the poisons in the water and the pollutants from the fuel.

A second approach is to try to find or develop new kinds of anaerobic bacteria which will be more efficient in their production of methane. It is also believed possible to find strains of bacteria which will produce hydrogen directly, instead of methane, which could lead to the biological fuel cells we mentioned. A third possibility is to use the catalytic activity of the enzyme, hydrogenose, which exists in algae, to cause the dissociation of water into hydrogen and oxygen.

Although these improved methods of photosynthesis cannot be considered as potential large-scale producers of power, it is clear that continued interest is warrented by the potential contribution to the organic waste disposal problem. It may be, in fact, that a less exotic photosynthetic process will again become an important source of energy. It is possible, of course, to make alcohol from sugar cane or grain. There have been times in recent history when industrial alcohol (as well as some popular beverages) have been derived from this source. In the war years, for instance, most industrial ethyl alcohol was made from molasses and other natural sources. When petroleum became cheap and plentiful again, it completely took over the market.

The rapid rise in the price of oil has brought fermentation-produced alcohol back to a nearly price competitive position. Alcohol from sugar cane, for instance, can now be produced for about 90 cents per gallon. Petroleum-derived alcohol, although it is price controlled at 66 cents per gallon (and scarce), sells at 1 dollar per gallon if bought outside the United States.

Sugar cane is an ideal crop for converting sunlight to sugar. It has the highest efficiency of solar energy conversion of any plant, 0.6 percent. It produces 4 tons of sugar (2 tons of alcohol or 625 gallons) per acre, and, including the pulp, 8 tons of carbohydrate per acre. If in Hawaii, for instance, one third of the molasses production were converted to alcohol, it would become unnecessary to import the 15 million gallons of gasoline that the state uses to run its agricultural machinery. Alcohol can be added to gasoline at up to 25 percent without any engine changes.

There are other similar possibilities; the 7 million bushels of grain which spoils each year in Nebraska could produce 20 million gallons of alcohol. The improved rubber plant is also an exciting possibility. It produces a pure hydrocarbon (not a carbohydrate containing oxygen which requires fermentation). It now yields about a ton of hydrocarbon per acre per year and the rubber growers predict that they could increase this yield to 3 tons per acre. It is quite possible that this plant could take over much of the raw materials market (the equivalent of about 730 million barrels of oil per year, 6 percent of the total energy budget).

It may well be that the rising prices of oil will drive us back to the land for energy and that large energy farms will bring needed wealth to the tropical areas of the world. We will now consider, however, a more direct energy farm.

Central Power Stations

While the use of large arrays of solar cells to convert solar energy to electricity do not appear economically feasible for a decade or so, there is growing interest in a simpler, thermal-collector system which may reach the pilot model stage within the next decade. The strongest advocates of this plan are the husband and wife team, the Meinels of the University of Arizona. They propose to trap and focus solar radiation by rows of transparent collectors, pump the heat to a central plant, generate steam, and produce electricity in the conventional manner. An artist's conception of the collector array is shown in Fig. 13-6, and a schematic of the collector operation in Fig. 13-7*a* and 13-7*b*. A proposed heat-transfer material is sodium, liquified by the heat. Temperatures as high as 1000°F could conceivably be reached, and an overall energy production efficiency of 30 percent has been estimated.

The heat-absorbing panel which is the heart of the collector, has, at its center, a hollow pipe covered with a selectively transparent coating similar to the silicon coats on the windows of modern office buildings. Silicon is an absorber in the wavelength range of the visible and ultraviolet radiation which is characteristic of sunlight, but is transparent to infrared (heat) radiation. If the silicon layer is put down on top of a reflecting coating on the pipe as in Fig. 13-7*a*, it acts as an absorber of incident energy. Normally, the pipe would emit heat in the infrared range as it became hotter, but since the silicon layer is transparent to infrared, the emitting surface at these longer wavelengths is the mirror coating, and mirrors are poor emitters. The pipe thus gains thermal energy readily but does not lose it easily. Energy loss is further restricted by placing the pipe in a vacuum.

FIGURE 13-6 An artist's sketch of the collectors of the Meinel Energy Farm. The heat is transferred to the power plant storage facilities by pipes carrying liquid sodium. (Courtesy of Helio Associates, Inc., Tuscon, Ariz.)

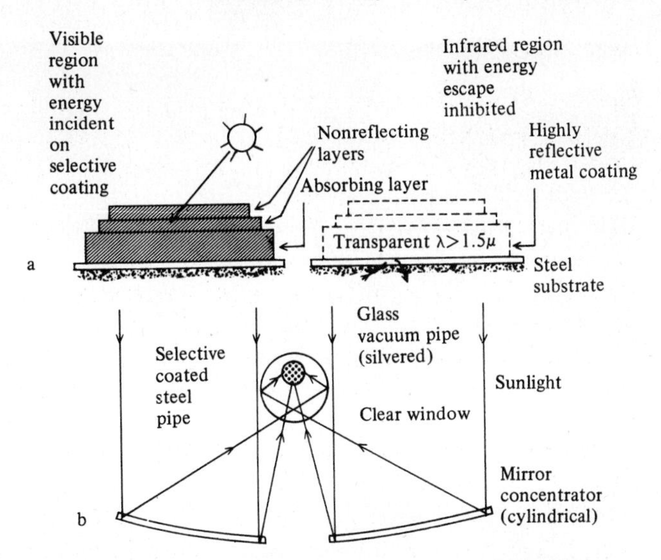

FIGURE 13-7 (a) The silicon layer absorbs energy from the incident visible radiation, but since it is transparent to infrared, the reflective metal coating becomes the infrared emitter and little energy is reradiated. (b) Cross-section of a collector. Since the glass pipe is silvered on the top and sides but transparent at the bottom, the light reflected from the mirror is trapped in the pipe. [Source: "Briefings Before the Task Force on Energy, Voll III, Committee on Science and Astronautics, House of Representatives, (Washington, D.C., 1972) p. 41.

Absorption alone will not provide enough energy to reach the temperatures near 1000°F which are necessary for efficient operation of the steam turbine. An additional collection area is provided by a curved mirror under the pipe which focuses radiation on the metal pipe (see Fig. 13-7*b*). The glass enclosure surrounding this metal pipe is silvered on the inside on all but the bottom surface, and the light from the mirrors is thus trapped in the glass pipe. After a reflection or two, it strikes and is absorbed by the metal heat-transmitting pipe.

The rest of the system is fairly standard. The heated material in the pipe flows to a storage tank (which has enough capacity to last through a couple of cloudy days) and then generates steam for the turbine. The overall system efficiency is expected by the Meinels to be 14 percent for their "early" system and 28 percent for their "mature" system. The almost 30 percent efficiency of the mature system compares very favorably with conventional electrical generating efficiency.

A most impressive statistic of this system is the area it occupies. The 120-mw power plant proposed will occupy 1 square mile of land in a region like Yuma, Arizona and produce 550,000 mw-hr of electrical energy. The relative size of the power plant and the collecting area is shown by Figs. 13-8*a* and 13-8*b*.

To produce half the 10 B mw-hr (which is the high projection of Chap. 11) for the consumption of electric energy in the year 2000) 20,000 square miles of land will be needed. That is a parcel of land 140 miles on a side. This is large, but distributed throughout the 6 to 8 sunshine states, the total impact would not be overwhelming. The Meinels call their system a *solar farm* to emphasize the comparison between setting land aside to convert solar energy to food energy and their planned conversion to electricity. Since some 500,000 square miles of land are now set aside

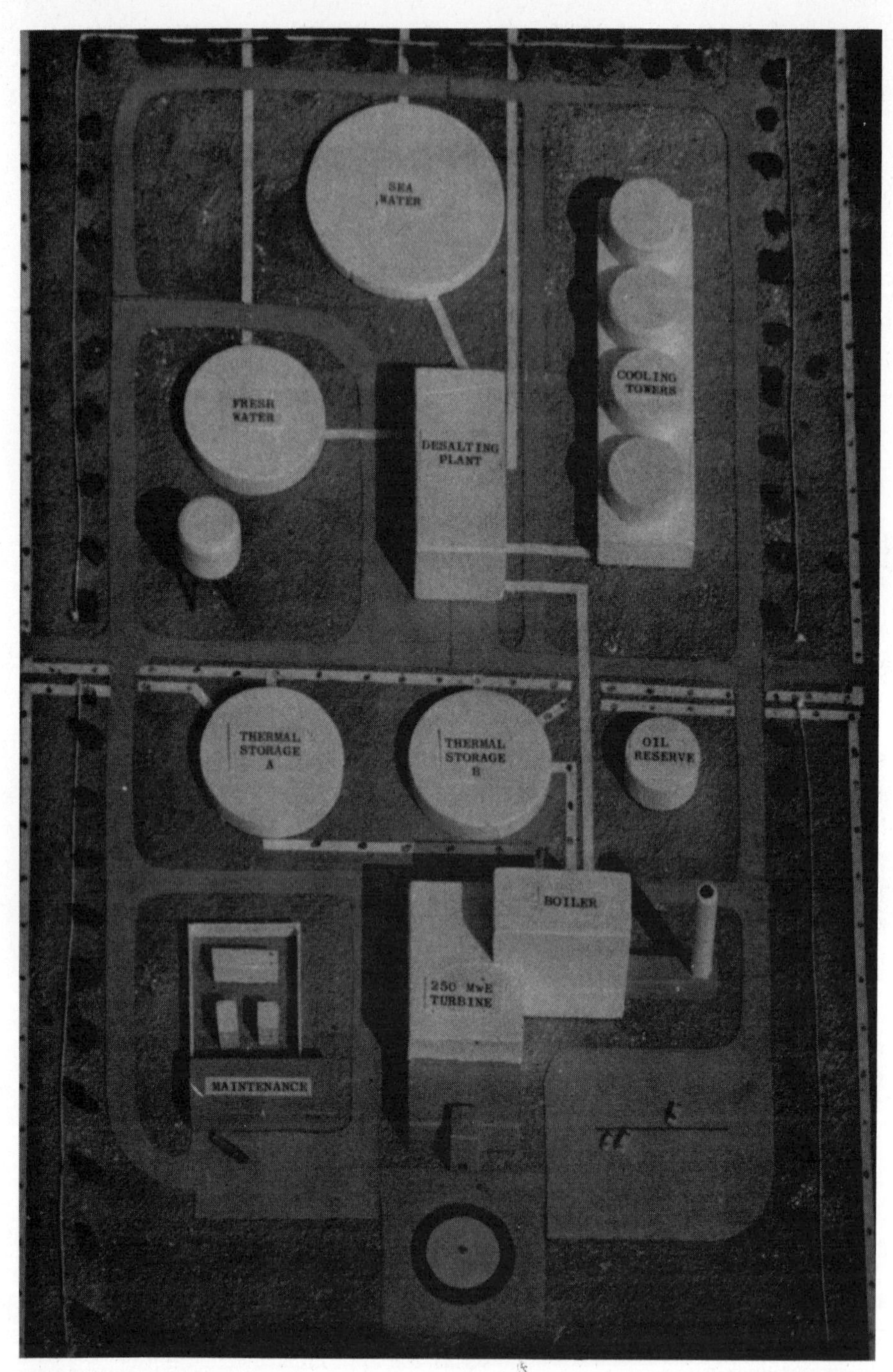

FIGURE 13-8 (a) A model of a 250-mw solar power plant. The components are as follows: (1) seawater; (2) freshwater; (3) desalting plant; (4) cooling tower; (5) thermal storage; (6) thermal storage; (7) oil reserve; (8) maintenance; (9) 250-mw turbine; (10) boiler. (Courtesy of Helio Associates, Inc. Tuscon, Ariz.)

FIGURE 13-8 (b) The total collecting area, approximately 2 square miles of land, for a 250-mw generating plant. (Coutesy of Helio Associates, Inc., Tuscon, Ariz.)

to grow the food (at less than 1 percent efficiency) which only provides 1 percent of our total energy consumption, the comparison makes its point.

As we have said, solar energy enjoys an advantage over most other forms: it can be used without unbalancing natural cycles. The solar farm demonstrates this beautifully. Under normal circumstances, the desert absorbs about 65 percent of the incident sunlight and reflects the remaining 35 percent. With the collectors in place, 95 percent of the incident radiation is absorbed but 30 percent of this (the generated electricity) is exported, and the remaining 65 percent is returned as leftover heat. Thus, the local balance remains the same. We should also point out that the same potential uses of left over heat exist for the solar plant as for any other thermal plant; desalination of water in one suggested use.

The environmental impact is lessened in other-ways. When we consider the choice between coal-fired steam plants such as Mojave and Navajo (see Chap. 9) and strip mining of the Black Mesa, the solar farms advantages are impressive. Navajo sheep or cattle could graze happily around the collectors (see Fig. 13-6).

The Meinels have received some National Science Foundation (NSF) research support to study the technical problems associated with the absorbing coatings, and are hopeful of forming a large group to begin the planning and the construction of a prototype plant which will show that their estimate of 5 to 6 mils per kw-hr electricity is within the range of possibility by the mid-1980s. They may well have started us on the way to much more efficient farming of the sun's free energy.

The Sea as Solar Collector

A central problem in utilizing solar energy on a large scale is that massive collection areas are needed. A crowded world may not be able to spare enough land area for such use. It is natural, therefore, to turn toward seas whose surfaces we put to minimal use.

Since the seas are warmed from the top rather than the bottom, they do not have the strong vertical mixing currents which are characteristic of the atmosphere. Their "winds" are the gentle, horizontal currents as in the Gulf Stream. As a result, there are significant temperature differences between the surface and the deeper layers. In tropical regions this difference may be as high as 20°C (36°F) between the surface and the 1,000-foot level. It is possible to conceive of a heat engine operating on this difference. Such an engine is shown in Fig. 13-9. In order to operate at such a low temperature it would have to employ some low boiling point liquid such as ammonia, as the working fluid. The small temperature difference would, of course, set a small

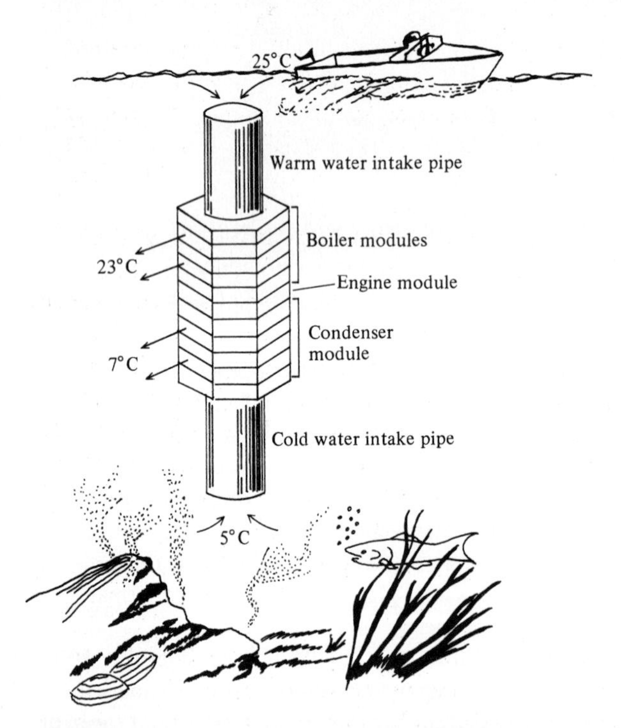

FIGURE 13-9 Artist's conception of a solar sea power plant which would float perhaps 200 feet below the surface of the water. (Source: C. Zener, "Solar Sea Power," *Physics Today*, **26**: 49, January 1973.)

upper limit; perhaps 5 percent on the thermal efficiency and practical efficiencies of 2 to 3 percent are all that can be actually expected.

The operation of the power plant is similar to the steam plant. Cool ammonia liquid at relatively high pressure is pumped to the surface region and is warmed by surface water flowing through a heat exchanger, becoming a high pressure vapor which can turn a turbine. The vapor, now at a somewhat cooler temperature is then pumped back down to the condenser where the cooler water condenses it to the liquid state.

The solar-sea power plant has several theoretical advantages. Neither high temperatures nor pressures are encountered; in fact, the hydrostatic pressure of the sea water could be used to balance the pressure in the boiler and condenser. The resulting system could conceivably have routine maintenance performed while running.

The problem of energy storage or transmission could be solved, it is suggested, by using the electricity to electrolyze water, with the resulting hydrogen and oxygen shipped or piped to shore. By conducting the electrolysis at great depths, the gases could be produced under pressure thus eliminating the costs of compressors.

The largest advantage is, of course, the enormity of the energy reservoir. The Gulf Stream, for instance, which brushes the Florida coast, could provide from its heated water an estimated 182 B mw-hr of electric energy, 75 times the expected United States demand by 1980. A group at the University of Massachusetts is presently preparing preliminary designs of a 400-mw solar plant to be located about 15 miles offshore from Miami.

Solar-sea plants could provide other useful products in addition to power; fresh water could be obtained by condensing water vapor and sea life might blossom in the nutrient-rich cold water pumped to the surface. In addition, other environmental effects seem to be potentially benign. Since the heat transferred from the warm surface to the deeper layers will be replaced by more solar absorption at the surface, the net effect would be a slight overall warming.

In spite of the size of the resource and the optimistic predictions of cost (165 dollars per kw or 3 mils per kw-hr), there has been no appreciable funding of research in this area. It may well be that the first step will be to develop the techniques of low-temperature turbines which are crucial to the success of these plants and useful, as well, as bottoming cycles on conventional plants, and for the development of the lower temperature geothermal sources. The proposals currently in the literature may be, as was the earliest proposal by the French physicist, D'Arsonval in 1881, before their time. The large thermal reservoir of the ocean will, however, continue to attract the attention of energy seekers and the prospect of less overall thermal pollution will bring the support of the environmentalists.

FUSION: THE BIG IF

The confluence of increasing demand and finite resources which were depicted in Chap. 11 left us with uncertain prospects of an energetic future: only the breeder reactors fueled by the Tennessee Mountains (or the Conway Granites of New Hampshire) gave promise of fuel for as long as thousands of years. In this Chapter we have seen that there are some promising technologies which may, sometime within the next 5 to 30 years, enable us to stretch the fossil fuel supplies and even tap into some of the energy which is continuously supplied to earth. The big "if" of the continuous sources is, of course, solar energy.

There is an equally big "if" among the depletable sources which we have not yet discussed: the fusion reaction which taps the nuclear energy of light nuclei. An understanding of the source of this energy begins with the plot of "binding energy per nucleon" versus "mass number" (see Fig. A5-2 in Appendix 5). As we point out in the discussion of that plot, energy is released when two light nuclei combine (fuse) to form a heavier one. The most productive nuclear reactions of this type are expected to be reactions involving the "isotopes" of hydrogen, deuterium (H^2), which has a neutron as well as a proton in its nucleus, and a radioactive form of hydrogen, tritium (H^3), which has two neutrons and a proton and decays with a half-life of about 12 years. The reactions of most interest are

$$H^2 + H^2 \longrightarrow H^3 + p + 3.2 \text{ Mev} \tag{13-1a}$$

$$H^2 + H^2 \longrightarrow He^3 + n + 4.0 \text{ Mev} \tag{13-1b}$$

$$H^2 + H^3 \longrightarrow He^4 + n + 17.6 \text{ Mev} \tag{13-2}$$

The meanings of these equations are simple; in Eqs. (13-1*a*) and (13-1*b*), two deuterium nuclei (H^2) combine to form either tritium (H^3) or an isotope of Helium (He^3), and, in both cases, energy is released (expressed here are millions of electron volts (Mev) (see Appendix 5 for further discussion). In Eq. (13-2) ordinary helium (He^4) is produced by the fusion of deuterium (H^2) and tritium (H^3), and a considerably larger amount of energy is released.

From these equations one can immediately see the major benefits of fusion as a source of energy. First, the radioactivity release is much less serious: only tritium among all the participating nuclei is radioactive. Second, the fuels (isotopes of hydrogen) are very plentiful since hydrogen is the most abundant element. Even though deuterium is a rare form of hydrogen (only one out of each 6,000 hydrogen atoms is H^2) the sea is a potentially huge source. If we can make the reactions in Eqs. (13-1*a*) and (13-1*b*) work, then 1 cubic km of seawater has energy content greater than all the fossil fuels on earth. Said in another way, the deuterium content of the oceans (only 1/6,000 of their water) will provide the energy equivalent that the oceans would provide if they were filled with gasoline—and 300 times as large. Here is energy potential stretching into our distant future: that's the good news.

The bad news is that we have not mastered this reaction, despite 400 million dollars spent during the past 25 years in the United States and 1 billion dollars in the world. The problem is not in obtaining pure deuterium; it is relatively easy to separate H^2 from H^1 since one is twice as heavy as the other. The problem is in bringing them together closely enough to make them stick, that is, closely enough to let the very short-range nuclear forces take over. Since the nuclei are electrically charged, they normally repel each other. To force them together takes energy; they have to climb an electrical hill, so to speak. That energy can be provided by a particle accelerator and these reactions have been studied in this manner. But for power generation, we must deal with large numbers of particles. The only practical way to give a large number of nuclei such energies is to heat them. There lies the catch; in order to assure that a sufficient number of nuclei have enough energy to cause fusion, the sample of deuterium must be at a temperature of at least 50 million °C—hotter than the temperature at the center of the sun. The only successful heating to that scale which we have so far achieved was by exploding the A-bomb, and, in this way, triggering the fusion reaction of the H-bomb.

The scope of the problem becomes clearer: to achieve fusion, solid deuterium or deuterium gas at an appreciable density must be held in a container and heated to temperatures at which any container will vaporize. Thus, means must be found to achieve this heating of deuterium, and yet keep it out of thermal contact with its container. There are two ways that this problem is being approached: by magnetic confinement and by laser bombardment.

There are several approaches using magnetic confinement undergoing active experimentation. In these processes the deuterium gas is decomposed into a "gas" of charged particles, a *plasma.* The energy to accomplish this is provided in several ways, usually by causing a large electric current to flow through the gas. Charged particles can be controlled by magnetic fields. Several different magnetic field shapes are being tried, a doughnut shape (the torus), a straight pipe with magnetic fields that turn the particles around at the ends (the magnetic mirror) or a pulsed field that tries to bring everything together at once with a single large burst of magnetic field (the theta pinch).

There are three parameters which must be controlled to make fusion work: the temperature, density, and confinement time. The deuterium has to be hot enough, dense enough, and hang together long enough to produce a self-sustaining fusion reaction. A crude criterion of feasibility is provided by the so-called *Lawson measure* which predicts that for a deuterium-tritium fuel mixture heated above the ignition temperature (50 million °C), the product of the density and pressure must be at least 10^{14} seconds – fuel ions per cm^3. Breaking this down, the rough criteria are that densities of 10^{14} to 10^{15} fuel ions per cm^3 (corresponding to pressures 10^{-4} or 10^{-5} times smaller than atmospheric pressure) must be held together for 0.1 to 1 second. While each of the three parameter (temperature, density, and confinement time) have been reached separately, no one experiment has yet been able to put them all together. The closest contender so far is one of the doughnut machines, the *ST Tokomak,* an American adaption of a Russian device. This machine has achieved a time density product of 3 to 6×10^{11} (but at a temperature well below ignition). It is expected, however, that the Tokomak principle will produce a successful controlled fusion reaction when larger machines are tested in the 1980s.

Achievement of the necessary temperatures, densities, and pressures, will demonstrate scientific feasibility; beyond that is the murky uncertainty of engineering feasibility. But before we explore that region, we must mention the second approach to fusion through laser bombardment.

Laser Fusion

In this approach the problems of density and confinement time are to be circumvented by bombarding a small pellet (about 1 mm in diameter) of frozen deuterium-tritium (frozen at 20° above absolute zero) with an energetic laser beam. The pulse of energy from the laser is poured into the pellet so quickly that its inertia holds it together while it heats to fusion ignition temperatures.

The laser fusion arrangement is shown schematically in Fig. 13-10. The target pellets drop one-by-one into a chamber and are struck by a focused pulse of laser energy. The pellet heats rapidly, fusion occurs, and the pellet explodes (with the force of 500 pounds of TNT). The amount of laser energy needed and the problems of holding the pellet together are minimized if several lasers are used and focused on the pellet from all sides. This technique produces an implosion, greatly increasing the pressures and temperatures in the interior of the pellet. The explosion products continue expanding until they strike the walls of the chamber where their kinetic energy is converted to thermal energy which can then be used to generate steam.

There are problems ahead for laser fusion also. The construction of a chamber which can hold up under the steady explosions is one of them. Laser power is a second requirement. The minimum estimated requirement is a pulse of 10^5 or 10^6 joules in 1×10^{-9} seconds. The largest laser in existence, a Russian one, delivers 600 joules in 2×10^{-9} seconds. A further problem is the cost of the pellet. Since each explosion will release energy, which, when converted to electricity is only worth about 1 cent, some extremely inexpensive means of pellet fabrication must be found.

Laser fusion is in its infancy and has generated much enthusiasm and greatly increased research support. Its proponents see it as providing power sources of conveniently small size (50 to 100 mw of electric generating capacity) allowing dispersion to industry, use in ship propulsion and for on-site power in urban areas. Its detractors don't believe it can be made to work. The answers may come first from the Soviet Union where a new, more powerful laser may be the first to cause this type of fusion to occur.

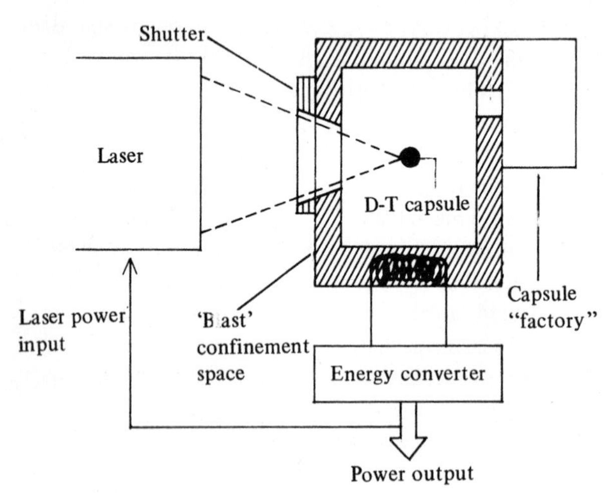

FIGURE 13-10 Schematic of a laser fusion reactor. The cavity would have to withstand explosions of about 500 pounds of TNT equivalent. It a new pellet were ignited each second, the thermal energy output would be about 200 mw. [Source: "Briefings Before the Task Force on Energy, Committee on Science and Astronautics, House of Representatives, Serial M, (Washington, D.C., 1971) p. 47.]

Fusion Reactors

It is conceivable that scientific feasibility of the fusion reactor may be demonstrated by the early 1980s. The problems, however, will be only beginning. While the fusion process does not have the huge radioactive inventory of fission to deal with, it has a new problem, high-energy neutrons. The large fluxes of these neutrons will cause structural damage, the seriousness of which is difficult to estimate. It is feared, however, that it may be impossible to find materials which will stand up under this radiation long enough to make commercial operation practical.

The large magnetic fields required will also be a problem; they will probably have to be provided by superconducting magnets operating at near liquid helium temperatures. Thus, the reactor will be 100 million degrees at its center and a few meters away will need to be near absolute zero. The magnets will also have to be shielded from neutrons.

As we have suggested earlier, the only fusion reaction in use for a considerable time will probably be the one described by Eq. (13-2) using tritium. This also poses problems. Tritium will no doubt be bred from lithium by the fast neutrons. The reaction is

$$Li^6 + n \longrightarrow H^3 + He^4 \qquad (13\text{-}3)$$

The reacting chamber will probably be surrounded by lithium liquified by the heat. A schematic diagram is shown in Fig. 13-11. Tritium is pumped from the lithium to be used as a fuel, and the hot lithium creates steam for a generator.

There are some drawbacks to reliance on the $H^2 + H^3$ reaction, which requires lithium for the production of H^3. Lithium is not a plentiful material and proved lithium resources would only provide energy equal to that available from all the fossil fuels. However, there has been no intensive search for lithium and the available

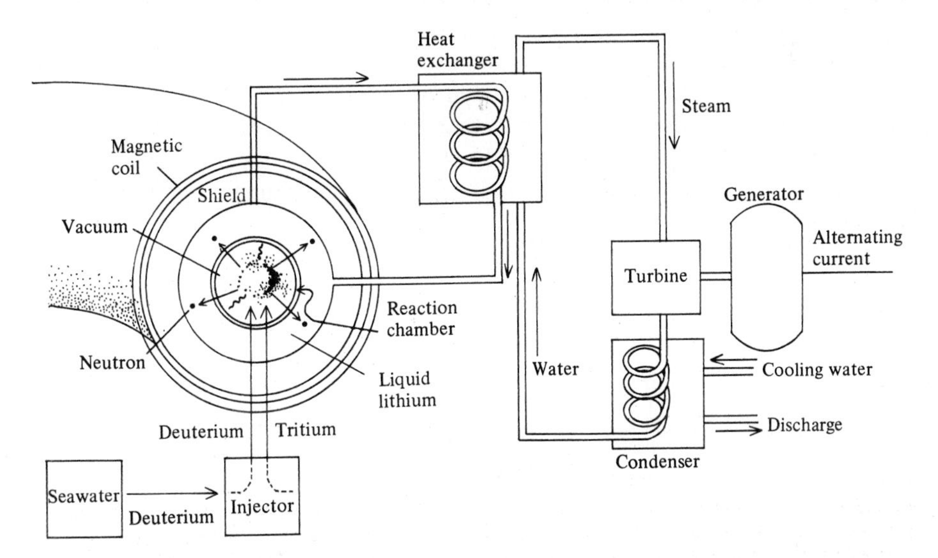

FIGURE 13-11 Artists' conception of a possible fusion reactor. The molten lithium serves not only as a heat exchanger but as fertile material for the breeding of tritium.

resources are probably great enough to power us for millions of years.

Fusion reactors might be used to increase our energy supplies in a different way. The existence of the large flux of high-energy neutrons might eventually make it possible for the breeder reactions, U^{238} and Th^{232}, (see Chap. 10) to take place and thus allow the fusion reactors to supercede the controversial breeder reactors as a source of fissile materials for light water reactors.

Because of the high temperatures, the fusion reactor could be very efficient, perhaps 50 to 60 percent. Since in the $H^2 + H^3$ reaction the products are charged particles, it may also be possible to separate them in a weak magnetic field which covers a large area in a fashion similar to that used in MHD. This direct conversion of kinetic energy to electric energy would be even more efficient, perhaps approaching 80 percent.

Environmental Effects

If the engineering problems we have discussed (and those we haven't such as the inflamability of hot lithium and the pounding in the pulsed magnets) are not enough to give pause, the environmental problem does. But first let's summarize the good news.

The environmental plusses are the absence of the danger of an explosion (there will not be enough fuel in the machine at any one time for a "critical mass" type of uncontrolled excursion) and the much lower radioactivity. There will be no radioactive fuel to be shipped back and forth and the only waste to be disposed of will be from reactor components made radioactive by neutron bombardment.

The fusion process offers a political advantage over the breeder reaction; it does not produce a bomb component like the breeder's Pu^{239}, and thus will add only a benign power component to the world. The high efficiencies will also reduce the amount of waste heat released into the water or air.

The environmental concern is centered on tritium: radioactive hydrogen that is part of the fuel in the $H^2 + H^3$ fusion reaction and a reaction product in $H^2 + H^2$ fusion. Tritium has both good and bad features. Compared to fission products like Sr^{90} or I^{131}, it appears almost innocuous. Its emission is a not very penetrating particle and its half-life is relatively short. In its gaseous form, tritium is not very active biologically.

The troublesome form of tritium occurs when it reacts with oxygen to form water, for water is biologically active. Tritium itself is very active in a different way; it is very difficult to contain; its small size compared to most other atoms allows it to diffuse (leak) through most walls. This diffusion and escape is particularly troublesome with metal walls. The fusion reactors, one fears, may have "radioactive B.O."

It is certain that careful design will be necessary to hold tritium leakage to acceptable levels. However, since the tritium is part of the fuel, there will be a double incentive to its recovery. Surrounding the whole fusion-reacting core in an evacuated enclosure from which the tritium is constantly pumped may be economically as well as biologically justified. The danger potential of tritium, however, will probably keep the fusion reactors (at least these which breed tritium from lithium) out of the cities until they are thoroughly understood. Nonethelesss, there is still hope that fusion reactors will turn out to be good neighbors and can eventually be placed in the cities where their "waste heat" will become "thermal enrichment."

When?

Within three years after the discovery that U^{235} could produce a chain reaction, Fermi and his colleagues had demonstrated one in a pile of graphite blocks under the University of Chicago football field, Stagg Stadium. It has now been more than 20 years since the physics of fusion has been well-understood, but its "Stagg Stadium" has not yet occurred. For this reason, prophecy is dubious. It does appear, however, that the increase in research support which the Energy Crisis has stimulated will allow the construction of bigger machines, and that from one or more of them we will obtain evidence of scientific feasibility sometime before the mid-1980s. As we have said, however, achieving scientific feasibility is the equivalent of getting our heads above water: the long swim to shore is still ahead.

In this explosive century we are accustomed to seeing all dreams come true. Any idea that is conceivable can eventually be implemented through genius and the magic of money. One has to be reminded, from time to time, that the inevitability of progress from conception to commercial realization is a belief, not a law of nature. The engineering difficulties, on whose solution the commercial future of fusion waits, are enormous. The optimists confidently predict pilot plants by 1990; the pessimists warn us that this future may never come.

SUMMARY

In this chapter we have examined some of the potential energy sources which may make energy abundant again. There are only two sources of a size that is not dwarfed by our rapidly expanding consumption: solar energy and energy from the fusion of hydrogen isotopes.

Solar energy is, of course, the major source of the earth's energy, at present providing food, and fuel, and the secondary sources, water and wind power. The size of this source and its daily delivery are causing us to turn more directly to it. There are no technological barriers to the immediate employment of solar energy for home heating and cooling and the new energy situation should cause rapid advances in that area. Large-scale conversion of solar energy to electric energy hinges on the development of more economical solar cells and on the mastering of the complexity of large-scale collection and thermal conversion.

There are other continuous sources which are beginning to attract attention. Of these, geothermal energy and improved use of photosynthetic energy show the most promise. Both of these are potentially benign environmentally and some of the proposals for improved photosynthesis, particularly those which use organic waste as fuel, hold out the promise of mitigating other environmental problems as well.

Another of the other continuous sources, wind power, could make large contributions, but there is at present little evidence of concrete interest in it or in solar-sea power which also needs exotic technology to realize its potential.

The other big "if" is fusion power, which could draw its fuel from the sea and cater to our needs for at least hundreds of thousands of years. The difficulties of this "if" are technological; it has not yet been possible to hold enough deuterium together long enough and get it hot enough. The optimists look to the demonstration of scientific feasibility by the mid-1980s but it is clear that even in such an optimistic view, fusion reactors will not make much of a contribution to our twentieth-century needs.

The visions that have been put forward in Chap. 13 have, perhaps, lifted some of the gloom that has been growing through the book. There is hope for the future. None of these techniques or sources offers us immediate relief in our present desperate state. It should be clear that for a while at least, we must value energy more highly than we have; we must look to an ethic of conservation. Some of the components of such an ethic are discussed in the next chapter.

14

Counting Calories: Strategies for Energy Conservation

The twentieth century has been a century of explosive exponential growth (see Chap. 5). We see this growth almost everywhere we look—in population and in the demands generated by families and by individuals.

We know, however, that exponential growth often violates the order of nature. We live on a finite planet; supplies are decreasing exponentially, and consequently, hostile environmental forces are rising. Sometime each of these exponentials must begin to turn over into the S or Sigmoid curve and growth must slow and stop. There are now before us, at least in the energy area, a multitude of symptoms which suggest that the 1970's are the turnover years and that the twenty-first century will be the "Sigmoid Century". In fact, as this chapter is being written, there is still an Energy Crisis; our supplies of gasoline, fuel oil, and electricity are running dangerously short. The central role that the fossil fuels and the energy from them play in our economy is being dramatically demonstrated. The effect of the embargo on oil shipments from the oil producing nations of the Middle East not only focussed our attention on the importance of gasoline to our economy, but also emphasized the importance of oil as a raw material. The shortage caused economic chaos not only in the trucking industry but also in the plastics industry, as its supply of petrochemicals decreased.

THE ENERGY CRISIS—TO BE CONTINUED

Looking at the United States and world energy picture in the mid-1970s, we can separate the Energy Crisis into three parts, or three crises of different time perspective. We have been in a critical period of shortage that was to some extent artificial, brought about by the oil embargo. It was focused on gasoline and heating oil, but extended, as we have said, along all the lines which connect petroleum to our economy. The response to this short-range crisis was equally short-range: enforced and harsh measures of conservation (50 mph speed limits, Sunday gasoline station closings, threats of gasoline and fuel oil rationing). Other less dramatic short-range measures were employed: priorities and allocations of petroleum products were established, importation was stepped up.

The oil which Secretary Kissinger spread on the troubled Mideastern waters is beginning to return to our ports in tankers. The embargo is over—but the Energy Crisis remains. This intermediate-range crisis which will occupy us through the remaining decades of the twentieth century will also be largely focused on oil, but difficulties with electric power production will compete for attention. We have already looked at the problems which will characterize this intermediate-range crisis: oil demand growth outstripping domestic supply, electric demand growth outstripping the electric power industry's ability to provide new generating capacity, and environmental insult growing with consumption. The measures which will allow us to meet and survive that crisis will be a mixture of economic, political, and technological ones. Some allocation may still be necessary; economic pressure may be applied to slow demand growth and increase production, and political accommodation to the oil producing nations will continue.

The intermediate time-range must also see a new energy conservation ethic. The nation must go on an "energy diet" in which all Calories are counted. Overall energy productivity must increase through improved efficiencies of conversion techniques (for example, through more efficient generation of electricity) and through improved overall system efficiencies in transportation, industry, and in the residential and commercial sectors. We will also look to technology to give flexibility; to give us oil and gas from coal, for instance, and to add to our total supplies by bringing in such continuous sources as geothermal and solar energy.

This mix of political, economic, and technological measures may carry us through the several decades of the intermediate crisis, but the energy situation will still be troubled. The long-range crisis of fossil fuel depletion will remain. In meeting this crisis, some political and economic problems will need attention; a balance between the energy and nonenergy uses of fossil fuels will have to be set and the competitive demands of the world's people will need refereeing. The major input, however, must be technological. We must find new and abundant sources.

It is our hope and expectation that the present short-term crisis will be over by the time of publication of this book (although another one may be upon us by then). We have in Chaps. 12 and 13 described the challenges that science and technology must meet if we are to survive the long range crisis. In this chapter, therefore, we will concentrate on the measures which must be applied to the intermediate crisis.

There is still a loudly expressed sentiment that the only way out of the crisis is to expand energy production. There are the "diggers and drillers" (as ex- Scretary of the Interior Udall has called them) who say "give us freedom from restraint and we'll provide the energy." But the solution to the growing scarcity of energy cannot be solely to "dig and drill." Production increases cannot come about overnight; it takes four or five years to put a new oil field into production, build an Alaskan Pipeline, or bring new coal mines into operation. And even if the constraints imposed by financial resources and by time were not there, the consequences of the "dig and drill" philosophy to our environment (offshore oil wells, strip mining of the Western coal fields, reduced emission controls) and to our economy (windfall profits, negative balance of payments) should provide alternate constraints. It should be clear that among the energy strategies which we apply during the next few decades some must increase the productivity of energy, eliminate the waste, and increase the amount of useful heat and work we get from the raw energy input.

INCREASING ENERGY PRODUCTIVITY

In this calorie-counting approach we must look at each sector of our economy and determine whether its allotment of energy is being used wisely. We must look at the system efficiencies and not only improve them (by insulating houses, for instance) but also by moving away from inherently inefficient ones (by decreasing the importance of the automobile in urban transportation, for instance).

In order to target our conservation strategies we turn again to the 1971 United States input-output energy flow diagram (Fig. 5-12) which is reproduced as Fig. 14-1. It is easy to spot the obvious inefficiencies. Of the 15.6 Q Btu of input energy which were channeled into the production of electricity that year, only 5.5 of them were actually converted to that important intermediate form. The remaining 10.1 Q Btu were lost mostly as waste heat which, since much of it was discharged into our streams, rivers, and lakes, produced the environmental problems discussed in Chap. 8. Of the 17.2 Q Btu of input energy to the transportation sector, only 2.5 Q Btu were actually used to move our cars, trucks, buses, airplanes; the remaining 15.1 Btu were wasted. Part of that waste created the Los Angeles smog.

We can identify waste in the other sectors: 4.3 Q Btu of the 16.3 Q Btu available to the household and commercial sector and 10.9 Q Btu of the 19.6 Q Btu input to industry were wasted. The much more efficient use of energy in these last two sectors is, of course, largely due to the form of energy required. In the generation of electricity and in transportation, mechanical energy is required; the energy must go through

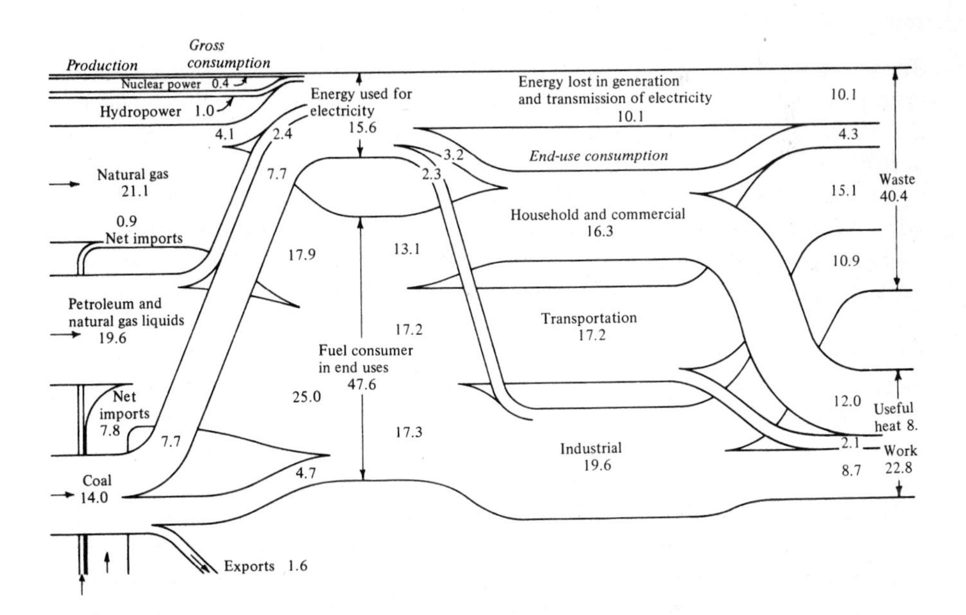

FIGURE 14-1 U.S. energy input-output. Nonenergy (raw material) use of fuels are excluded. The overall efficiency is about 41 percent. (Courtesy of Earl Cook, Texas A & M University.)

the thermal bottleneck we examined in Chap. 4. In the latter sectors, household, commercial and industry, 69 percent of the energy is used in the form of heat which is efficiently available from the chemical or nuclear fuels.

We see, however, even from this rough analysis that none of the sectors of our economy are free from inefficiency; each can stand examination. With our recognition of the inherent waste in electric power generation, we will be particularly alert for ways of reducing the consumption of electric energy.

MORE PER GALLON

With a utilization efficiency of about 15 percent and a growing appetite that now accounts for 25 percent of the United States energy consumption, transportation is the logical place to begin our category-by-category conservation study.

Transportation has two basic functions: to move people and to move things. We can further break down these categories geographically into intercity transport and urban transport. In Chap. 5 (Fig. 5-14) we have already examined the distribution of the energy consumed in transportation between urban passenger transport (UPT), intercity passenger transport (ICPT), intercity freight transport (ICF), and urban freight transport (UF). In the same chapter we also provided a breakdown within the three categories (ICPT, UPT, ICF) of the percentages of passenger-miles or ton-miles carried by the various transport modes: automobile, truck, railroad, and the like (see Figs. 5-15, 5-16, and 5-17). Those earlier figures gave us our targets for energy conservation by indicating where the large amounts of energy were being consumed. The first target is passenger transport.

Moving People

The job of moving people around in this country consumed about 60 percent of the energy expended in transportation in 1970 and 15 percent of the total energy consumed that year. Although the boundaries are overlapping, we will look separately at ICPT (trips of 100 miles or more) and UPT.

From Fig. 5-14 we see that UPT uses the most energy. Figure 14-2, however which compares the passenger-miles accounted for in these two types of passenger transport, shows quite another story; ICPT produces the larger number of passenger-miles per year. This comparison already provides a hint of inefficient energy usage.

From a comparison of the rates of growth of passenger-miles (Fig. 14-2) and energy consumption (Fig. 5-14) we find an even stronger hint of inefficient usage. The number of passenger-miles traveled in ICPT and UPT increased by 120 percent and 130 percent respectively during the period 1950-1970. During this same period, however, energy consumption increased by 150 percent and 170 percent in the two categories. The greater growth of energy consumption points, in both cases, to declining efficiency.

We will use as an indicator of efficiency the quantity, energy intensiveness, (EI), which was defined in Chap. 5 as the number of Btu per passenger-mile or ton mile used by a mode of transportation. We showed the comparative EIs for several passenger modes in Fig. 5-19. Figures 14-3*a* and 14-3*b* show the changes in the EI for the most important modes of ICPT and UPT during the period 1950-1970, along

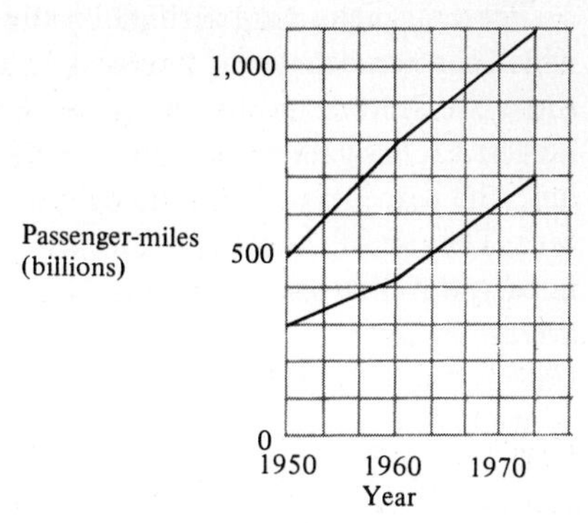

FIGURE 14-2 The number of passenger-miles accounted for by ICPT has grown more steeply than those of UPT. (Source: Eric Hirst, "Transportation Energy Use and Conservation Potential," *Bulletin of the Atomic Scientists,* **29,** November 1972.)

with the average EIs of ICPT and UPT for that period. What we already suspect is apparent in these two figures. In both types of passenger travel, the EI has been increasing—more rapidly in the UPT sector than in ICPT. These data begin to give some insight into what has been happening. It is clear from Fig. 14-3*a,* for instance, that the train (whose EI has been more than cut in half over the past 20 years) cannot have been playing a significant role in ICPT. From Fig. 14-3*b* we are led to suspect that the automobile dominates UPT.

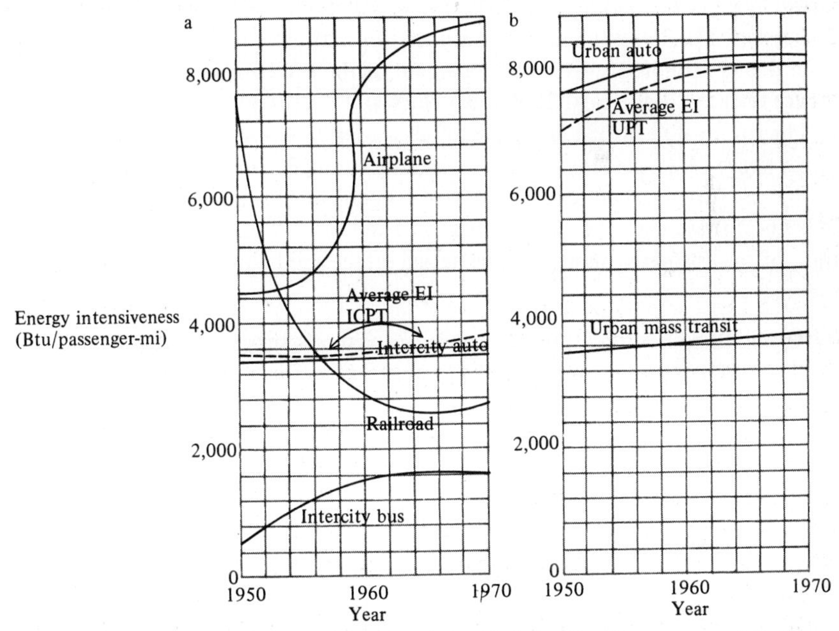

FIGURE 14-3 (a) The El of the airplane has almost doubled over the 20-year span while that of rail transport has decreased by about the same factor due to the switch to diesel. The average El factor is about that of the ICPT automobile. (b) The urban automobile has shown an increasing El over the 20-year span and the average El is approaching that of the automobile. (Source: Hirst, "Transportation Energy," *op. cit.*)

These suspicions are verified by the data in Figs. 5-17 and 5-18 of Chap. 5, in which the percentages of the total passenger-miles carried by the various modes in these two sectors are shown. In the ICPT sector, the airplane, the mode with the highest EI, has gained passengers at the expense of the two low EI modes, rail and bus. The percentage of people travelling by automobile has hovered around 90 percent. The rise of the average EI between 1960 and 1970 for ICPT, therefore, reflects, in part, this increasing importance of air travel (see Fig. 14-3*a*). The approach of the average EI to that of the urban auto indicates the decreasing importance of the bus transport in UPT (see Fig. 14-3*b*).

We have now all the data necessary to investigate the effects of strategies to reduce energy consumption. In the terms we are using we can affect savings either by reducing EIs of the various modes or by shifting some of the passengers from high to low EI modes. Let us briefly examine some specific strategies and compute the estimated savings. We will rank these in rough order by the amount of disruption they would produce in the system—by the ease of implementation.

Emergency Measures. The emergency measures taken by the Nixon Administration during the embargo consisted both of efforts to lower the EI for ICPT by lowering speeds, and to lower the total number of passenger-miles travelled by making weekend recreational driving difficult. The expectations of savings were based on the following.

All cars get more miles per gallon at lower speeds. Typical performance data are shown in Fig. 14-4. A drop from 70 mph to 50 mph increases gas mileage by about 20 percent. The Department of Transportation estimate of the savings expected from the 50 mph limit was about 5 percent of the total automobile fuel used in ICPT (lower than the theoretical 20 percent since all travel is not at high speeds).

The anticipated savings from the Sunday sales ban on gasoline were based on the rough figure of 25 percent gasoline consumption on Saturdays and Sundays. A ban on Sunday driving, therefore, might save 10 percent or so of the total. The net effect of both of these may have been as high as 1 million barrels of oil per day, about a third of the estimated shortage. The actual effect will be measurable only well after the fact.

Increasing Load Factors. Another nondisruptive strategy for energy conservation would be to increase the number of passengers per vehicle since none of the various

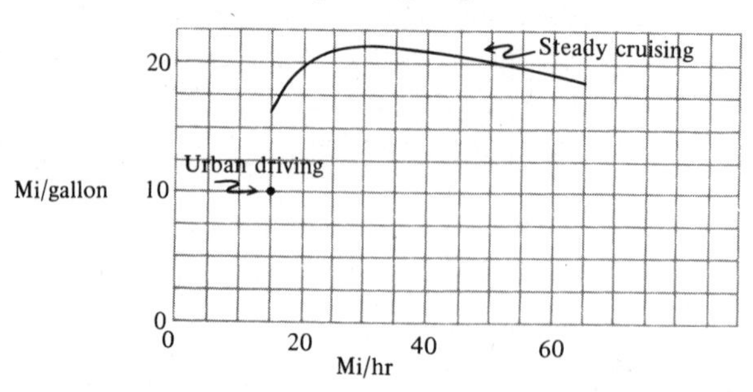

FIGURE 14-4 The miles travelled per gallon of gas reaches a peak at 30 to 40 miles per hour and then drops steadily as the speed is increased. (Source: "A Report on Automobile Fuel Economy," Environmental Protection Agency.)

modes of ICPT or UPT are operating near their capacity. The average percentage of capacity at which a transport mode operates is called the *load factor.* For automobiles in UPT, the load factor is 28, meaning that an average 5 passenger car operates in urban transport at 28 percent capacity; it carries 1.4 passengers. It can easily be shown that the EI is proportional to the reciprocal of the load factor (1 over the load factor), thus increasing the load factor lowers the EI.

Car pooling, for instance, is a mechanism for increasing the load factor. As an example of the size of energy savings possible, if the load factor for automobiles in UPT were to be increased 10 percentage points from 28 percent to 38 percent (still less than half full) the EI would decrease from 8200 Btu per passenger-mile to 8200 $\times$ 28/38 or 6100 Btu per passenger-mile, a savings of 2100 Btu per passenger-mile. At the 1970 level of UPT (690 billion passenger miles by auto) this amounts to a savings of 1.45 Q Btu of energy, or 8.8 percent of the total energy consumed in transport.

With the data which has been presented, we can compute the savings from this strategy applied to other passenger modes. Some representative values are displayed as total Btu and as energy percentages in Table 14-1. We see that the changes in automobile load factors, since the automobile dominates the passenger-mile statistics, produce the largest relative changes.

There is some concrete evidence to back up these conjectures. In 1971, following a year of heavy losses, three airlines, TWA, United and American, were permitted, under Civil Aeronautics Board (CAB) supervision, to meet together and plan reductions in their schedules between some of the largest cities. The result of the overall reductions in flights increased the load factors of these three airlines from the industry average of 39 percent to 54 percent. The reported savings were 85 million dollars in operating costs and 120 million gallons in fuel. It is estimated that industry-wide cooperation to increase laod factors to the CAB's goal of 55 percent, would save 800 million gallons of fuel. If we convert jet fuel to its energy value of 130,000 Btu per gallon, that 800 million gallons is 110 T Btu. This figure for the total savings is a little more than half the estimate shown in Table 14-1, reassuring us that our calculations are reasonable.

Improving Engine Efficiency. Although the internal combustion engine has dominated the transportation market since the early days of this century, its efficiency

Table 14-1 Energy Savings from Load Factor Increases in Passenger Travel

Transport mode	*LF change*	*Savings per billion Savings per billion passenger-miles*	*Total savings at 1970 levels*	*Percent of U.S. total energy*†*, 1970*
UPT, automobile	28*-38	2.16×10^{12} Btu	1.45×10^{15} Btu	2.2
UPT, automobile	28-50	3.60×10^{12} Btu	2.48×10^{15} Btu	3.7
UPT, bus	20*-30	1.27×10^{12} Btu	0.27×10^{15} Btu	0.4
UPT, bus	20-50	2.28×10^{12} Btu	0.49×10^{15} Btu	0.7
ICPT, airplane	40*-50	1.68×10^{12} Btu	0.20×10^{15} Btu	0.3
ICPT, airplane	40-60	2.80×10^{12} Btu	0.32×10^{15} Btu	0.5

*These are approximate present levels.

†Taken as 67.4×10^{15} Btu as in Hirst (see footnote 1).

record has not been a distinguished one (see Chap. 4, Fig. 4-3). A viable conservation strategy would be to improve on its efficiency by increasing the miles per gallon achieved with our now precious fuels.

We know that there are great differences in the miles per gallon ratings of different cars. Some of these were shown in Table 4-3. Increases in engine efficiency, however, would come under the heading of disruptive changes—as far as Detroit is concerned. They have resisted change with remarkable success. Change may come, nonetheless; the growing strong preference of the public for small cars (American Motor sales up 33 percent and General Motors sales down 42 percent in January 1974) will cause a lowering of the overall EI of passenger transport.

Improvements in other forms of transport are also slow in coming. It will take a radical new engine, perhaps a turbine, steam, or electric engine, for example, to improve bus transport. If, however, urban buses (EI of 3,800) were to be replaced by subways or similar rapid transit systems whose EIs are half that of the buses', significant savings could be expected.

The savings which could be achieved by improvements in engine efficiencies are shown for various modes of passenger transport in Table 14-2. Again, we see that improvements in the efficiency of the automobile engine offer us the largest potential savings.

Changing Transport Patterns. We have so far made our estimates within the framework of the mix of passenger transport modes shown in Figs. 5-15 and 5-16. As we commented in the discussion of those figures, one of the obvious inefficiencies of our passenger transport system is that it uses high El modes. Energy savings could be obtained by shifting some passengers from high EI to low EI modes while maintaining the same total passenger-miles.

Whether this is a disruptive or a nondisruptive tactic depends on the local situation as well as on the size of the change contemplated. It is quite clear, for instance, that neither buses nor trains could take over all ICPT. It would however, be possible within the system to turn half the air ICPT over to buses and trains. The same argument holds with respect to urban traffic; the mass transit systems of most cities can handle only a small percentage of the total UPT now handled by automobiles. We

Table 14-2 Energy Savings from Changes in Efficiency of Fuel Use in Passenger Transport

Transport mode	*Efficiency change (percent)*	*Savings per billion passenger-miles*	*Total savings at 1970 levels*	*Percent of U.S. total energy*, 1970*
UPT, automobile (10-15 mpg)	50	2.74×10^{12} Btu	1.90×10^{15} Btu	2.8
ICPT, automobile (15-20 mpg)	33.3	1.19×10^{12} Btu	1.16×10^{15} Btu	1.7
ICPT, automobile (15-30 mpg)	100	1.70×10^{12} Btu	1.66×10^{15} Btu	2.5
UPT, bus, subway	100	1.90×10^{12} Btu	0.04×10^{15} Btu	0.5

*Taken as 67.4×10^{15} Btu as in Hirst (see footnote 1).

have computed in Table 14-3 the savings which would result from some relatively modest changes in the mix of passenger travel modes.

One of the shifts we show is, perhaps, more difficult to achieve than the others: a shift (in this instance, of only 5 percent) of UPT passenger-miles from the automobile to bicycling and walking. The EIs of these modes can be taken as zero in this calculation since they depend on food energy and not on petroleum. The data of Table 14-3 recommends such a shift; at the 1970 level the saving would amount (at 6 M Btu per barrel) to 46.6 million barrels of oil, which is 4.3 percent of the oil imported in 1970. In addition to energy savings, this shift would also benefit the health of the participants and reduce pollution all the way along the line.

The gains of these various changes are not as large as some of the others we have seen, but are still significant. The savings are directly proportional to the percentage shifts; if, for instance, 50 percent of air ICPT were shifted to trains, five times as much energy, amounting to 650 T Btu, 1.2 percent of the United States grand total of energy would be saved.

Moving Things

The other 40 percent of the total energy used by transportation in 1970, 6.74 Q Btu in all, was expended as fuel in moving freight of all kinds. Included as freight are not only the packages and boxes of Railway Express but also all the vast raw and manufactured material of our economy. The range of materials (and the speed of their delivery) stretches from the oil flowing slowly through the nation's pipeline network to the lobsters air-freighted from Maine. We will conduct our search for potential saving in this area in the same manner as we followed in our analysis of passenger transport. We will look at growth in energy consumption and in ton-miles (the freight equivalent of passenger-miles), at the EI of the various modes, and at the mix of modes in actual use. With this information in hand we can consider conservation strategies.

Figure 5-14 showed the growth in ICF. Unfortunately, we do not have sufficient data on UF to make it a part of this study. The data in Fig. 5-14 for UF energy consumption is an estimate arrived at by subtracting the energy consumed in ICPT, UPT, and ICF from the total energy spent for transportation in those years. What is left will include, therefore, all the other transportation energy uses not taken into ac-

Table 14-3 Energy Savings from Shift of Passenger Traffic from High to Low EI Mode

Transport mode shift	*Savings per billion passenger-miles*	*Total savings at 1970 level*	*Percent of 1970 U.S. total energy*
20% UPT, auto–bus	4.4×10^{12} Btu	0.61×10^{15} Btu	0.9
5% UPT, auto–bicycle, walking	8.2×10^{12} Btu	0.28×10^{15} Btu	0.4
20% ICPT, auto–bus	1.8×10^{12} Btu	0.35×10^{15} Btu	0.5
20% ICPT, auto–train	0.6×10^{12} Btu	0.12×10^{15} Btu	0.2
20% ICPT, air–train	5.6×10^{12} Btu	0.13×10^{15} Btu	0.2

count: recreational and passenger boating, general aviation, and subways. In comparison to energy consumed in transporting freight, this amount is small. We see from these data that although the ICF traffic (in ton-miles) increased steadily, by 64 percent over the 20-year period from 1950 to 1970, energy consumption actually decreased by 12 percent over the same period. The mystery of this accomplishment is easily dispelled by Fig. 14-5 which displays the EI of the various important modes of ICF, and by reference to the percentage of the ICF carried by each mode (Fig. 5-17). The drop of the EI for trains from 3200 to 670 Btu per ton-miles as a result of the shift from steam to diesel overcame the rising EI of the other forms of transport and the increasing importance of these other forms. The fastest growing mode is air freight, with the highest EI of them all. In freight transport, as in passenger transport, we seem to be trading energy economy for speed and convenience.

Strategies for Conservation in Freight Transport. Let us now examine the nondisruptive and the disruptive—the easy and the hard—strategies open to us in ICF. In theory, the same strategies are available to us here; we can lower EIs or shift freight from high EI to low EI modes. In practice, freight traffic is quite different.

That difference was indicated in the uneven application of the emergency controls of late 1973. While passenger automobiles were urged to drive at 50 mph, truck traffic was allowed 55 mph. This difference was justified—incorrectly—in terms of miles per gallon efficiency; in fact, truck engine efficiency is also higher at lower speeds. The real justification was that if trucks are required to deliver the same tonnage per day, and if speeds are reduced, more trucks are required on the roads. Since it turns out that truckers are paid by the miles travelled, it is economically important (to them) that they travel fast. The fact that truckers acted together to protest (and that a few big trucks could block any highway) quickly won them concessions that virtually eliminated them from the controls.

There are other obvious inflexibilities in freight transport; you can't ship new

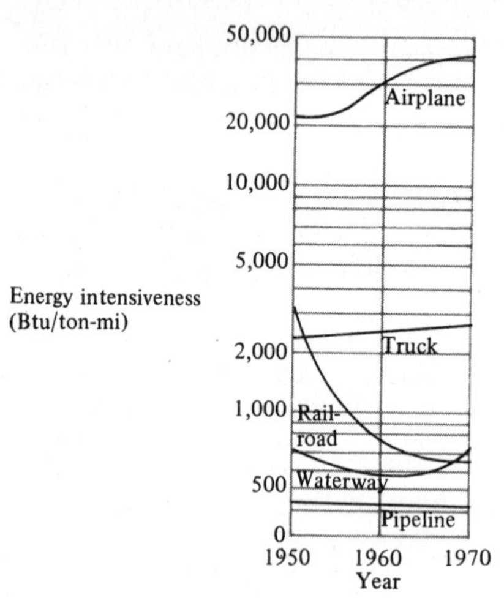

FIGURE 14-5 Only rail frieght transport has shown a decreasing El (shown here in a semilogarithmic display) While that of the airplane has doubled. (Source: Hirst, "Transportation Energy," *op. cit.*)

automobiles or household furnishings by pipe, for instance, nor can you ship (economically) natural gas or crude oil by truck or train. In spite of all this inflexibility, we can still apply some of the conservation strategies, with greater or lesser amounts of disruption. Perhaps the least disruptive would be to change the mix of modes, in other words, to shift some truck freight to train, and air freight to truck or train. The savings which would accrue from 20 percent shifts are shown in Table 14-4: even though air freight ton-miles are relatively small, the great reduction in going from the airplane's EI of 41,000 to 670 Btu per ton-mile for rail, results in a noticeable savings. Given the expected growth, such a change could become even more important. If present trends continue, ICF by air is expected to amount to 1 percent of the total in 1980–34 billion ton-miles compared to the 0.33 billion ton-miles in 1970. The same 20 percent shift in 1980 will save ten times as much energy.

The other strategies employed in our theoretical efforts to conserve energy in passenger transport are not available to us in freight transport; load factors are already relatively high (few trucks or trains run near empty) and significant improvements in engine efficiency are far in the future. There are probable efficiency improvements to be made in the urban freight area, perhaps by adding on some freight transport to mass transit. This needs study. It may also be possible to increase the use of low EI pipelines (shipping coal by slurry, for instance) and waterways. These will not, it appears, be of help to us in the intermediate range future we are examining.

A Conservation Scenario for Transportation

The data we have assembled in Tables 14-1, 14-2, 14-3, and 14-4 show us which strategies are most productive in savings and which are least productive. In order to gain some idea of reasonable expectations for total savings, we have calculated the total and percentage savings at the 1970 level and mix of transportation traffic. What is of more interest is to select some of the most promising of these strategies and apply them to the projected 1980 level and mix of transportation traffic in order to set a real goal for conservation in this area. To set this goal we need to know what modes of transportation are expected to carry how many ton-miles or passenger-miles in 1980. These data, taken from a projection by Hirst[1] are collected in Table 14-5. They represent an extrapolation of past trends.

The strategies which will give us significant savings in 1980 will be those which show large savings per passenger- or ton-mile and are applied to transportation modes

[1]E. Hirst, "Energy Consumption for Transportation in the U.S.," Oak Ridge National Laboratory-NSF Environmental Program ORNL-NSF-EP-15, March 1972.

Table 14-4 Energy Savings from Shift of Freight Traffic from High to Low EI Mode

Transport mode shift	*Savings per billion ton-miles*	*Total savings at 1970 level*	*Percent of 1970 total energy*
20% ICF, truck–train	2.03×10^{12} Btu	0.85×10^{15} Btu	1.3
20% ICF, air–truck	38.3×10^{12} Btu	0.03×10^{15} Btu	–
20% ICF, air–train	40.3×10^{12} Btu	0.03×10^{15} Btu	–

Table 14-5 Projections of Energy Consumption by The Various Modes of Transport 1980*

Mode	*Percent of total*	*Passenger-miles* $\times 10^{12}$	*Total energy* $\times 10^{15}$ *Btu*
ICPT			
Auto	85	1445	
Air	13	220	
Bus	1.5	25	
Rail	0.5	20	
Total		1710	8.37
UPT			
Auto	97	1370	
Bus	3	40	
Total		1410	6.97
ICF			
Rail	37	890	
Truck	21	510	
Waterway	16	390	
Pipeline	25	600	
Air	0.4	10	
Total		2490	2.62
			17.96
Total transportation energy (including urban freight, etc.) in 1980			21.55
Total U.S. energy consumption, 1980			88.1

*Hirst, "Energy Consumption for Transportation," *op. cit.*

which are expected to be important in 1980. The selection we have made and the savings they would bring about in 1980 are summarized in Table 14-6. We see that the application of these various strategies could reduce the 1980 total energy by almost 12 percent and cut the energy consumed by transportation almost in half, saving 1.7 billion barrels of oil. Referring back to Fig. 11-9, we see that this is 31 percent of the oil that is expected to be imported in 1980 under the Case 4, (low demand) assumptions of the NPC.

The projected savings are obviously significant; but are the strategies realistic? Here we are on less certain ground; political, economic, cultural factors enter. Most of the savings come from changing our reliance on the automobile and from making the automobile, when we do use it, more efficient. The biggest increment comes from increasing the automobile load factor from 28 to 50 percent; this means that five passenger cars will carry, on the average, 2.5 people. Such an increase can be achieved by car pooling, both formal, employment-related car pooling, and informal car pooling for shopping and recreation. Can it be done? We can only look to past experience. During World War II gas rationing, the average occupancy went from 2 passengers per car, prior to July 1942, to 2.44 passenger per car by December 1942 and 2.66 passengers per car by March 1943. Change, we see can occur.

Before leaving Table 14-6 we should remark that this is just one selection of

Table 14-6 An Outline of Combined Strategies for Energy Conservation in Transportation, 1980

Strategy	Total savings in 10^{15} Btu	Percent of 1980 total transport	Percent of U.S. total, 1980
UPT			
Increase auto load factor to 50%	4.95	22.9	5.6
Increase bus load factor to 50%	0.97	0.5	0.1
Shift 20% urban auto to bus	1.21	5.6	1.4
Shift 5% urban auto to bicycle and walking	0.56	2.6	0.6
ICPT			
50% increase in ICPT auto efficiency	1.66	7.7	1.9
Shift 20% ICPT auto to bus	0.52	2.4	0.6
Shift 20% ICPT air to train	0.25	1.2	0.3
Increase ICPT air load factor to 50%	0.63	2.9	0.7
ICF			
Shift 20% ICF truck to train	0.22	1.0	0.3
Shift 20% ICF air to train	0.08	0.4	0.1
Total	11.05	47.2	11.6

strategies. It is quite possible that by 1980 other more productive strategies will be in sight. Out-and-out gas rationing could force reductions. Technology may help us more rapidly than we anticipate. New efficient mass transit systems may take over large percentages of urban traffic, for instance, or a functioning personal television-telephone circuit may drastically reduce the amount of air passenger-miles businessmen pile up going to and from meetings. It would seem the better part of wisdom, from our mid-1970 viewpoint, however, to begin to actively stimulate some of the strategies indicated in Table 14-6 as insurance, while we await these and other technological miracles.

The Electric Automobile

We cannot leave the subject of transportation energy and its conservation without mentioning the electric automobile, that promise of technology which has been with us, and is with us, but whose performance, like that of the old Brooklyn Dodgers, brings forth the defense, "wait until next year." What may next year's electric automobile bring us?

The attractions of the electric automobile are many. It does not produce air pollution as does the present automobile, it is quieter, and is inherently more efficient than gasoline-powered cars. The advantages of no pollution and high efficiency appear most impressive until one remembers the power plant out there providing the electricity–inefficiently–and with its own pollution.

A realistic comparison of the electric automibile with the present one must be done in terms of system efficiences as discussed in Chap. 4. This comparison for the two types of automobiles is given in Table 14-7, using the data on the gasoline-powered automobile earlier shown in Table 4-4, and some of the electrical energy

Table 14-7 Comparison of System Efficiency of Gasoline and Electric-Powered Automobiles

	Efficiency of step		*Cumulative efficiency*	
Step	*Gasoline*	*Electric (coal)*	*Gasoline*	*Electric*
Raw fuel production	83*	96		
Transportation of raw fuel	97	97	81	94
Generation of electricity		33		31
Transportation of electricity		90		28
Battery		80		22
Engine conversion	25	90	20	20
Mechanical	70		14	
Transmission	70	90	10	18
Overall efficiency			10	18

*96 percent efficiency for production of petroleum, 87 percent efficiency for refining to gasoline.

system data of Table 4-1. We see that the electric automobile, at present generating-efficiencies, is half again as efficient as the gasoline-powered automobile.

The attractiveness of the electric automobile is further enhanced by the fact that future improvements in the efficiency of electric power generation seem much more certain than improvements in the efficiency of the gasoline engine. From the point of view of pollution, it is also easier to control the emission from a few central power plants than from millions of individual automobiles.

There are, of course, practical difficulties with presently available electric automobiles and we will briefly mention these later. The next big question concerns our ability to generate the electric power needed for this conversion.

In 1968 there were 101 million motor vehicles registered in the United States: 83.7 million automobiles, 17 million trucks, and 350,000 buses. The average horsepower was 60; converted to electrical power units this is 12.6 M mw, 46 times the 1968 electrical generating capacity of 276,000 mw.

Fortunately, most of the motorized horsepower is not used. If we compare fuel use in 1968, we find that motor vehicles used 10.7 Q Btu while electric utilities generated 4.5 Q Btu. The energy actually used for motor vehicles was, therefore, only a little more than twice the electric energy generated.

We have not taken into account, however, the greater efficiency of electric automobiles, 65 percent versus 12 percent if we only compare the operating efficiency (the last four lines of Table 14-7). It has been estimated by Netschert[2] that the actual energy requirement of an average automobile is 1200 Btu per mile, and, of trucks and buses, 2800 Btu per mile. At 65 percent efficiency the energy required of all the 1968 cars, trucks, and buses would have been 2.4 Q Btu, half the generated energy.

If we set even more modest goals and focus only on the urban automobiles, the total energy needed is reduced to about 1 Q Btu, 22 percent of the electricity generated in 1968. We could throw in half the trains for an addition 0.12 Q Btu, and

[2] B. C. Netschert, *Bulletin of the Atomic Scientists,* 29, May 1970.

provide this energy by an overall increase of 25 percent of the 1968 generating capacity (an increase which has already been exceeded since that year). In the use of this capacity there would be the additional advantage that battery charging would take place, for the most part, during late night, off-peak hours.

Thus, electrified transportation seems possible and would result in significant fuel savings as well as lowered pollution. Its arrival waits more on its own technical difficulties than on a shortage of generating capacity. Improvements must be made in the power/weight ratio of batteries, so that a car can be light and fast enough to operate, at least part of the time, at suburban freeway speeds.

MORE COMFORT PER KILOWATT

Man's first use of nonfood energy, fire, enabled him to make his living space more comfortable. In spite of the enormous amounts of energy now used in transportation and industry, heating and cooling our residental and commercial buildings still accounts for about a third of our total energy.

In Chap. 5 we looked, in some detail, at the various end uses of energy in these two sectors and at the amounts of fuel and electricity which provided that energy. We will briefly review that data in order to identify the appropriate targets for conservation strategies.

Again, as we did in the study of transportation, we will concentrate on intermediate-range strategies: conservation practices which could conceivably be implemented by 1980. Longer-range strategies could, of course, grow out of the ones we will suggest (a conservation practice, if it is economically sound, should be continued) but the mix of strategies for the 1990s should have an input from advanced technology as well.

Keeping the Home Fires Burning—Low

We will look first at the residential sector for conservation targets. Figure 14-6 displays the important end-use consumption in this sector and the sources of that energy. We see that heating and cooling are the big consumers and that natural gas and oil are the most important sources, followed closely by electricity. We also charge to the residential sector, the energy loss in the conversion of primary energy (coal, oil, or natural gas) to electric energy.

Lowering the Thermostat. There is an important emergency strategy suggested for this sector which is analogous to the speed limit lowering in the transportation area. American homes are kept hot in the winter and cool in the summer—often cooler in the summer than in the winter. Since the heat transferred (in or out), whether it is by radiation or by conduction through the walls, is greater for greater temperature difference, energy can be saved by reducing this difference. In the average American home it is estimated that 1°F change in the daytime thermostat setting—down in the winter and up in the summer—reduces fuel consumption by 3 to 4 percent. Nighttime reductions are similar. A 5°F change in the "comfort temperature" during daytime and a 10° change at night would reduce fuel usage by 15 to 20 percent. We

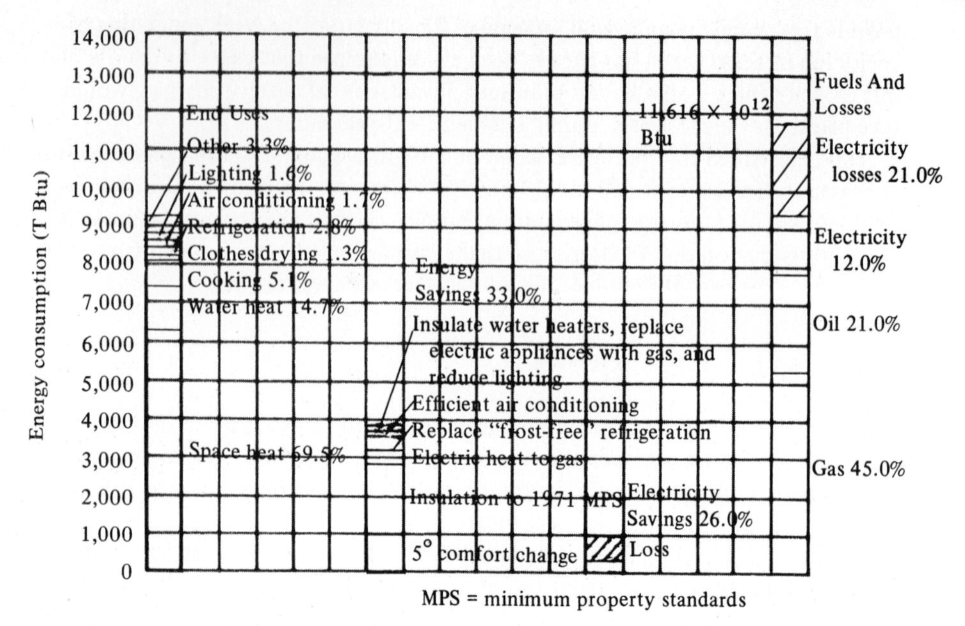

FIGURE 14-6 The 1968 residential energy consumption by end use and fuel. The conservation strategies which are described in the text would have reduced energy consumption by 33 percent and electrical energy consumption by 25 percent.

show these and other savings calculated at 1968 energy consumption levels in Fig. 14-6.

Most expert opinion is that indoor health is improved at these lower winter temperatures (except in the case of the aged); this is a strategy which should continue. It is not an inconsequential saving: 15 percent of the 1968 space heating total is 1000 T Btu, equivalent to 165 million barrels of oil, about 15 percent of our oil imports in 1968.

Insulation. A strategy which would complement thermostat lowering is that of better insulation. It has been shown by experiment, both real and simulated, that a significant amount of energy in the residential sector is being used to heat and cool the outdoors.

Some perspective on the size of the "heat leak" is obtained from the testimony of an insulation manufacturer, G. J. Haufler[3] before the Congressional energy conservation hearings. Haufler reported on studies which showed that an "average home" (1,600 square feet of floor space, located at the climate of St. Louis) uses, if completely uninsulated, 1,600 gallons of fuel oil per year for heating. Fully insulated (6 inches of insulation in the ceiling and 3.5 inches in the walls) reduces the consumption to 900 gallons per year, a 45 percent savings.

From the point of view of the consumer, the strategy of insulation pays off in

[3] Testimony by Mr. George J. Haufler, Executive Vice President of the "Certain-Teed Products Corp." Printed in *Energy Conservation,* Hearings before the Subcommittee on Consumer Economics of the Joint Economic Committee Congress of the United States, November 19, 1973, pp. 54-79.

dollars as well as Btu. In this same report it is claimed that such a house could be completely insulated, while under construction, for 325 dollars and that this would save 175 dollars per year in fuel costs.

The attic is a prime target for insulation. Half the heat is lost there, and, since it is usually accessible, a reasonably adept homeowner can insulate his attic himself. The cost, 100 to 200 dollars, will be paid back from savings in fuel bills in two to five years (depending on climate) and in colder climates (Minnesota, for instance), the payback period can be as short as a year (although it is hard to imagine uninsulated houses in Minnesota).

Similar results have been reported in other studies. The Department of Housing and Urban Development (HUD) studied the possibilities of conserving residential energy in a "typical single family house" in the Baltimore-Washington area.[4] This "typical house" was two stories, cost 30,000 to 35,000 dollars, had 1,500 square feet of floor space, used natural gas for heating, had some insulation in the walls and ceiling, but no storm windows. Natural gas was used for heating and for an ornamental outdoor gaslight which used 10 percent of the total gas. Pilot lights for the gas appliances consumed another 8 percent. The ornamental light is an immediate target for conservation. We will comment on the pilot light later.

The largest energy leak came from what is called *air infiltration,* or air circulation between the inside and outside caused mainly by leaky doors and windows. While some ventilation must be provided, most houses have much more than needed. In terms of total *primary energy,* (which takes into account the losses in the generation and transmission of the electricity) the addition of storm doors and windows to such a house results in a 12 percent saving of the energy used for heating and cooling. They also found that the house was underinsulated, and that adding 12 percent more increased the savings to 21 percent.

A third study which produced generally similar results was that of Moyers[5] at Oak Ridge, who has studied the effects of the changes in insulation and infiltration reduction brought about by the new FHA *Minimum Property Standards* (MPS). There is, of course, no "national standard" for insulation, but it is a good guess that few homes are built to more stringent standards than those put out by the FHA. In 1972 the FHA reduced the MPS from its previous level (set in 1965) by about 25 percent, which required additional insulation. (The MPS is set in terms of allowable heat losses.) Moyers found that, averaged over typical climates of Atlanta, New York, and Minneapolis, these improvements would reduce the heat loss by 29 percent in gas heated homes, and 19 percent in electrically heated homes (which already had stricter insulation requirements).

The Moyers study went beyond the FHA regulations and determined the *economic optimum* levels of insulation, the amount of insulation which yields the maximum economic benefit to the homeowner (at 1970 to 1971 fuel prices). If insulation were to be brought up to these levels, savings of 43 percent for gas heated homes, and 41

[4]Testimony reported in *Conservation and Efficient Use of Energy,* Part I, No. 14, Joint Hearings on Government Operations, June 19, 1973, p. 188.

[5]J. D. Moyers, "The Value of Thermal Insulation in Residential Construction Economics and the Conservation of Energy," Oak Ridge National Laboratory Report, ORVL-NSF-EP-9, December 1971.

percent for electrically heated homes, should be achievable. An average savings of 42 percent in all residential space heating in 1970 would have amounted to 2.68 Q Btu, 29 percent of the total residential consumption in that year.

The Inefficiency of Electric Heat

In Chap. 5 we commented on the rapid growth of electric resistance heating and the reasons for that growth. In resistance heating, current flows through a high resistance wire located in the baseboard, floor, and/or ceiling and the electric energy is converted into heat. This is a highly efficient conversion process, but, as noted in Chap. 4, the system efficiency of electric energy production is low, 30 percent or less compared to about 60 percent for gas. As an indication of the size of the potential savings, let us compute the amount of energy needed if all the houses which were electrically heated in 1968 had been heated instead with gas.

From Table 6-9 we find that in 1968, 3.4 million electrically heated homes used 48 B kw-hr of electricity. We will take as a rough "heat rate," 10,500 Btu per kw-hr, for that year. This conversion factor is based on the average amount of energy spent in the power plant to generate electricity and also includes a 10 percent loss in transmission and distribution. The 48 B kw-hr were, therefore, produced by the expenditure of about 500 T Btu of primary energy.

The gas furnace is 60 to 70 percent efficient. Using the lower figure, the energy equivalent of 48 B kw-hr of electricity, 164 T Btu (at an energy conversion of 3412 Btu per kw-hr), would have totalled with gas furnaces, 273 T Btu (164/60 = 273). Thus, the conversion of those electric resistance heaters to gas would have saved 227 T Btu, or 3.4 percent of the heating total that year.

One might question a conservation strategy that increases the use of natural gas, a fuel that is experiencing shortages. However, natural gas is used to produce electricity, 3245 Q Btu worth in 1968, more than ten times the heating requirement we have just computed. It would seem an obvious improvement to burn this gas directly to provide heat rather than wasting two thirds of its energy value in converting heat energy to electric energy and then back to heat energy. We will see, in the following sections, other instances in which an electric to gas conversion offers the possibility of energy conservation.

Heat Pumps. In Chap. 4 we pointed out the inherent advantages of heat pumps. The energy used in electric resistance heating is high-temperature energy. As we emphasized by the waterwheel analogy (Fig. 4-8), this high-temperature energy could be better used to do work and move heat energy from the outside in during the winter, and from the inside out during the summer. Unfortunately, the heat pump has a high capital cost and has, so far, a poor maintenance record. In 1970 only 11 percent of the electrically heated homes used them. The increasing cost of fuel and the renewed developmental efforts being expended on this device may increase its attractiveness.

In may even be that gas will get into the act here as well. There is considerable interest in the gas industry in developing heat-activated heat pumps which would use the absorption-desorption cycle described in the discussion of solar cooling (see Chap. 13).

Other Residential Conservation Strategies

From Fig. 14-6 we see that water heating is the next end use to be examined for energy leaks, or for the possibility of improved efficiency of utilization.

First the leaks: an electric water heater consumes about 4,500 kw-hr per year. This energy goes into heating the water and making up for the losses through the tank walls and from the piping. Only the loss through the walls seems a practical target. In his paper on energy conservation, Moyers[6] also reported on studies of the effects of additional insulation on water heaters. They found that in a 50-gallon electric water heater with 2 inches of factory installed insulation, 20 percent of the energy (960 kw-hr) is wasted. An additional 3 inches of insulation (which pays for itself in fuel savings in a couple of years) saves 350 kw-hr, or 7.8 percent of the total consumption of water heaters.

In 1968 the 14.5 million electric water heaters used 65.4 B kw-hr of electricity. The additional insulation would have saved 5.1 B kw-hr. When we save electrical energy, however, we get better than a 2 for 1 return because of the losses in its generation. Using the heat rate of 10,500 Btu per kw-hr, the total savings would have been 53.5 T Btu. Insulation will bring about savings on gas water heaters as well. There were 41.2 million gas water heaters in use in 1968. A savings of the same amount, 1.2 M Btu per unit (the Btu equivalent of 350 kw-hr), would have amounted to an additional 49.2 T Btu. The total savings from the additional 3 inches of insulation would have been 103 T Btu, or 0.9 percent of the residential total.

We have available to us in water heating the same "drastic" strategy as we considered in space heating, the replacement of electric water heaters by gas. The 65.4 B kw-hr of electricity consumed by the electric appliance converts to 687 T Btu of primary energy. An equivalent gas heater uses 27.2 M Btu. The total consumption of an additional 14.5 million gas heaters would have been 394 T Btu a savings of 293 T Btu, or 2.5 percent of the residential total.

Air Conditioning Inefficiency. The strategies we have already described, will bring about savings in air conditioning energy as well as in space heating by reducing air filtration and adding to the insulation. The economically optimum amount of insulation of the Moyers report will reduce the electricity requirements for air conditioning by about 15 percent. In terms of primary energy, this amounts to 71 T Btu, or 0.6 percent of the residential total for the 45 B kw-hr of electric air conditioning in 1968.

There is, however, the potential of even greater savings. In recent studies of air conditioner cooling efficiency, it has been found that models of the same cooling capacity range in cooling efficiency from 4.7 Btu per watt-hr to 12.2 Btu per watt-hr, a factor of almost 3. Of course, the price differs also: the less efficient units are generally less expensive. Again, the choice is between lower initial cost and lower cost over the operating lifetime.

Moyers estimates that an improvement in the average efficiency of air conditioners, from the value of about 6 Btu per watt-hr characteristic of 1970 to 10 Btu per watt-hr (a value still lower than several commercially available models attain) would

[6] Moyers, J. D. *Ibid.*

have reduced the 45 B kw-hr to 27 B kw-hr, a savings of 18 B kw-hr or 190 T Btu of primary energy.

Cooking and Freezing. There are other candidates for conservation strategies from which we will get smaller returns. The refrigerator is one. We have not as yet seen studies of the returns to be expected from increased insulation, but there is no doubt that savings will be possible. What is clear, however, is that the "frost-free" refrigerator, a relic of the "free energy age" whose walls are periodically heated to remove the frost, is a prime conservation target. This now popular refrigerator uses about 60 percent more energy than does the standard model. In 1968 about two thirds of the total were of this type. Thus, two thirds of the 73.2 B kw-hr total for refrigeration could be charged to them. A 60 percent savings amounts to 310 T Btu or 2.6 percent of the residential total.

To be consistent we should also advocate replacement of the electric ranges and clothes driers by gas devices. There were 23.6 million electric ranges and 15.2 million electric clothes driers in use in 1968. They consumed a total of 43.1 B kw-hr of electricity—453 T Btu of primary energy. Gas clothes driers use 9 M Btu per unit (50 percent of it in the pilot light) and gas ranges, 10.6 M Btu per unit per year on the average. Replacement of the electric appliances by the gas ones would have resulted in the consumption of 382 T Btu for a savings of 71 T Btu.

In the case of clothes driers, we could consider the entire consumption a target to be replaced by solar energy of the old-fashioned type—outdoor clotheslines.

We are obviously now left only with the small consumption devices to look at and cannot hope for significant savings. Lighting uses about 46 B kw-hr of electricity; a 10 percent reduction would save 48 T Btu of primary energy. If the estimated 8 percent of the natural gas consumption that goes into pilot lights were saved by replacing them with electric lighters (or matches), that would save an additional 370 T Btu.

It is argued by the gas industry, however, that pilot light energy is not all wasted. In the water heater, for instance, it contributes energy, making up for heat losses. In any home which is being heated (rather than cooled) the pilot light helps in the job. The realistic "savings" are thus considerably lower than 8 percent of the total.

The "energy saved" columns in Fig. 14-6 present a summary of the results which would have been achieved in 1970 by application of the measures, big and small, that we have suggested. We see that an overall savings of 33 percent of the residential total and 5.6 percent of the United States total has been described. More important, consumption of electric energy—and its inherent losses—has been reduced by 70 percent.

We cannot make detailed 1980 energy conservation estimates for the residential area as we did for transportation. The projections of the growth of space and water heating, of the number of new refrigerators, and so on have not been assembled. We can report one study, however, that has attempted to estimate energy savings realizable by 1982 through improved insulation and storm doors and windows.

The National Mineral Wool Insulation Association[7] undertook a study in 1972

[7] National Mineral Wool Insulation, Inc., "Impact of Improved Thermal Performance in Conserving Energy," March 1973.

which considered the expected growth in the housing market and the effects of the new FHA standards. They project a total demand of 11.0 Q Btu for residential heating and cooling in 1982. (This compares with 7.1 Q Btu in 1968). It is their estimation that by insulating all new housing built during that period to the FHA standard, by upgrading the insulation of existing houses and adding storm doors and windows, a maximum of 2.8 Q Btu, or 26 percent of the total, could be saved. What is perhaps more interesting are their "reasonably attainable objectives," listed below:

1 3 of every 4 new units constructed from January 1973 on have proper thermal treatment.

2 1 of every 4 existing family homes, constructed prior to 1973 have had improved insulation.

3 1 of every 8 single family homes constructed prior to 1973 add storm windows and doors.

If these objectives are attained, the 1982 savings would be 1.5 Q Btu, 14 percent of the 1982 total. A total of 8.2 Q Btu would be saved over the ten-year period. They further compute that this program of insulation would save 17.2 billion dollars, three times its cost of 6.4 billion dollars.

The 1.5 Q Btu saved in 1982 is about 2 percent of the anticipated 95 to 100 Q Btu of total energy consumed in that year. This figure can surely serve as a conservative goal, and we can hope to increase the savings above that figure by the other measures we have described.

Changing Life-Styles. We cannot leave our discussion of energy consumption in the residential sector without mentioning some "disruptive"—but interesting—changes that are taking place on a small scale. These are the return to simplier life-styles among the young, symbolized perhaps by the *Whole Earth Catalog* and the commune. While energy conservation is not the goal of this movement toward simplicity, it is a result.

We will look only at some representative statistics and comparisons in an article by Michael Corr and Dan McLeod, "Home Energy Consumption as a Function of Life Style."[8] The authors compare, for instance, the energy consumption in various types of baths. The results are shown in Table 14-8. The differences are impressive. While one doesn't really expect the "Guatemalan Sponge Bath" to become popular in this country, these comparisons serve to remind us that energy can be conserved by different bathing practices and to emphasize the contributions which can be made by solar water-heating.

The authors also present some interesting data on home heating, recommending more attention to efficient fireplaces and reminding us of the great efficiency of heating and cooling in primitive structures such as igloos and adobe huts. In an igloo, for instance, the indoor temperature at the sleeping level remains at about 35°F even with outside temperatures of -20°F. In adobe huts, inside temperatures hover at a comfortable 80°F while outside temperatures climb to 140°F. Again, one does not

[8]Source: Corr and McLeod, *op. cit.*

Table 14-8 Energy Use and Style of Bathing*

Bath style	*Hot† water (gal)*	*Resources Btu*	*Persons served*	*Btu per person*	*Comment*
Big tub bath	30	35,000	1	35,000	Assumes tank temperature increase of 87°F and 62% efficient gas water heater.
Small tub bath	15	17,500	1	17,500	
15 min shower	30	35,000	1	35,000	
5 min light shower	5	5,850	1	5,850	
American clothes Washing machine cycle	24	27,700			Doesn't count operation of washing machine motors, etc.
Wood-fired Japanese bath	46	86,300	5	17,300	Maximum bath temperature is about 114°F; $\Delta F = 61°F$.
Wood-fired Japanese bath + one tub laundry	46	86,300 -27,700 58,600	5	11,700	Doing tub wash after bath saves 27,000 Btu.
Japanese bath at gas efficiency	46	41,200	5	8,250	At 896 Btu/gal raised from 53°F to 114°F.
Japanese bath at gas efficiency with 1 tub laundry	46	41,200 -27,700 13,500	5	2,700	
Electric Sauna and cold showers		49,200 (space heat)	5	9,850	300 ft³ room with 7.5 kw-hr heater operating 63% of 1/25 hr.
Electric Sauna and 1 hot shower per person	25	29,300 (water) 49,200 (space) 78,500	5	15,700	
Electric sauna and 1 hot shower for 2 people	10	11,700 (water) 47,200 (space) 60,900	2	30,400	
Sauna at gas heating efficiency (62%) and cold dips		21,650 (space)	5	4,340	The ratio of electric space heating to that of gas is 0.273/0.62 = 0.44
Sauna at gas heating efficiency (1 hot sponge bath per person)	5	21,650 (space) 5,850 (water) 27,500	5	5,500	
Sauna at gas heating efficiency (1 quick hot shower for 2 people)	10	21,650 (space) 11,700 (water) 33,350	2	16,700	
Sponge bath	1	1,170	1	1,170	
Gautemalan sponge bath	1	solar	1	solar	Bucket of water is left in sun for 6 hr.

Outdoors: 3 30 gallon water tanks in small roof "greenhouse"π	30	solar	2	solar	Yields 30 gal of water at 120°F for 2 tub baths with warm water left over for morning chores. Tanks are painted black; Functional 8 to 9 months in California (varies with altitude).

*Source: Corr and McLeod, *op. cit.*

expect that there will be a building boom in either of these. These figures however, may stimulate some architects and homeowners to collaborate on more imaginative as well as efficient home design in the future.

Since many young (and some older) Americans are attracted to communal living, energy comparisons of this style with the more common urban life are of interest. Corr and McLeod studied 12 communal households in the Minneapolis area. The results in terms of power consumption are shown in Table 14-9. The energy savings of these groups are considerably understated by these data. Not only do they use less power, but they get much more use out of appliances and automobiles, for example, and thus save considerable amounts of manufacturing energy as well.

For all its energy-saving attractiveness, we do not, of course, seriously present these changes as part of a "scenario for 1980." The movement of a small percentage of our youth in these directions, however, may signal some change in attitude toward consumption which will eventually bring the exponential curves down.

CONSERVATION IN THE COMMERCIAL AREA

The great diversity within the commercial sector makes the identification of specific strategies much more difficult than it was in the case of the transportation and resi-

Table 14-9 Comparison of Energy Consumption in Commercial Households*

	Comparison	*Sample communes*		
	Average	*Low*	*Average*	*High*
Natural gas/100 ft^3/mo/person (Minneapolis)	33	17	23	30
Electricity/kw-hr/mo/person (Minneapolis)	247	17	39	66
Gasoline/gal/yr/person (U.S.)	285	60	181	444
	Reduction by percentage			
	Low			*Average*
Natural gas reduced by	49%			30%
Electricity reduced by	93%			84%
Gasoline reduced by	79%			36%

*Source: Corr and McLeod, *op. cit.*

dential sectors. That there are, however, significant amounts of energy to be saved in that area cannot be questioned. The percentage distribution of end uses in 1968 is displayed in Fig. 14-7. As in the residential sector, space heating leads, followed by air conditioning, refrigeration, and water heating. In this conservation study we are ignoring the nonenergy end uses "asphalt and road oils" which are assigned to this sector. The fuel mix is somewhat different; electricity acounted for about 11 percent and the losses in generating electricity added 23 percent more. Coal provided 6.5 percent of the energy, oil 27.5 percent, natural gas 21 percent and the nonenergy uses, the remaining 11 percent.

We recognize some familiar conservation targets in this breakdown: space heating with its leaks through poor insulation and excessive air infiltration, air conditioning which is affected by the same leaks, and refrigeration and water heating where insulation and replacement (electricity by gas) might make contributions.

The immediately applicable strategy of lowering the thermostat in the winter and raising it in the summer should produce significant results here also. At the previously quoted rate of 3 to 4 percent fuel reduction for each 1°F lowering (or raising in the summer), a 6°F change (down to 68°F and up to 74°F) would save 18 to 24 percent of the fuel consumed for space heating and air conditioning. The total used for these two purposes in 1968 was 5.3 Q Btu. If we take a 20 percent savings as being realistic, the rather slight reduction in comfort which these changed thermostat settings would cause would be balanced by a saving of 1.06 Q Btu, 12 percent of the total energy consumed in the commercial sector and 1.6 percent of the United States total.

The replacement strategies, at least when applied to the 1968 levels of space heating, will not be as productive since electric space heating did not contribute importantly to the commercial area in 1968. We will not be able to apply a simple "frost-free versus standard" formula to refrigerators since commercial refrigerators come in all shapes and sizes. There are, no doubt, savings to be made in refrigeration, as anyone who has been chilled by the open refrigerators in the supermarket can believe,

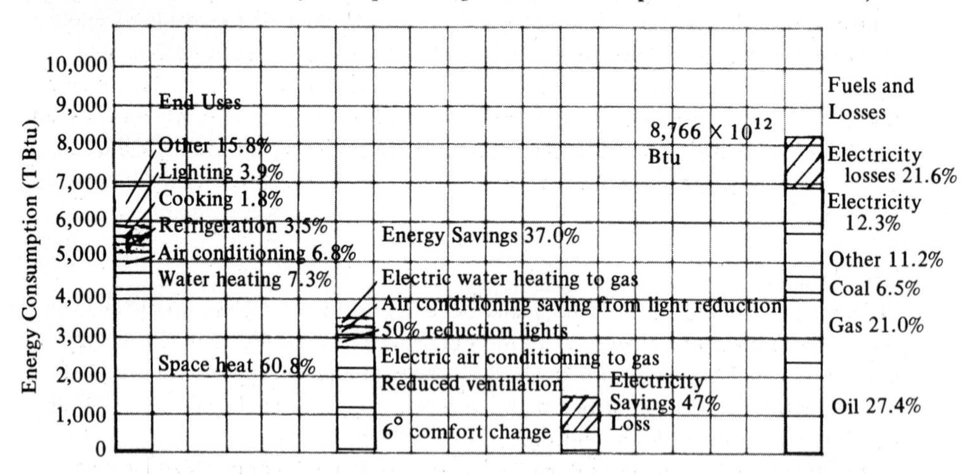

FIGURE 14-7 1968 commercial energy consumption by end use and fuel. The conservation strategies described in the text would have reduced energy consumption by 37 percent and consumption of electric energy by 47 percent.

but quantification of such savings awaits further studies. To compensate somewhat for the loss of old targets we will discover a new one, lighting, which takes a larger share of energy here. We will also consider a new strategy—the *total energy concept.*

Stopping the Heat Leaks

"Sample field observations indicate that the state of insulation and draft sealing in existing commerical buildings is not significantly different from that in existing residences."[9] Air infiltration is an especially troublesome leak in these buildings, accounting for 25 to 50 percent of the energy used in heating and cooling. It appears that under present construction practices the ventilation of the entire building is designed to the levels needed in special areas such as toilets, kitchens, and conference rooms, and is thus larger than needed, perhaps by as much as a factor of 4. Reduction of excess ventilation and control over the ventilation to the critical areas so that it is only used when needed, could reduce fuel needs by 15 to 25 percent. At the 1968 levels of consumption this would amount, on the average (20 percent), to another 1.06 Q Btu, and an additional 12 percent reduction in the commercial total.

Savings through improved insulation will be similar, in percent, to those we described in the residential sector; a 29 percent savings if all buildings were brought up to the 1971 FHA standards, and 43 percent if they were insulated to the economically optimum amount. Since it is not likely that existing commercial buildings will be reinsulated, we will have to look to stricter standards on new buildings for realistic savings.

Water Heating

Commerical use of hot water heaters is an area which needs investigation. Since we know of no study of their insulation specifications, we can speculate that their relatively larger size has resulted in better insulation and lower percentage losses. It is realistic here, however, to apply the electric-gas replacement strategy since about one fourth of the commerical water heaters are electric. In the base-line study of 1968 consumption which we have been using (see Chap. 5) electric water heaters used about 79.5 B kw-hr to heat 116 billion gallons of water. In terms of primary energy this is 229 T Btu. Gas water heaters use about 1040 Btu per gallon at their efficiency of 64 percent. Thus, they could have heated the 116 billion gallons at a cost of 121 T Btu, saving 108 T Btu, or 1.2 percent of the energy charged to the commercial sector.

Lighting

Illumination of buildings accounts for about a quarter of all the electricity sold in the United States, and, therefore, about 4 percent of the total commercial energy use. In metropolitan areas with large amounts of commercial space like New York, the percentage of electricity used in illumination may be much higher; estimates as large as 65 percent are given by Consolidated Edison.

Lighting levels are, in general, guided by the minimum recommended lighting

[9] C. A. Berg, "Energy Conservation through Effective Utilization," National Bureau of Standards, NSSIR 73-102, February 1973.

standards of the Illuminating Engineering Society (IES). Over the last 15 years the lighting levels recommended by the IES have been increased by a factor of 3 or more. In 1952, for instance, the recommended level of classroom lighting in New York City was 20 footcandles. This was raised to 30 in 1957 and to 60 in 1971. In libraries an even greater increase has taken place, from 20 footcandles in 1952 to 40 in 1957, 50 in 1959, and 70 in 1971. The theory behind these increases is one of "undifferentiated lighting"; it is the belief of the IES that the most comfortable lighting environment is one in which there are no contrasts greater than 3 to 1.

The physiological basis for this belief seems weak since the human eye adjusts easily to much greater contrasts. But, as a result of building design stemming from this thinking, our buildings are brightly lighted throughout their entire floor area including walkways and other perimeter areas in which daylight itself could produce adequate illumination.

It has been estimated[10] that energy consumed by interior lighting could be reduced to as low as 36 percent of the original expenditure by more careful design and placement of fixtures, by circuiting perimeter lighting separately in order to use natural light whenever possible, and by turning lights out when not in use. Even a 50 percent overall reduction would save 39.2 B kw-hr of the 78.4 B kw-hr which the 4 percent figure assigns to commercial lighting. In terms of primary energy this amounts to 412 T Btu, 4.7 percent of the total commerical energy.

Air Conditioning

The reduction in energy used for illumination has a secondary conservation effect: it reduces the energy needed for air conditioning in the summer by perhaps as much as 10 percent. In 1968 gas air conditioning accounted for 97 T Btu and electricity for 107 B kw-hr (1125 T Btu) for a total of 1222 T Btu. The 10 percent reduction amounts to 122 T Btu.

It is also of interest here to examine the consequences of substitution of gas air conditioning (30 percent efficient) for electric air conditioning (50 percent efficient). The 107 B kw-hr amounts to 366 T Btu of energy used in cooling. Since gas air conditioning is less efficient it would use 50/30 times more, or 620 T Btu. In terms of primary energy, this is still a savings of 500 T Btu, 5.7 percent of the commerical total.

Total Energy Plants

A somewhat controversial entry into the energy conservation picture is the *total energy plant*, a package designed by the gas companies to challenge central power station generation of electricity. The concept is rather simple; instead of buying electricity from the public utilities, the larger customer (an apartment complex, industry, or school) installs a gas engine generator set and generates his own electricity. The savings come from the utilization of exhaust heat that the on-site locations allows.

Some of the problems this approach runs into are illustrated in the example of a school in York, Pennsylvania. As reported by R. E. Roushey, the engineer in charge,

[10] P. B. Stein, *Energy and Human Welfare,* (Riverside, N.J.: Macmillan, 1974) Vol. III, Chap. 2.

a study was made of the needs of a school which was to receive heavy year-round use. The installation of a total energy system was recommended; the initial cost was estimated at 150,000 dollars with a less than five-year payoff. The school board accepted this recommendation. This, according to Roushey, upset the local utility:

> . . . the local power company solicited the support of other neighboring power utilities and wrote letters to all of the board members threatening them with surcharge suits and enumerating many unfounded claims. We countered these claims as ridiculous and unfounded. The power company then applied for and received a PUC rate reduction which was approved in less than 2 weeks for a lump energy charge of 11 mills per kilowatt-hr with no demand charges and included in the package, free installation.
>
> Our original report had verified a 4.1 year payout including interest and with their new rate this jumped to 4.8 years. The board again unanimously confirmed their original decision and approved the project for total energy.
>
> The school total construction cost was $23.52 per square foot at a time when all electric roof top unit construction was averaging above this amount. The project has been in continuous operation for over 6 years at an average monthly cost as predicted of less than $2,000. It has its own power generating and recovery system and over 500 tons of air conditioning.[11]

Mr. Roushey also provided the comparison of his school with a similar all-electric one nearby (see Table 14-10).

In the total energy system the use of waste heat can allow an overall efficiency of 70 to 80 percent instead of the 30 percent central power station efficiency. There are drawbacks. The actual thermal conversion efficiency of the total energy system, since it is a smaller unit, will be generally lower than that at the larger plant. The larger plant is thus able to get more energy out of the fuel. As some of the technological improvements we anticipated in Chap. 12 come into being, the edge now claimed by the on-site system will lessen.

The other problems pointed to by some critics are those of maintenance and flexibility. The smaller units will require a certain minimum of maintenance as will

[11]Letter from R. E. Roushey, engineer to Hon. F. B. Dent, Secretary of Commerce, printed in *Conservation and Efficient Use of Energy,* Part I, No. 14, Joint Hearings on Government Operations, June 19, 1973, p. 220.

Table 14-10 Comparison of Total Energy Plant with All-electric Plant*

Criteria	*Total energy*	*All-electric*
Construction cost	$ 23.52	$ 22.91
Pupil capacity	784	600
Percent air-conditioning	100	40
Size square feet	106,400	65,666
Occupancy hours per week	100	47
Average monthly energy cost	$ 1,489	$2,794
Auxiliary energy	None	†
Energy cost square feet per month (cents)	1.4	4.2

*Source: Letter from R. E. Roushey, engineer, to Honorable F. B. Dent, Secretary of Commerce. Reprinted in *Conservation and Efficient Use of Energy,* Part 1 No. 14, Joint Hearings on Government Operations June 19, 1973, p. 220.

†Emergency power (propane).

the associated equipment. There is some question as to how the skilled manpower will be made available to the small establishments. Flexibility of fuel use is also an advantage to the larger station; they can switch from oil to gas to coal, for instance, if necessary, while the on-site units are limited to one fuel.

Some of these questions will be answered by the operation of the total energy plant being built by HUD to service a housing development in Jersey City: "Operation Breakthrough." This plant is expected to show a 34 percent savings in fuel over those of a conventional system which utilizes a central oil furnace and purchased electricity. The total energy plant, it is estimated, would use only half the primary energy of an all-electric system. This plant is now in operation and the measured savings will soon become available for comparison with these estimates.

We have just suggested several strategies for energy conservation. They differ greatly in the ease with which they could be applied. Lowering the thermostats and reducing lighting are relatively easy. It should also be possible to gain the efficiencies of reduced air infiltration without too much remodeling. On the other hand, the strategies of better insulation and the replacement of electric services by gas are more difficult. We can, at best, hope for their employment in new buildings built between now and 1980.

The savings at 1968 levels and forms of use amounted to 37 percent of the total commercial energy consumption and 4.8 percent of the United States total (see Fig. 14-7). Electrical energy consumption in the commercial sector was reduced 47 percent by the indicated strategies.

What are reasonable 1980 savings? Again, we lack the necessary detailed projections on which to base our estimates. We can set a lower limit by quoting an industrial study, in this case Shell Oil Company's "The National Energy Problem: Potential Energy Savings."[12] Shell projects a consumption of 8.9 Q Btu by the commercial sector in 1980. Of this amount, 850 T Btu, or 9.5 percent of the total, could be saved by insulating new buildings to the new standards, revising the venilation, resetting the thermostats in the way we have suggested, and reducing the lighting to 65 percent of the present levels. The 850 T Btu saved would be about 1 percent of the projected 1980 total energy.

ENERGY CONSERVATION IN INDUSTRY

The industrial sector is the largest consumer of energy, accounting for 41 percent of the total consumption in 1968. Thus, it is reasonable to look for the potential of large savings there. The search for savings, however, is immediately frustrated by the great diversity of the energy-consuming operations within that sector and by the overall lack of specific data on amounts of energy used and efficiencies of use in specific operations.

The 1968 breakdown of energy consumption in this area is reviewed in Fig. 14-8 by industry grouping and by end use, along with the percentages of energy supplied by the various fuels. We see that heat is again the largest end use, accounting for al-

[12]Shell Oil Company, "The National Energy Problem: Potential Energy Savings," Public Affairs, P. O. Box 2463 Houston, Texas 77001, October 1973.

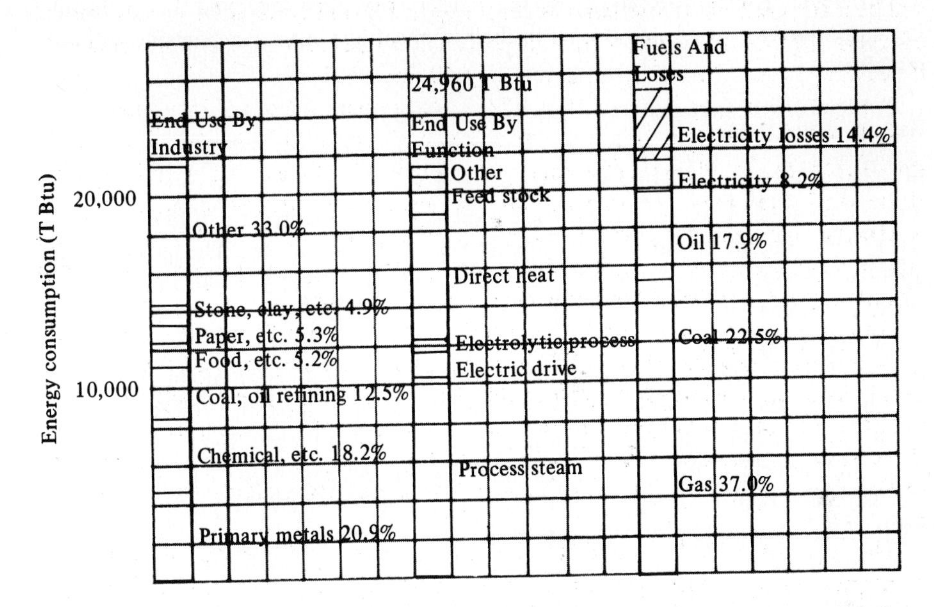

FIGURE 14-8 1968 industrial energy consumption by end use, industrial category, and fuel. Heat accounts for 70 percent of the end use.

most 70 percent of the total. The energy-intensive industries are those producing primary metals and chemical.

In the National Bureau of Standards report noted earlier it is stated that savings of as high as 30 percent are attainable in industry.[13] Although the actual strategies are not specified, a separate communication to industry does provide some general suggestions.[14] In addition to the obvious ones, such as turning off heat in unoccupied space, lowering thermostats, improving insulation and reducing venilation, several other conservation strategies were suggested. Among these were the recycling of waste (drain oil, sawdust) as a fuel supplement, recovering some of the waste heat from furnaces for preheating and other uses, and improving furnace efficiencies with automatic controls.

As part of a special effort to reduce electric energy consumption it was suggested that lighting be reduced overall, that the overall electric load efficiency be studied, and that efforts be made to shift as much of the load as possible to off-peak hours. This could be accomplished by starting air conditioning and heating early in the morning before work and shutting them down before closing, for instance, as well as by shifting working hours themselves.

In these suggestions we see a mixture of easy and hard strategies. Reduction of ventilation losses, repairing of insulation, weather stripping, reduction of lighting,

[13]C. A. Berg, *op. cit.*

[14]Frederick B. Dent, Secretary of Commerce, "A Message to American Industry." Reprinted in *Conservation and Efficient Use of Energy,* Part I, No. 14 Joint Hearings on Government Operations, June 19, 1973, p. 214.

and lowered thermostat settings are immediately applicable here. The adjustment of load to demand, the efforts to change the efficiency of plant load, and the rescheduling suggestions are somewhat deeper in their implications but still achievable without remodeling.

Recycling and Reuse

In addition to improved utilization efficiency in the direct use of energy in industry, it would be possible to bring about significant savings of energy, and of raw materials, by recycling more of the energy-intensive products of industry. The most obvious opportunities for the application of this tactic are in the primary metals industry, where 21 percent of industrial energy was consumed in 1968.

As an example of what could be gained by recycling efforts in this area, we will look at two examples; automobile manufacture and the production of aluminum. According to Hirst,[15] the 8.82 million automobiles manufactured in 1968 required about 1000 T Btu for their production, or about 110 M Btu per automobile. This is in rough agreement with the 4,000 to 6,000 kw-hr in Chap. 4. In Chap. 4 it was also estimated that recycling of metal in the automobile can save 10 percent of the input energy. In addition, it was pointed out that if car manufacturers were to produce automobiles that lasted longer, much more significant energy savings could be achieved. Such change in manufacturing strategy will be a disruptive strategy, however, as it would shift manpower from assembly lines to maintenance.

The aluminum industry is another prime target in the recycling effort. The production of 1 ton of aluminum requires about 75 M Btu, about two thirds of it as electrical energy. To reprocess aluminum scrap requires 8.5 M Btu ton, one ninth as much. In 1969 the aluminum industry consumed 285 T Btu in producing 3.8 million tons of aluminum and 7 T Btu in producing 860,000 tons of aluminum from scrap. Thus, only little over 2 percent of the total production came from recycled materials. If an additional 10 percent of the demand for aluminum had been satisfied by recycled material the savings would have been 25 T Btu. Since two thirds of this energy saved would be electric, the primary savings would have been greater. Two thirds of the 25 T Btu corresponds to 4.9 B kw-hr. At the 1968 heat rate of 10,500 Btu per kw-hr this amounts to 52 T Btu of primary energy to which we must add the remaining one third nonelectric energy for a total of 68.8 T Btu saved.

Aluminum production is one of the fastest-growing industrial categories. The production capacity of United States plants increased by 50 percent from 1960 to 1968. Increased use of recycled aluminum thus offers increasing savings in the future.

A similar example exists in the paper industry. Pulp made from recycled waste paper consumes only one fourth the energy of pulp made from virgin wood. The percentage of waste paper used per pound of pulp, however, decreased during the past decade from 28 percent in 1957 to 21 percent in 1967. In 1967 some 94 billion pounds of paper were produced at an average of about 11,700 Btu per pound. A ten percentage point increase in the amount of waste paper used from 21 to 31 percent would have saved about 1000 Btu per pound or about 94 T Btu, a significant saving, 9 percent of the industry's total.

[15] Hirst, *op. cit.*

For an estimate of what the realistic 1980 savings are for the industrial sector we are again dependent on the Shell Oil Company report.[16] Their conservative prediction is that 1.7 Q Btu out of a 1980 industrial total of 26.8 Q Btu could be saved, 6 percent of the industrial total, and about 2 percent of the United States total in that year.

ENERGY CONSERVATION BY LOWERING DEMAND

We have concerned ourselves so far with the results of conservation strategies. We have asked and answered the question: In what ways can we increase the efficiency of energy utilization and how much energy can be saved by this increased efficiency? We must turn now away from the ends to the means. How can these desired ends be brought about most effectively?

Before we discuss the motivating strategies which have been suggested, we must examine the end itself—energy conservation—in a broader context. We must be sure we know why we want to conserve energy and be aware of all the other cross-linkages in the economy which will receive their own signal when the tag marked "energy" is pulled.

Reasons for conserving energy change as we move our attention along in time from the immediate short-range Energy Crisis to the intermediate one, and then to the long-range one. In the short view we must conserve energy because we do not have enough of it to continue running our economy as we have been. Lowered thermostat settings and speed limits were designed to make sure that there was enough fuel in the winter of 1973 to 1974 to keep the houses warm and the cars and trucks running.

The reasons for energy conservation in the intermediate-range crisis are somewhat different. They are derived not from actual shortages but from the rising costs needed for investment in more utility plants and Mideast oil purchases, as well as the social and environment costs measured in terms of acres strip-mined, beaches polluted, and air fouled. There is enough fuel of all kinds to last us several decades, but its cost is increasing. The goal of conservation is to make sure that our energy is spent wisely.

We cannot at this stage see into the long-range crisis when the fossil fuels will become too scarce to be burned, and, consequently, we cannot project conservation into that era. We can only insist that reasonable percentages of the economic gain we obtain from energy expenditure in the next decades be plowed back into energy research and development to make certain that those new sources are ready when needed.

As to the cross-linkages within the rest of the economy, a careful examination of these are beyond the scope of this book. We will point to some of the more obvious ones. The Depression, as we have seen from earlier data, saved a considerable amount of energy. Similarly, the recession, or at least the decline in industrial productivity which we experienced from summer 1969 to spring 1972 saved, it is estimated, 8 Q Btu of energy.[17] It is obvious that all energy savings are not "good" by definition.

[16] Shell Oil Company, *op. cit.*

[17] I. M. Stelzer, in testimony before the Energy Conservation Hearings of the Committee on Interior and Insular Affairs. Printed in *Energy Conservation,* Part I, Serial No. 93-7 (92-42) March 22 and 23, 1973.

In the same light, we have pointed to increases in energy consumption, those caused by automobile emission controls, for instance, or by the choice of a cooling tower over straight-through cooling, which are wise additional expenditures. Nonetheless, it does seem clear that for the next decade, and probably the one beyond that, energy conservation strategies of the kinds we have described in this chapter can be important components of the national energy policy, along with increased production and technological advances. It is thus important that we examine briefly some of the mechanisms available to us for motivating the implementation of those strategies.

The Carrot and the Stick

Among the several motivating mechanisms available to bring about energy conservation are two which could be classified under the preceding heading: the carrot supplied by the market place, and the stick of government regulation and taxation. If we let the various forms of energy reach their natural prices, free from present distortion, the efficient uses we desire will naturally occur, we are assured. Homeowners will add insulation because they will save money as well as energy by doing so; small cars, car pools, and mass transit will play more important roles in transportation because they will be economical alternatives; gas heaters will replace electric ones for the same economic reasons.

Taxation, the stick, will have to be used to force energy producers and consumers to add the social and environmental costs of energy into the price of their products. The air, land, and water can no longer be considered free dumping-grounds; the costs of pollution must be "internalized," in economists' jargon, they must be added to the bill. In the same manner the user of the coal must pay for the black lungs of the coal miners and for reclamation of strip-mined land. The tax is not to be seen as punitive; it must accurately reflect the damage, and, by doing so, provide an economic incentive to improve coal mining practice, reduce pollution, put waste heat to beneficial use, and, generally speaking, make energy conservation and environmental improvement "profitable."

"Expensive Except for the Price." This quote from the testimony of Laurence Moss, Vice President of the Sierra Club[18] makes the point that the present pricing system is not working, it is not giving accurate signals to the consumer. Because of such subsidies as the depletion allowance and the other tax write-offs–including the royalities paid to foreign governments–gasoline is priced below the actual cost of producing it. When the tremendous subsidies to automobile transportation such as highways are added to the picture, our enormous waste of energy in transportation becomes less surprising.

Similar distortions occur in the electric utility industry. A great change has been taking place there. The most desirable way to operate a "natural monopoly" such as a utility is through marginal cost pricing, that is, by establishing a price which reflects the cost of the last unit produced. But the electric power industry has been a decreas-

[18]Laurence Moss, in testimony before the Energy Conservation Hearings of the Committee on Interior and Insular Affairs. Printed in *Energy Conservation*, Part I, Serial No. 93-7 (92-42) March 22 and 23, 1973, p. 225.

ing cost industry. The last unit produced (from the big new generator) has been cheaper than the average unit cost. If the price were set to the cost of this unit, the utilities would lose money. In order to compensate for this and to continue to take advantage of these decreasing costs, the regulatory agencies have allowed discriminatory pricing (different prices to the home and industry, for instance) and the so-called *declining block rates.* In this rate structure the charge to each class of customer decreases as consumption increases; in the first "block," electricity may cost 2.5 cents per kw-hr, the next 2 cents, and so on. Thus, demand is increased, the production system is enlarged, and the average costs to all decrease.

It now seems apparent that electric power generation is no longer a declining cost industry. Rather, it appears that all costs—costs of fuel, of labor, of capital—are increasing. Not only are costs rising, but the economies of scale are no longer there; the efficiency decrease with time which we noted in Chap. 6 has flattened out; the larger plants are not significantly more economical than the last group built. Increased demand, particularly increased peak demand, means that older less-efficient units are called into service along with the inefficient peaking devices such as diesel engines and gas turbines described earlier.

Discriminatory pricing and the declining block structure are now anachronisms. They are designed to create demand. Many economists are calling for the real application of marginal cost pricing, not only in the electric power industry but the other energy industries as well. If natural gas sells for 45 cents per million cubic feet, on the average, when it costs 1 dollar per million cubic feet to produce it from the new wells, then the consumer is not getting the message. He increases his demand for cheap gas and requires more of the expensive gas to be produced. If he had to pay 1 dollar per million cubic feet, he would use gas more efficiently and make more realistic demands.

There are obvious problems with pricing at the marginal cost since the gas from old wells and electricity from new generating plants is really cheaper and windfall profits would obtain from overall price increases; but there are ways to handle these. Taxation is one—raising money, perhaps, for research and development of new energy sources which have not had their depletion allowances and other subsidies to attract investment. It has also been suggested that something like a *reverse declining block* structure be applied to the rates of electric energy; that the first block, which would be large enough to take care of typical residential need, would be priced well below the marginal costs, to make sure that the poor aren't penalized. The big users would pay the high prices.

The reliance on the marketplace, on marginal cost pricing to lower demand, hinges on what is called the *price elasticity* of energy: will higher prices lower demand? It is clear that the consumption of individual fuels is price sensitive (see discussion of the electric case in Chap. 11). It is not as clear that energy itself as a commodity will show these same effects. Observation of consumer practices in energy-poor countries, however, suggests that energy consumption is responsive to price. In any event, it does seem reasonable to price energy at its real cost and see what happens.

Consumer Information. Along with a pricing policy which reflects the real costs must come a new level of consumer energy-awareness. The eventual success of the we have examined will depend on a reorientation of consumer think- :osts to life-cycle costs. We have already pointed to examples of the g consequences of this. A well-insulated house costs more, but the more than make up the difference in four or five years; a more ef- ioner has a higher initial cost but pays for itself. We have described ı the lower initial cost has been used to promote energy (wasteful electric resistance heating, for instance) and the barrier that higher initial costs has set against the development of solar heating.

If the American consumer is to turn from energy-wasteful to energy-conserving pricing, he will need information. He will have to have accurate data on the fuel economy of his car, on the heating and cooling costs of the insulation options available to him, and on the furnaces and air conditioners he might buy. Each appliance he buys should have its energy cost labelled so that he can calculate its real cost over its period of use. This may take federal regulations and assistance, as well as the cooperation of industry and trade organizations. It involved the consumer in the conservation decision through his pocketbook rather than forcing him through rationing and allocation. In the long run, it seems the surer way.

SUMMARY

In this chapter we have surveyed some of the conservation strategies which could be employed to reduce energy consumption. We concentrated our attention on strategies which might be implemented in this decade to reduce consumption and thereby reduce energy importation and the environmental destruction that unhampered growth will bring.

We looked first at the transportation sector, which is our least efficient end-use category, and found that the most productive strategies called for decreased use of the automobile for urban transportation and of the truck for intercity freight. We also saw the desirability of increasing load factors in all of transportation. We outlined strategies which could cut the 1980 projections for transportation's energy consumption in half, and lower the total by 12 percent.

In the residential area the most profitable conservation target was home space heating, in which lowered thermostat settings, more insulation, and decreased air infiltration resulted in large savings. The conservative estimate here was for a savings of 14 percent of the residential total and 2 percent of the national total. We can add to these savings some fraction of the savings obtainable by switching from electric resistance heating and water heating to gas, by using more efficient air conditioners, and abandoning frost-free refrigerators. These and the other techniques we listed, if rigorously applied, would have saved 33 percent of the residential total in 1968. Since many of them will be even more appropriate in 1980, it is not too unrealistic to hope for 25 percent success which would save an additional 8 percent of the residential total and, perhaps, an additional 1.5 percent of the United States total. Conservation in the residential sector could be reduced overall it appears, by 3.5 percent.

The commercial and industrial areas are much more diverse, and savings are correspondingly more difficult to quantify. In the commercial area, however, we described strategies for saving on space heating and cooling, and lowered illumination levels which could reduce the 1980 projected commercial consumption by 9.5 percent and total United States consumption by 1 percent. We described some of the general areas of conservation in industry that might bring 6 percent reduction in that area and a 2 percent overall savings. Thus, it appears not unreasonable to hope for an overall reduction of 18 to 20 percent in all categories.

It is important to point out that many of the strategies are aimed at reducing the consumption of electric power. In particular, they are designed to avoid the very wasteful step of turning fuels to heat energy to electricity and then back to heat. Each kilowatt-hour of electric energy saved is 10,000 Btu of primary energy conserved for other purposes.

Finally, we looked briefly at some of the changes in the economic system which could provide "pocketbook" motivation for conservation. In particular, we passed along the urgings of economists, that energy prices must be freed of the distortions of of subsidies (such as the depletion allowances) and should also be forced to include the social and environmental costs their production incurs.

We put forward a belief that if consumers receive true signals on energy costs and are provided with accurate information on the energy consumptions of their purchases (homes, automobiles, or electric toothbrushes, for instance), they would make wise, that is, energy-conserving, decisions. It is, after all, on such a belief in the efficiency of accurate information, that this book itself rests.

15

The Road to 1985

We began this book with the description of an energy-rich future. The 1985 which we sketched has twice the present electrical generating capacity; 40 percent of this capacity is nuclear. The breeder reactor has begun commercial operation, opening up plentiful U^{238} as a resource, while fusion has begun its move out of the laboratory and into the pilot plant stage. The other fuels are also in plentiful supply; oil and gas shortages are a thing of the past, although 40 percent of the oil and 20 percent of the natural gas are imported. Coal is again a major fuel; it is not only burned in the generation of electricity, but is also the raw material for new fuels such as solvent refined coal, and liquid and gaseous fuels as well.

The 1985 we pictured is a possible future. It lies within the range of options that the facts of this book encompass. Is it the probable or preferred future? That is another question—a question of goals, values, and mechanisms. We must now turn, briefly, in this closing chapter, to that question and the ones which derive from it: How do we select our preferred future? How do we get there?

We have been marshalling throughout this book as thorough a description as we could put together of energy supply and demand during the next few decades, and the environmental consequences of the production and consumption of that energy. It is from that list of existing resources, new technology, and new sources of energy, that the real picture of the future will be constructed. The chapters on environmental impact are there to guide us in our selections.

The next step would be to choose, to make value judgments, to weigh, for instance, the need of coal for electrical power production against the cost of gutting our Western prairies. Before we can rationally examine these options, however, there is one more demand, and one further resource—money—which must be examined.

THE COSTS OF ENERGY

All through this book we have emphasized that the cost of energy, after many decades of decrease, has started to rise. In a very real sense we have apparently come to the end of another frontier, this one, the frontier of free energy. From now on we will have to be prepared to pay more for it. Much of the cost increases will be brought about by the exponential growth of demand. We need more oil, so we will have to import it. We need to quadruple our electrical generating capacity by 1990, so we must enter on a crash program of plant and facilities construction. In a secondhand way, that same demand has brought us to the point of no return in our use of the environment. We are overloading many of its important cycles. We now must begin to pay for the use of the environment as a dumping ground; it is no longer free. We must also begin to pay for repair of existing damage.

It should be clear that these costs (energy prices, oil and gas importation, new energy investment, and costs of environmental protection and repair) are not independent. All are part of the increase in the cost of energy, and environmental protection, in particular, adds to the requirement for investment capital. We will, however, for the sake of simplicity, treat each one more or less independently.

Energy Prices

In Chap. 6 we emphasized that the steadily decreasing cost (in constant or current dollars) of electric energy has been a major cause of the rapid growth in its demand. The electric power industry has been a "decreasing cost" industry (see Chap. 14). In such an industry there are strong economic incentives to increase consumption, by preferential rate structures, for instance.

Figure 15-1 shows the cost of the fossil fuels (at the source) in both constant and current dollars. Since the constant dollar price is perhaps the surest measure of cost we see that all the fuels have been bargains, and in the case of oil and coal, increasingly good bargains throughout most of the 1960s.

The bargain days, however, are over. Between October 1973 and January 1974, for example, the posted (tax reference) price of Middle Eastern crude oil rose from $3.01 to $11.65 per barrel. This is only one component, but the major one, of the total price. All in all, the F.O.B. price of oil from that area probably rose $5.50 to $6 per barrel during those four months.

The sensational prices of 17 to 22 dollars per barrel which were quoted during the embargo were not typical, but represented desperation prices paid by small firms and utilities for oil that either was not embargoed or was slipped through the embargo.

The costs shown in Fig. 15-1 are for energy at its source, mine or wellhead. What is of more concern to us are the costs to the consumer. These data are provided in Fig. 15-2. These "average residential prices" are in current dollars, and the Consumer Price Index (CPI), set at 100 for 1972, is provided by comparison. The CPI provides a measure of the cost of all consumer items. What we see from Fig. 15-2 is similar to the picture presented in Fig. 15-1: bargain days in the 1960s coming to an abrupt end. In the 1970s all the fuels and electricity are rising more sharply than the CPI. The increases in the three fuels are particularly startling.

All predictions are that these increases will continue. The NPC projections of the average percentage increase in prices of fuels over the period 1970-1985, are[1]

Oil at the wellhead:	up 60 to 125%
Gas at the wellhead:	up 80 to 250%
Coal at the mine:	up about 30%
U_3O_8:	up about 30%

[1] A Summary Report of the National Petroleum Council's Committee on U.S. Energy Outlook, *U. S. Energy Outlook,* (Washington, D.C.: National Petroleum Council, December 1972).

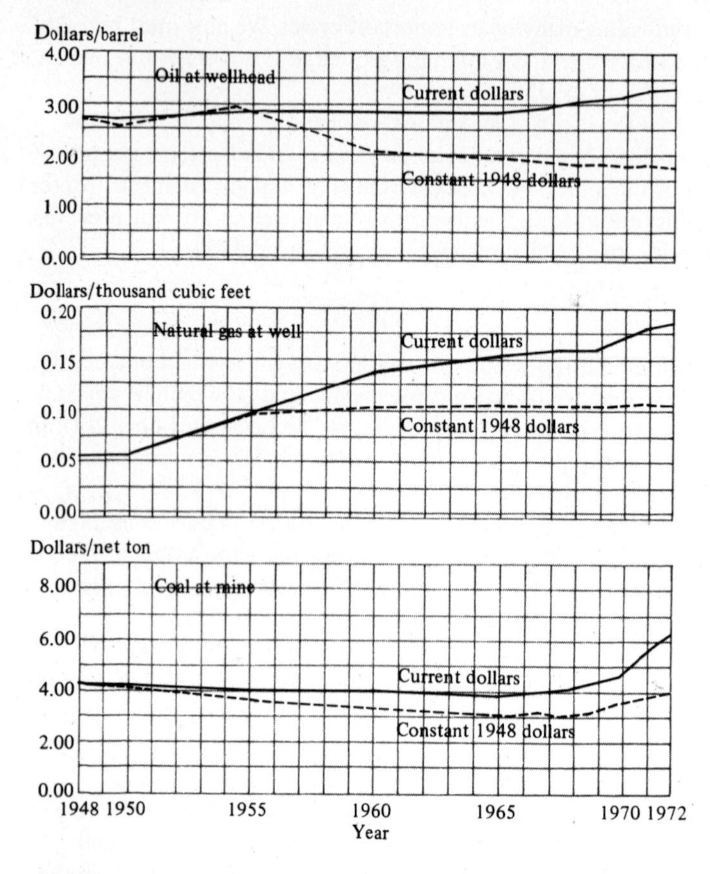

FIGURE 15-1 Average value of oil, gas, and coal in current and constant dollars. The cost of energy from the fuels has dropped in the case of oil and coal and remained more or less constant for natural gas since the mid-1950s. Even in current dollars energy was a bargain until the beginning of this decade. (Source: *Resources of the Future, Annual Report,* 1973.)

These price increases are in terms of 1970 collars; they are not escalated by inflation. They are production prices and do not take into account increased costs of distribution, transportation, or environmental protection costs.

We are so newly entered into this period of increasing prices that it is difficult to tell what the long-range effects will be. In particular, we do not yet have sufficient data to say whether the increasing prices will stimulate conservation practices. The experience with the electric utilities during the conservation measures of 1974 cause us to view even this as a mixed blessing. Voluntary conservation of electrical energy in many states has been so successful that the utilities profits are falling below acceptable levels and several of them are now asking for additional rate increases to make up for these losses.

It is likely that the traditional market pressures will have some effect, that rising prices will cause the fuels to be used more efficiently, and, at the same time, make the energy market more attractive, encouraging inventiveness and investment in new sources. Nevertheless, we face a costly decade or two.

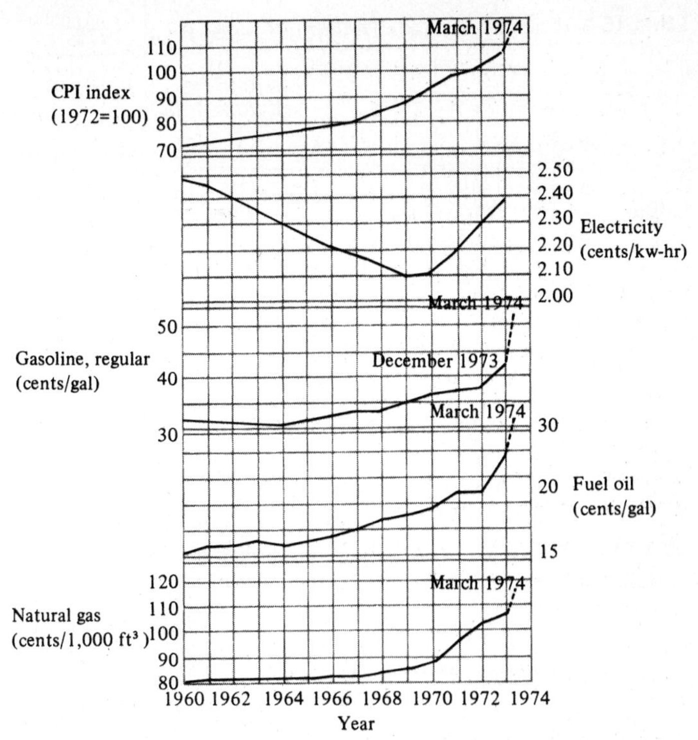

FIGURE 15-2 The average residential prices of energy (including taxes), with the exception of electricity, rose slowly over the last decade. In the last few years, however, these current dollar prices have turned sharply upward. (Source: U.S. Bureau of Mines.)

THE IMBALANCE OF ENERGY TRADE

An important cause of the rising prices is the insufficiency of domestic petroleum supplies. As we summarized in Chap. 11 we can expect to be importing in the best projection (the NPC Case 1) 30 percent of our oil and 12 percent of our natural gas in 1980; these percentages are 66 percent and 18 percent in the worst projection (Case 4). This imported energy will not come free. In 1970 the United States payment for imported oil was 3.4 billion dollars and for gas, 0.2 billion dollars. That same year we exported oil and coal which earned 1.5 billion dollars; thus, the net deficit from energy purchases alone was 2.1 billion dollars, almost half the 1970 over-overall balance of payments deficit of 4.7 billion dollars.

A glimpse at the future outflow of dollars is provided in Table 15-1 in which the 1975 and 1985 balance of trade deficit for energy fuels is summarized for the two extreme cases, 1 and 4 of the NPC study. We see that in 1975 the projected deficit is already 4 to 6 times the 1970 deficit and could, in the worst case, go as high as 15 times the 1970 deficit by 1985.

There is rising concern over the effects of these huge dollar exchanges in the financial market. In a recent study T. R. Stauffer, of Harvard's Center for Middle Eastern Studies, examined the effects of the 40 billion or so dollars per year surplus oil money

Table 15-1 Balance of Trade Deficit in Energy Fuels, 1975*

Type of fuel	1975 (billion dollars) Case 1	1975 (billion dollars) Case 4	1985 (billion dollars) Case 1	1985 (billion dollars) Case 4
Oil imports (delivered)†	10.9	14.6	5.4	29.1
Natural gas and LNG imports	0.5	0.5	4.9	5.4
Total energy fuels imports	11.4	15.1	10.3	34.5
Oil exports	(0.4)	(0.4)	(0.4)	(0.4)
Steam coal exports	(0.2)	(0.2)	(0.3)	(0.3)
Metallurgical coal exports	(1.3)	(1.3)	(2.1)	(2.1)
Total energy fuels exports	(1.9)	(1.9)	(2.8)	(2.8)
Total energy fuel deficit	9.5	13.2	7.5	31.7

*A Summary Report of the National Petroleum Council's Committee on U.S. Energy Outlook, *op. cit.*

†Including synthetic gas feedstocks.

flowing into the oil-producing countries. With interest and existing balances, he projects a total surplus of about 300 billion dollars by 1980. As large as this is, the total financial market of the industrial nations seems capable of absorbing it. The seven major industrial countries absorbed 350 billion dollars in 1971 and the 1974 estimate is 450 billion dollars. The total size of the financial market in these countries was 4 trillion dollars in 1971 and should be 6 trillion dollars in 1974. Stauffer's conclusions were:

> With respect to the industrialized countries, the general issues may be summarized as follows
>
> *1* The one-time inflationary contribution of higher oil prices is about 2 percentage points.
>
> *2* Benighted bilateralism—that is currency devaluations or trade promotion programs—invites retaliation and is not fruitful.
>
> *3* "Oil deficits" can be offset over the next several years only through compensatory capital flows, *not* on trade account.
>
> *4* The "oil surpluses" will constitute about 7 to 10 percent of new debt issuance in the OECD countries and by 1980 might accumulate to 2 to 4 percent of global financial markets (excluding real estate).
>
> These higher prices do entail real transfers of real resources of purchasing power from oil consumers to oil exporters, but the global economy does not appear to be untowardly burdened. The relatively small magnitudes facilitate adjustment. The key question, therefore, is not one of magnitudes, but of whether the surpluses can be rechanneled back to the debtor countries with minimum friction.[2]

These surplus dollars and the high prices of oil they represent will have more serious effects on the underdeveloped countries, a cause of international concern. These increases in energy cost will probably cause increases in the costs of raw materials for these countries. An enlightened and redesigned foreign aid program with larger input from the oil exporters will also be needed.

[2] T. R. Stauffer, "Oil Money and World Money: Conflict or Confluence?" *Science,* **184**: 321, April 1974.

Demands for Investment Capital

In Chap. 6 we pointed out that the electric utility industry, if it is to carry out the quadrupling of capacity projected for the period 1970-1990, will require 400 to 500 billion dollars in new investment capital. To put this in perspective, the United States financial market in 1971 absorbed 145 billion dollars in new money (loans, for example). The total size of the 1971 United States market was estimated at 2.73 trillion dollars or 2,730 billion dollars. The incremental increase due to utilities investment is thus about 20 percent of the 1971 total. Spread over the 20-year period the requirements are thus large, but not impossible to attain.

The increases in production which are required for the other fuels will also make their own demands on capital. Not only will money be needed for oil and gas drilling, for instance, but for pipelines and tankers, ports, refineries and all the other ancillary activities associated with production, transmission, and distribution. To the direct costs of traditional fuel production we must add the cost, for instance, of gasification plants and of the facilities to process oil shale.

A summary of the cumulative capital requirements of energy industries over the period 1971-1985, from the NPC Report[3] is given in Table 15-2. Only the two extreme cases (1 and 4) are considered. These data include an estimate for the utilities investment over that period which is about one half the FPC estimate for the period 1970-1990 we quoted earlier. The estimates of Table 15-2 do not include capital needs for marketing or distribution; the latter is a significant omission in the case of electricity and gas.

It is interesting, but not surprising, to note that the capital requirements for the optimistic Case 1 are almost 100 billion dollars greater than the pessimistic Case 4 projections. The requirements for tankers and terminals, however, in the Case 4 projection, which relies on heavy importation, is 23 billion dollars as against 2 billion dollars in the Case 1 projection. To handle the projected imports we must not only build super tankers, but deepwater port facilities to unload them.

The totals, again in the 400 to 500 billion dollar range are, in comparison with the total investment figures we quoted earlier, not overwhelming. Again, the story is: "We can have the energy, but we will have to pay for it."

Paying for Pollution

There is a fourth category of real financial need that will make unprecedented demands on our capital in the next two decades, the cost of pollution control and abatement. The Council on Environmental Quality (CEQ), in its fourth annual report (1973), has analyzed these costs in some detail and breaks the overall costs down under four headings: (1) damage costs, due to illness and property damage; (2) avoidance costs, which take into account the costs of traveling further to find a clean swimming beach, for instance; (3) transaction costs, the costs of making and enforcing regulations; and (4) abatement costs, to reduce pollution through sewage treatment plants or emission control devices, for example.

Firm data are difficult to come by in most of these areas. In Chap. 7, for instance,

[3]A Summary Report of the National Petroleum Council's Committee on U.S. Energy Outlook, *op. cit.*

Table 15-2 Summary of Cumulative Capital Requirements U.S. Energy Industries, 1971-1985*

Type of fuel	Supply cases (billions of 1970 dollars) 1	4
Oil and gas		
Exploration and production	171.8	88.0
Oil pipelines	7.5	7.5
Gas transportation	56.6	29.5
Refining†	19.0	38.0
Tankers, terminals	2.0	23.0
Subtotal	256.9	186.0
Synthetics		
From petroleum liquids	5.0	5.0
From coal (plants only)	12.0	1.7
From shale (mines and plants)	4.0	0.5
Subtotal	21.0	7.2
Coal‡		
Production	14.3	9.4
Transportation	6.0	6.0
Subtotal	20.3	15.4
Nuclear		
Production, processing, enriching	13.1	6.7
Total all fuels	311.3	215.3
Electric generation, transmission	235.0	235.0
Water requirements	1.1	0.7
Total energy industries	547.4	451.0

*A Summary Report of the National Petroleum Council's Committee on U.S. Energy Outlook, *op. cit.*

†Based on maximum U.S. requirements, some of which may be spent outside the U.S.

‡Cases 1 to 4 do not include capital requirements for coal production for synthetic fuels. These requirements in billions of 1970 dollars are as follows: Case 1: 2.0; Cases 2/3: 0.8; Case 4: 0.3.

we quoted a CEQ estimate of air pollution damage amounting to 16.2 billion dollars in 1968. In this latest report, they also provide an estimate of 24.9 billion dollars as the damage cost from air pollution in 1977 if there were to be no pollution control. Water pollution damage, on the other hand, has not been estimated. It is apparently the case, however, that recreational needs (swimming, fishing, boating) are dominant here. Transactional costs are often ignored, and while they do not fall into the billion dollar range, they are considerable (see Table 15-3).

The major costs are, of course, abatement costs. Major items are emission controls on power plants and automobiles, water treatment, and solid waste disposal. As we see in Figs. 15-3, 15-4, and 15-5 in which the expenditures of state and municipal governments, the federal government, and private industry, are presented, the costs are rising swiftly. From these figures we can also see the relative amounts of these expenditures in the various pollution categories.

The complete CEQ projection of the costs of pollution control is given in Table 15-4. This table includes capital as well as operating and maintenance costs and projects 274.2 billion dollars as a cumulative total over the period 1972-1981.

Table 15-3 1972 Federal Transaction Cost of Environmental Progress (millions of dollars)*

Program	*Air*	*Water*	*Land*	*Other*	*Multi-media*	*Total*
Research and development	136.5	70.2	26.3	82.3	32.8	348.1
Planning	2.2	19.8	0.2	3.5	1.2	26.9
Monitoring and surveillance	29.0	30.8	4.4	18.4	2.0	34.6
Administration, standard setting, enforcement	82.2	72.8	0	16.6	27.4	199.0
Other	15.0	92.5	7.5	83.4	4.5	202.9
Total	264.9	286.1	38.4	204.2	67.9	861.5

*Source: Office of Management and Budget, "Back-up Information for Special Analysis S, Federal Environmental Programs" (unpublished report of the Office of Management and Budget, 1972).

This is a staggering amount of money, but several considerations soften the blow of this estimate. These are the costs of control; we must remember, as well, that the damage we are trying to abate is also large. The total cost of control in 1971, 11.3 billion dollars, for instance, is only a little more than half the estimated damage *from air pollution alone* in 1968. A second perspective is given by a comparison with the total GNP over the same period; the 274.2 billion dollars is only about 2.5 percent of the projected total GNP.

Who Pays? The payment for this environmental bill will be spread in different ways through the income ranks of our society. A breakdown of the costs between the public and private sectors is given in Fig. 15-6. The private sector will pay about two thirds of the total bill.

There is, indeed, a concern that the environmental bill not be loaded onto the

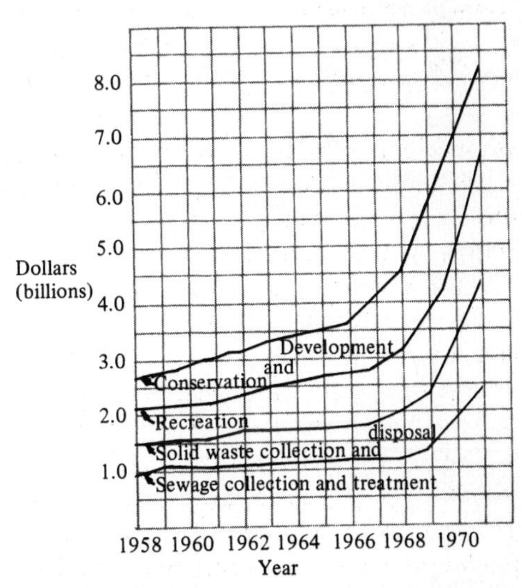

FIGURE 15-3 Most of the state and municipal environmental money from 1958 to 1971 has been devoted to waste disposal and recreation. (Source: "Environmental Quality" Fourth Annual Report of The Council on Environmental Quality, 1973.)

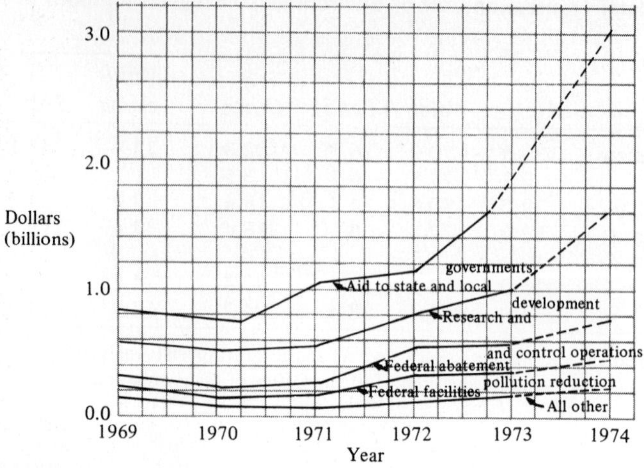

FIGURE 15-4 Federal pollution abatement expenditures, much of it channeled directly to state and local governments, has doubled since 1968. (Source: "Environmental Quality," *op. cit.*)

backs of the poor. They will not benefit from many of the control measures as much as the rich. They do not, for instance, use the recreational facilities as intensively as do the rich and are not as concerned (although they are more affected) with air pollution. Generally speaking, the taxes, through which the public sector finances these control measures, are progressive, that is, the rich pay more. The private expenditures are regressive since they are added to basic services and prices; consequently, the poor pay a higher share. Assessment of the overall impact of the *additional costs* in 1976

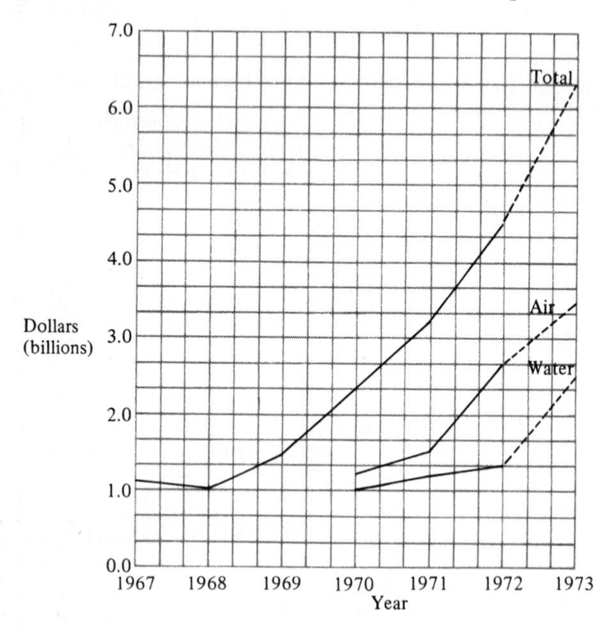

FIGURE 15-5 Industrial investment in pollution control has quadrupled since the passage of the environmental protection acts of the late 1960s and early 1970s. (Source: "Environmental Quality," *op. cit.*)

Table 15-4 Estimated Total Pollution Control Costs (billions of dollars)*

Pollutant/medium	1971			1981			Cumulative–1972-1981		
	O&M†	Capital costs	Total annual costs	O&M†	Capital costs	Total annual costs	Capital invest-ment	O&M†	Total annual costs
Air pollution									
Public	0.2	<0.05	0.2	1.0	0.2	1.2	1.4	7.1	8.4
Private									
Mobile	1.1	<0.05	1.2	6.2	4.3	10.5	27.1	39.1	58.8
Stationary	0.4	0.3	0.7	4.2	1.5	5.7	11.4	27.0	38.4
Total	1.7	0.3	2.1	11.4	6.0	17.4	39.5	73.2	105.6
Water pollution									
People									
Federal	0.2	NA	NA	0.2	NA	NA	1.2	2.8	NA
State and local	1.2	3.8	5.0	2.6	7.0	9.6	47.2	20.0	76.9
Private									
Manufacturing	0.4	0.3	0.7	2.2	1.5	3.7	12.3	15.8	27.5
Utilities	0.2	0.1	0.3	1.6	0.9	2.5	6.8	10.9	16.5
Feedlots	0	0	0	<0.05	<0.05	<0.05	0.2	<0.05	0.2
Construction sediment	<0.05	NA	NA	<0.05	<0.05	<0.05	0.3	<0.05	0.2
Total	2.0	4.2	6.0	6.6	9.4	15.8	68.0	49.5	121.3
Noise									
Commercial jet aircraft	NA	NA	NA	NA	NA	NA	(0.4-1.6)	NA	NA
Radiation									
Nuclear power plants	NA	NA	NA	<0.05	0.2	0.2	1.2	<0.05	1.0
Solid waste									
Public	1.0	0.2	1.2	1.7	0.4	2.1	2.6	13.8	16.3
Private	2.0	<0.05	2.0	3.1	0.1	3.2	0.3	25.2	25.5
Total	3.0	0.2	3.2	4.8	0.5	5.3	2.9	39.0	41.8
Land reclamation									
Surface mining	NA	0	NA	0.8	0	0.8	0	4.5	4.5
Grand total	6.7	4.7	11.3	23.6	16.1	39.5	112.0	166.2	274.2

*Source: Office of Management and Budget, "Back-up Information for Special Analysis S, Federal Environmental Programs" (unpublished report of the Office of Management and Budget, 1972).

†Operating and maintenance costs.

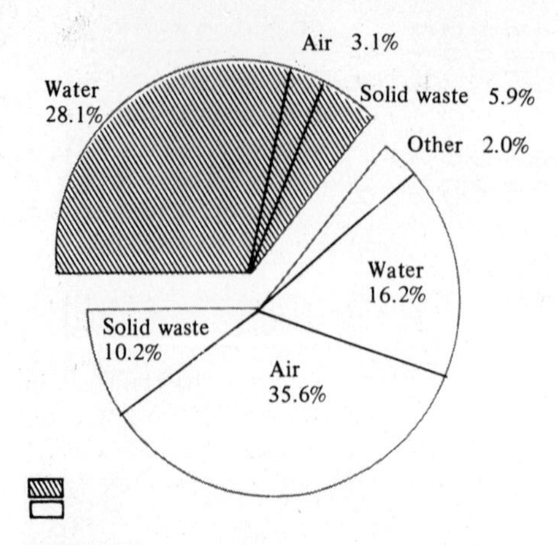

FIGURE 15-6 Distribution of total environmental expenditure from 1972 to 1981. The private sector carries about two thirds of the total environmental expenditures and the largest of that goes to air pollution abatement and control. Most of the public expenditure is earmarked for water pollution, largely for the upgrading of sewage treatment. (Note: Figures do not total due to rounding.) (Source: Environmental Quality," *op. cit.*)

as a function of income is given in Fig. 15-7. The conclusion is that the overall impact is slightly regressive and the poor pay a slightly larger share—an unfortunate, but not unusual result. In the light of this analysis, the suggestion (Chap. 14) of a change in electrical rate structure which lowers the cost of the first "block" of energy, seems even more reasonable.

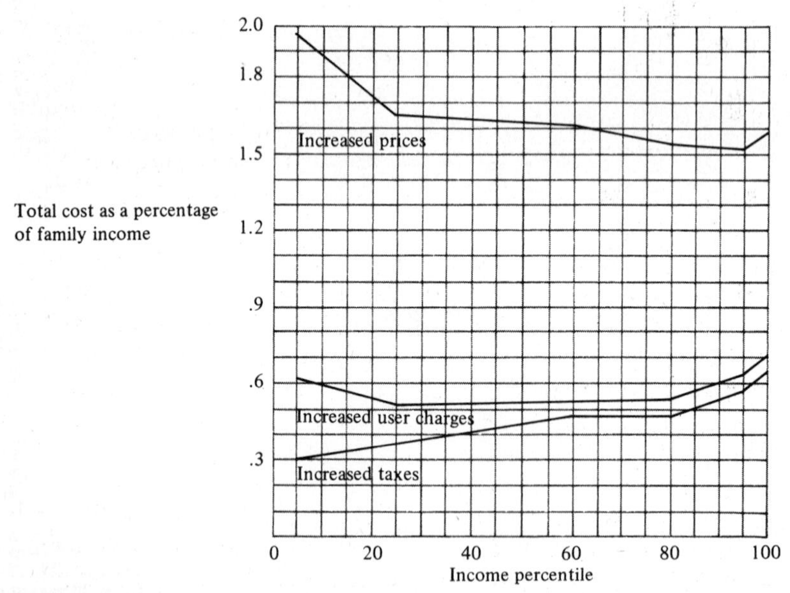

FIGURE 15-7 Incidence of incremental pollution abatement costs as a percentage of family income, 1976. Increased prices add significantly to the pollution costs paid by low-income families while a slightly greater percentage of the costs borne by taxes are paid by the rich. The user charges add higher percentages at the extreme income ranges. (Source: "Environmental Quality," *op. cit.*)

The Energy Cost of Pollution Control

The "diggers and drillers," as well as others who speak against any limits on energy growth, remind us constantly that we will need energy and more energy to clean up our environment. It will be useful, therefore, as we attempt to assess our overall needs, to try and estimate just how much energy that will take.

There have been several estimates made of these energy needs. Eric Hirst, of the Oak Ridge Environmental Program, has recently published a fairly detailed analysis[4] which considers five types of control: (1) automobile emission controls; (2) secondary level sewage treatment; (3) solid waste management; (4) air pollution control at stationary sources; and (5) waste heat dissipation with cooling towers at power plants.

All of these control techniques consume energy. In order to meet 1976 standards the automobile emission controls cause an estimated fuel penalty of 15 to 30 percent. (We compared this loss to other losses in Chap. 4.) Hirst assumes a 20 percent loss.

It also takes energy to treat waste water. Such treatment is conventionally categorized into *primary treatment,* which consists of screening sedimentation, sludge digestion and disposal; *secondary treatment,* by "activated sludge" in which the biological processes of assimilation and degradation are carried out in a controlled manner; and the advanced *tertiary treatment,* which produces essentially clean, reusable, water as an effluent. (At present, 10 percent of sewage is untreated, 30 percent receives only primary treatment and 60 percent secondary treatment. Tertiary treatment is rare; it is used, for instance, to preserve Lake Tahoe.) The EPA goal is for secondary treatment of 90 percent of urban waste. The Hirst estimate is that 290 T Btu would have been needed for secondary treatment of *all* waste water in 1970, 0.47 percent of the 1970 energy total. Tertiary treatment would have doubled this figure.

Energy is required for solid waste treatment, mainly for transportation to land fills. Energy is available from solid waste if it is burned to produce steam for electricity, or if the metals are recycled.

Energy is also needed to run the air pollution control devices on stationary sources, the electrostatic precipitators which remove the particulate matter and the scrubbers, and the other devices which remove the sulphur oxides (see Chap. 7). The electricity needed to operate a 95 percent efficient precipitator is about 0.1 percent of the power plant output, while an 80 percent efficient removal (limestone scrubbing) of the sulphur oxides would require 4 percent of the output. Hirst estimates that 90 percent efficient particulate removal and 70 percent efficient sulphur oxides removal on all producing sources would take 26 B kw-hr per year or 560 T Btu at 1970 levels. (The Btu total includes some heat energy.) To take possible stricter controls and other sources into account, Hirst raised this estimate by 50 percent to 39 B kw-hr and 840 T Btu, 1.2 percent of the 1970 total energy.

The energy cost of waste heat dissipation comes mostly from the lowered efficiency of the turbine (due to the fact that the condensers are not cooled as much as in straight-through cooling and, therefore, T_{out} is higher in the efficiency equation). Some energy is also consumed by fans and pumps. Hirst estimates that if 50 percent of all power plant capacity in 1970 had been equipped with wet cooling towers, the

[4] E. Hirst, "Pollution Control Energy Costs." Presented at the American Society of Mechanical Engineers Annual Meeting, Detroit, Michigan, August 1973.

electricity consumed would have been 14 B kw-hr, 1 percent of the total 1970 electricity consumption. In terms of primary energy, this would have amounted to 160 T Btu, 0.2 percent of the 1970 total.

Table 15-5 summarizes the energy cost of these controls in terms of both primary energy use and of electrical energy, and compares them with the 1970 energy consumption data. These hypothetical 1970 measures would have increased total energy consumption by about 4 percent and the consumption of electricity by about 6 percent. These figures are in general agreement with other estimates; the NPC, for instance, estimated about 4 percent for 1980 controls and Marc Ross, in a study undertaken for the Ford Energy Policy Project[5] estimated (in the preliminary report) that 3 percent of the total energy will be needed for pollution control.

We cannot leave this subject of the energy cost of pollution control without commenting on some additional energy-saving measures. As we have already pointed out, some control measures *produce* energy rather than consume it and, in a rational future, these would be included. Hirst has included some of these in his study. If 10 percent of the solid waste in 1970 had been incinerated and the heat used to generate steam for electricity, 25 B kw-hr of electrical energy, or 1.8 percent of the total consumed, could have been produced. This would have contributed 0.4 percent of the total primary energy that year.

Hirst also considered the energy savings from recycling. If one third of the United States products of steel, aluminum and paper, had been obtained from recycled scrap instead of virgin material, we would have saved 440 T Btu, 0.7 percent of the 1970 total and 23 B kw-hr of electricity, 1.7 percent of the total electric energy in 1970. If these savings had been taken into account, the pollution control needs would have been lowered to 2.6 percent of the 1970 total energy and the electric energy to 2.3 percent of the 1970 total.

It appears from these studies that environmental control will not require an overburdening fraction of either our financial or our energy resources.

[5] Marc Ross, "Exploring Energy Choices," a preliminary report of the Energy Policy Project, Ford Foundation, 1974.

Table 15-5 Energy Use for Hypothetical 1970 Environmental Qualtiy Measures*

Control measure	*Primary energy (T Btu)*	*Percent of total*	*Electrical energy (B kw-hr)*	*Percent of total*
Auto pollution control	1100	1.63		
Secondary sewage treatment	290	0.43	25	1.8
Solid waste management (Landfilling)	75	0.11	3	0.2
Control of stationary services	840	1.25	39	2.8
Waste heat dissipation (Cooling towers)	160	0.24	14	1.0
Total	2465	3.66%	81	5.8

*Source: Hirst, "Pollution Control Energy Costs," *op. cit.*

THE ROAD TO 1985

Within the next few years this country will set out, guided either by conscious new decisions or by the inertia of past practices, on the road to 1985. How will that road be chosen? We return now to the question of goals and mechanisms.

The process of selecting goals and establishing the mechanisms for their achievement is a three-step process. The first step is to describe the options; it involves a description not only of the specific goals (more air conditioning, urban transportation by electric automobile, and so on), but also a description of the entire ecological web that unites these specific goals. The second step involves values, the cost-benefit judgments. With this step the goals are selected, and, in the selection of these goals, all the costs we have described in the preceding chapters must be taken into account and their payment arranged. The third step is to select and bring into being the mechanisms of change which will cause us, as a society, to arrive at those goals. This book has, of course, been designed to contribute importantly only to the first of these steps: to describe the energy options and the web of environmental consequences which connect them. From this perspective we will look briefly to the other steps, but first a review of the description.

Energy and the Environment: A Summary

The major fact which should have emerged from the previous chapters is that for the next 50 years, at least, this country's pursuit of wealth and well-being will not be energy-limited. There are enough Btu stored in the fossil fuels which still underlie our part of this continent to carry us that far, even if we allow our consumption of energy to continue to grow at the same exponential rate of the past few decades.

We have, however, reached, or at least come close to the limits of the capacity of the environment to absorb the damage caused by the production and consumption of that energy. As we saw in Chap. 11 the fossil fuel resources exist in amounts which are inversely proportional to their environmental threat. Natural gas, the cleanest of the fuels, is in shortest supply, and coal, which has the most environmental impact, is the most plentiful.

For every million tons of coal strip-mined we wreck 100 to 200 acres of land. (This assumes production at 5,000 to 10,000 tons per acre, which is about the national average.) The 300 to 500 million tons which will be strip-mined per year this next decade will cost us, therefore, 30,000 to 100,000 acres of ruined land per year. The other half of this coal, if produced underground, is costly in human terms, bringing on death, injury, and disease. Underground mining, as presently practiced, also damages the land and water through acid water drainage.

Coal is troublesome at the other end of its energy journey also. We have described the air pollution damage from the sulphur oxides. Without proper emission controls, it is estimated that the sulphur oxides from each 10 million tons of coal burned in a densely populated area causes 2,000 or so additional cases of respiratory diseases in children under 5 years of age, and 2,000 early deaths in people over 60.[6] There is also

[6]Ross, *op. cit.*

a new worry over the effects of very small particles which escape not only the electrostatic precipitators but also the defenses of the respiratory system. These particles lodge in the lungs and carry injurious substances such as sulphuric acid to this sensitive area. It may, in fact, turn out that burning oil is as hazardous as is burning coal in this respect.

Oil is not trouble-free, in any event. Its hazards come at all stages in its journey from well to combustion. The new rush to tap the oil under the continental shelf will lead to more of those rare but disastrous spills such as the one which occurred in Santa Barbara; perhaps this time it will happen off Atlantic City. We will also pay in damage as well as in dollars for the huge imports of oil we anticipate in this next decade. We will need supertankers and deep ocean ports. The capacity, of a supertanker is about 250,000 tons or 1½ million barrels of oil. (Presently the largest supertanker, the *S. S. Globtik,* has a capacity of 476,292 tons. The Torrey Canyon, whose wreck caused such a spectacular oil spill, had a deadweight of 118,285 tons.) The estimated importation of 3.5 million barrels per day in 1980 will require two of these ships to dock and discharge their oil somewhere in this country every day. It will not only be the spectacular wreck (which will, no doubt, occur), but also the accidental and purposeful dumping and the spills during transfer which will add part of this oil to our sea and foul our coastlines. Some rough figures on the amount we can expect are given in Table 15-6 which also includes the spills anticipated from the Outer Continental Shelf (OCS) development and from tanker shipments at the Trans-Alaskan Pipeline (TAPS).

If, instead of supertanker shipments from the Middle East, we look to the mountains of oil shale in the Colorado Green River basin, we put oil alongside coal as a land use threat, for we must then strip-mine these mountains. As we have said, the 1 million barrels per day of oil from shale, which is the 1985 target, will require the mining and disposal of about 1.5 million tons of waste rock per day. The reclamation of that beautiful country will be more difficult than in the rain-watered Appalachians.

Even after we have the oil and it has arrived at its point of conversion, it still has

Table 15-6 Quantities of Potential Oil Spills and Discharges into U.S. and World Waters (barrels per day)*

Source	*Major accidents*	*Minor and moderate spills*	*Ballast or waste water treatment*	*Total*
Outer Continental Shelf	280-1060	52	18	350-1,130
Trans-Alaskan Pipeline Shipment	68-384	6	26-540	100-416 (ballast treatment) 614-930 (100% load-on-top treatment)
Imports				
U.S. waters	68	3	0	71
World's waters†	384	6	540	930-2,100

*Source: Energy Policy Project, "Exploring Energy Choices," (Washington, D.C.: Ford Foundation, 1974).

†All (in barrels per day) likely to be spilled and discharged for each 2,000,000 barrels per day produced and transported.

much potential for damage. Burned to produce electric power, it can produce sulphur oxides or the particulate pollution we have just mentioned. Converted to gasoline and burned in our almost 100 million automobiles, it causes the photochemical smog we discussed in Chap. 7.

There are other forms of environmental damage that all the fuels share. They all produce heat and carbon dioxide when burned. When they are burned in power plants much of that heat is wasted and dumped into rivers, lakes, and oceans. In the Midwest, Southwest, and North Central Plains areas, in particular, there is much concern that the quantity of cooling water needed will tax the supply capacity. We are also watching carefully to see what effect the slow buildup of carbon dioxide in our atmosphere will have.

The transmission and distribution of energy in our country, along with its conversion, have caused problems of land use. We have seen the projected increase in the number of generating plants (see Fig. 1-5). Half of these plants will be coal burning and require 300 to 400 acres apiece. All will need siting near a lake, river, or the ocean and thus degrade recreational land, to say the least. If, instead, we turn to cooling towers we can anticipate local climatic changes. The transmission lines (500,000 miles in 1990) add their land use problems, as do the new refineries which must be built, the coal gasification plants, the nuclear fuel reprocessing plants, and so on.

The nuclear plants, which offer us some relief from the air pollution and land use problems of coal burning plants, bring different problems of their own. Their deadly wastes must be stored for at least a few centuries. We are faced, for the first time in mankind's history, with the necessity of guarding and monitoring hazardous materials for several generations. In addition, the use, in the high temperature gas reactors and the breeders, of a fuel from which an atom bomb can be constructed, makes the world that much more vulnerable to the insanity of fanatics. And behind all this is the unknown threat of catastrophic accident.

The Question of Values

The "fourth law of ecology," in Barry Commoner's words, is: "There is no such thing as a free lunch."[7] Throughout this book, and particularly in the summary of this last chapter, we have tried to total up the price of our energy menu. It will be costly, but so are lunches; nonetheless, we must continue to consume energy. We must make the energy policy decisions which are necessary to the continued growth of our society.

We must decide, for instance, on how much energy we will demand, on the reliance we will place on each of the various fuels and on nuclear energy to satisfy these demands. We must decide what level of environmental protection we are willing to pay for. We must decide what recreational and aesthetic values we are willing to set aside, and how deeply we are willing to commit our future generations to the hazards of radioactive wastes, reactor accidents, and the new and various dangers from plutonium. It has been the primary goal of this book to assist the reader through the first step, the description of options. We have provided as comprehensive a cost analysis

[7] Barry Commoner, *The Closing Circle,* (New York: Knopf, 1972).

as is possible within the limits of these pages and our present knowledge. This second step, however, the cost-benefit judgement and the decisions on goals, is not one that we can guide. It is a responsibility not of the scientist, but of the citizen.

The distinction between scientist as guide and scientist as advocate is one that is often blurred. It is not easy for the public to determine, for instance, when Erlich the Scientist stops explaining the causes and dangers of overpopulation, and when Erlich the Advocate begins to press for a particular course of action. There are, in fact, two quite different attitudes among scientists about their public role. The first of these, the "Voice of the Dolphin" role,[8] is that taken by the highly respected "scientist's lobby," the Federation of American Scientists, and by the informal alliances of well-known scientists who occasionally buy newspaper ads, to inform the public of critical situations, contribute testimony at nuclear reactor hearings, and perform other "public services." The fact that important scientific voices are raised on each side of these scientific-political questions (Pauling versus Teller in the nuclear weapons controversy and Seaborg versus Gofman and Tamplin on nuclear reactors, for instance) reduces, and, in fact, threatens, the public impact of science.

A different interpretation motivates the "science information movement," for which the Scientist's Institute for Public Information[9] serves as a prototype. These scientists see themselves as partners only in the public enterprise. In the cost-benefit analysis of possible courses of action it is their duty (as well as privilege) to provide a clear and objective evaluation of the costs: describing what is certain and admitting what is uncertain. The other half of the calculus, the benefit analysis, involves decisions on values and value judgments which are not within the realm of science.

In the preceding chapters we have tried to stay within these limits, to stick with "the facts ma'am, nothing but the facts." Our bias and preferences have, no doubt, come through at times: we were, perhaps, somewhat overly critical of the growth of consumption in the electrical industry; perhaps a bit too worried about nuclear energy; maybe overzealous in pressing for a reduction in urban automobile transportation. But we tried.

Facts, however, are not enough. There must be action; the reader may now be asking, "What should we do?" This question deserves an answer. We will, therefore, in these closing paragraphs, shed the cloak of objectivity and set down, briefly, some of the possible courses of action that seem to us to hold the most optimistic promise for the future.

Mechanisms: What We Should Do

We have neither the space nor the talent and experience to set forth a new "United States Energy Policy."[10] From the several years of study that have accompanied

[8] Named after the book of that title by the late Leo Szilard. In that story the scientists subtly take over the decision-making responsibility and lead the world to peace and prosperity.

[9] This organization is a confederation of local groups of scientists and citizens who inform themselves and others on socio-political problems with scientific content. It is the publisher of *Environment* Magazine and has offices at 68 E. 30th St., N. Y., N. Y.

[10] The results of the Ford Foundation Energy Policy Project will begin to appear in print in Fall 1974 and Spring 1975 and should provide much stimulation of policy discussion.

the writing of this book, however, we have reached two major conclusions: (1) we need real energy options beyond fossil fuels and fission; (2) we must enter into a determined program of energy conservation.

What specific actions can be suggested for the fulfillment of these goals?

More Options. We have come into the energy-short 1970s with only one alternative to the fossil fuels: nuclear energy. Because of this it has been difficult to rationally examine the pros and cons of the reactor; and since not even the reactor was ready to take over when the Middle Eastern sheiks and the environmental disposal services raised their prices, we have found ourselves in a "sellers' market."

The blame for this lack of planning must be shared. The Federal Government does not have a carefully and intelligently constructed energy policy. In particular, it has not properly discharged its responsibility to lead the way by financing the necessary research and development. The history of research and development in this past decade is shown in Table 15-7. All bets until the 1970s were on the nuclear entries. This nuclear dominance has been extended not only through financial support but also through optimistic projections of plentiful, cheap energy. The publicity campaign, with its predictions of cheap nuclear-produced electricity, has effectively killed off all but the most determined exponents of alternative sources of power: wind power, solar power, geothermal, and the like. It is only now, as we begin to see that 2 cents per kw-hr nuclear electric energy is an illusion, that proposals for solar energy farms,

Table 15-7 Federal Energy Research and Development (R & D) Funding, FY 1969 through FY 1975 (million dollars)*

Item	*FY 1969*	*FY 1970*	*FY 1971*	*FY 1972*	*FY 1973*	*FY† 1974*	*FY‡ 1975*
Coal resource development	23.3	30.4	49.0	76.8	85.1	164.4	426.7
Petroleum and natural gas	13.5	14.8	17.5	23.8	18.7	19.1	41.8
Nuclear fission							
LMFBR	132.5	144.3	167.9	237.4	253.7	357.3	473.4
Other civilian nuclear power	144.6	109.1	97.7	90.7	152.8	173.2	251.7
Nuclear fusion							
magnetic confinement	29.7	34.3	32.3	33.2	39.7	57.0	102.3
laser-pellet	2.1	3.2	9.3	14.0	35.1	44.1	66.3
Energy conversion with less environment impact	12.3	22.9	22.8	33.4	38.4	65.5	178.5
General energy R & D	3.0	4.2	8.7	15.4	8.1	28.8	59.8
Conservation including transportation	N/A	N/A	N/A	N/A	32.2	65.0	115.7
Solar	N/A	N/A	N/A	N/A	4.0	13.8	50.0
Geothermal	N/A	N/A	N/A	N/A	4.4	10.9	44.7
Total	361.0	363.2	405.2	524.7	672.2	999.1	1810.9

*Source: Energy Policy Project, *op. cit.*

†Estimated

‡Proposed

Note: Data for FYs 1969-1972 are not exactly comparable to FYs 1973-1975, as additional programs have been included.

solar-sea plants, and wind-driven generators are once again receiving serious attention. Our narrow vision has, however, cost us 20 years of development time. We must now, as an informed and concerned public, insist that the government continue and expand its proper role of financing these uncertain areas. It is from them that we will fuel our future.

We must also place some of the blame for our present condition on the energy companies (for the most part, these are oil companies). In their search for profit above all, they have not served us well. In particular, it appears that they have not used the subsidies we have given them, such as depletion allowances and tax incentives, to increase our supply of inexpensive energy. Instead, they have made us increasingly dependent on foreign oil and used their profits, too often, in nonenergy related financial enterprises in this country. It is now time to reexamine their role and their regulation.

Energy Conservation. Along with a determined and adequately financed search for energy options, we must embark on a program to use the energy we have as efficiently as possible. This thrust must come along several lines: we must change the way in which the marketplace deals with energy; we must change our own national consciousness regarding energy; and we may need to make some specific changes in our energy-related technology. In addition to these three, we can lower our consumption of energy by closing new ecological cycles with more recycling of resource materials. We will now describe some of the necessary components of each of these four thrusts.

The *marketplace* can be used to reduce our demand for energy. What will be important to achieve is a price structure which includes all the costs. In the past we have subsidized the consumption of energy in two very important ways: (1) we have, with depletion allowances and declining rate structures, for instance, made it possible to sell energy at less than its real cost of production; and (2) we have not charged ourselves for the use of the environment in producing that energy and in disposing of the waste products of its consumption.

In Chap. 14 we described some of the suggestions for pricing energy more realistically. In particular, we should expect to pay what it costs to produce, transport, and distribute it. The environmental costs must be added in a different manner; pollutant control and abatement is not profit-making and so a structure of taxes and penalties must be built which put environmental costs on the same basis as the other costs. This is much preferred, the economists assure us, to a system of regulations and prohibitions. The marketplace, with its built-in feedback, choices, and incentives is our most flexible mechanism for carrying the message that energy is not to be wasted.

We must also develop a *conservation ethic*; we must (and have already begun to) bring about a change in our view of energy similar to that which the closing of the land frontier brought to our view of land. The "dig and drill" approach is as outmoded in the energy business as "slash and burn" is in agriculture. We must educate ourselves to view energy as a scarce commodity, to prize gasoline economy over horsepower, to look at fuel consumption rather than just the cost of air conditioning and heating plants. We must temper comfort with conservation when we set thermostats, and convenience with conservation when we travel. We must, in other words, go beyond a willingness to "pay the price" for energy. We must, in some instances, treasure it beyond price.

With this new energy consciousness may well come specific *changes in technology* away from some of those which are energy-intensive. As we have pointed out in earlier chapters, much of the increase in energy consumption by industry has resulted from the substitution of energy-intensive products such as plastics, synthetic fabrics, and aluminum for the natural and energy-economic materials such as paper, wood, cotton, and wool, and for less energy-intensive materials such as steel or brick. A new energy ethic, reinforced by realistic marketplace pressures, may cause us to return to a higher proportion of these substances to clothe, house, and serve our needs.

That these substitutions may have economic motivation as well, has begun to gain support. A California hospital[11] has reported the savings of several dollars per day in the average patient's bill, by replacing disposable items, food trays and dishes, napkins, aprons, and even hypodermic (throw-away) needles, by those which can be reused. That part of this savings is due to more reliance on manpower (and womanpower) in this era of high unemployment, is also significant. Extension of this approach to schools and businesses, and a general retreat from the disposable life is attractive from both an energy and a valid waste-management point of view.

The fourth component of this determined effort to conserve energy is the *closing of ecological cycles.* We can reduce our need for more energy, and, at the same time, alleviate some pressing waste disposal problems, by returning used materials to the beginning of their use cycle as raw materials.

Recycling can save large amounts of energy, not only in the cases of steel, aluminum, and paper which we have described, but in oil, for instance, where a significant fraction of our needs could be met if used lubricating oil were cleaned and reused, or, if in the bottling of beverages, we were to make returnable bottles the preferred choice.

To encourage recycling we must also look to the marketplace for incentives and disincentives. We must examine the depletion allowance here also, for it makes the use of virgin pulp in papermaking, for instance, more profitable than the use of scrap paper. There are also tax inequities (as well as the host of oil additives) which work against the recycling of oil. Finally, the preferential freight charges which now favor the transport of raw materials over scrap and recycled materials must be changed.

We can, in this expansion of recycling, go beyond the return of used materials resources and use our growing pile of waste itself as a source of energy. In Chap. 12 we described some of the available technology for burning waste for steam-electricity, and for creating even gaseous and liquid fuels from it. All these recycling techniques pay out in two ways: they reduce the amount of raw materials we must take from our diminishing resources, and thus diminish the environmental damage from their production. In addition, they diminish the present overload of the environmental disposal systems. It will take changes in our economy, in our laws and regulations, and in our consumer habits to bring these new energy economics into practice. The pressure for these changes must, in the end, come from the political public.

[11] Alan Nadler, M.D., 830 Scenic Dr., Modesto, California, private communication.

SUMMARY

Much of this book was written during the tense fall and winter days of 1973 to 1974. The houses were cold and the gasoline lines long. The summer sun and the lifting of the Middle Eastern embargo have warmed us again, and the lure of vacations put the automobiles back on the roads. The twin crises of energy and environment are, however, still very much with us, and the most important battles are still before us.

The gravest danger of the Energy Crisis, which has so dominated our media during the past year, is that it will swing the pendulum of public support away from environmental concerns. That swing has already begun. The Alaskan Pipeline controversy was resolved on the basis of energy priority; energy considerations were an important part of the argument for postponing automobile emission controls and for removing air pollution restrictions on the burning of high sulphur coal. We are also the target of a determined and well-financed advertising campaign by the energy companies. Environmental protection is not one of the high priorities of the "diggers and drillers," that priority can only receive support from a concerned public. It was this public concern that forced the environmental revolution of the 1960s whose triumphs are now threatened.

It has been our major goal in this book to look at energy as a part of our environment, not as a threat to it; to show, as well as present knowledge permits, the various ways and the costs of satisfying our energy needs and, at the same time, preserving the environment which supports all aspects of our life. We are all on the road to 1985 together. What that year, and the few years until the close of this century can be for mankind, finds dual symbolism in the picture of 15-8, the sun and the sea. We can see in this scene the sun setting on a polluted sea and on a short, energy-rich episode in man's long history. We can also see in clear, unpolluted air, a rising sun which, with the sea, promises eons of plentiful energy. Both of these futures now lie before us.

FIGURE 15-8 The energy of the future may come from the sun through solar heating, from the the generation of electricity, and from the sea's rich resources of deuterium for fusion. (Source: Robert Greenler, University of Wisconsin.)

APPENDIX 1

A Vocabulary Guide to the Mathematics

Our intent in this book is to tell the full story of energy in as simple and straightforward a way as we can. We want to make the important facts about the production and consumption of energy and their environmental consequences accessible to anyone. We have, therefore, made a determined effort to limit the use of mathematical language.

It is not always possible, however, to talk in meaningful way about energy without using some numbers. In our discussion of supply-and-demand considerations, for for instance, numbers are essential; in fact, very large numbers are used to compare domestic supplies of oil with demand, for we are dealing with billions of barrels of oil, and quadrillions of Btu. (The various units of energy and power are examined and defined in Chap. 3 and Appendix 2.) We will use this appendix, therefore, as a vocabulary guide to these large numbers.

A second area of presentation in which we will need some simple mathematics is that of growth. The Energy Crisis is a crisis of growth. In its most elementary form, it is caused by demand growing faster than supply. We will not be able to discuss it with satisfactory crispness and precision unless we can find simple ways to characterize that growth. We will not be satisfied with recounting the past and describing the present; the real payoff comes from projecting some of the past and present trends into the future in such a way as to illuminate the choices which are before us. For that exercise we need to understand annual rates of growth, for instance, and doubling times. We will translate those terms from mathematics to English in this vocabulary guide.

A third way in which mathematics will intrude into our discussion is pictorially. Much of the data we wish to present is most efficiently shown graphically. Many of these graphs will be cast in the same famílar form as the stock market's ups and downs or the cost-of-living curves. The vertical axis which measures the increase (or decrease) of the quantity being studied will have a linear scale, that is, each vertical unit (each vertical inch on the scale, for instance) will measure the same amount of increase or decrease.

Unfortunately, we cannot handle all our data with linear scales. The range of growth in the per capita consumption of electrical energy for instance, is too great. The graph must go from 49 kw-hr per person per year in 1900 to more than 8,000 kw-hr per person per year in 1970. This is an increase of 160. To handle such large changes we will use a less familar vertical scale, the semilogarithmic scale and we will use this appendix to prepare you for that also.

There is an even more important reason for plotting much of the energy data on a semilogarithmic scale. Much of the growth we will be looking at is of a special kind known as exponential growth. A semilogarithmic presentation of this kind of growth makes it much easier to understand, for a quantity growing exponentially rises as a straight line in a semilog presentation. The vocabulary guide will interpret this pervasive exponential curve for you.

These ideas will constitute our main emphasis in this guide. We will end with some discussion of the phrases and sentences of mathematics, proportionality and equations, and their use. We will aim only at a rudimentary facility. Our intent is not to teach mathematics but only to translate the basic terms.

A CODE FOR LARGE NUMBERS

There is no way that we can escape the necessity of using large numbers in this book, for in this country millions of people use millions of tons of coal to generate trillions of kilowatt-hours of electricity. Since 1 trillion is written 1,000,000,000,000 we can justify turning to a code on the grounds of space-saving alone (although there are other reasons). As we have just indicated, 1 trillion is represented by the number 1 followed by 12 zeros. Our code will be based on this, on powers of 10. We are familar with the operation of squaring a number and know that 10^2 is 100. The 2 is called the power and 10^2 is "ten to the second power." Familar powers of ten are: $10^2 = 100$, $10^3 = 1000$, and $10^4 = 10{,}000$. We note that the power is equal to the number of zeros to the right of the 1; hence, our code is born. (We do have to add, by fiat, that $10^0 = 1$).

Table A1-1 summarizes the large numbers which appear in this book, expressed

Table A1-1 Large Numbers

Power of ten notation	*Word*	*Abbreviation*
10^0	one	na*
10^1	ten	na
10^2	hundred	na
10^3	thousand (kilo)†	k
10^4	ten thousand	10k
10^5	hundred thousand	100k
10^6	million (mega)†	M
10^7	ten million	10M
10^8	hundred million	100M
10^9	billion	B
10^{10}	ten billion	10B
10^{11}	hundred billion	100B
10^{12}	trillion	T
10^{13}	ten trillion	10T
10^{14}	hundred trillion	100T
10^{15}	quadrillion	Q

*na, no abbreviation.

†The terms in the parentheses are the common prefixes which we will use, kilowatts or megawatts, for instance, meaning thousands or millions of watts.

first as a power of 10, then as a word, and finally, in the last column, as the abbreviation which we will use throughout the text.

As an example of the way we will indicate such numbers, the total energy consumed in the United States in 1973 was 75,600,000,000,000,000 Btu or 75 quadrillion, 600 trillion Btu. We will write this in power of ten notation as 75.6×10^{15} Btu. The power, in this case 15, is the number of zeros to the right of the decimal point. We could write it equivalently as 7.56×10^{16} Btu or $75{,}600 \times 10^{12}$ Btu. However, to further simplify this notation, we will use the abbreviations listed in the last column of Table A1-1. This figure for total energy will then be either 75.6 Q Btu or 75,600 T Btu.

We stop our table at a quadrillion, 10^{15}, even though some of our numbers (the total amount of heat energy available from coal resources, for instance) are as large as 10^{18}. We will just write that as 1,000 Q.

The power of ten notation can take us in the other direction also, toward smaller and smaller numbers. For these numbers we use negative powers; 10^{-1} is one tenth, 1/10, or 0.1, and 10^{-2}, one hundredth, or 0.01, and so on. The code applicable here is also obvious. The negative power is the number of decimal places to the *left* of the first nonzero number; 10^{-1} is one decimal place (tenths), 10^{-3} is three decimal places (thousandths). Using this notation we write a number such as 0.00595 as either 5.95×10^{-3}, 59.5×10^{-4}, 595×10^{-5}, and so on. Small numbers, however, are not of the same general importance as are large numbers in this book, and we have not bothered to set up an abbreviation code for them.

SEMILOGARITHMETIC GRAPHS

Whenever a quantity to be displayed changes by more than a factor of 10 during a particular period we are examining, we begin to lose detail in presenting it on a linear graph. Figure A1-1, for instance, shows the growth in per capita consumption of electrical energy in this country over the period 1890-1970. In order to contain the very large numbers characteristic of consumption in the past decade we must choose a scale whose divisions are so large that they tend to mask the details of the fluctuations in demand in the early part of this century, when the demand was much smaller. We see more of these details, on the other hand, in Fig. A1-2, where the same data is plotted on a graph in which the vertical scale is logarithmic.

The basic idea behind the logarithmic scale comes from the powers of ten notation we have just discussed. If a quantity spans a range of sizes from 10 to 10,000, for instance, and you are interested in seeing the details of its fluctuations when its size is between 10 and 100 as well as when its size is between 100 and 1,000, then you have to allot the same vertical space on your scale to the 10 to 100 range as you do to the 100 to 1,000 range. In Fig. A1-2, you will notice that the vertical distances from 10 to 100, 100 to 1,000, and 1,000 to 10,000 are all identical. The logarithm is the appropriate mathematical quantity to accomplish this. You remember (perhaps) that the logarithem X of a number N "to the base ten," is the value of X that satisfies the equation

$$N = 10^{X}$$

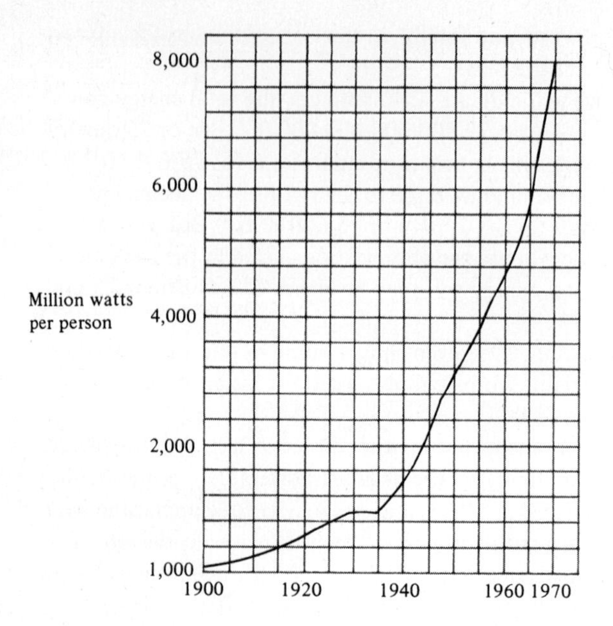

FIGURE A1-1 The per capita consumption of electricity has increased by a factor of 160 since 1900.

that is, the power to which 10 must be raised to equal the number N. Without worrying about the details of logarithms, which we will not need, we can see from this equation that the logarithm of 10 is 1 (since $10^1 = 10$) of 100 is 2 (since $10^2 = 100$) and of 1,000 is 3 (since $10^3 = 1{,}000$). Thus, if we plot as equal distances along a vertical scale the *logarithms* $X = 1, 2, 3$ rather than the numbers, we can obtain a scale that greatly compresses the data. All the numerical values between 1 and 10 fall into the first division, those between 10 and 100 into the second, and so on. The vertical scale in Fig. A1-2 is laid out in this fashion; equal distances are equivalent to equal logarithms rather than equal numbers.

EXPONENTIAL FUNCTIONS

There is an even more important reason for using the semilog presentation in this book (and elsewhere) and that is because of the great simplicity it offers in displaying quantities that are increasing in size, exponentially, with time. Such quantities are of great importance in our discussions of energy.

A quantity grows exponentially when the amount of its increase depends on its size. A familar example is a savings account. If you have N dollars at 6 percent interest per year, then after a year you have $N + 0.06N$, after another year $N + 0.06N + 0.06\,(N + 0.06N)$, and so on. Figure A1-3 shows the growth of a 1,000 dollar savings account at 6 percent per year on a linear scale. This is an exponential curve; the 1,000 dollars has grown to over 10,000 dollars in the 40-year period.

This same data is plotted on semilog graph paper in Fig. A1-4. The exponential growth curve on a semilog plot is a *straight line.* It can be shown, of course, that the

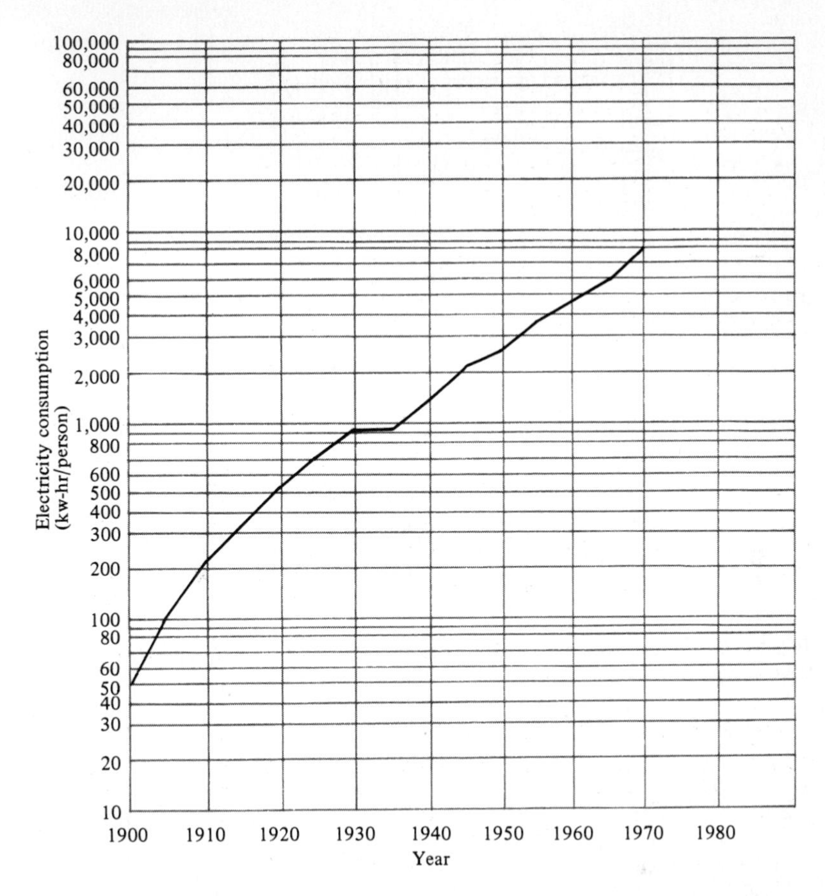

FIGURE A1-2 The data of Fig. A1-1, when plotted on a semilogarithmic scale, are somewhat easier to interpret. The straight line sections indicate exponential growth.

relationship between logarithms and exponentials is such that this has to be, but we will not take the time to develop that here. What we need to know is why a straight-line representation of an exponential is important and how it can be interpreted.

A straight line is the preferred representation of a growing quantity because it is straight. Thus, one can easily project it; one can see where it is going. We can project the savings account in Fig. A1-4 and see that it reaches 30,000 dollars after 58 years. We cannot project a curve such as that shown in the linear plot of Fig. A1-3 with such accuracy. A straight line has the significant advantage that it is determined by two points and can then be extended in either direction by that simple tool, the ruler. We are, therefore, at a considerable advantage, in dealing with these exponential curves of growth, when we plot them on semilog graph.

As we have said, a quantity grows exponentially when its amount of increase depends on its size. We should now state that condition more exactly. The example we have been using is the savings account and the *rate* of increase; the amount of increase per year depends on the size of the account at the beginning of that year. In mathematical terms this can be written

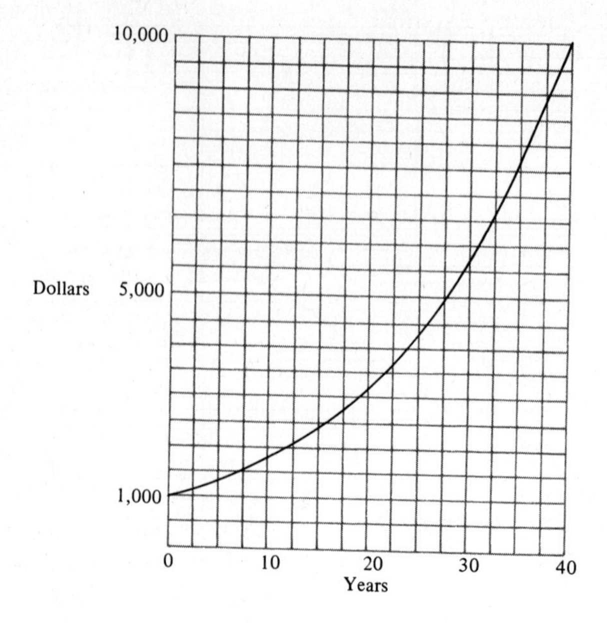

FIGURE A1-3 The growth of a savings account at 6 percent interest per year serves as an example of pure exponential growth, here plotted to a linear vertical scale.

$$\frac{\Delta N}{\Delta t} = rN \qquad \text{(A1-1)}$$

which reads: the change in N, (we use the symbol Δ to stand for "the change in") over the time interval Δt (the change in t), is equal to some constant r times N (where N is the size of the savings account at the beginning of the time interval). In this example the constant r is, of course, the yearly interest, 6 percent per year. This relationship tells us that the amount of growth per year, $\Delta N/\Delta t$, is dependent on (proportional to) the size N.

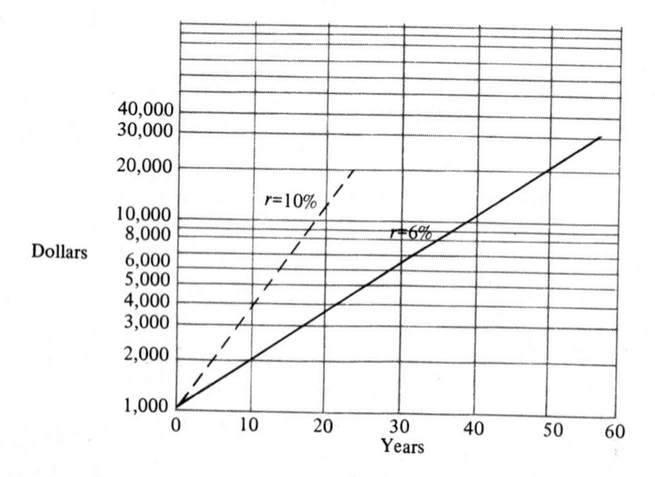

FIGURE A1-4 On a semilogarithmic vertical scale, an exponential curve becomes a straight line. The annual rate of growth (or the interest) determines the steepness of the curve.

There are many quantities of importance to us which grow in this way. The population does, for example, since the number of children born per year depends on the number of people. For less obvious reasons, energy consumption, pollution, and many others are growing more or less exponentially.

Doubling Times and Rates of Growth

We can see from the example and from Eq. (A1-1) that the constant r is of great importance in characterizing exponential growth; it determines how rapid the growth will be. In the case of a savings account, r is the interest (6 percent per year in our example). Generally it is the *annual rate of growth,* the annual percentage increase. If r is large, the straight line slopes up more steeply than if it is small. Thus, a savings account with $r = 10$ percent will follow the dotted curve of A1-4 (and the account reaches 10,000 dollars much more rapidly). The annual rate of growth is thus a most important quantity and we will often refer to it in the book's discussions of growing quantities.

A second and related measure of growth is the doubling time, the time that is needed for the growing quantity to double in size. This is also constant, as we can see from Fig. A1-4. The savings account at 6 percent interest doubles in 12 years, doubles again (goes from 200 to 400 dollars) the next 12 years and continues doubling every 12 years. The doubling time, therefore, is 12 years, a constant. The doubling time for the account at 10 percent interest is shorter, about 7 years.

We may, in fact, take as the most characteristic feature of an exponential curve of growth, that it has a constant doubling time. That time is an obviously important quantity. The fact that the doubling time of electrical energy consumption is 10 years is, as we point out, causing many of the difficulties of the Energy Crisis. Ten years is a short time in which to build the necessary generating and other facilities.

One final comment on the exponential curve of growth, is that the annual rate of growth [r in Eq. (A1-1)] and the doubling time are related. We will not derive this relationship (although it is a simple exercise in mathematics). An approximate statement of it is that

$$T_{\text{double}} = \frac{70 \text{ years}}{r\ (\%\text{ per year})}$$

In words, the doubling time in years (T_{double}) is equal to 70 years divided by the annual rate of growth in percent per year. We will leave it to the reader to verify this for the growth curves of Fig. A1-4.

The ease with which the doubling time, and, therefore, the annual rate of growth, can be read off of the straight-line, semilog plot of an exponential is a further reason for using that kind of a representation.

We should not leave this discussion of doubling time without emphasizing what an enormous change a doubling is. Let us take as our example the total consumption of energy in this country. This is increasing at an annual rate of about 3.5 percent. From Eq. (A1-2) this means that the doubling time is about 20 years (70 years/3.5 = 20 years). The interpretation of that statement is that during the past 20 years we have consumed as much energy as we did in our entire previous history, and that in the next 20 years we will again consume an amount of energy equal to the total previous consumption.

We can also get some visceral feel for the relation between the annual rate of growth and the doubling time by using an example from inflation. If a sack of food cost 10 dollars in 1960 and the rate of inflation is 7 percent per year, then the doubling time is 70 years/7 = 10 years, and that same sack of food will cost 20 dollars in 1970. Its price will have doubled in 10 years.

PROPORTIONALITY AND EQUATIONS

Our final translations from the mathematics are concerned with the phrases and sentences of mathematics. In the text we often use the phrase: "is proportional to." We can define exponential growth as that type in which the rate of growth of a quantity "is proportional to" the size of that quantity. Using the symbols we defined in Eq. (A1-1) we write this as

$$\frac{\Delta N}{\Delta t} \propto N$$

where the symbol $\propto$ means "is proportional to." What does this statement mean? It means that if you double N then you double $\Delta N/\Delta t$; if you cut N in half, you cut $\Delta N/\Delta t$ in half, and so on. Other examples would be the area of a circle which is proportional to the square of the radius of that circle, (Area $\propto r^2$) or the energy available from gasoline, which is proportional to the amount of gasoline you have (Energy $\propto$ gallons of gas).

A statement of proportionality tells us the nature of the dependence, for instance, that a circle of twice the radius of another has four times its area. It does not allow us to compute the area knowing the radius. For that we need an equation. To make an equation out of the area proportionality, we need a constant, which we will symbolize by k, and write

$$\text{Area (circle)} = kr^2$$

The constant k provides the numerical value. If we measure the area in square centimeters (by counting all the square centimeters inside a circle drawn on graph paper, for instance), measure the radius in centimeters, and square it, we can compute k by rewriting the equation

$$k = \frac{\text{Area}}{r^2}$$

We find, of course, in this example, that k is the constant π, 3.14.... In our earlier savings account equation the constant was r, the annual rate of interest. It is this constant that makes the units on either side of the equation equivalent. In Eq. (A1-1) the units of $\Delta N/\Delta t$ are dollars per year. On the other side, N has the units dollars. Thus, r is a pure number (6 percent, without units) *per year*.

In the energy from gasoline example, to make an equation out of that proportionality we must write it as

$$\text{Energy} = k \times N \text{ (gallons)}$$

The actual numerical value of k depends on the units we desire for energy; if we want Calories, then k = 33,000 Calories per gallon; if we want Btu, then k = 130,000

Btu per gallon. If we desired to find the amount of energy in Calories available in a tank, say 20 gallons, of gasoline, we would use the equation

$$\text{Energy (Calories)} = 33{,}000 \frac{\text{Calories}}{\cancel{\text{gallons}}} \times 20 \cancel{\text{gallons}}$$

$$= 6.6 \times 10^6 \text{ Calories}$$

Note that we have treated the units, gallons, like numerical quantities, and cancelled them, leaving only Calories on the right as units to equal the Calories on the left. This is a handy way of checking to see if we are using the correct units for k.

In Appendix 2 we provide all the relevant units of power and energy (and their definitions) as well as the conversion factors which allow for their interchange. From the numbers given, for instance, we can determine that the conversion factor between Calories and Btu, which are both units of energy, is

$$\text{Number of} \frac{\text{Btu}}{\text{gallon}} = K \times \text{Number of} \frac{\text{Calories}}{\text{gallon}}$$

$$K = \frac{\text{Number of Btu/gallon}}{\text{Number of Calories/gallon}}$$

$$= \frac{130{,}000 \text{ Btu/}\cancel{\text{gallon}}}{33{,}000 \text{ Calories/}\cancel{\text{gallon}}}$$

$$= \frac{130{,}000}{33{,}000} \frac{\text{Btu}}{\text{Calorie}}$$

$$= 4 \text{ Btu/Calorie}$$

Again, the cancelation of units guides our computation. With this technique and the tables in Appendix 2 the mystery of the units will unravel.

APPENDIX 2

Force, Work, Energy, and Power

What is energy? It has been defined in everyday terms as "the force that sets things in motion," or "the capacity to do work." Both of these definitions suffer from the same weakness: they are circular, they define energy in terms of words (force, work) which are themselves undefined.

If definitions are couched in familiar terms and if precision is not too important, this sort of circularity may not greatly impede communication. In science, however, where verbal statements usually have mathematical expression, precision of definition is of primary importance. To gain this necessary precision, terms such as energy, force, work, and the like are defined "operationally," that is, by the operation necessary to observe or measure them. In this appendix we will make some such definitions, review the relationships between these terms, and provide some of the quantitative expressions which allow us to represent them by numbers. We will also give, in the last sections, some representative examples of energy and power and provide a table of conversion factors between the various units in which energy and power are measured.

FORCE

We will begin with a definition of force. We have an intuitive understanding of this concept; we might say "we have some feel for it." With force defined we will link force and energy and go on to define work and power.

What is a force? A rough but operational definition is: a force is a push or a pull. This is not yet satisfactory, for we cannot translate that definition into numbers. The next question, then, is: What does a force do? We can think of three situations, at least, in which forces are involved: (1) static situations in which there is no motion (a person holding a suitcase or Atlas holding up the world); (2) situations in which there is steady motion (a horse pulling a wagon, an engine pushing a car); or (3) situations in which the motion is changing, either in speed or in direction (a rock falling, a car rounding a curve).

It was Newton, the great nineteenth-century English physicist who gave us convincing evidence and mathematical statement of the fact that a net force is only associated with the last of these, the change in motion, and that in the other examples the total force is zero. In all examples of no motion or of steady, straight-line motion, it can be shown that several forces are applied and that their sum is zero.

There is an opportunity to present much traditional physics here but we will reluctantly set most of it aside and try to stay on the trail of energy. We will only point out that force is a *vector,* that two numbers are needed to specify it:[1] the *strength* of the force (in pounds, for instance) and the *direction* of the force (up, down, north by northeast, and so on). Since both these quantities need to be taken into account, the addition of forces is a little complicated and we will not get into it here. One can easily believe, however, that two equal and opposite forces acting on the same object can cancel each other out, leaving zero net force. There are other geometrical arrangements of forces which can also sum to zero.

Much of Newton's original inspiration in his study of the relation of force to the change in motion came from the careful observation of the motion of a falling body (from watching an apple, we are told). As far as he could tell, the force of gravity (the earth's pull on an object) was constant, independent of its height above the earth's surface. When an object falls, its speed or velocity[2] is not constant: it increases continually as it gets closer to the earth. A constant force can not be proportional to a changing velocity.

The increase or the change of velocity is called *acceleration.* Mathematically, the acceleration is

$$a = \frac{\text{change in velocity}}{\text{time interval}}$$

(This is not a precise mathematical statement, and we get into trouble if the velocity is fluctuating too rapidly, but it will do for us here.)

When Newton measured the acceleration of the falling body, (this change in velocity divided by the time interval), he found that it was constant. This suggested that the force (a constant) might be proportional to the acceleration (also a constant). A large number of further experiments confirmed that this was the correct relationship, not just for gravitational forces, but for all forces. Newton proposed, therefore, that force was proportional to acceleration. In symbolic forms

$$F \propto a$$

If one quantity is proportional to another, it is also true that one quantity is equal to some numerical constant times the other, in this case, that

$$F = C \times a$$

Where F is the force, a is the acceleration, and C is a constant. This equation is a more powerful statement than the proportionality. If we can determine the numerical value of C, for instance, then we can compute a for any value of F.

Since force and acceleration are different physical quantities (pounds and feet per second per second, for example) the constant provides the matching that allows us to have the same units on either side of the equation. Its size is important; if it is big, then a small force can produce a big acceleration, and vice versa.

This last statement gives us a strong clue as to what the constant in the force-acceleration equation might be. What is it that determines the acceleration of an ob-

[1] Quantities requiring only one number are called scalars. Temperature, power, and energy are scalars, while force, velocity, and acceleration are vectors.

[2] In keeping with convention we will use speed to mean the magnitude (feet per second or miles per hour) and velocity to describe the vector which has direction as well as magnitude.

ject acted on by a force? It is the *mass* of that object, or, we would say less precisely, its "heaviness." We know that a person pushing a Volkswagen can produce more acceleration than he can pushing a Cadillac. The constant in the force-acceleration equation is the mass, which we will symbolize by m. Thus

$$F = ma$$

What is mass? We identified it with heaviness, but it is more than that; it is a measure of the quantity of matter in an object. It could be identified with the number of atoms in that object. It is an unchanging property of the object and, thus, the same anywhere in space. The mass of an astronaut is the same on earth and on the moon. (It is, however, different from the more familar quantity, weight; we will clear up that confusion in a moment.)

Since mass is a fundamental property, we need a way to measure it. (Counting the atoms would be cumbersome and we would still need to know the mass of one atom.) *Measure* means, operationally, "compare with a standard." (We measure lengths by comparing them with a ruler.) In the scientific system the standard mass is the *kilogram* (kg), a cylinder of metal of carefully specified dimensions. It is divided into the subunits, *grams*; there are a thousand grams in a kilogram. The usual way of comparing masses with this standard is shown in Fig. A2-1. We use a balance. We put the unknown mass on one side, and put enough grams (or kilograms) on the other side to balance it. Counting the grams (or kilograms) needed to balance determines the mass.

We have thus provided an independent way to determine the mass of an object. We can now use the relationship expressed by the Eq. (A2-1) to find a numerical value for force. We will use the falling body experiment as an example. If we take a kilogram mass, drop it, and measure its acceleration, we find a constant acceleration of 9.8 meters per second per second. We can now operationally define force with Eq. (A2-1), $F = ma$. It is defined because a prescription for its measurement is provided. The unit of force we chose is called a *newton* and has the dimension of mass times acceleration, or kg × meters/second^2.

$$1 \text{ newton} = 1 \text{ kg} \times 1 \text{ meter/second}^2$$

The force on a 1-kg mass dropped to earth is, therefore

$$F = 1 \text{ kg} \times 9.8 \text{ meter/second}^2 \quad \text{or}$$

$$F = 9.8 \text{ newtons}$$

The gravitational force on a mass at the surface of the earth is a familiar one; it is called the *weight* of that object. Thus, the weight of the standard kilogram is 9.8 newtons. Weight in the more familiar "English" system (soon to disappear, we hope) is given in pounds. We can determine the relation between newtons and pounds from our knowledge of the weight and mass of water. A liter of water has a mass of 1

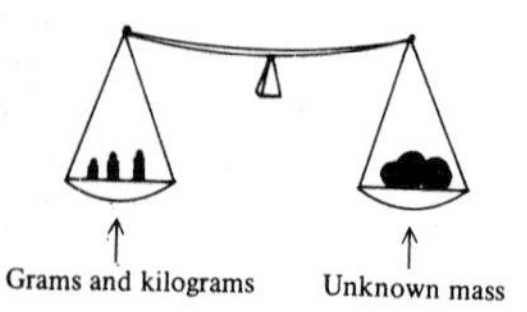

FIGURE A2-1 We measure mass by comparing with standard grams and kilograms on a balance. The mass of the standard is chosen by international scientific agreement.

kilogram and, therefore, a weight of 9.8 newtons. We also know that a liter is about a quart. Since "a pint's a pound the world around" and there are about 2 pints in a liter, 2 pounds are about equal to about 10 newtons. More precisely

$$1 \text{ pound} = 4.45 \text{ newtons}$$

We have now identified and given units to force and associated it with acceleration. Eq. A2-1 is our first important law and one of the two important characteristics of force. The other is its "pairness."

The "Pairness" of Forces

Forces always occur in pairs. Another way of saying this is: to exert a force you have to push or pull on something and when you do the something pushes or pulls back. Newton said it: for every action there is an equal and opposite reaction.

The evidence for this is all around us. It is most obvious in static cases; you push on a wall, the wall pushes back. You stand on a scale, your weight pushes down, the spring in the scale pushes back up.

It is this "reaction" force which is responsible for motion. You push your foot against the ground and it pushes back: you walk. An automobile tire turns, pushing against the street, the street pushes back and the car moves. This is tricky stuff and you may want to go into it a little more deeply. If the forces are equal and opposite then why don't they balance to zero? The answer is that they act on *different* objects, the tire pushes on the road, the road pushes on the tire. We sketch the static case and that of the walking man in Figs. A2-2 and A2-3. In the first of these there is no motion, no acceleration. If we are to believe Eq. (A2-1) this means that the total force, on either the man or the wall, must be zero. How can this be? We see from Fig. A2-2 that there are *two* forces acting on the man: the reaction of the wall to his push and an oppositely directed resisting force at his feet, supplied by the friction between his feet and the floor. If the force he applies to the wall is greater than the possible frictional resistance, then the forces don't balance and he moves. (It can happen.) As Fig. A2-2 shows, at both points of application the forces occur in pairs, but only two of them act on the man and thus contribute to his motion (or lack of it).

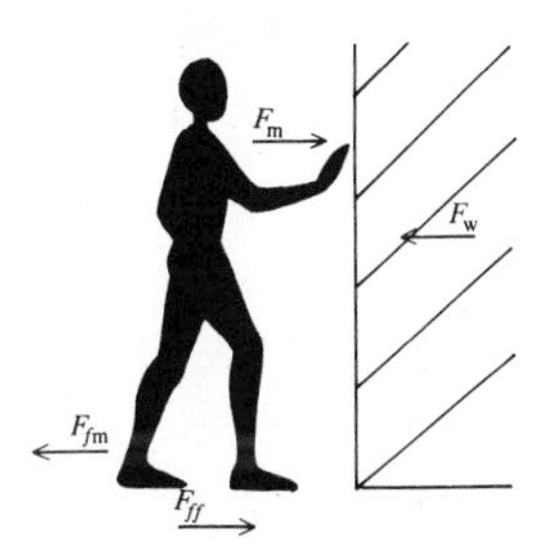

FIGURE A2-2 In the static case of a man pushing on a wall, the two forces *on the man* are the opposing force of the wall F_W and the frictional force of the floor F_{ff}. If they are equal and opposite, there is no net force on the man and he does not move.

FIGURE A2-3 The up and down forces cancel, so there is no motion in that direction. The man pushes horizontally also, and the floor pushes back. Since this is the only horizontal force acting *on the man,* it causes motion.

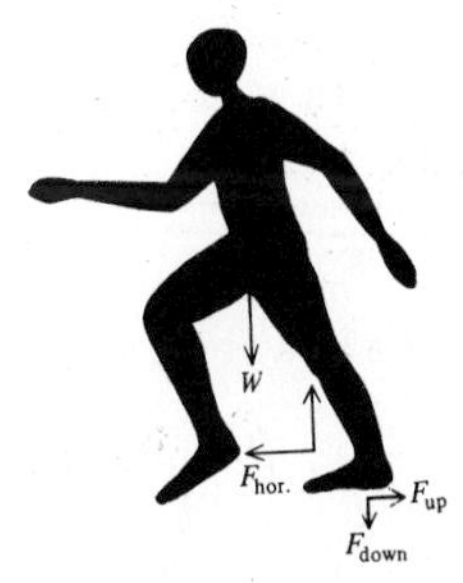

We find the forces paired in the example of a man walking (see Fig. A2-3). He pushes straight down on the ground, the ground pushes up. The upward force F_{up} is just enough to support his weight W: he does not move up or down. In addition to the up-down forces, he pushes backwards against the ground with his leg muscles, and the ground pushes in the opposite direction, forward. There is no force to cancel this one. Therefore, since F is not zero, a cannot be zero and he moves.

A third example of the pairing of forces occurs in the case of the falling apple. If the apple has a mass m the earth is pulling on it with a force $F = mg$, where g represents the (constant) acceleration of gravity, 9.8 meters per second^2. Where is the other force? The object is pulling on the earth with an equal force f, but in the opposite direction. Since f, by Eq. A2-1, is equal to $m_e a_e$, where the subscript e means "of the earth," we see that the acceleration a_e of the earth will be very small, since m_e is so large.

The Gravitational Force

We have already referred several times to the force of gravity. What is this force? Why does an object have weight? We have defined the weight of an object as the force with which it is pulled toward the center of the earth. We have so far talked only about this pull at earth's surface. The force of gravity, however, is a universal force. It is a property of mass. Any two masses are pulled toward each other by a force proportional to their product and inversely proportional to the square of their separation. Mathematically, this is stated as

$$F_{grav} = \frac{Gm_1 m_2}{r} \qquad \text{(A2-2)}$$

where m_1 and m_2 are the masses, r the separation between their centers, and G is a constant which takes care of the units and is a measure of the strength of the force.

The force of gravity is a very weak force. If, for instance, you took two 1,000-kg spheres (about 2,200 pounds apiece), 1 meter apart (about 3 feet), the force between them is 6.67×10^{-4} newtons, which is about the size of the weight of a fly. It takes objects the size of the earth (or at least a mountain) to show appreciable forces of gravity. On a universal scale, where one deals with stars and galaxies, the gravitational force is extremely important. We also see from Eq. (A2-2) that as a star or a galaxy contracts and r gets smaller and smaller, the force becomes larger, compressing the gases to the enormous densities we find at a star's center.

We should not leave the gravitational force until we have reconciled Eq. (A2-2) with the earlier statement that the gravitational force on an object at the surface of the earth is independent of its height above the earth's surface. Substituting g for a in Eq. (A2-1), the force on mass m at the earth's surface is

$$F = mg$$

We now rewrite Eq. (A2-2) as

$$F = \left(\frac{Gm_e}{r^2}\right) \times m$$

Remember that while, precisely speaking, r is $r_e + h$ (where r_e is the earth's radius and h is the distance of the object above the earth's surface) it is permissible to write $r = r_e$ since r_e is so much larger than h. Therefore,

$$F \cong \left(\frac{Gm_e}{r_e^2}\right) \times m$$

All the terms in the parenthesis are constant. By putting in the appropriate numbers we can show that

$$g \cong \frac{Gm_e}{r_e^2} = 9.8 \text{ meters/second}^2$$

at, or near the earth's surface, and, therefore, F (at the earth's surface) = mg, and does not depend on h.

The Electric Force

Gravitational attraction is the most obvious of the forces, but the weakest. The electrical force is not so obvious, but it is much stronger than gravity and much more important to life.

The mathematical equation defining the electrical force is quite similar to the one defining the gravitational force. Any two electric charges will feel a force which is proportional to the product of the strength of the charges and inversely proportional to the square of the separation between them. The mathematical statement is

$$F_{\text{elec}} = \pm k \frac{q_1 q_2}{r^2} \qquad \text{(A2-3)}$$

where q_1 and q_2 are the strengths of the charges, r the separation between them, and k (like G) is a constant which takes care of the units and is a measure of the strength of the electrical force. This is as always a paired force; each charge feels it.

In spite of the similarity in form, an important difference between this expression and that for gravitational force is signaled by the $\pm$. This gravitational force is, as far as we know, always attractive (pulls mass together); the electrical force is attractive if the charges are unlike (positive and negative) and repulsive if the charges are the same (both positive or both negative).

Another important difference between the electrical force and the gravitational force is the strength. We can get an idea of this difference by comparing the gravitational attraction between two protons separated by some distance r, with the repulsion due to their electrical charge at this same separation. Taking the ratio of Eq. (A2-2) to Eq. (A2-1)

$$\frac{F_{\text{elec}}}{F_{\text{grav}}} = \frac{kq_p^2}{Gm_p^2}$$

When we put in the values for k, G, q_p (the charge on the proton) and m_p (the mass of the proton) we find that

$$\frac{F_{elec}}{F_{grav}} = 1.2 \times 10^{36}\dagger$$

The electrical repulsion between two protons is 10^{36} times larger than the gravitational attraction. For an electron with 1/2,000 the mass of the proton but with the same strength of charge, the ratio is even larger.

The electric force is strong enough to hold atoms and molecules together, and it does. Most of the forces which are familiar to us, the forces of springs and rubber bands, surface tension, friction, adhesion, and the like are different manifestations of the electrical force. The gravitational force is responsible for the shape of the universe as a whole; the electrical force shapes the immense diversity of form at the scale of size from man to atoms.

The Nuclear Force

The third member of this triumvirate of basic forces is that one which holds the nucleus together. As soon as it was determined that the nucleus, the tiny dense core of the atoms, was composed of electrically charged protons and the electrically neutral particles, neutrons, it was clear that there had to be a force other than the gravitational and electrical ones to hold the protons together. (The protons, since they are alike, strongly repel each other electrically and gravity is too weak to overcome this.)

We cannot write down a simple equation for the nuclear force; we know less about it than we do about the other two. We do know that it is very short-range; it does not reach much beyond the 10^{-12} cm size of the nucleus. It is attractive through most of its range, and it must be much stronger even than the electrical force, for the nucleus is very strongly bound together. (Actually, there seems to be more than one nuclear force having quite different strengths and properties, but this statement applies to both the "weak" and the "strong" nuclear interaction.) That is about all we can say.

Force Fields

With these descriptions of the three kinds of forces we have subtly altered our concept of a force. "Push or pull" implies contact, but we see that these three forces operate without contact. The more sophisticated description must be made in terms of a "field of force." An electric charge or a mass alters the space around itself. When another charge or mass is brought into that region of space it "feels" a force, even though it is not "touched." We call this alteration of space a force field.

Equation (A2-2) and (A2-3) describe how those forces depend on the separation

†The appropriate values are

$$k = \frac{9 \times 10^{9}\ \text{newton meter}^2}{(\text{coulomb})^2}$$

$$G = 6.7 \times 10^{-11}\ \frac{\text{newton meter}^2}{(\text{kg})^2}$$

$$q_p = 1.6 \times 10^{-19}\ \text{coulomb},\quad m_p = 1.7 \times 10^{-27}\ \text{kg}$$

and on the strengths of the charges or masses. We see, from this description, that things never touch. The firm grip you have on your pencil while writing is caused by the electrical repulsion between atoms in your fingers and those in the pencil. From the point of view of those atoms, there are miles of empty space between your finger and the pencil.

ENERGY, WORK, AND POWER

The concept of a force is one which is, more or less, intuitively obvious; we can equate it with pushes or pulls which we feel in our muscles. The concept of energy is more abstract. We do have some intuitive feeling about energy; it causes us to avoid stepping in front of speeding cars and to feel nervous under suspended pianos. We know that in both cases there is a capability of damage, a capability to rearrange matter. We extend the same sort of ominous feeling to lighted firecrackers, bent bows, water behind a dam, and with some prodding to a lump of coal or a battery. In all these examples a familar object has had something added to it which makes it different, the car because it is speeding, the piano because it is lifted, the bow because it is bent, and so on. This added quantity is energy.

Kinetic and Potential Energy

In our examples we have already provided a basis for dividing energy into its two main types: the energy that is there because of motion, kinetic energy, and the energy that is stored in a motionless object, potential energy.

Kinetic energy was "discovered" in the study of collisions between masses. (Spherical pendulums on long strings were used.) In the study of such collisions the Dutch physicist, Christiaan Huggens, in the seventeenth century, found that the quantity $\frac{1}{2}mv^2$ was "conserved," that is, it did not change; it was the same before and after the collision. More speecifically, if a mass m_1 had a velocity v_1 before collision and v_3 after collision (as in Fig. A2-4) and a second mass m_2 had velocities v_2 before and v_4 after collision, he found that

$$\underbrace{\tfrac{1}{2}m_1v_1^2 + \tfrac{1}{2}m_2v_2^2}_{\text{Before}} = \underbrace{\tfrac{1}{2}m_1v_3^2 + \tfrac{1}{2}m_2v_4^2}_{\text{After}}$$

A quantity that is conserved in an interaction is usually an important quantity. It provides a handle on the interaction. It gives an equation (like the previous one) which can help to understand what is happening. For instance, if the masses and velocities before the collision and one of the velocities after the collision are measured, the other velocity is uniquely determined. This quantity, $\frac{1}{2}mv^2$ was thus worthy of naming and, after several temporary and inappropriate designations, it was finally

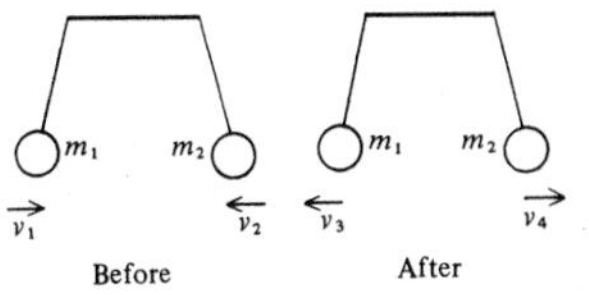

FIGURE A2-4 In the experiment of the colliding pendulums, it was found that the *sum* of the kinetic energy ($\frac{1}{2}mv^2$), before and after the collision, was the same.

called kinetic energy. The kinetic energy of an object of mass m moving with a velocity v is

$$\text{Kinetic energy} = \tfrac{1}{2}mv^2 \tag{A2-4}$$

Where does this kinetic energy come from and where does it go? The answer leads us to the definition of potential energy and to the relationship between force and energy. To answer it we go back to our experiment with falling objects.

If we drop an object of mass m from a distance h it will have a kinetic energy $\frac{1}{2}mv^2$ just before it hits the ground. Where did that energy come from? The conclusion forced on us is that the energy must be associated with its height above the earth. We know its velocity will be greater if we drop it from a greater height. From this conclusion we immediately see the source of the energy; we put it there by lifting the object.

How was this energy stored in the lifted object? What caused the storage? The necessary ingredients were a force, for the pull of gravity had to be overcome if the object was to be lifted, and a distance over which the force acted.

The stored potential energy is, in fact, just the product of this force times the distance h through which the object was lifted. We can show this by the following argument. The kinetic energy is $\frac{1}{2}mv^2$. If we know h, however, we can calculate v^2, the velocity of a falling body. You may remember from your high school physics that the velocity of a body falling from a height h is

$$v = \sqrt{2gh} \quad \text{or}$$
$$v^2 = 2gh$$

where g is the acceleration of gravity, 9.8 meters/second^2 which we defined earlier.[4]

If we now substitute that expression for v^2 into Eq. (A2-4) we obtain

$$\tfrac{1}{2}mv^2 = \tfrac{1}{2}m(2gh) \quad \text{or}$$
$$\tfrac{1}{2}mv^2 = mgh \tag{A2-5}$$

On the left we have an expression for kinetic energy, on the right an expression involving the mass and height to which the mass was lifted. This term must also be an expression for an energy (since it is equal to an energy). It fits our expectation for gravitational potential energy since it depends on h, g, and m. More important, it is the product of the force of gravity at earth's surface, mg, and the distance of lift, h.

Force, Work and Energy

In lifting an object of mass m a distance h above the earth's surface, we increase its potential energy by mgh, that is

$$\text{The change in potential energy} = mgh \tag{A2-6}$$

To store this potential energy we needed to provide a force mg to overcome the gravitational pull. This force was in the direction of motion, in the direction of increasing h. As a matter of fact, to lift the object we had to apply a force slightly larger than mg and give the mass a little kinetic energy of motion. This extra kinetic en-

[4]The derivation of this can be found in the early chapters of any elementary physics text.

ergy is dissipated in some way, as heat in your arm when you stop lifting for instance. It is part of the heat tax on storage energy referred to in the discussion of the Second Law (see Chap. 3).

In every case of stored energy which we will investigate we will find the same factors: a force must operate against an opposing force. The application of this force over a distance is what we call *work.* It is this work which produces the change in potential energy which occurs in lifting the object. It is also work which produces the change in kinetic energy when the object is dropped. In that case, the force is the gravitational force itself, acting over the distance h pulling the mass back to earth. In the first instance we did work, in the second the earth's gravity did the work.

The computation of the work, the product of force times distance, is relatively easy when the force is constant and in the same direction as the motion. If, however, either of these conditions is altered, the computation of the work becomes more complicated.

The violation of the requirement that the force be along the direction of motion is the easiest one to handle. As we have said, force is a quantity of the type called a vector, which means that it has both a magnitude (a size) and a direction. There are rules for the addition, subtraction, and multiplication of vectors analogous to those which govern simple numbers. We will not need most of those laws for our discussion here, but will invoke one of them: that any vector can be represented by two perpendicular components. The meaning of this statement can be made clearer with Fig. A2-5, where the force exerted on the walking man's foot is examined in detail. This force F is the ground's reaction to the foot's pressure (shown as the dotted F in Fig. A2-5). It is a single force and both supports the man and causes him to move. It can be replaced by the two perpendicular components shown F_H, the horizontal force and F_V, the vertical force. The effects of these two forces acting together are idential to the effect of the single force. We can, therefore, look at the contribution of both of these component forces to the work. Since work is the product of the distance moved times the force *in the direction of the distance moved,* only the horizontal component F_H can contribute to the work. There is no motion (at least in our simplified version of walking) in the vertical direction and, therefore, no work accomplished. All F_V does is support the walker's weight. The work, which here goes into kinetic energy of motion, is thus

$$W = F_H \times D$$

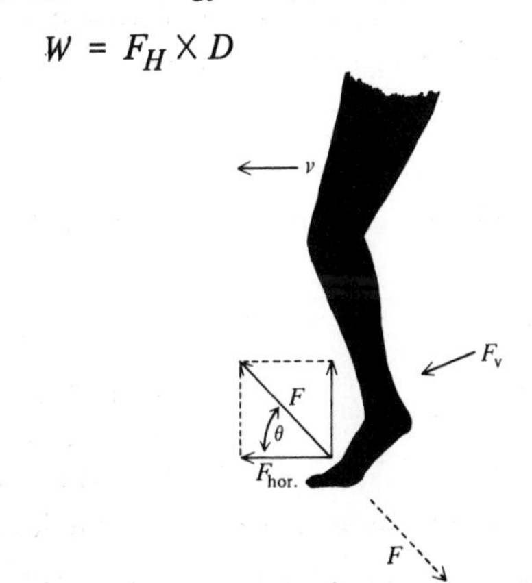

FIGURE A2-5 The single force F, which is the ground's reaction to the push of the foot, can be "resolved" into two perpendicular components: the horizontal force F_h which causes motion, and the vertical force F_V which supports the weight.

where D is the distance moved. Looking again at Fig. A2-5 and using some simple trigonometry we can write this more generally as

$$W = FD \cos \theta$$

where F is the total applied force, D is the distance moved (it is of course a vector also, with magnitude and direction), and θ is the angle between the directions of F and D. We will almost always deal with cases of parallel F and D for which $\cos \theta = 1$.

The second complication, a force which varies with distance (rather than a constant one), is more difficult to deal with. A graphical analysis of the storage of gravitational energy suggests how this might be handled. Figure A2-6 plots the force mg along the vertical axis versus the distance (in this case, the height). The force is a constant and thus plots into a horizontal straight line. The work mgh is, as we can see, represented on this graph by the *area under the curve of force versus distance* from the point where the motion started to the point (h in this case) where it stopped.

We can use this same relation to determine the work accomplished by a force which varies with distance. Figure A2-7 plots a force that is directly proportional to the distance, and is thus a sloped straight line as shown. To find the work done by this force we break the distance up into little intervals such as d. The work done over this small distance is the average force operating at that distance times d. But this is just the area of the small shaded rectangle. We can get the total work by performing this operation over and over for the requisite number of small d's. Since the sum of the areas of all the small rectangles from the beginning point O to the end point D is just the total area under the curve of force versus distance, this graphical method still produces the work. This is a general statement.[5] The work done by a force which varies with distance is found by determining the area under the F versus D plot, over the distance moved.

We have been careful to point out that the area under the force-distance curve gives us the work, and, therefore, the increase in potential energy; we cannot find the total potential energy in this way. The total potential energy is arbitrary. In the case of the lifted object we have been discussing, and whose F versus D curve we showed in Fig. A2-5, we arbitrarily set potential energy at $h = 0$. Perhaps $h = 0$ is measured from the floor of a laboratory on the third floor of a building; the mass still has some gravitational potential energy at this point, for it could fall to ground level; some potential energy still remains even at the surface of the earth, for in theory, if not in practice, it could "fall" into the earth's center. Thus, we can arbitrarily set potential energy = 0 at any chosen reference point. In gravitational examples we usually chose the place where the falling object will stop.

[5] We have provided only a suggestive argument here. A rigorous proof can be given using the laws of calculus.

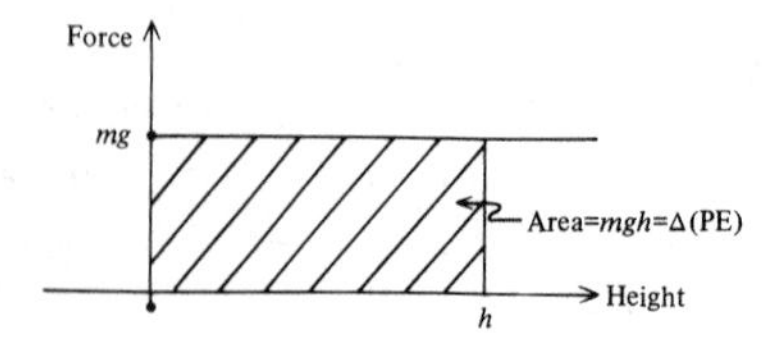

FIGURE A2-6 When the force is constant, it is easy to see that the work done by the force mg operating over the distance h is just the area under the f versus h curve from zero to h. This work changes the potential energy.

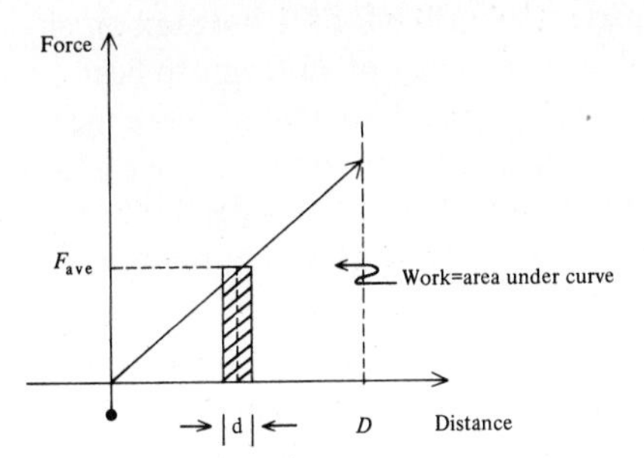

FIGURE A2-7 When the force varies with distance, the work done by a small motion d is the area of the small rectangle $F_{ave} \times d$. Since the total area can be very well approximated by the sum of these small rectangles, the total work done by the force operating over the distance D is the area under the curve from zero to D.

From the force-distance plot we can compute, by finding the area for the appropriate distance, the *work* done to change the potential energy. This general method of finding work will prove useful later on when we investigate the work done by the pressure of a gas (Appendix 2 and Chap. 4).

Work as the Change in Energy

We will be much concerned in this book with energy conversions. Since we know from the First Law of Thermodynamics that energy is neither created nor destroyed, its use must be through conversion. Work is the name we give to a large class of those conversions, those which are to, or from, mechanical forms. Thus, we do work in lifting an object (mechanical to gravitational potential). We obtain work from an electric motor (electric kinetic energy to mechanical kinetic energy) or from a heat engine (thermal energy to mechanical kinetic energy). Work is characterized, as we have seen, by the operation of a force over a distance. If the system supplies the force, as in the falling body case or in the very different case of the battery discussed in Appendix 4, then work is done *by* the system. If the force is supplied externally, by a man lifting the objects or a battery-charger putting energy into the battery, then work is done *on* the system. The final result of the investment of work, on or by a system, is a change in the energy of at least that part of the system on or by which the work is done.

Work Against Dissipative Forces

We have so far considered work done in storing energy. Much work is also done against dissipative forces, particularly so in the case of "processing work (see Chap. 3). We will not try to either list or describe all the various important forces of this type, only to make some general statements about them. Most of these dissipative forces oppose motion. The most common example is friction, caused by the adhesive forces between solid surfaces in contact. The result of forcing an object against this force is to convert work to heat energy. Another example is air resistance, a complicated resistance to motion which depends on the velocity of the moving object. Again, the kinetic energy is converted to heat energy. Other important dissipative forces are that one due to electrical resistance (see Chap. 6 and Appendix 4), and the more sophisticated forces which oppose the motion of electric conductors in magnetic fields.

These dissipative forces are, for the most part, responsible for the heat tax on energy conversions and for the eventual downhill run of energy of all forms to heat energy (see Chap. 4). It is also interesting, perhaps, to remark that all of those mentioned above, as disparate as they may seem, are essentially variations on the electric force; evidence again of the dominance of this force in the affairs of man.

Units

Our operational definition of work and energy is not complete until we have discussed units. These tell us the size of the "measuring stick" we are using, whether we have a foot ruler or a yardstick, to carry out the analogy. Since work is defined as the change in energy, both work and energy will be expressed in terms of force times distance. In the English system, force is in pounds and distance in feet; a common unit of work (or energy therefore is foot-pounds. A 200-pound man lifting himself a vertical distance of 20 feet by walking up a flight of stairs changes his gravitational potential energy (does work) equivalent to 200 × 20 = 4,000 foot-pounds.

In the metric system, force is in newtons and distance in meters; the unit of work is newton-meters which is called a *joule,* after the Scottish physicist, James Joule, who did much to establish the Law of Conservation of Energy. Using the earlier relation, 1 pound = 4.45 newtons, and remembering that 1 foot = 0.305 meters, the work, 4,000 foot-pounds, just computed, is equal to 5,429 joules.[6]

Since work can bring about changes in other forms of energy, and, conversely, work can be derived from conversion of other forms of energy, other units are often employed in the discussion of work and energy. The most important of these are the units by which heat energy is measured, the Calorie and the British thermal unit (Btu), and the electrical energy unit, the kilowatt-hour. This latter unit is obtained by multiplying a unit of power by a time. The special nature of heat energy is discussed in Chap. 4 and in Appendix 3. The concept of power and its relation to work and energy follows in the next section of this appendix. Brief definitions of all these units are given, however, in Table A2-1 and specification for their conversion provided in Table A2-2.

Power

In the preceding sections we have discussed work, both storage work and processing work. We have not yet considered how long it might take to accomplish that work. This consideration brings us to a discussion of the concept of power. Power is a measure of the rate at which work is done, and since work produces a change in energy, we could call power the flow of energy.

The quantitative definition of power is, therefore, the following: if we do an amount of work W during a time interval t the power is given by

$$P = W/t$$

Units of power are thus work per unit time; for example foot-pounds per second or joules per second. Conversely, work is easily found as the product of power and time, $W = P \times t$.

The most familiar unit of power is the horsepower, which is roughly the rate at

[6] 4,000 foot-pounds × 0.305 meters/foot × 4.45 newtons/pound = 5.429 newton-meters.

Table A2-1 Commonly Used Units of Energy and Power

Units	*Definitions*
ENERGY UNITS	
Calorie (Cal) (or kilocalorie)	The amount of heat energy needed to raise the temperature of 1 kg of water 1° Centigrade.
Btu (British thermal unit)*	The amount of heat energy needed to raise the temperature of 1 lb of water 1° Fahrenheit.
foot-pound (ft-lb)*	The energy required to lift a 1 lb weight 1 ft.
joule	The energy supplied by a force of 1 newton applied for 1 m.
watt-hour	The energy supplied by 1 watt of power in 1 hour (see definition of watt under *Power*).
kilowatt-hour (kw-hr)*	10^3 or 1,000 watt-hours.
horsepower-hour (hp-hr)*	The energy supplied by 1 hp of power in 1 hour (see definition of horsepower under *Power*).
POWER UNITS	
horsepower (hp)	A time rate of energy expenditure of 550 ft-lb per sec.
watt	A time rate of energy expenditure of 1 joule per sec.†
kilowatt (kw)*	10^3 or 1,000 watts.

*Denotes those units most frequently used.
†Most often used to measure electrical power. See Appendix 4 for definition in electrical terms.

which a horse can work for an extended time. The definition now has been made more precise; 1 horsepower is defined as 550 foot-pounds of work per second. Thus, in the example of the man climbing the stairs, if he made it up the 20-foot flight in 10 seconds, he would have operated at just under 1 horsepower. If, however, he raced up them in 5 seconds, his output would have been almost 2 horsepower.[7]

The other important unit of power is the watt, the name given (in honor of James Watt, who contributed so much to the understanding of the steam engine) to the power unit, 1 joule per second. It is applicable to any measurement of power; 1 horsepower, for instance, is equivalent to 750 watts of power. The watt, and its larger brother, the kilowatt (1,000 watts) are most often used, however, to measure electric power. In Tables A2-1 and A2-2 the units of power and their conversion factors are also given.

ENERGY AND POWER IN PERSPECTIVE

In this final section of Appendix 2 we present three tables with representative data on comparative energy and power levels of many natural and man-made systems.

[7] *a*

$$\frac{4{,}000 \text{ foot-pounds}}{10 \text{ second}} = \frac{400 \text{ foot-pounds}}{\text{second}}$$

$$\text{horsepower} = \frac{400 \text{ foot-pounds}}{\text{second}} \times \frac{1 \text{ horsepower}}{550 \text{ foot-pounds/second}} = \frac{8}{11}\text{horsepower} \cong 0.7 \text{ horsepower}$$

b

$$\frac{4{,}000 \text{ foot-pounds}}{5 \text{ second}} = \frac{800 \text{ foot-pounds}}{\text{second}}$$

$$\text{horsepower} = \frac{800}{550} = \frac{16}{11}\text{horsepower} \cong 1.5 \text{ horsepower}$$

Table A2-2*a* Conversions: Energy Units

In one	*There are*	
Cal	4	Btu
Cal	3.1×10^3	ft-lb
Cal	4200	joules
Cal	1.16×10^{-3}	kw-hr
Cal	1.6×10^{-3}	hp-hr
Btu	0.25	Cal
Btu	780	ft-lb
Btu	1055	joules
Btu	2.9×10^{-3}	kw-hr
Btu	3.9×10^{-3}	hp-hr
ft-lb	3.2×10^{-4}	Cal
ft-lb	1.3×10^{-3}	Btu
ft-lb	1.4	joules
ft-lb	3.8×10^{-7}	kw-hr
ft-lb	5.1×10^{-7}	hp-hr
kw-hr	860	Cal
kw-hr	3.4×10^3	Btu
kw-hr	3.6×10^6	joules
kw-hr	2.7×10^6	ft-lb
kw-hr	1.35	hp-hr
hp-hr	640	Cal
hp-hr	2.5×10^3	Btu
hp-hr	2×10^6	ft-lb
hp-hr	2.68×10^6	joules
hp-hr	0.75	kw-hr

The first of these, Table A2-3, emphasizes that energy is a commodity which can be bought and sold and gives some comparative prices of the energy from various foods and fuels. Work is also purchased; one buys a book or a car or a house without inquiring how long it took to build it. We pay for power, however, by giving the fast worker a higher salary, by buying the automobile with the bigger engine or employing the higher capacity printing press.

Table A2-4 compares the amounts of energy involved in different activities. These range from the extremely small amounts of energy carried by the photons of light, to the enormous daily energy output of the sun. There are several points of interest

Table A2-2*b* Conversions: Power Units

In one	*There are*	
kw	1,000	watt (joules/sec)
kw	740	ft-lb/sec
kw	0.24	Cal/sec
kw	0.96	Btu/sec
kw	1.3	hp
hp	550	ft-lb/sec
hp	0.75	kw
hp	0.18	Cal/sec
hp	0.71	Btu/sec

Table A2-3 Energy Values for Some Fuels and Foods (Retail Prices)

Source	*Energy (Cal/kg)*	*Energy (kw-hr/lb)*	*Cost (per 1,000 kw-hr)*
Coal (stove coal)	7200	3.86	$ 5.20
Fuel oil	10,800	5.72	4.30
Natural gas	11,000	5.86	5.46
Gasoline	11,530	6.10	10.00
Alcohol (denatured)	6400	3.42	10.25
Alcohol (Scotch, 80 proof)	2580	1.37	2,920.00
Bread	2660	1.42	220.00
Butter	7950	4.20	200.00
Sugar	4100	2.16	68.00
Beef steak (sirloin)	1840	0.97	1,640.00
Electricity	—	—	20.00

that should be noted. By comparing the photon energies with the energy required by photosynthesis we see that photosynthesis must be a multistep reaction since a single red or blue photon does not have the energy required. We also see that ultraviolet light could tear the molecules of life apart. Further along, the table illustrates the

Table A2-4 Some Representative Energy Data

Energy Source	*Calories*	*Btu*
Energy in 1 photon of red light (λ = 7,000 A)*	6.8×10^{-23}	27.0×10^{-23}
Energy in 1 photon of blue light (λ = 4,000 A)*	11.8×10^{-23}	46.8×10^{-28}
Energy converted (per reaction) in photosynthesis	19.0×10^{-23}	75.4×10^{-23}
Energy in 1 photon of ultraviolet light (λ = 2,250 A)*	21.2×10^{-23}	84.2×10^{-23}
Energy (gravitational) of 1 lb of mass at 1 mi above sea level	~2	~8
Energy to melt 1 lb of ice (at 0°C)	36	140
Energy to evaporate 1 lb of water	245	970
Energy (chemical) released by exploding 1 lb of TNT	520	2,065
Energy (chemical) released by burning 1 lb of wood	1,250	4,965
Energy (chemical) released by burning 1 lb of sugar	1,860	7,385
Energy (chemical) released by burning 1 lb of coal	3,300	13,100
Energy (chemical) released by burning 1 lb of gasoline	5,250	29,840
Energy needed to manufacture an automobile	5.2 M	20.6 M
Energy needed to send Apollo 17 to Moon†	1.42 B	5.64 B
Energy (nuclear) released by fusing 1 lb of deuterium	37 B	147 B
Energy (nuclear) released by fission of 1 lb of U^{235}	137 B	544 B
Energy equivalent of 1 lb of mass ($E = mc^2$)	9.8 T	38.9 T
United States daily energy consumption (1970)	47 T	189 T
World daily energy consumption (1970)	140 T	556 T
Energy needed to boil Lake Michigan‡	~400 T	~1,600 T
Solar, earth's daily total at top of atmosphere	3.6×10^{18}	14.3×10^{18}
Total energy in fossil fuels§	$\sim 2.5 \times 10^{19}$	$\sim 10 \times 10^{19}$
Sun's daily output	7.1×10^{28}	28.2×10^{28}

*λ = wavelength, A = angstrom, 10^{-11} m.

†Courtesy Marshall Space Flight Center.

‡5.8×10^{12} kg raised from 25°C to 100°C.

§M. King Hubbert, "Energy Resources," Publication 100-D, Committee on Natural Resources, National Research Council, (Washington: National Academy of Sciences, 1962).

great differences between gravitation, chemical (electrical), and nuclear energies.

In Table A2-5 some representative power data are shown. The very efficient machinery of insects, birds, or fishes shows up in their small power levels (birds routinely fly hundreds of miles nonstop). By comparison, man's power needs for transportation are enormous and increase with speed, as the data on airplanes shows.

Table A2-5 Some Representative Power Data

Source	*Amount (kw)*
Bird flying (humming bird)*	0.0007
Dog running fast*	0.06
Human at rest*	0.09
Fish swimming (dolphin)*	0.21
Horse working steadily (1 horsepower)	0.75
Human (top athletic performance)*	1.7
Per capita power, United States (1973)	40
Automobile at 60 mph‡	100
Automobile at top acceleration†	150
Bus‡	3,000
Commuter train‡	21,000
707 Jet‡	180,000
SST‡	11.9
Fossil fuel consumption in urban area (per mile^2)	625,000
Electrical power for New York City (peak demand)	7.35 M
United States electrical power consumption (aver)	1.86 B
World power consumption (1970)	6.0 B
Solar power involved in photosynthesis	40 B
Incoming solar power (total)	173 T

*Schmidt-Nielsen, "Locomotion: Energy Cost of Swimming, Flying, Running," *Science,* **177**: 222, 1972.

†4,735 lb automobile accelerates to 60 mph in 11 sec.

‡"Lost Power," *Environment,* April 1972.

APPENDIX 3

Heat and Heat Engines

Most of us are aware that mechanical energy—work—can produce heat. If you hammer on a nail, the nail gets hot. If you rub your hands together, causing a force (friction) to operate over a distance, you notice that your hands heat up. But the full realization that heat was a form of energy, and that mechanical work was convertible into heat energy, and heat convertible to work at a fixed exchange rate, was slow in coming to physicists. The eventual understanding was one of the greatest accomplishments of nineteenth-century physics.

The early physicists thought of heat as something which flowed from a hot body to a cold body, as water flows downhill. They considered it a fluid, "caloric", stored in the body. It moved of its own accord to a cooler body which, having less caloric, had room for more. They also postulated that caloric could be released by mechanical means, such as hammering, scraping, or grinding. The early heat engines, the steam engine, in particular, were thought to work like waterwheels: heat flowing from high to low temperature could be tapped for work like water flowing downhill.

That this theory was insufficient was slowly realized by several physicists in the early nineteenth-century. Two different but interesting observations are worth reporting. One was made by a German M.D., Julius Mayer, completely untrained as a physicist, the other by an American physicist, Benjamin Thompson, who left the country for political reasons, became the Bavarian Count Rumford and found himself at one time supervising the boring of cannon barrels for the Bavarian Army.

Mayer's intuition, that work and heat were connected, came while he was working as a ship's doctor in tropical Java. While treating bleeding patients there, he was struck by the bright red color of their blood, much redder than the blood of patients in Germany. He knew enough about blood to recognize that this redness indicated a high oxygen content in the venous blood. The body functions were, at that time, sufficiently well understood that he could correctly ascribe this higher oxygen content to the fact that in a warm country the body did not need to produce as much heat as in the colder Europe and; since heat was produced by oxidation (burning) of the blood sugar, the reduced need of heat allowed more oxygen to be returned by the veins. He carried this line of thinking one important step further. He knew that a man, by working, could produce heat, warming himself or his surroundings. He saw the logical inference that the mechanical work also must have come from oxidation, ultimately from the food. He became convinced of the *equivalence* of work and heat and spent his life proving this and convincing others.

Count Rumford arrived at the same conclusion from an entirely different experience. He was in charge of supervising the tedious operation of boring the barrels of cannons. This was accompished by horses walking around and around turning a large drill which ground away at the barrel hole. An enormous amount of heat was generated in the process. The caloric believers explained this as the release of heat from the metal in the grinding process. But Rumford did some experiments and made some measurements. He used a blunt drill which did not cut out much metal and found that he still "released" enormous quantities of heat and kept on "releasing" it as long as the horses turned the equipment. He measured the amount of the metallic dust removed and found, for instance, that after 960 turns on a cannon blank weighing 113 pounds, its temperature had risen from 60°F to 130°F and that only 2 ounces of metal had been removed. In other experiments he showed that the heat released was independent of the metal removed. The "calorics" believed that all that heat had come from that small amount of metal. Rumford didn't, and looked to the tiring horses as the ultimate source of the heat.

As with all theories in physics, experiment is the final arbiter, and the caloric theory fell because of experimentation. It was finally laid to rest by Joule, a Scottish physicist, who measured the exact equivalence between mechanical work and heat energy with a device by which a falling weight turned a paddle wheel in water and caused a small temperature rise.

MEASURING HEAT

In order to describe exactly what Joule was able to prove and to interpret its meaning, we need to sharpen our understanding of the terms we have been using. We start with heat and temperature. That they refer to different quantities is a commonplace observation; if we put a large pan of water and a small pan of water over two identical heat sources, gas flames, for instance, we can easily believe that the same amount of heat is going into each pan, but observe that the water in the small pan comes more quickly to a high temperature than that in the large pan.

Temperature

Hot and cold are terms we identify with the concept of temperature (which is related to, but certainly not the same as heat). Even under the misconceptions of the caloric theory, the early physicists had the quantitative concept and the means of measuring temperature. We can, by our sense of touch, distinguish between hot and cool items and even rank a series of objects from hottest to coolest. But what science needs is numbers: a thermometer on which a temperature scale can be based. To make a thermometer we need to find a substance with a property that changes in a uniform way with temperature. It was early recognized that the expansion of the volume of a liquid is such a property. Mercury, the liquid metal, is an ideal substance. An amount of mercury is confined in a glass tube of small cross-section; as the mercury is warmed, it expands and the top of the column moves up the tube; as it is cooled, the column lowers. We can use the changes in height to indicate temperature.

The first step is to calibrate it, that is, to choose certain reference temperatures, mark the thermometer at these points, and divide the distance between these reference temperatures into a number of degrees. There are two common scales now in use, the Fahrenheit scale and the Centigrade (also called the Celsius) scale. For the Fahrenheit scale, the reference temperatures originally chosen were zero degrees Fahrenheit, (0°F) as the temperature of a mixture of salt, ice (and water) and 100°F as the temperature of the normal human body. The temperature interval between these two points was divided into 100 degrees.

The choices of reference temperatures of the Fahrenheit scale were not fortuitous; neither temperature is precisely determined. The choices for the Centigrade scale were more successful; zero degrees Centigrade (0°C) was established at the freezing point of water and 100°C as the boiling point (both at atmospheric pressure). There are 100 Centigrade degrees in-between these two points. The Fahrenheit scale was eventually standardized to the same points so that now, in degrees Fahrenheit, freezing is at 32°F and boiling at 212°F.

The relation between these two scales is shown in Fig. A3-1. From this figure it is clear how to convert from one scale to another. There are 180°F between the freezing point and the boiling point of water, and 100°C between the same reference points. Thus, the desired Centigrade temperature will be represented by 100/180, or 5/9, as many degrees as the Fahrenheit interval. The interval is 122° - 32°F, or 90°F. The desired Centigrade temperature is, therefore

$$X = 5/9\,(^\circ F - 32) \quad \text{or} \tag{A3-1a}$$

$$X = 5/9 \times 90$$

$$= 50^\circ C$$

To find a Fahrenheit temperature given a Centigrade one, we work backwards. If the centigrade reading is 80°, then Y is $9/5 \times 80$, or 144° Fahrenheit degrees above the freezing point. The Fahrenheit temperature is, therefore, 144 + 32 = 176°F. Symbolically

$$Y = 9/5\,(^\circ C) + 32 \tag{A3-1b}$$

$$= 9/5\,(80) + 32$$

$$= 176^\circ F$$

There is more to say about temperature; the mercury thermometer, or any thermometer, only works over a certain range of temperature. At -38.8°C, mercury, for instance, becomes a solid and cannot perform as a thermometer. We will pick up this topic again, but for the time being, with a temperature measuring device at hand, let us turn to a discussion of heat.

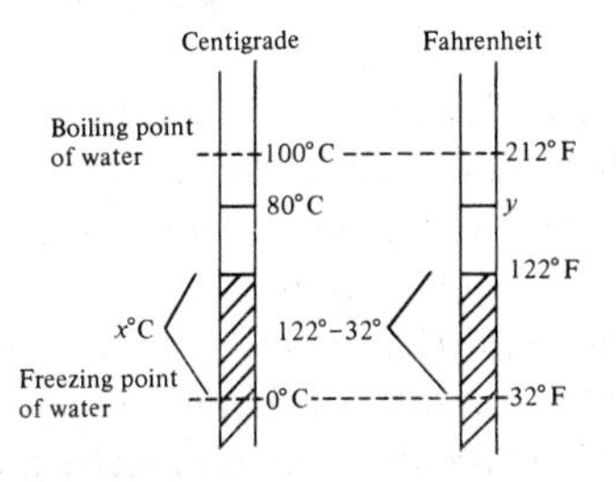

FIGURE A3-1 In the conversion of Centigrade to Fahrenheit and vice versa, 180°F equals 100°C and the difference in zeros must be taken into account.

Units of Heat Energy

Mayer and Rumford argued persuasively that heat was a form of energy. In the next few sections we will try not only to make this identification meaningful, but to get down to the molecular level and become familiar with this strange form of energy. In fact, even under the ancient auspices of the caloric theory, we can set up a system for measuring the amount of heat transferred from one body to another.

If we heat an object many things can happen: it can melt or boil, expand, change color, catch fire: it may even become incandescent, or explode. The most common result, however, is for its temperature to rise. We can use this phenomena to define a unit with which to measure heat energy. The problem is to isolate a quantity of heat. It doesn't come in handy chunks like mass or electric charge. We can best investigate it by the classic method of mixtures. Suppose we mix together 1 kg of water at 60°C and 1 kg at 40°C. The resulting 2 kg have a temperature of 50°C. We can perform the mixing at different temperatures and with different masses. Table A3-1 shows some typical data. The results follow a predictable pattern; some heat energy is removed from the hot water, lowering its temperature, and the same amount is added to the cooler water, raising its temperature. If we let Q_h and Q_c stand for these two amounts of heat energy, we are saying that $Q_h = Q_c$. It can further be deduced from these data, that the amount of temperature change brought about by the addition or removal of heat energy depends on the amount (the mass) of the water. We can put all this together in one expression. For water

$$Q_h = M_h \text{ (mass)} \times (T_h - T_f)$$

$$Q_c = M_c \text{ (mass)} \times (T_f - T_c)$$

where T_h, T_c, and T_f are the temperatures of the hot and cold water, and the final temperature of the mixture and M_h and M_c are the mass of hot and cold water. Since $Q_h = Q_c$ from energy conservation we then obtain the equation

$$M_h(T_h - T_f) = M_c(T_f - T_c) \qquad \text{(A3-2)}$$

We will leave it to you to verify that Eq. (A3-2) does explain all the data of Table A3-1.

The method of mixtures gives us one way to isolate a quantity Q of heat energy by relating it to the temperature and mass of a given amount of water. We can now define it. The most familiar unit of heat energy is the Calorie (often called the kilocalorie). *It is the amount of heat energy which will raise the temperature of 1 kg of*

Table A3-1 Comparison of Temperatures Obtained When Mixing Specific Quantities of Water at 60°C and 40°C

Mass of hot water (kg)	*Temperature of hot water (°C)*	*Mass of cold water (kg)*	*Temperature of cold water (°C)*	*Final temperature (°C)*
1	60	1	40	50
1	60	2	40	33.3
2	60	1	40	53.3
2	60	2	40	50
2	60	3	40	48
3	60	2	40	52

water 1°C. (A precise definition would also include the temperature at which the Calorie is measured, but such precision is not necessary for our purposes.)

This Calorie (capitalized to distinguish it from the small calorie which raises the temperature of 1 gram of water 1°C) is the same unit used by nutritionists. The heat energy equivalent of the normal daily food intake is about 2000 Calories.

Because most of the data in this book comes from engineers, most of it is given in terms of an engineering unit, the Btu which comes from the English System of units (pounds, feet, quarts) rather than the metric system. *One Btu is the heat energy needed to raise the temperature of 1 pound of water 1°F.* To establish the relation between Calories and Btus one must use the conversion factors between degrees F and degrees C, and between pounds and kilograms. The result, rounded off to the nearest decimal point, is

1 Calorie = ~4 Btu (actually, 3.96 Btu)

Water is a natural substance for a standard; it is readily available, easily handled, and easy to heat, to cool, and to mix. With the definition fixed, it is now possible to investigate the heating of other substances. The question we seek (through experiment) to answer is: How many Calories are needed to raise the temperature of iron (or lead and copper, for example) 1°C? This amount is called the *specific heat.* For water we have defined it as one Calorie/kg°C. It turns out that water has a high specific heat; the values for other materials shown in Table A3-2 are all less than 1.

HEAT AS ENERGY

We now have the quantitative tools to report on Joule's experiment. He put a paddle wheel in a closed and insulated container of water and turned the paddle wheel by letting a weight fall. The gravitational potential energy of the weight was converted into kinetic energy of the paddle wheel and then into heat energy in the water. By measuring the temperature rise of the water and paddle wheel, and using the appropriate specific heats, he was able to show that it took 4,200 joules of mechanical energy to produce 1 Calorie of heat energy, that is

1 joule = 4200 Calories

He did many different experiments to verify this. He found not only that this rate

Table A3-2 Specific Heats of Various Materials

Material	*Specific heat (Cal/kg °C)*
Water	1
Ice	0.5
Steam	0.5
Iron	0.12
Copper	0.09
Paper	0.4
Glass	0.2
Rock	0.2
Lead	0.03
Gold	0.03

of exchange (1 joule = 4200 Calories) held at different temperatures and for different amounts of energy, but that the same "exchange rate" held when a different form of energy was converted to heat energy. He sent a measured amount of electrical energy, for instance, through a heating element immersed in water, measured the temperature rise, and found that 1 joule of electrical energy also converted to 4200 Calories.

The fact that 1 joule of mechanical or electrical energy converts to the same amount of heat energy can, in fact, be taken as the final proof of the correctness of our interpretation of the phenomenon of heat and heat flow as energy. If we heat a substance, we give it energy; if we put a hot substance next to a cool one, energy will flow from the hot to cold. This understanding is a big step. There are two more steps between us and a thorough appreciation of the unique role of heat in the energy picture. The first of these is to put a magnifier on nature and find out just exactly in what form this heat energy appears, what its connection with temperature is, for instance. The second is to investigate the laws which govern the conversion of other forms of energy to heat and, more importantly, the conversion of heat to mechanical energy. We turn first to the molecular picture of heat.

Heat as Motion

The recognition that heat is a form of energy was a large step for physics. It doomed the caloric theory and opened the way to a mechanical interpretation. The early experimenters began to suspect that heat involved some kind of motion. It was only after a thorough picture of the atomic and molecular structure of matter became available that it was possible to specify *what* was moving. We now recognize that the kind of energy we call heat is actually the motion of the atoms or molecules of the hot substance. Heat energy is, in fact, the average kinetic energy of the molecules (or atoms) of the substance.

This picture of heat energy allows us to interpret several heat-related phenomena; conduction, for instance. *Conduction* is one of the three ways by which heat is transferred from one region to another. (The other two important forms of heat transfer are *convection,* in which mass moves; the rise of hot water in a kettle is an example, and *radiation*, in which energy is carried away in wave form, as infrared radiation, for example.) Conduction occurs between objects in contact; the kinetic (heat) energy of the hot substance is transferred to the cooler one by collisions between atoms or molecules. On the average, the more energetic molecules lose energy in these collisions to the cooler less energetic ones. (In fact, the electrons play an important role here and it is for that reason that good conductors of electricity, copper and silver, for instance, are also good conductors of heat.)

This interpretation also allows us to understant the "changes of phase" such as melting and boiling. To see how this interpretation proceeds, we will describe at a molecular level what happens when we carry water through its various phase changes to steam. We plot the temperature against the input heat energy in Fig. A3-2.

We start with ice, at a temperature somewhat below its melting point. When we begin to warm it, that is, when we begin to put heat energy into it, the molecules begin

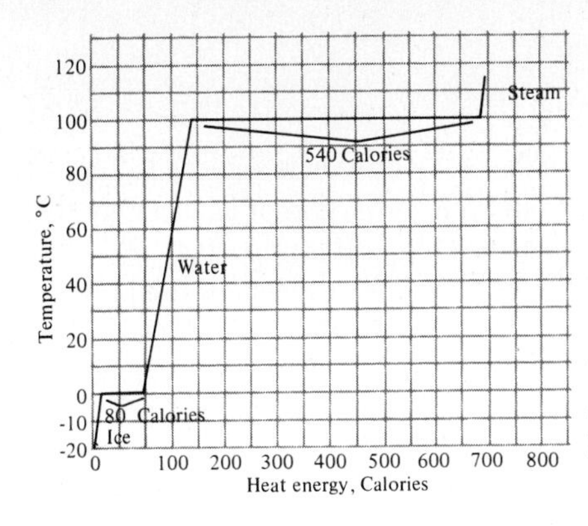

FIGURE A3-2 Heat input versus temperature rise for changing ice to water to steam. Relatively large inputs of energy are needed to change ice to water and water to steam.

to move more rapidly, but are still held by electrical forces in a regular structure; they each move back and forth about a fixed point. At the melting temperature, however, these molecules have received enough energy that they begin to overcome those forces, as if they were snapping springs, and now move more freely under weaker forces. As we continue to provide heat energy, the temperature no longer rises; it remains at the melting point, 0°C, and more and more springs are snapped, allowing the molecules the relative freedom they have as a liquid. The amount of heat energy needed for this is formidable. As we saw in Table A3-2, it required 0.5 Calories to raise the temperature of 1 kg of ice 1°C. To *melt* a kilogram of ice, however, requires 80 Calories. The same is true of other solids: a large input of heat energy is required to melt them.

We now have water at 0°C. If we continue to provide heat energy we can raise its temperature, 1°C for every Calorie we put into a kilogram. After we have expended 100 Calories in the kilogram, it is at 100°C and a second *change of phase* takes place. The kinetic energy of the molecules has been rising steadily; their velocity has been increasing and they have made excursions further and further apart. At the boiling point they begin to overcome the remaining forces that hold them together and to fly free of the liquid into the surrounding air, as steam. Again, the temperature holds constant as we pour in energy and all this energy goes into snapping the liquid bonds. The energy required to give the water molecules their ultimate freedom is even greater than the melting energy, 540 Calories per kg. The steam can continue to be heated up and up and no further phase changes take place; the average kinetic energy of the molecules just continues to increase.

We can take these processes in the reverse direction with the results you should expect. As we begin to take heat out of our super-heated kilogram of steam, it drops in temperature about 2°C for every Calorie removed (see Table A3-2; the specific heat of steam is 0.5 Calories per kg°C). When it reaches 100°C, however, we have to remove 540 Calories from it to make it condense into water. This is, of course, why steam heating is so effective; each kilogram of steam introduced into the radiator releases 540 Calories upon condensing to water.

After it all has condensed, we can cool it further, removing 1 Calorie for each 1°C drop. At the freezing point, however, we must remove the 80 Calories to cause it to solidify. Ice is a good cooling agent for just this reason. Put an ice cube in a glass of water and it will take 80 Calories per kg out of the water in order to melt: 1 kg of ice can cool 80 kg of water 1°C, or cool 20 kg, 4°C or 4 kg, 20°C, etc.

Hot Gases

While there is much physics to be learned from the study of liquids and solids, our aim is toward an understanding of heat engines and the conversion of heat energy to mechanical work. In that conversion it is the gaseous state that is important. We will confine ourselves in the remainder of this appendix to an investigation of the properties of gases.

This choice has certain benefits: the most important properties of gases can be developed on the basis of a very simple model. We will only deal with single atom (or molecule) gases, helium for instance, and gases like it, in which the energy can appear in only one form, motional kinetic energy of the molecule, and in which potential energy changes are small enough to be neglected. We will be working, therefore, with the "ideal gas" laws. These laws, however, give quite satisfactory descriptions, for our purposes, of real gases.

The identifying feature of a gas is that it has no definite volume; it expands to fill any container. Let us consider a container such as the piston-cylinder arrangement in Fig. A3-3. In our simple gas there are only three properties of importance: the pressure P the volume V and the temperature T. What is the relationship between them?

We first must define pressure. How does it show itself? In the arrangement in Fig. A3-3*a*, we know that the piston, which has a certain weight W, will compress the gas to a certain volume V and, at that volume (and temperature), the gas will exert a force which balances W. This force is provided by the pressure of the gas. What is pressure? It is defined as a force per unit area, $P = F/A$. We use the concept of pressure when the force is not exerted at a single identifiable point (by a rope or a rod, for instance) but, instead, forces are exerted at all points over a surface, a condition that occurs when we are dealing with gases or liquids. In the example of Fig. A3-3*a*, therefore, if the cross-sectional area of the piston is A, the total force exerted by the gas is $P \times A$ and it must equal the weight W if the piston is balanced.

The difference between a pressure and a force is perhaps best shown by the difference between the pressures in bicycle tires and automobile tires. Two bicycle tires support 200 pounds when inflated to 60 pounds per square inch. Four automobile tires support 4,000 pounds when inflated to 25 pounds per square inch. We leave it to the reader to find out what *area* of these tires must be in contact with the ground to produce the necessary forces.

How does a gas exert pressure? Remembering our simple model of the gas—molecules constantly in motion, colliding with each other and the walls—we can answer. The molecules collide also with the piston, uncountable numbers of them per second. When an object bounces off another object it exerts a force on it. Each molecule that collides with the piston exerts a force on the piston (see Fig. A3-3*b*). The effect of the steady barrage of molecules is, therefore, an average force on each unit area—

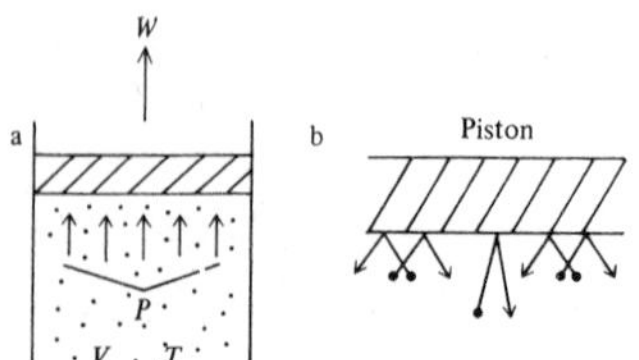

FIGURE A3-3 *a*. The pressure P acts over the entire area of the piston producing a force $P \times A$ which equals the weight W of the piston. *b*. The force P per unit area is caused by the molecules bouncing off the piston.

a pressure. The total force is, as we have already said, the product of this pressure and and the total surface area.

The total pressure exerted by a gas depends on the number of molecules, on the mass of each molecule (heavier molecules can exert more force), and on their velocity (the faster they are going, the more force they can exert upon impact). With this information we can use our model to investigate the relation between pressure and volume.

Suppose we hold the temperature constant and reduce the volume by half, what happens to the pressure? We need here to anticipate a future result, that the temperature is related to the kinetic energy of the molecules, and that holding it constant means that the velocities do not change (since kinetic energy = $\frac{1}{2}mv^2$). With that restriction, it is easy to see what must happen to the pressure. What we have done is to increase the density of the gas, that is, double the number of molecules per cm^3. This will, of necessity, double the number of molecules hitting a given area of the piston (or any other surface) and, therefore, double the pressure. The relation between pressure and volume at constant temperature is, therefore, that pressure is proportional to the reciprocal of the volume

$$P \propto I/V \text{ or} \qquad \text{(A3-3}a\text{)}$$

$$PV = \text{const}$$

that is, the product of the pressure and the volume at a fixed temperature is a numerical constant. This relation between P and V, which we have argued for from our model, was discovered, in fact, long before the molecular model was conceived. It is known as Boyle's Law, discovered by the physicist Robert Boyle in 1660.

From our model it is not as easy to find the relation between pressure and temperature. Let us first turn to experiment. Suppose we enclose a gas (again an ideal one such as helium) in a container of fixed volume. We insert a pressure gauge[1] into it, and then investigate the change of pressure at different temperatures. We place the container in boiling water, measure the pressure, then in a mixture of ice and water (0°C), then in other mixtures of known temperatures (alcohol and ice at -30°C for instance). If we plot these data, we get the interesting curve indicated in Fig. A3-4. The relation between pressure and temperature is linear. It is not proportional however; doubling T does not double P (as can be seen by checking values of P at

[1] Pressures can be measured by allowing them to push against a known area and measuring the total force in some conventional manner, with a spring for example.

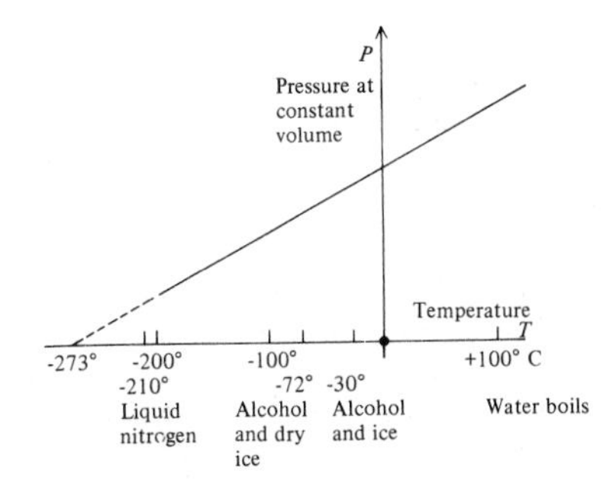

FIGURE A3-4 If the pressure is measured for a constant volume of gas at different temperatures, it produces a straight line which extrapolates to $P = 0$ at $T = -273$°C.

$T = 50°C$ and $T = 100°C$). The curve has another interesting feature: the pressure seems to go to zero if we extend the straight line backwards to $T = -273°C$.

If we were to repeat this experiment with other gases,[2] we would find the same features, a linear relationship with $P = 0$ at $T = -273°C$. This suggests that $T = -273°C$ is some kind of absolute zero, a limiting temperature. It also suggests that, if we construct an absolute temperature scale with $T = 0$ at absolute zero (and, therefore, $T = 273°$ at 0°C, 373 at 100°C, and so on), P will be proportional to T with this temperature scale

$$P \propto T \text{ or} \qquad \text{(A3-3}b\text{)}$$
$$P/T = \text{const}$$

Most of the temperature-containing relations indicated here and in Chap. 4 will involve absolute temperature, that is, a scale with $T = 0$ at absolute zero. There are two such scales in use: the Kelvin scale, which is based on the Centigrade scale and for which water freezes at 273°K, and boils at 373°K, and the Rankine scale, based on the Fahrenheit scale for which 0°F is 460°R, and water, therefore, freezes at 492°R and boils at 672°R.

Absolute zero is mentioned in Chap. 4 in the discussion of the efficiency of heat engines. It is an important consequence of a detailed understanding of molecular motion. For now, however, let us just accept it as a fact of nature.

The result of our investigations so far is to find that if we express temperatures on an absolute scale, the relationship between our three quantities (which we get by combining Eqs. A3-3*a* and A3-3*b*) is

$$PV/T = \text{const}$$

A more precise statement is

$$PV/T = Nk \qquad \text{(A3-4)}$$

where k is a constant and N the total number of molecules in the volume V.

All we need to know about gases (and much more) is contained in this equation. We see that an increase in T at a constant V causes an increase in P, and vice versa (the vice versa explains why an automobile tire gets hot when you pump it up). The most important consequence from our point of view is the information this relation can give us about the molecular meaning of temperature.

Temperature and Kinetic Energy

What we are seeking is a relationship between temperature and the motion of the molecules. Equation (A3-4) relates temperature to pressure and volume. We can obtain the relation we are seeking if we can relate the motion of the molecules to P and V. We already have most of the facts needed to accomplish that. Let us review them.

1 Pressure depends on the number of molecules per unit volume; the more mole-

[2] Such a device as we have described can be calibrated and used as a thermometer. These "constant volume" thermometers have more extensive range of temperature than the mercury thermometer.

cules in a given volume the more that can strike the piston (in the wall) per second. Thus

$$P \propto N/V$$

2 Pressure depends on the mass of the molecules since massive molecules exert greater force on striking the wall than do "lighter" ones. Thus

$$P \propto m$$

3 Pressure depends on the speed with which molecules hit the piston or wall; fast-moving molecules produce a greater impact and thus greater force.

$$P \propto v$$

4 Pressure also is dependent on the speed in an additional way: for a given number of molecules per unit volume the number of impacts per second also depends on the speed; as the speed increases molecules can come from further away, in a given amount of time. Thus, the volume that molecules can be drawn from per second is larger. This makes the number of impacts per second larger.

We can put all these dependencies together and obtain.

$$P \propto \frac{N}{V} \times M \times v \times v \quad \text{or}$$

$$PV \propto Nmv^2$$

We recognize mv^2 as a kinetic energy term (kinetic energy = $\frac{1}{2}mv^2$). In an actual gas the molecules do not all have the same velocity and we replace this by the average kinetic energy which we will label E_{ave}. The relationship then becomes

$$PV \propto NE_{ave} \tag{A3-5}$$

Comparison of Eqs. (A3-4) and (A3-5) gives us the connection we are seeking. The absolute temperature T must be proportional to the average kinetic energy, that is

$$T \propto (\text{kinetic energy})_{ave} \tag{A3-6}$$

It is because this relation between T and E_{ave} is checked by experiment that we have confidence in our simple model of a gas and the ideal gas laws.

HEAT ENGINES AND THE "PERFECT CYCLE"

In Chap. 4, which this appendix accompanies, we have shown how work can be done by a hot gas. We will briefly review that description here and add to it the important dimension of the P versus V curve.

In the first of these examples work is done at constant pressure (see Fig. A3-5*a*). Let us examine the work done by moving the piston the small distance Δd. The pressure remains constant at P since the piston moves; the work done is thus the force, $P \times A$, (where A is the area of the piston) times the distance Δd. Thus

$$\Delta W = P \times A \times \Delta d$$

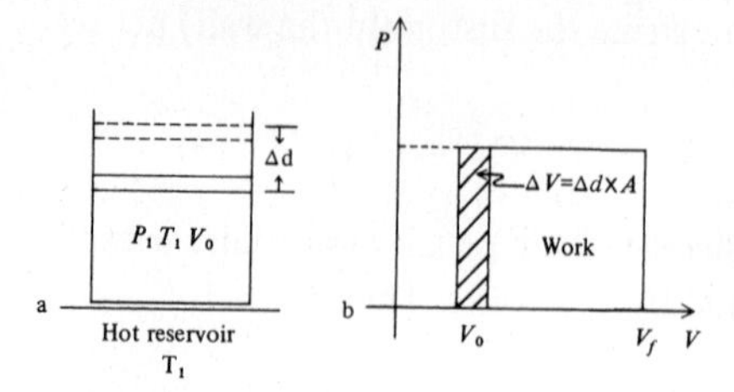

FIGURE A3-5 *a*. The heat energy from the reservoir causes the piston to rise but the pressure remains constant. *b*. The work done in moving the piston a distance Δd is $PA \times d$ or $P \Delta V$, the area of the cross-hatched rectangle. The total work is the total area.

But $A \, \Delta d$ is the change in volume, ΔV, thus

$$\Delta W = P\Delta V$$

indicated by the small cross-hatched rectange of Fig. A3-5*b*. Since $PA\Delta d$ can also be interpreted as $F \times \Delta d$, we can, as in Appendix 2, find the work from the area under the *F*-*d* curve which, in this case, is the same as the area under the *P*-*V* curve from V_0 to V_f.

In this example, which is also examined in Chap. 4, heat energy is put into the gas from an external source and the molecular motion of the gas converts this to mechanical work. It is also possible to derive the energy from the internal energy of the gas without having an outside source. We will briefly develop that example with the aid of Fig. A3-6, which shows a cylinder-piston arrangement on an insulated stand (no heat energy in or out).

Suppose the piston in Fig. A3-6 were light enough that the force of the gas molecules, P_0A, could lift it. A balance is obtained by adding a weight W to the piston as shown. If the weight is now removed, the upward force of gas pressure is greater than the downward force of the piston's weight. The piston moves up until it is stopped by the cylinder top or because the gas pressure is no longer great enough to move it.

Again work has been done; a force (pressure times area) has operated over the distance Δd. Where did the work come from in this case? This time there was no inflow of heat energy. (The cylinder rested as an insulated stand.) The inescapable conclusion is that the work came from the internal energy of the gas; that for each amount of work W obtained, the internal energy of the gas was reduced by an equal amount.

This conclusion leads to a second one. If the internal energy of the gas has been reduced, then the temperature has been reduced also [see Eq. (A3-6)]. Thus, in this case all three properties of the gas, P, V, and T are changing.

In spite of the greater complexity, we can still define the work in terms of a *P*-*V* diagram as shown in Fig. A3-7. As the piston moves the small distance, Δd of Fig. A3-7*a*, the pressure does change but we can approximate the work by using an average pressure, P_{ave} of Fig. A3-7*b*. As Δd becomes smaller and smaller (that is, as the width of the small rectangle of Fig. A3-7*b* becomes less and less) this becomes a bet-

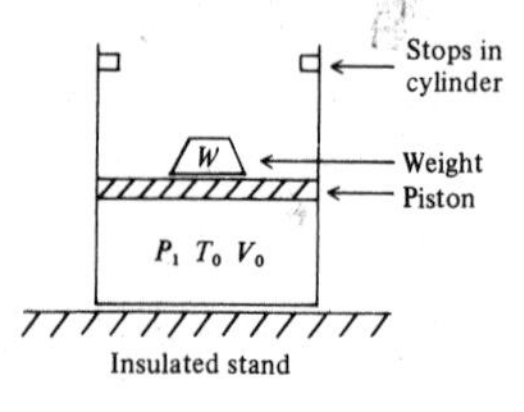

FIGURE A3-6 If we remove the additional weight W which is needed to balance the force exerted on the piston by the hot gas, work will be done on the piston (it will move upwards) without any input of additional energy.

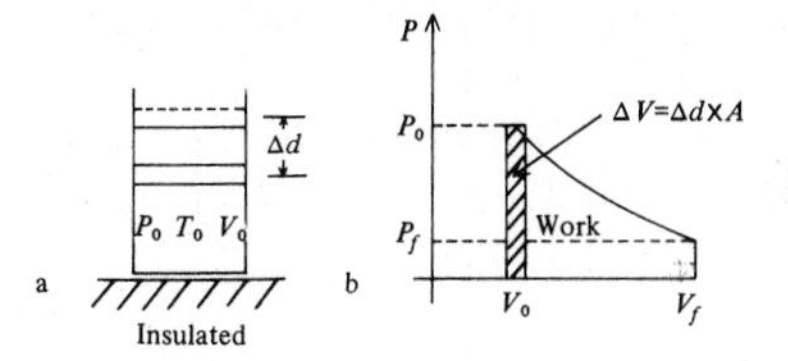

FIGURE A3-7 For a pressure that varies with change in volume (as in Fig. A3-6) it is still true that the work done is the area between the horizontal axis and the curve of P versus V.

ter and better approximation. Thus, the work done during the small movement Δd is still well approximated by

$$\Delta W = P\Delta V$$

that is, by the small rectangular area of Fig. A3-7*b* and the total work is still given by the total area under the P versus V curve.

To study an engine, we must now replace these one step operations with cyclic operations. In an engine, if work is to be produced continuously, the working fluid must be returned to its original state. It was the genius of the young French physicist Sadi Carnot (1796-1832) that added the necessary ingredients to this analysis. He pointed out that the only important processes in the cycle were the absorption of heat energy from a hot reservoir, the performance of mechanical work, and the rejection of a lesser quantity of heat energy to a cooler reservoir. The cycle should not depend on the physical properties of the working substance since this is returned to its initial state at the end of each cycle.

Carnot also was the first to recognize that the thermodynamically perfect engine would be reversible or, as he stated it

> Whatever amount of mechanical effect it can derive from a certain thermal agency, an equal amount be spent in working it backwards, an equal reverse thermal effect will be produced.

A reversible process is one in which the energy flows can be turned around and caused to return the energy converting device to its initial state. Many engines can be reversed; a generator can be reversed and run as a motor. For the process to be reversible, however, in the Carnot sense, the electric energy generated by the generator should, if used to run it as a motor, provide the same mechanical energy as originally turned the generator.

Real engines and real processes are not reversible; their irreversibility is insured by losses to friction, leakage, and so on; in other words, they have to pay the heat tax. Carnot realized, however, that by designing a theorectically perfect, reversible engine cycle he could obtain a description of the theoretical limits on engine efficiency.

The Perfect Cycle

It was easy to hypothesize a frictionless piston-cylinder, but Carnot realized that he needed thermal reversibility as well. Since heat energy does not flow from cold to hot objects, all heat energy transfers had to be between substances at the same temperature (no more impossible than frictionless motors). The four-step cycle he studied is shown schematically in Fig. A3-8.

The working fluid can be steam, air, or some other gas; it is not important. In step

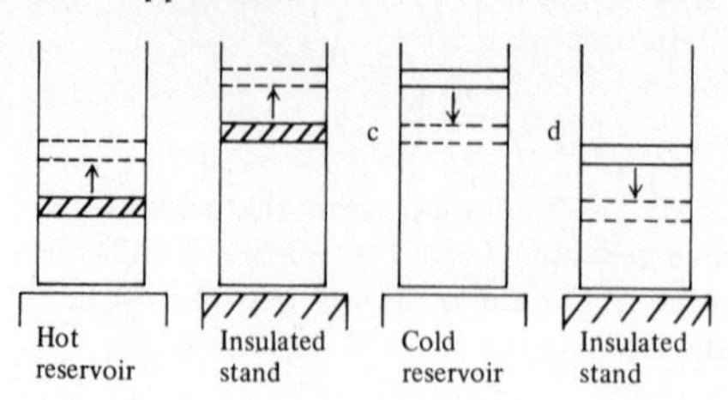

FIGURE A3-8 The four steps of the "perfect" or Carnot cycle.

(*a*) the gas is compressed and at a temperature T_1, it then expands doing work, but heat energy flows into it keeping the temperature constant at T_1. This is shown as curve 1-2 on the plot of P versus V in Fig. A3-9. At the end of this step (point 2) the cylinder is removed from the hot stand and placed on an insulated stand [step (*b*)]. The gas continues to expand doing work at the expense of its internal energy and its temperature drops to the exhaust temperature T_2. The corresponding curve on the P versus V plot is 2-3. It is now put in contact with the cold reservoir at temperature T_2 [step (*c*)]. The piston now compresses the gas, doing work on it, (driven by a flywheel, for instance) and heat energy is caused to flow into the colder reservoir. This is the energy-wasting step, but it is clearly necessary if the piston is to get back to its initial condition so that it can repeat the cycle. The curve which represents this is 3-4 in Fig. A3-9. The last step (*d*) is to move the cylinder again to the insulated stand and allow the compression to continue. During this step, 4-1 in the diagram, work is done on the gas, but since no heat flow is allowed, this work increases the internal energy and, therefore, the temperature. At point 1, the end of this cycle, the pressure has returned to its original value and the temperature is at T_1; the process can be repeated.

As we showed in earlier examples, the area under the P versus V curve is a measure of the work done. Thus, the work done *by* the gas is the area under the combined curve 1-2-3; the work done *on* the gas (during compression) is the area under 3-4-1. From Fig. A3-9 it becomes clear that the output work, the heat energy which has been converted to usable work, is given by the cross-hatched area of the closed curve 1-2-3-4-1 as this area is the *difference between* the area under 1-2-3 and that under 3-4-1.

We can compute the efficiency of this engine. The efficiency is given by

$$\text{eff} = \frac{W_{\text{out}}}{Q_{\text{in}}} \times 100\%$$

Heat energy Q_1 was put in during step (*a*) and some heat energy Q_2 was rejected during step (*c*). Since this is an ideal machine (no friction or heat leakage) the work done must be the difference between these two (since energy must be conserved). Therefore

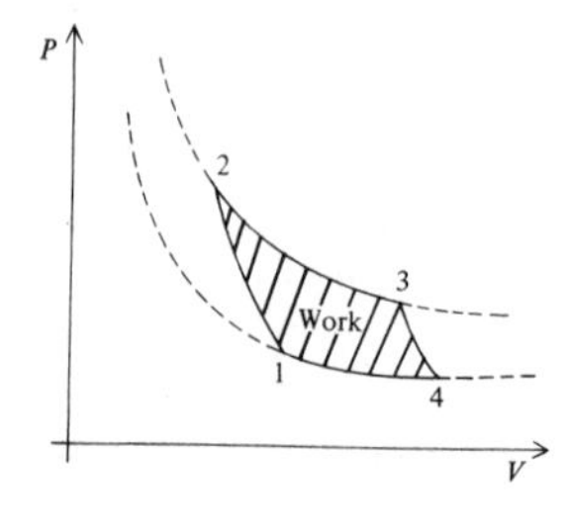

FIGURE A3-9 A plot of P versus V for the Carnot cycle.

$W_{out} = Q_1 - Q_2$ and the efficiency is

$$\text{eff} = \frac{Q_1 - Q_2}{Q_1} \times 100\% \quad \text{or}$$

$$= 1 - \frac{Q_2}{Q_1} \times 100\%$$

The next step is a big one. We won't derive it, but only argue for its plausibility. The question is: What can the heat input and output, Q_1 and Q_2, depend on? There is only one property available, the temperature of the reservoirs. We have no other one available. Since we have seen earlier that the amount of heat transferred depends on the temperature difference, we set the Q's proportional to the T's, that is, $Q_1/Q_2 = T_1/T_2$, and write

$$\text{eff} = 1 - \frac{T_2}{T_1} \times 100\% \tag{A3-7}$$

where T_2 is the exhaust temperature and T_1 the input temperature.

The fact that this cycle is reversible, that it can be run backwards using the work to lift heat, is due to its ideal nature; there are no losses due to friction, leakage, and the like. It is this reversibility that makes it the "perfect engine." This was stated by Carnot in the following terms

> Of all heat engines receiving heat from the same constant temperature source and rejecting heat to the same constant temperature receiver, none can be more efficient than the reversible engine.

It is this fact, that a reversible engine has the highest possible efficiency, that makes it useful for study. All real engines operating through the same temperature difference will have lower efficiencies.

The proof of this statement takes the following form. Suppose there were another engine operating between the same temperatures that was *more* efficient than the reversible one. It would accomplish an amount of work W with *less* input heat energy than Carnot's engine, that is, if Q_R is the input energy required for Carnot's reversible engine and Q_X that for the hypothetical, more efficient engine, then for the same work output, Q_R is greater than Q_X.

We could, therefore, hook the more efficient engine up to the reversible engine so as to run it in reverse as a heat pump (see Fig. A3-10) and, by putting Q_X into engine X, deliver the larger amount of heat energy Q_R back to the hot reservoir. We would

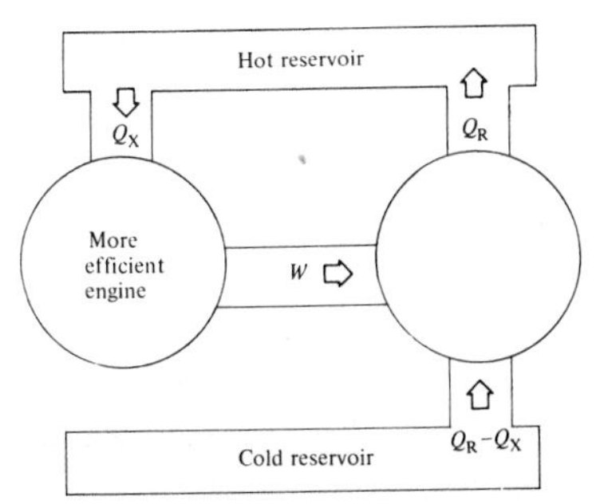

FIGURE A3-10 A hypothetical "more efficient engine" driving a Carnot engine.

thus have constructed a device that took an amount of heat energy $Q_R - Q_X$ from a low-temperature reservoir and delivered it to a high-temperature reservoir with no input of work. Such a device violates that statement of the Second Law of Thermodynamics that we have given in Chap. 4. Since this is impossible, it is consequently impossible for any engine to be more efficient than the reversible engine.

ENTROPY: IT'S A DOWNHILL RUN

The ideal heat engine, whose efficiency we just computed, is perfect because it is reversible. This means that, driven in one direction, (say 1-2-3-4-1 in Fig. A3-9), a quantity of heat Q_1 is absorbed from a source at temperature T_1; some of it is converted to work W and the remainder Q_2 is exhausted at temperature T_2. Driven in the opposite direction, the same amount of work W put into the system extracts the same amount of heat energy Q_2 from the cooler source and transfers it, along with the work (which has been converted to heat energy) to the hotter source. It is clear that such an engine cannot exist in practice, for some of the heat energy will leak away and some of the work must be used to overcome friction.

Since some energy is lost for useful work in an irreversible process, the nineteenth-century physicists looked for a "property" of the process (like temperature, pressure, or energy) which would provide a measure of the departure of a process from reversibility and thus signal the decreased availability of the energy involved in the process. Let us "make this perfectly clear." In the case of a bucket of water poured on a waterwheel, the energy has not changed in total amount, but while the quantity remains the same, we know that somehow the "quality" of that energy has changed, for it can no longer be used to power the waterwheel. The sought after measure of the "quality" of the energy was first proposed in 1854 by Rudolf Clausius, a German physicist. He reasoned that only in the ideal "reversible" process was the quality of the energy unchanged; in such a process, therefore, he looked for some term which remained constant. Identifying this term was simple; he took the relationship which we have just used in deriving Eq. (A3-7), rearranged it as $Q_1/T_1 = Q_2/T_2$, or $Q_1/T_1 - Q_2/T_2 = 0$. The ratio of the amount of heat energy added to or removed from a system, to the temperature at which the heat is added or removed is the property he sought. He called it *entropy*.

Changes in Entropy

Like potential energy, it is difficult to assign an absolute value to entropy and we always deal with changes of entropy. If we choose S as our symbol for entropy, then the change in entropy ΔS as a quantity of heat ΔQ is added to or removed from a large source of heat energy at constant temperature T, is

$$\Delta S = \Delta Q/T$$

How does this measure the availability of energy? Consider the reversible cycle in Fig. A3-8. If it is an ideal "reversible" process then, as we have shown, ΔS_1, for the removal of heat from the hot source, equals ΔS_2, for the exhausting of heat to the cooler source, since $Q_1/T_1 = Q_2/T_2$. But suppose such a cycle were irreversible; how would this irreversibility show itself?

We compare the two cycles schematically in Fig. A3-11. In the reversible process $\Delta S = \Delta S_2 - \Delta S_1 = 0$ since $\Delta S_2 = Q_2/T_2, \Delta S_1 = Q_1/T_1$. In the irreversible process an amount of heat energy q is lost to friction and leakage and is not available for work. The entropy change ΔS_1 of the hot source is Q_1/T_1 as before. The entropy change of the cold source, however, is $\Delta S_2 = (Q_2 + q)/T_2$, which is larger than before. Thus, the overall entropy change is

$$\begin{aligned} \Delta S &= \Delta S_2 - \Delta S_1 \\ &= \frac{Q_2}{T_2} + \frac{q}{T_2} - \frac{Q_1}{T_1} \\ &= \frac{q}{T_2} \end{aligned}$$

For all irreversible processes ΔS is positive.

That this heat energy is lost is clear when we try to reclaim it. The work available to run the engine backwards is $W'' = W - 2q$, since another q is lost to friction. This work can lift only $Q_2' = Q_2 - 2q$ from the cold source. The total heat returned to the hot source Q_1' (including the work which is converted to heat energy and adding the loss q which returns to the hot source) is (see Fig. A3-11)

$$\begin{aligned} Q_1' &= W'' + q + Q_2' \\ &= (W - 2q) + q + (Q_2 - 2q) \\ &= W + Q_2 - 3q \end{aligned}$$

since

$$\begin{aligned} W + Q_2 &= Q_1 \\ Q_1' &= Q_1 - 3q \end{aligned}$$

The heat returned to the hot source is thus less by $3q$ than the amount put in. The condition ΔS positive is thus an indication of the degradation of energy from an available to a less available form. When we take heat energy from a hot source and use it to do work some of it is lost for any further work. Even in the direct transfer of heat energy from a hot to a cooler source, entropy increases. If we remove Q from a source at T_1 (see Fig. A3-12) and deliver all of it to the cooler source at T_2, the entropy change is

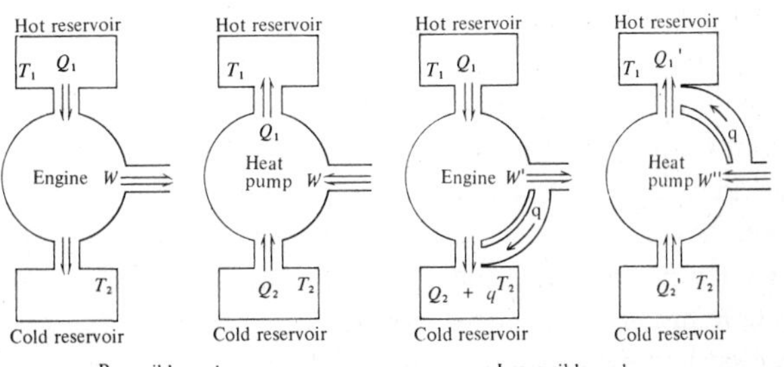

FIGURE A3-11 The losses q due to friction and leakage make it impossible, in the irreversible cycles, to restore all of the heat energy to the hot reservoir with the heat pump.

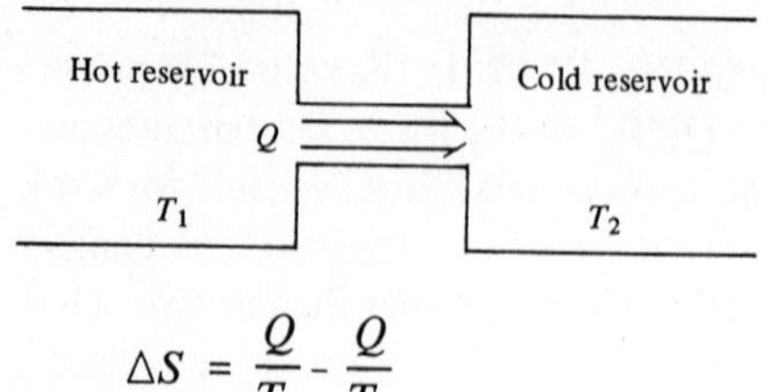

FIGURE A3-12 If we allow heat energy Q to flow from a hot to a cold reservoir, the entropy increases since $\Delta Q/T_2$ is larger than $\Delta Q/T_1$.

$$\Delta S = \frac{Q}{T_2} - \frac{Q}{T_1}$$

and as T_2 is less than T_1, ΔS is positive and entropy increases.

Since there cannot be a net transfer of heat energy unless there is a difference in temperature, it is clear that in all naturally occurring energy transfers or conversions, the entropy of the whole system always increases. This leads to a new way of stating the Second Law

> In any energy conversion or transfer, within a closed system, the entropy of the system increases.

Order and Disorder

If all the concept of entropy did for us was to allow us to restate the Second Law in this way, its introduction here would not be justified. It is capable of a broader interpretation having to do with order and disorder. To discuss this interpretation we have to introduce the concepts of a "state of a system" and the "probability of a state." By state of a system we mean a complete description of the system in terms of all the relevant variables. The state of a pair of dice can be described by the sum of the two numbers they represent, six, for instance. The probability of that state is proportional to the number of ways in which six can be achieved: 1-5, 2-4, 3-3, 4-2, 5-1. This probability can then be compared with the probability of a 2 or a 12, for instance, which can only be achieved one way. The ratio of these probabilities will be 5/1.

The states to which the concept of entropy apply are states of a collection of molecules, of the gas in the cylinder of a heat engine, for instance. Here the relevant variables are pressure, volume, and temperature, and the probabilities are expressed in terms of these variables. Entropy, it can be shown, depends on these probabilities. (The entropy of a state is, in fact, a constant times the natural logarithm of the probability of that state.) Thus, when a gas goes from a less probable to a more probable state, the entropy increases, while a change from a more probable to a less probable state means a decrease in entropy.

You can immediately see the direction of this argument. Reversible processes take place between states of equal probability, as in an exchange of heat energy between two sources of equal temperature, but no energy is transferred or converted in such a process. The energy conversions which do occur spontaneously, the irreversible processes, are obviously processes that go from less to more probable states, for that is the meaning of probability.

It is here that order and disorder come in. Let us look again at a simple example, say in the tossing of five pennies. Suppose they all come up heads; that is a highly ordered state and much less probable than the less ordered state of two heads and three tails. The latter state can occur in more ways than the former. Another more relevant example is the state of a gas in which all the molecules are in one half of a container and the other half, sealed off by a partition, is empty. If a hole is made in the partition, the gas will reach a state in which both halves are equally filled. This is less ordered than the state half-full, half-empty—and more probable. That it is more probable, is clear if you estimate the odds of all the molecules reentering the original half at one time. Thus, the statement that "in any naturally occurring process in a closed system, the entropy increases," is equivalent to the statement that "in any energy conversion within a closed system, order decreases."

Another way to see this connection between entropy and order is to consider a mixture of ice and water at 0°C or $T = 273°K$. If we add an amount of heat Q to this mixture, some ice melts and the entropy is increased by Q/T. If we remove the same amount of heat, the entropy decreases, $\Delta S = -Q/T$ and ice is formed. Ice is a more ordered substance than water; its molecules are arranged in a crystalline form. Thus, the decrease of entropy (formation of ice) is connected with an increase in order and vice versa.

The example of the decreased entropy of ice in comparison with water illustrates another important point. In the formulation of the Second Law in terms of entropy, we emphasized that entropy always increases in an interaction within *a closed system.* In part of that system it may decrease, in the ice cube trays in a refrigerator, for instance, but in the entire system—refrigerator plus room—it will increase. The heat energy Q_1 removed from the ice at a low temperature T_1 will cause an entropy *decrease* of $\Delta S_1 = -Q_1/T_1$ but outside the refrigerator a larger amount of heat Q_2 will be delivered to a hot source and when the inevitable losses are taken into account, the entropy increase, $\Delta S_2 = Q_2/T_2$, will be greater than the decrease. Thus, we can always buy order, but at the expense of energy.

The changes in entropy in terms of energy degradation, and of order going to disorder, can both be seen in the example of burning coal. A lump of coal has a certain amount of chemical potential energy stored in it because carbon atoms and oxygen atoms are in a higher state of potential energy as separate atoms than as carbon dioxide. Thus, energy is released when coal is burned. But there is also order; the carbon atoms in coal are arranged in a highly ordered crystalline structure. After burning, the carbon dioxide is a hot gas and the molecules move in random, disordered motion. In going from order to disorder of structure the entropy has increased also. The gain to us was that we did not have to search for carbon atoms in their normal low concentration, but found them, instead, at 75 percent or so concentration in coal. This saved the energy we would have to have spent to concentrate carbon into a fuel. Thus, we make use of both potentials, of energy and order, and by this use, increase the total entropy of the universe.

The concept of entropy is a powerful one. It was called "time's arrow" by Eddington, for it is the single physical quantity which enables us to keep track of the passage of time. As the universe ages, its entropy increases. On the earth as well, as we discuss in Chap. 4, the real crisis is not a shortage of energy, for energy does not disappear. The crisis is in the steady buildup of entropy as we burn our fuels, fission our uranium, and even as we use up our (highly ordered) concentrations of iron, copper, and other raw materials. The end is a long way off, but entropy's arrow points toward heat and disorder.

APPENDIX 4

The Nature of Electricity

Among the three intermediate forms of energy (thermal, mechanical, and electrical) shown in Fig. 3-3, Chap. 3, electrical energy is unique. Thermal and mechanical energy are important end uses in our society (in fact, the most important end uses) as well as intermediate forms. Electrical energy, however, is truly intermediate; it is energy in transit and must be converted to another form to be used.

Electric energy is a form of kinetic energy; it is due to the motion of charged particles and is clearly different from the electric potential energy or chemical energy. Electric potential energy is stored in atoms and molecules by rearranging their electronic structure, or putting it more crudely, by pushing and pulling electrons into new positions against the electric forces which hold both atoms and molecules together. Electric kinetic energy is obtained when electric charges are set into motion by electric forces just as mechanical kinetic energy is due to mass in motion.

THE VOCABULARY OF ELECTRICITY

In this appendix we will introduce the vocabulary of electricity; we will define terms such as charge, current, voltage, and resistance with which electrical energy and power are discussed. We will also provide descriptions, at a very basic level, of the electrical devices: batteries, generators, motors, and transformers all of which are important in the production, consumption, storage, and transmission of electrical power. This appendix is intended to provide a background to Chap. 6 in which the role of electric power in our industrial society is discussed.

Electric Charge

What is electric charge? It resists verbal definition as tenaciously as does mass or any other of the fundamental quantities of science. It must be defined operationally. Electric charge is what gives rise to electric forces, electric currents, and the like, whose effects can be measured, just as mass gives rise to the gravitational force and to inertia and momentum. Electric charge is the source of electric force and is defined quantitatively through Eq. A2-3 of Appendix 2.

Electric charge differs qualitatively from mass in two important ways: (1) there are two kinds of electric charge and the electric force can, therefore, be either repulsive or attractive; (2) electric charge is "quantitized": it comes in chunks.

The two kinds of charge are called positive and negative. They could have been called black and white, or male and female, or by any other set of names which underlined their oppositeness. Positive and negative are convenient since, in combination, the two can cancel each other–hence the lack of charge of most of bulk matter.

The most important carrier of negative charge is the electron, the elementary particle which is present in the outer shell of the atom and which determines its chemical properties. The most important positive charge carrier is the proton, an elementary particle which is 1,840 times as heavy as the electron: 1.67×10^{-24} grams of mass to the electrons 9.11×10^{-28} grams. The proton and the uncharged but massively similar neutron, make up the nucleus, the core of the atom. It is the positive charge of the proton which gives the nucleus its positive charge, and it is the attraction between the positively charged nucleus (approximately a sphere of 10^{-12} cm diameter) and the outer electrons (in a spherical shell of about 10^{-8} cm diameter) which holds the atom together.

The proton and the electron have equal amounts of charge, even though there is the fundamental difference in quality (positive and negative). A series of experiments have shown that this amount is, in fact, the smallest amount of charge that exists in nature, and that all other charges are multiples of this 1,2,3,4,. . . electron charges.

As we have said, bulk matter is usually electrically neutral; each atom contains equal numbers of electrons and protons. Charges can be moved, however, and it is possible to remove one kind of charge from a piece of matter and deposit it on another. You have, no doubt, done this by sliding your feet across a deep pile rug, removing electrons by friction, and collecting them on your body. The resulting spark attests to their presence. If charges are separated in this fashion, large attractive forces can be established between the two oppositely charged pieces of matter. The force can be large in spite of the small size of each charge because of the enormous number of atoms or molecules in an ordinary chunk of matter. In one cup of water, for instance, there are 14.6×10^{24} water molecules; if an electron could be removed from each water molecule, the cup of water could exert a force on a similarly charged cup of water a meter away which was enormous, some 10^{36} times greater than the force of gravity between those two cups of water, greater, in fact than the gravitation force of the sun on the earth. The presence of large forces and millions and millions of small charge carriers makes electricity a much more convenient way to transport kinetic energy than to put bulk matter in motion as in pulleys and axles. (Transmitting electric kinetic energy by wires is also more convenient, but not cheaper, than shipping chemical potential energy by rail, pipeline or boat.)

The electron is the most important charge carrier. It is the electron whose movement from the rug to the shoes leads to the shock at the metal door knob. It is the electron whose movement through a copper wire constitutes the usual electric current. The mobility of the electron depends, of course, on the strength of the forces which hold it in place. In solid material the outermost electrons are bound to the lattice, the regular and repeating structure of the atoms which make up the solid material, rather than to the individual atoms. This is just another way of saying that the outermost electrons are subject to forces from neighboring atoms. In certain materials, and the metals are the prime examples, these electrons are essentially unbound as far as the atom is concerned, that is, they are bound to the material; they do not fall off, but they are not connected to any particular atom. These are the free electrons which can be moved through the metal very easily by the application of an electric force. In metals like copper, silver, and aluminum there is approximately one "free" electron per atom: these are the *conductors* of electricity. In other materials, quartz and glass, for instance, the number of free electrons is much smaller—much less than one per atom on the average—and it is difficult to make many of them move; these are the *insulators.* Even in insulators, however, many of the electrons are loosely bound (the forces are weak) and they can be removed if given a little energy, as, for instance, by a shoe scraping across a deep pile rug; both the shoe and the rug are insulators, but the spark attests to the movement of electrons.

Measuring Charge

Charge can be measured in many different units. It could be measured in electron charges—a charge of 1, 2, or 10^{24} electrons, for instance. The relative smallness of this charge unit makes it impractical; we normally have to deal with too many electrons. The unit we will use is the *coulomb*, named after Charles Coulomb, whose experiments led to the measurement of electric charge. The coulomb is a large unit of charge; the spark that zips from your finger to the knob carries perhaps 10^{-6} coulombs of charge, a bolt of lighting has only about 20 coulombs of charge. The charge on one electron is 1.6×10^{-19} coulombs or, conversely, it takes 6.25×10^{18} electrons to make one coulomb of charge.

Electric Current

The reason for choosing such a large unit becomes clearer when we talk of electric current, or the transport of electric charge. It is easiest to define current in the case of electrons moving through a wire as in Fig. A4-1. The current, normally designated i, is the amount of charge passing through the cross-section A in one second. The common unit of current, the ampere (often shortened to amp) is one coulomb per second. Symbolically, therefore, through any cross-section A, the current is

$$i = \Delta q/\Delta t \tag{A4-1}$$

where Δq is the amount of charge passing through A in time Δt. Therefore, one ampere of current can consist of 6.25×10^{18} electrons passing through A in one second.

Current can be caused by the movement of other charged particles, ions (atoms or molecules with electrons added or taken off) moving in gases or in liquids are examples. The direction of positive current is arbitrarily taken as the direction which positive charges move. Therefore, the electron current is a negative current. A little thought will convince you that a negative current flowing in one direction is the same as a positive current in the opposite direction. We will be dealing mostly with electron currents in this chapter.

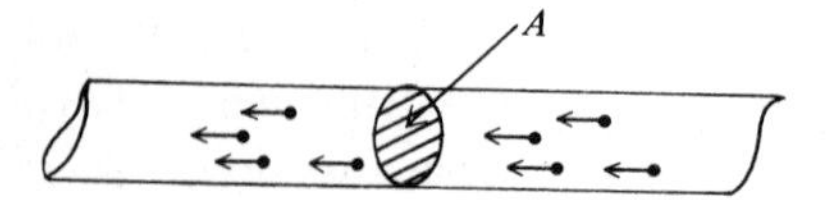

FIGURE A4-1 If n electrons pass through the cross-section A of a wire in 1 second, then the current, $i = \Delta q/t$ is $n \times 1.6 \times 10^{-19}$ amperes.

Voltage

Another familiar term is voltage. Most people know that electric circuits in house wiring are labeled 110 volts. The volt is a unit of potential difference, a term closely related to potential energy. We will define it by analogy to the gravitational case.

The increase in potential energy caused by lifting a mass m from the surface of the earth to a point some small distance h above it is

$$\text{Potential energy} = mgh$$

Alternatively, we say that an amount of work $W = mgh$, is performed in lifting the mass the vertical distance h. The change in potential energy—the work done on the mass m—depends on the mass. We can, however, define a quantity, potential difference, as the work per unit mass and remove this dependence on mass. Thus

$$\text{Potential difference} = \frac{W}{m} = hg$$

This potential difference depends only on the height h and the constant g (the acceleration of gravity). It does not depend on the mass and, therefore, describes the effect of the gravitational field (the region in which there are gravitational forces) rather than the interaction between a particular mass and that gravitational field.

Electric charges are moved about by electric forces. If charges are moved *against* the attraction or repulsion of electric forces, work is done on them and their potential energy is thereby increased. If electric charge is allowed to move from a point of high electric potential energy to a lower one, work is done by the electric field. The electric potential difference between two such points can be defined in a manner similar to the gravitational case, as the work divided by the charge. Symbolically,

$$\text{Potential difference} = \frac{W}{q}$$

Thus, electric potential difference is independent of charge and depends only on the electric field. The volt, the most common unit of electric potential difference (named after the Italian physicist Allessandro Volta who invented the first battery) is defined as

$$1 \text{ volt} = \frac{1 \text{ joule (of work)}}{1 \text{ coulomb (of charge)}} \tag{A4-2}$$

If you have a 12-volt battery, therefore, this means that a positive charge at the positive terminal is at a higher potential energy than it would be at the negative terminal. That charge will "fall" in going from the positive (+) to the negative (−) terminal, performing 12 joules of work for each coulomb of charge that travels around the connecting circuit. The battery, on the other hand, does 12 joules of work in lifting a coulomb from the negative to the positive terminal internally. This work comes at the expense of the chemical energy of the battery materials. (We will discuss the battery at some length in a later section.)

To say that a potential difference exists between two points in an electric circuit is to say that the charge at one point will have a greater potential energy than at the other. If left free to move, positive charge travels from positive to negative potential and negative charge in the reverse direction. In crude terms, since work is force times distance, the larger the voltage over a given distance, the larger is the force available to push (or pull) the charge through that distance. In this sense, one can say that voltage measures the "force" available to push charges through a circuit.

Electric Current in Wires

The most important medium in which charged particles move, from our point of view, is metal, the copper and aluminum wires that carry electric current. It is the free electrons, as we have said, which constitute the current in a wire. We visualize these electrons moving in the following way: a copper wire is used to connect the terminals of a battery. The positive terminal is at low potential for the electron; it is "downhill." Electrons at that end of the wire move on to the positive terminal leaving that region of the wire positively charged. Free electrons from the region next to this one move in to fill the vacancy and leave their own vacancy behind them. In this manner the free electrons in the metal are all set in motion toward the positive terminal.

It is obvious that the circuit must be completed somehow. If the electrons just piled up on the positive terminal they would soon be numerous enough to repel any other electrons which tried to leave the wire and the current would cease to flow. It is the job of the battery to somehow "lift" these electrons from the positive terminal up to the negative terminal where they can enter the wire at the other end.

There is an enormous number of atoms, and thus free electrons, in even a small copper wire. Therefore although one ampere of current requires that 6.25×10^{18} electrons move through a cross section of wire in one second, there are so many electrons available that none of them have to move very fast. It is interesting to put numbers with this assertion. In each 64 grams of copper there are 6×10^{23} atoms. The density of copper is 8.9 grams/cm^3. A typical current-carrying wire, # 18 wire, has a diameter of 0.1 cm. Therefore, in 1 cm of this wire there are 6.5×10^{21} electrons.

To find their speed, we have to ask, "How long a piece of # 18 wire will be needed to hold 6.25×10^{18} electrons?" One electron per atom in a piece of wire of that length will have to move through the cross-section in 1 second. A 1-cm length holds 6.5×10^{21} electrons. It will, therefore, take a piece $(6.25/6.5) \times (10^{18}/10^{21})$ cm long, which works out to 0.00095 cm long or about 1/1,000 of a cm. The speed of the electrons need be only 0.001 cm/second: a very low speed.

When a piece of wire is connected between battery terminals there is a potential difference set up immediately throughout its length. Stated another way, there is an electric field in the wire and, at any point in the wire an electron feels a force which causes it to begin to move. In contrast to the relative slowness with which the electrons move, this field is set up very rapidly, essentially with the speed of light. When you switch on a light, electrons all along the circuit begin to move, current begins to flow and the light comes on immediately, even though it will take a long time for an electron from the switch region to finally reach the light itself, if it ever does.

THE RELATIONS BETWEEN CURRENT AND VOLTAGE IN ELECTRIC CIRCUITS

In the previous section we introduced the basic concepts of current flow in conducting wires, identifying it with the movement of the free electrons, and implying, at least, a continuity condition. If you put in a certain number of electrons as a current at one end of the wire, you expect the same number to flow out at the other end. Electrons do not pile up anywhere along the wire; if they did, they would quickly establish an electric field which would oppose the movement of any more electrons into that space. Conversely, for current to flow in a circuit it must be complete: there must be an electrical connection between all points.

In this section we wish to look further at what happens to an electron current in typical electric circuits. In particular, we want to define that quantity, resistance, which provides the electrical friction through which the heat tax operates.

As we have already shown, the minimal components of an electric circuit are a source of potential difference to cause the electrons to move, and wires to provide the electrons whose motion make up the current. We have so far considered only the constant potential difference such as is provided by a battery. A voltage (as we shall begin to designate potential difference) of the type put out by a battery is called a direct current (dc) voltage. As shown in Fig. A4-2, such a voltage plotted against time maintains a constant value and direction (that is, the positive terminal remains the positive terminal).

Direct current voltages have certain important applications in automobile electrical systems, flashlights, and electroplating, for example, but from the point of view of total electrical use, the more important type is alternating current (ac) voltage. As is suggested by the name, an ac voltage changes in both magnitude and direction, in other words, the polarity of the output (+ or −) regularly changes. Later we will describe how generators (or alternators as they are called) produce such a voltage.

An ac voltage has the form, as a function of time

$$V = V_{max} \sin \omega t$$

which is shown in Fig. A4-3. The parameter ω determines how many times a second an ac voltage changes from positive to negative. In the United States system, ac voltages "cycle" 60 times a second, that is, ω is 60 cycles per second and the polarity is reversed 120 times per second. As Fig. A4-3 shows, it is no longer possible to talk of "the voltage" (or "the current") in the ac case as both these quantities vary with time. What is usually specified is the root-mean-square (rms) value, which is the roughly equivalent dc value as far as its heating effect is concerned. The rms value is $2/\sqrt{2}$ or 0.707 times the maximum value. Thus, the 110-volt house current we speak of, has a maximum value of 155 volts.

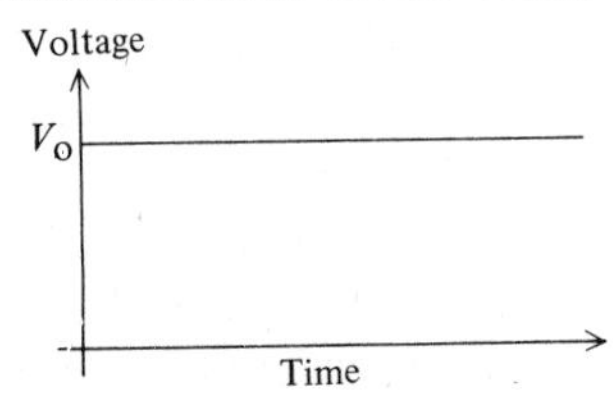

FIGURE A4-2 The dc voltage maintans a constant value V and constant polarity.

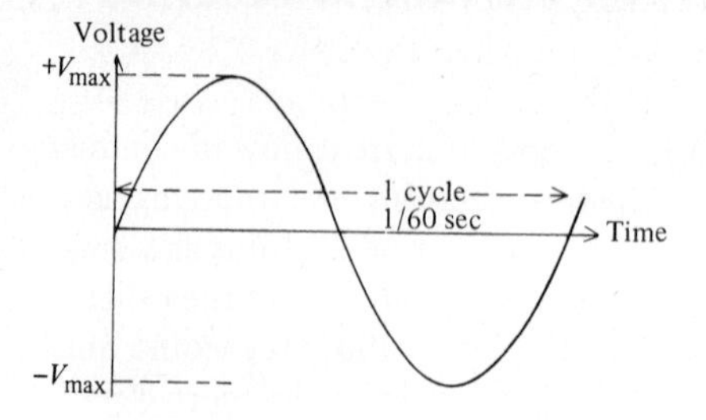

FIGURE A4-3 The ac voltage goes from $+V_{max}$ to $-V_{max}$ and repeats the *sinusoidal* variation 60 times a second.

Resistance

The most important, and certainly the simplest, of the resistances to current flow is the ohmic resistance within the wires. We have identified electric current in wires with the movement of free electrons. When you try to move a charged particle (electron, proton, or ion) through any medium (gas, liquid, or solid) the motion will be resisted. There are, after all, other atoms about and their electrons will be attracted or repelled by the moving charge; they will take up some of its energy just as a moving pool ball loses energy by collision with other pool balls. This lost energy contributes to the motion of the atoms of the transmitting medium and, therefore, heats it up, the heat tax again. The motion of the free electrons through the metal is thus very erratic. In the first place, they are moving randomly at relatively high speeds due to the temperature. The applied voltage gives them a small component of motion in the direction of the electric field (along the wire). This velocity produces a small drift of about 0.001 cm/second in that direction. Thus, the electrons are bouncing wildly about, but, on the average, drifting in the direction of the electric field.

The speed of this drift depends upon the number of collisions which interrupt the electron's drift and on the force of these collisions. In the metals such as copper or aluminum which are used for wire, and in several other materials, this drift speed depends only on the strength of the electric field. Since the voltage applied is proportional to the field in this important class of conductors the current i and the applied voltage V are related by

$$\begin{aligned} V &= i \times \text{constant} \\ &= iR \end{aligned} \tag{A4-3}$$

where R is a constant which takes into account the resistance to motion which we have just described. It also depends on the cross-sectional area of the wire and its length. R is called the resistance of the conductor. The common unit is the ohm named for yet another pioneer in electricity, George Simon Ohm, a German high school teacher whose experiments led to Eq. (A4-3). If V in Eq. A4-3 is one volt and i is 1 ampere, then R is 1 ohm.

It is this resistance that accounts for much of the loss of electrical energy in moving it from place to place. Those electrical devices whose purpose is to convert electrical energy to heat energy, space and water heaters, toasters, irons, and the like depend upon high resistance wires for the conversion. As we noted in Chap. 4, this conversion can be 100 percent efficient.

The definition we have just given for resistance holds true whether the applied voltage is ac or dc. The electrons are light enough to respond easily to the alternating field produced by an alternating voltage. Thus, the current in a resistive circuit has the same form and the same time relation (it peaks when the voltage peaks) as the ac voltage indicated in Fig. A4-3.

Capacitance and Charging Current

Resistance is the simplest of the circuit elements; it provides a straightforward resistance to motion and does not change the time relationship between the current and voltage. The next circuit element, the capacitor is more complicated.

From the viewpoint of dc voltage, a capacitor is a device for storing electricity. A typical capacitor consists of a pair of metal conducting plates separated by an insulator, which may be air. This arrangement is represented in Fig. A4-4*a*. When a dc voltage is established across the plates, electrons begin to move in the circuit, leaving the metal plate, which is connected to the positive battery terminal, and piling up on the other plate. Charges are therefore stored on the plates, held there by the attraction between the positive and negative charges on either side of the insulating space. The amount of charge Q which is stored depends on the voltage applied across the insulating space; this is a measure of the attractive force between the charges. It also depends on the size of the plates, their separation, and the insulating properties of the material between the plates.

It is of interest to consider the current that flows in establishing this charge on the plate. Immediately upon applying the voltage, a large current flows as the charges redistribute themselves. As the charge builds up on the plates, it becomes more and more difficult to put additional charge on them. The current as a function of time is of the form shown in Fig. A4-4*b*. It is large at first, and then drops rapidly towards zero. This is a transient current of short duration (fractions of a second typically). In a dc circuit, a capacitor is an open element; there is no steady current flow. It is used to store electrical energy. By connecting one plate to the other—"shorting" or discharging the capacitor—this energy can be released for any desired purpose, for instance, to create a large spark.

If the applied voltage is ac, the response is altogether different. Because the ac voltage alternates 60 times a second, the charges must flow back and forth in the circuit from plate to plate. Thus, in a capacitive ac circuit, a real current flows (except across the insulating space where it is more difficult to analyze what happens).

One of the major causes of energy loss from transmission lines is this charging current. The line is at a high voltage, 300,000 to 500,000 volts, while the ground

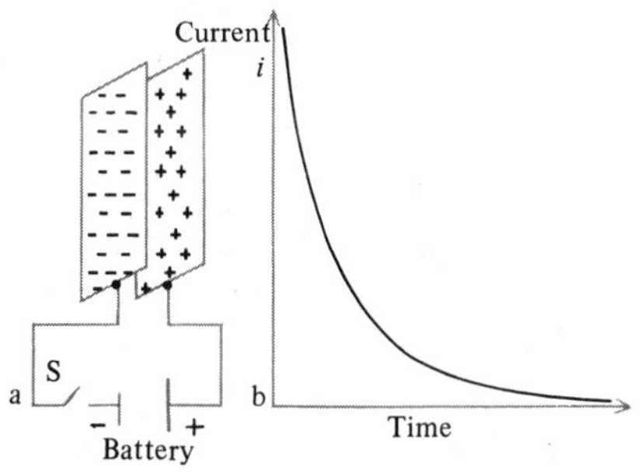

FIGURE A4-4 *a*. The battery maintains a potential difference across the separated plates which holds + and − charges in place. *b*, A current flows when the switch is closed, but if the voltage is dc it quickly dies out.

underneath it is at zero voltage. Thus, the line and the ground act as the plates of a capacitor and part of the current in the line is used to charge this long, narrow capacitor. This charging current increases the total current carried by the line and therefore increases the i^2R losses. In underground line the effect is much more pronounced since the separation between the line and the ground is much less and the capicitance is therefore greater; in fact, in underground lines at 350,000 volts, when the length becomes 25 miles, *all* of the current goes into charging and none is left for power transmission.

It is possible to compensate for this with a third type of circuit element called an inductance. We will not describe this here. They have the opposite effect as capacitors and in series with them make the line act more or less as a purely resistive line. This compensation every 25 miles is expensive, however, and adds greatly to costs. The lack of charging current is one of the major advantages of dc transmission lines (see Chap. 6).

Electric Power

The final addition to the basic vocabulary of electricity is electric power. Power is work per unit time. Since potential difference (voltage) is work per unit charge, and current is charge flow (charge per unit time) power is dimensionally the product of volts times current.

$$\text{Power} = \text{volts}\left(\frac{\text{work}}{\cancel{\text{charge}}}\right) \times \text{current}\left(\frac{\cancel{\text{charge}}}{\text{time}}\right)$$

$$= \frac{\text{work}}{\text{time}}$$

This is the definition for the dc case since the voltage and current do not change with time. In the units we are using, power (P) is

$$P = \text{volts}\left(\frac{\text{joules}}{\cancel{\text{coulomb}}}\right) \times \text{amps}\left(\frac{\cancel{\text{coulomb}}}{\text{second}}\right) \qquad \text{(A4-4)}$$

This unit of power is the watt defined earlier (see Appendix 2). Thus

$$1 \text{ watt} = 1\,\frac{\text{joule}}{\text{second}}$$

We can also combine the expression for electric power in terms of volts and amperes with the definition of resistance [Eq. (A4-3)] and learn an important fact about electric power transmission. If we want to deliver a certain amount of power through a transmission line then, since $P = iV$, we have a choice. We can deliver the power at high current and low voltage, or the same amount of power at high voltage and low current; the product determines the power. To force a current i against the resistance R of the transmission line, requires a potential difference $V_R = iR$. Thus, an amount of electric power iV_R is used up just in transmitting the current i through the line; it is "lost" as heat. This lost power depends only on the current, that is

$$P_{\text{lost}} = i \times V_R = i \times iR$$

$$= i^2R \qquad \text{(A4-5)}$$

In fact, it increases as the square of the current. It is less wasteful, therefore, to ship electric power at high voltage and low current.

Generally, the transmission lines which crisscross the country from power plants to the metropolitan areas where the power is used, are high-voltage transmission lines with transmission voltages ranging from 230,000 up to 765,000 volts. Experimental lines at as high as 1 million volts are contemplated. (We discuss the transmission of electric power in more detail in Chap. 6.) This i^2R loss, which increases with the length of the line (R increases with length), is one of the factors which makes it expensive to ship electricity long distances. Since this loss is a conversion of electric kinetic energy to heat, it is again an example of the irreversibility of energy flow. This heat cannot all be turned back into electricity.

Power in the ac case is more complicated. Dimensionally, it is still the product of the voltage V and the current i. But now, however, both the voltage and current are varying in time and may be "out of phase," that is, may peak at different times. This occurs because the time relationships between current and voltage depend, in the ac case, on the nature of the circuit. If the current and voltage are badly out of phase, if they peak at different times, then a considerable amount of power can be wasted. This waste occurs because the power delivered is given by some sort of average product of the varying current and voltage. If they are very much out of phase, the current may have to be fairly large to give enough power. The i^2R losses, however, depend directly on the current. The power companies are, therefore, most anxious to reduce this phase mismatch whenever possible.

In the usual home use of ac electric power, however, this phase mismatch is not very important; most of the home uses are resistive, and power is given by the simple product, Vi. In the watt-hour meter on the dwelling wall, the power Vi runs a synchronous motor, a clock. The speed of the motor is proportional to the average power. The number of revolutions in a given time thus measures the energy used during that time. You can gain an idea of how the power use in your household varies by observing the different rates of revolution of the turning disk at various times of the day.

THE BASIC ELECTRICAL DEVICES

In the previous sections we have introduced the vocabulary of electricity and the basic circuit elements. We will now look at the four most important electrical devices: (1) the battery which produces dc voltage; (2) the generators and alternators which convert mechanical energy to dc and ac electrical energy; (3) the motors which accomplish the reverse conversion; and (4) the transformer which is the key element in the transmission of ac electrical power. We will look first at the conversion of chemical energy to electrical energy in the battery.

Batteries

The most important devices for producing electricity in terms of total power are the generators and alternators which convert mechanical energy into electrical energy. For certain uses, however, especially where a small portable source of power is needed, the battery, which converts chemical energy to electrical energy, is the only choice.

A battery works on the preferential exchange of ions between a metal and an acid. A typical example is provided by the zinc strip dipped into a dilute solution of zinc sulphate ($Zn^{++}SO_4^{--}$) shown in Fig. A4-5*a*. Some of the zinc ions from the strip dissolve into the solution; they go into the solution as doubly charged ions (Zn^{++}) leaving the electrons on the strip and thus giving the strip a negative charge. This process is soon stopped. The negatively charged zinc strip soon attracts as many Zn^{++} ions as it loses and a state of equilibrium is reached. There is a potential difference between the negatively charged metal strip and the solution which depends on the concentration of the zinc sulphate. This is not a battery, however; it needs another half. We can provide this by a similar arrangement of a copper strip in a copper sulphate ($Cu^{++}SO_4^{--}$) solution (see Fig. A4-5*b*). The same thing happens, copper ions (Cu^{++}) go into solution, but not as readily as in the zinc case; consequently, the potential difference between the strip and the solution is less.

To make a battery, the first step is to connect the two strips (see Fig. A4-5*c*). Electrons now flow from the zinc strip to the copper strip since the zinc strip is at a higher negative potential. The electrons neutralize Cu^{++} ions at the copper strip leaving a surplus of SO_4^{--} ions at the copper strip, matched by the same number of Zn^{++} ions clustered around the zinc strip. Again, due to the buidup of these ions, the electron flow soon stops. The final step is to make the wall shown in Fig. A4-5*c* porous, so that the ions can pass through it. Then, SO_4^{--} ions and the Zn^{++} ions can flow together and neutralize making the electrically neutral $ZnSO_4$ again. Under these circumstances, there is a potential difference of about one volt between the zinc and the copper strips and a steady flow of current can be achieved.

Where did the energy come from? We could see its source directly if we were to put the zinc strip into the $Cu^{++}SO_4^{--}$. Immediately, the Zn^{++} replaces the Cu^{++} and heat is released, about 50 Calories for every 65 grams of zinc. The Zn^{++} is apparently more tightly bound to the SO_4^{--} than is the Cu^{++} and energy is released when it replaces the Cu^{++}. What the elaborate arrangement of the battery accomplishes is the slowing down of the reaction by making it take place in two steps: one at each electrode, as the strips are called. Under these circumstances, the energy is released as electric energy by the flow of electrons, rather than as heat energy. It is a similar process which occurs in the fuel cell as we see in Chap. 12.

Actual batteries operate on this same principle, but with different electrodes and solutions. In the familiar flashlight dry cell, for instance (which is not really dry), the negative electrode is the zinc outer container, the positive electrode, a carbon rod in the center, and the electrotyle, (as the solution is called) is soaked up in some porous material, sawdust, or powdered carbon. The reaction is similar: Zn^{++} goes in-

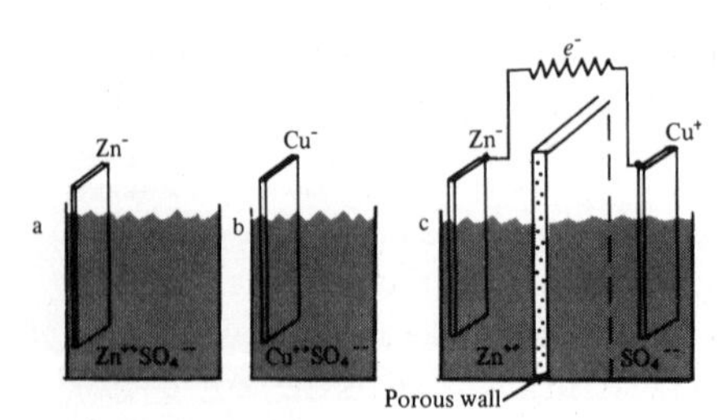

FIGURE A4-5 The compontents of a battery.

to solution, electrons flow and neutralize the positive ions which the Zn^{++} displaced. The potential difference is about 1.5 volts.

The more interesting battery, from an energy point of view, is the familiar lead storage battery. Here the negative electrode is lead and the positive electrode is lead peroxide (PbO_2) in a paste on a lead plate. The electrolyte is sulphuric acid. The reaction is complicated, but the net result is that for each two electrons that flow, one molecule of lead peroxide is changed to insoluble lead sulphate at the positive plate and an atom of lead is changed to the same insoluble substance at the negative plate. To achieve this, two molecules of sulphuric acid disappear from the electrolyte and are replaced by two molecules of water. Thus, the discharge of the battery is indicated by the weakening of the electrolyte. This can be measured by checking the density of the electrolyte with the familiar specific gravity device.

The importance of the storage battery is that by forcing current through the cells in the opposite direction the reverse process takes place, lead sulphate disappears, lead and lead peroxide are regenerated along with sulphuric acid, and the cell is returned to its original condition.

Thus, batteries have two important and unique properties; they convert chemical energy to electric energy, and they can store electrical energy by reconverting it to chemical energy. The only other device accomplishing the chemical-to-electric-energy conversion are the fuel cells, still in the experimental stage.

For large-scale energy storage, batteries must compete with pumped-storage hydroelectric energy. At the present, batteries are too costly. General Electric Company estimates[1] for instance, that in order for a 60 percent efficient battery to be competitive with pumped storage, it must store energy at a capital cost of $7.80 per kw-hr, and for a 90 percent efficient battery at $10.30 per kw-hr. For comparison, the cost of energy storage in present lead storage batteries (which are about 60 percent efficient) is in the range of 20 to 70 dollars per kw-hr. We can rule lead batteries out on another count; there is just not enough lead to use it in batteries in this way. There is much work under way, however, to develop more efficient batteries in the competitive cost range and progress can be expected.

Electrical to Chemical Energy

In the battery, chemical energy is converted to electrical energy. In recharging the storage battery, however, electrical energy is converted to chemical energy. There are other important uses of this same process. One is electroplating, in which, for instance, a copper electrode and one of another metal are put into a solution of copper sulphate. When a potential difference is applied across the electrodes, copper ions (Cu^{++}), attracted to the negative electrodes, leave the solution, capture electrons, and are deposited on the other electrode, while ions from the copper electrode go into solution to replace them. Other metals like silver, gold, and chronium can be plated out in this way.

Another important electric-to-chemical-energy conversion is the similar process of electrolysis. In this process electric energy is used to break up a compound and

[1] Cristopher, Hamlen, and Simms, "Batteries and Fuel Cells in Future Energy Handling," General Electric Report No. 72CRD255, September 1, 1972.

the two components of the compound are collected at the electrodes. Two examples will serve to illustrate it.

Aluminum is refined by electrolysis. Aluminum is very abundant in the earth but not as a pure metal. It is usually found in the form of an oxide such as aluminum oxide (Al_2O_3). To refine aluminum from a solution of this material, electric current is passed through it. The electrical energy then separates the aluminum and oxygen and the aluminum is carried to one of the electrodes, also made of aluminum, where it collects (plates out). Large currents are needed and, therefore, large amounts of electric power; in fact, it takes more than 16,000 kw-hr of electricity to produce a ton of aluminum. It must be remembered that this energy is stored as chemical energy in the aluminum and oxygen. When they again recombine in the "rusting" of aluminum, this energy is given up (slowly) as heat.

In the electrolysis of aluminum the stored chemical energy is not the desired end product. The electrolysis of water, however, is a reaction in which this energy storage may soon be of commerical importance. When an electric potential difference is applied across two electrodes in water, the water molecules separate. The H^+ ions are attracted to the negative electrode and the O^- to the other. It takes energy to decompose the water molecules and that energy is stored in the separated hydrogen and oxygen atoms. When they come back together through the oxidation of hydrogen, that is, burning with oxygen, this energy is released as heat. It is being suggested that the electrolysis of water may become an important way to store electric energy. The hydrogen produced is usable as a fuel; it has the additional advantage that the residue of the burning–the ashes, so to speak, is water. The prospects of large-scale use of hydrogen fuel is discussed in Chap. 12.

We now return to the discussion of ways to generate electric energy.

The Magnetic Force

The phenomenon of magnetism, that certain kinds of iron (first found near the ancient city of Magnesia in Asia Minor) attract iron to them, is as old as the early indications of electricity. The slow growth in the understanding of magnetism through the centuries was separate from, and parallel to, similar growth in the understanding of electricity. In 1820, however, a Danish physicist named Hans Oersted discovered than an electric current produces a magnetic field. We now recognized that magnetic force is a relativistic effect, that it depends on the motion of the charge.

We will not deal with origins of the magnetic force here, interesting though they are, but will concentrate, instead, on its properties as they apply to the conversion of energy. To say that a magnetic field exists in a region of space is to say that at each point in that region, there is a magnetic force of a certain strength and direction. Magnetic fields exist around magnets; a typical one is the uniform field between the North and South Poles of the specially shaped magnets of motors and generators. Magnetic fields can also be established by coils of wire; the solenoid is a familiar example. In both these cases the field originates in moving charges: in the current of the solenoid and in spinning electrons in the iron magnet. Similarly, the magnetic force acts only on moving charges and can thus exert a force on another magnet, on a current-carrying conductor, on a stream of moving charges (as in the electron beam of a television set), or on a single moving charge. To illustrate the peculiar nature of this force, it is sufficient for us to consider a single charge moving with velocity v in a region of magnetic field.

Figure A4-6*a* portrays this simple event. There is a magnetic field between poles labeled N and S. We indicate this field symbolically by B, and use the arrows to show its direction from N to S. The charge, an electron, moves through the field as shown, at right angles to B with a velocity v.

The magnetic force on a moving charge in comparison with other types of forces is a weird one. It is always at a right angle to the direction of motion of the charge. It is, therefore, *a force that can do no work* since work is only done by that amount of force which acts parallel to (along or against) the direction of motion (see Appendix 2). The magnetic force on the electron is not along the direction of motion v, but instead, as Fig. A4-6*a* shows, it is perpendicular to both v and B. It is perpendicular, therefore, to the plane formed by B and v. Another view of this is provided in Fig. A4-6*b*.

The other property of the magnetic force which is of importance is that it depends on the angle between v and B. It is a maximum when v is perpendicular to B as in Fig. A4-6; it is zero when v is parallel to B. If the electron in the example were directed along B from N to S or S to N, there would be no magnetic force.

The numerical magnitude of the force also depends on the amount of charge; it is stronger if the particle is, for instance, a helium nucleus with two proton charges, than if it is only the singly charged electron. We can put all this together and write

$$F_{\text{mag}} = qv\text{B} \sin\theta \qquad \text{(A4-6}a\text{)}$$

where q is the numerical value of the charge (in coulombs, for instance) and θ the angle between v and B. Since we will only deal simple cases for which $\sin\theta$ is either 0 or 1 (v either parallel to or perpendicular to B) we will usually work with this relation in the form

$$F_{\text{mag}} = qv\text{B} \qquad \text{(A4-6}b\text{)}$$

Equation (A4-6*b*) specifies only the numerical magnitude of the force F_{mag}. We have specified its direction already; it is perpendicular to both v and B, and, therefore, is perpendicular to the plane created by v and B. (In Fig. A4-6*a* this plane was the plane of the paper.)

The magnetic force is *always* perpendicular to v. It cannot, therefore, cause a

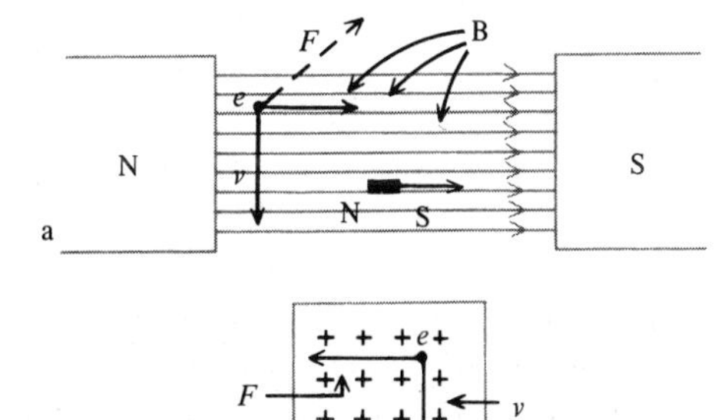

FIGURE A4-6 *a*. An electron moving with a velocity v at right angles to the magnetic field B feels the force F at right angles to v and B. *b*. An end-on view, the + symbols represent a B field into the paper.

charged particle to speed up or slow down (which is equivalent to the statement that it can do no work). All it can do is to cause a particle to curve. If the field is strong enough, in fact, it can bend the electron or other particles around in a circle. It is this force which sends particles chasing themselves in the great particle accelerators like cyclotrons and "cosmotrons." All we need to know, however, in order to explain how electric power is generated and motors run, is that the force is perpendicular to v and B.

A simple electric power generator is shown in Fig. A4-7, a "U" made of copper with a cross piece of length L which can slide on it. This contraption is placed in a magnetic field perpendicular to the plane of the paper, one which could be formed by a North Pole of a magnet above and South Pole below the paper. As we have said, the copper atoms in the wire align themselves so that there is a loosely bound electron which we've called a free electron, for each atom. If the cross piece is moved from position a to b these electrons are being physically moved in the magnetic field. If we focus on the cross piece in Fig. A4-8, we see that an electron in the wire will have a force exerted on it at right angles to its direction of motion, v, and to the direction of B, that is, along the wire. If it were not for the rest of the circuit, the electrons would pile up at the end of the wire, but the rest of the circuit allows them to flow, as shown by the arrows in Fig. A4-7. Since the same force acts on each free electron along the wire, there is a uniform electric field in the wire and a potential difference across its length. Moving a conductor at right angles to a magnetic field establishes a potential difference across the length of the conductor.

The existence of a potential difference means work is being done since potential difference is work per unit charge. How is work being done? Let's look again at this wire and now consider the effect of the current caused by the potential difference which flows in it, as in Fig. A4-9*a*. Current is charge in motion. These moving charges, and, therefore, the wire (since they are in the wire) will experience a force because of their motion. As a comparison between Figs. A4-8 and A4-9*b* shows, this force will be in a direction *opposite* to the direction in which the wire was originally pushed. If you are to continue to generate the potential difference, you have to overcome this "induced" force. The mechanical work which is being converted to electrical energy is the work involved in pushing the wire against this force over the distance through which the conducting wire is moved. If it were not for this opposing force, energy conservation would be violated, for electrical energy would be created without the input of work.

We have just stated that the motion of a conductor in a magnetic field sets up a potential difference along the conductor—an "induced" potential difference, or voltage. This voltage causes a current to flow, and *the action of the magnetic field on*

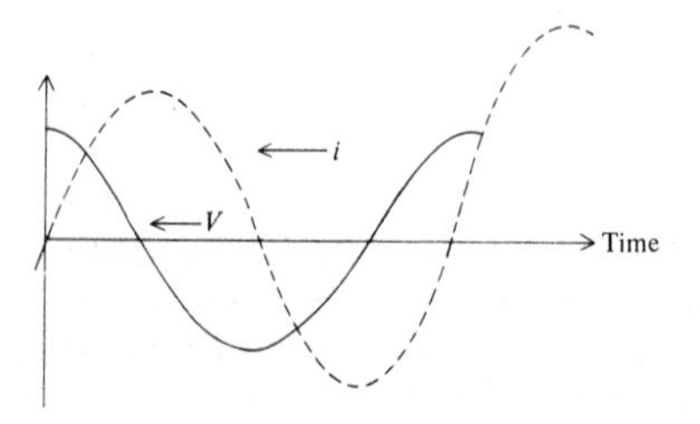

FIGURE A4-7 A simple generator. If the wire cross piece is moved as shown, the free electrons it contains will feel a force causing them to move up the wire and create the current *i*. The magnetic field B is into the paper.

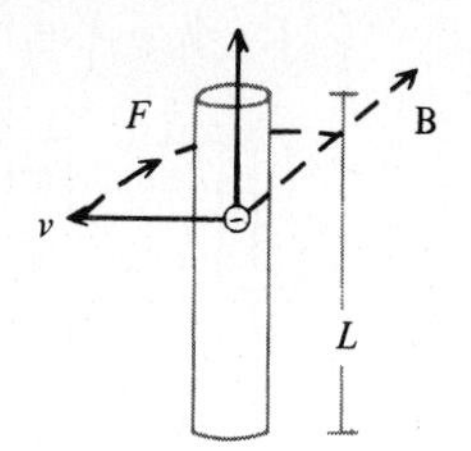

FIGURE A4-8 A magnified view of v, B, and F for an electron in the cross piece of Fig. A4-7.

that current opposes the motion which established it. This is a special instance of a very general law of electromagnetism, *Lenz's Law*, which states that

> the potential difference induced in a conducting coil by a changing magnetic field through that coil is always in a direction to oppose that change.

As we have shown, Lenz's Law is required if energy is to be conserved.

Electric Motors

We have shown the basic mechanism by which mechanical energy is converted to electrical energy in a generator or alternator. Let us now look at the reverse of that process, the conversion of electrical energy to mechanical energy in a motor. It is, in fact, the same process, only the sequence of events is different. We start with a wire carrying a current as in Fig. A4-9*a*. There will be a force on the moving charges and on the conductor that carries them, causing the conductor to move in the direction of that force. Thus, we have the beginning of a motor. But immediately as it moves, a potential difference will be induced and a current will flow opposite to the current which caused the original motion. That it opposes it is clear if you note that in Fig. A4-9*b* the force F moves the wire in the opposite direction to the direction in which it was moved in the example of Fig. A4-8. Thus, the potential difference generated by motion in this direction will be opposite to the one that caused current to flow in the "generator" we have been describing. This is just Lenz's Law again. To operate the motor you have to force current to flow against that potential difference. You must therefore supply an external electric potential difference, forcing charge through the motor against the induced potential difference. This is the source of the electric kinetic energy (potential difference times charge equals work or energy) to be converted to mechanical energy. In both cases, the generator and the motor, there is some i^2R heat tax since the currents run through wires which have resistances. This can be very small, however, and both motors and generators can be very efficient, 90 percent or more.

We have thus far presented the fundamental features of motors and generators. Practical devices are, of course, more complicated. The next step toward a practical generator is to rotate a coil in a magnetic field (see Fig. A4-10*a*). If we look at the

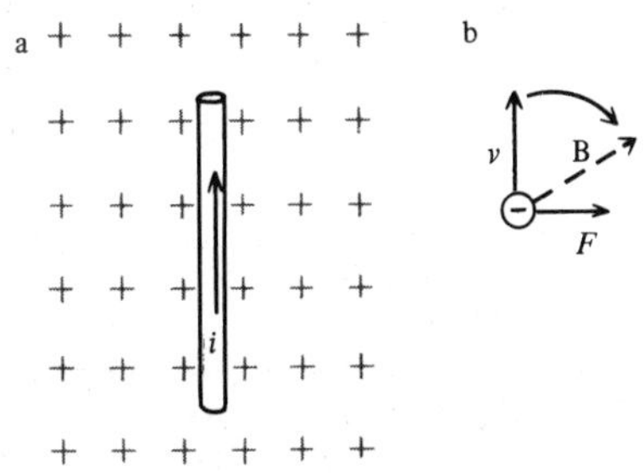

FIGURE A4-9 Current i in a wire oriented as shown in (*a*) (magnetic field B into paper), results in a force as shown in (*b*) on the electrons which constitute this current.

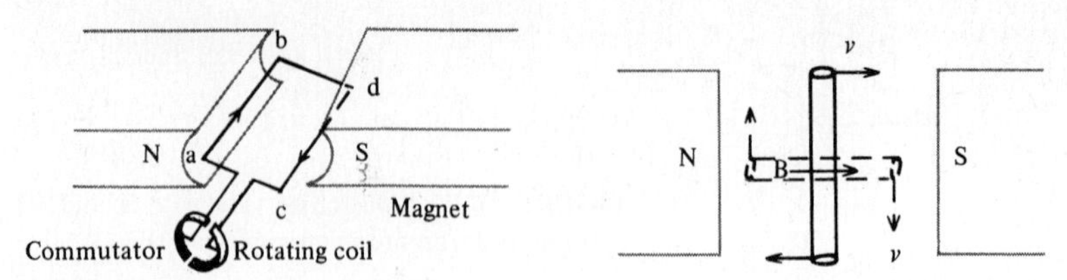

FIGURE A4-10 *a.* If the commutator is split as shown, the voltage is unidirectional dc as in Fig. A4-11. If it is solid, the voltage is ac as in Fig. A4-3. *b.* The force on the electrons in the long sides, ab, dc of *a* is zero when the velocity *v* is parallel to B and a maximum when it is perpendicular to B.

rotating coil from along its axis as in Fig. A4-10*b*, we see that the potential difference is induced only on the long sides (*ab, dc* in Fig. A4-10*a*) since only on these sides is the induced force *along* the wire. Since the force depends on the angle between the motion of the electrons and the direction of B on the sides it takes the form

$$F = qv \text{B} \sin \theta$$

The potential difference produced by the rotating coil will have this same $\sin\theta$ dependence (it is caused by this force) and it will take the form

$$V = V_{max} \sin \theta$$

shown earlier as Fig. A4-3, with $\theta = \omega t$.

The generation of a dc potential difference is a bit more complicated, for it is now necessary to arrange for connections which reverse the voltage every half-cycle. This is accomplished by the split ring or "commutator" shown in Fig. A4-10*a*. As can be determined by following the coils in their rotation, the output voltage picked up by these connectors is always in the same direction and has the time dependence shown in Fig. A4-11. As the number of coils is increased and set at angles to each other around a cylindrical "armature," other series of voltage curves identical to A4-11 are produced, but with peaks shifted to different times as in Fig. A4-12. These then add to form an almost constant unidirectional dc voltage.

Transformers

As we have seen in the section on electric power, it is least wasteful in terms of i^2R losses to transmit electric power at very high voltages.

At either end of the transmission line, however, the voltage requirements are different. At the plant, power is generated at 2,000 to 30,000 volts; it is used in the

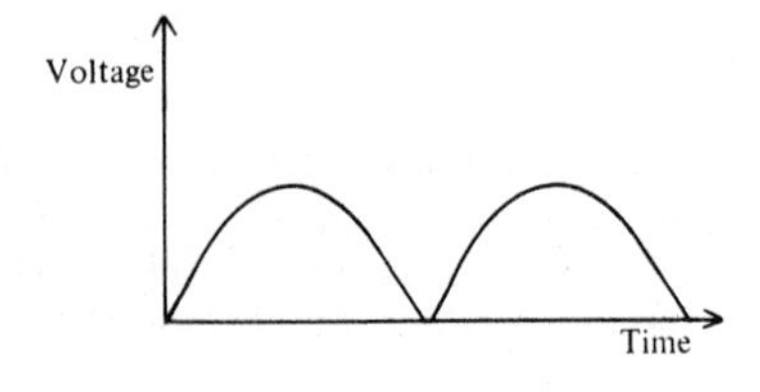

FIGURE A4-11 The split rings or commutator of Fi. A4-10*a* reverse the voltage each half cycle.

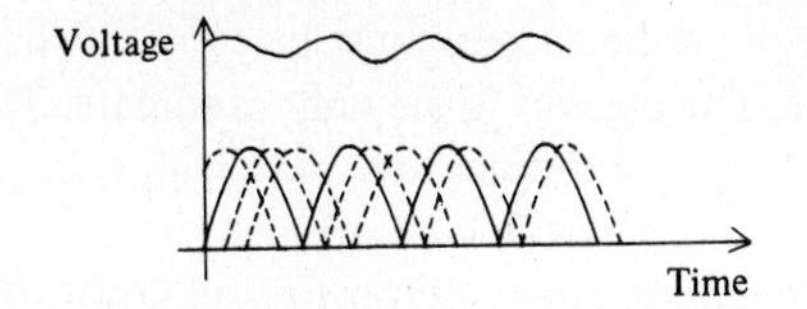

FIGURE A4-12 If many coils are wound on an armature, the cumulative effect is to produce a nearly constant unidirectional voltage.

home at 110 or 220 volts.[2] To make this system work there has to be some convenient and efficient method to increase or decrease voltages. Fortunately, there is a device, called, appropriately enough, the transformer, which accomplishes this.

The transformer is another application of Lenz's Law: the current induced in a conducting coil by a changing magnetic field through that coil is always in a direction to oppose that change. In earlier examples the change in the magnetic field was brought about by the motion of the coil turning in the field. It is also possible to produce a similar effect by changing the magnetic field itself. This is easily accomplished by changing the current which is producing that magnetic field.

Figure A4-13 is a schematic drawing of a transformer. The important components are the two sets of coils wound around opposite sides of a continuous "core" of iron. Suppose we first consider a low voltage—say, 100 volts—applied to the side with the single coil, the primary winding. Current in a coil wrapped around iron sets up a magnetic field in the iron. (This is how electromagnets are built.) The magnetic field goes all around the iron core (as shown by the B lines) and specifically, it passes through the many coils (we have drawn 10) on the other side. This is the secondary winding.

If the 100 volts which are applied from the left is ac, what is the nature of the field in the iron? It is also sinusoidal like the voltage (see Fig. A4-3), starting at a maximum, going to zero, reversing its direction, and going to a maximum in the new direction. What will be the effect of this changing magnetic field in the secondary coils? A changing current in the primary coil sets up a changing magnetic field in the iron. By Lenz's Law that changing magnetic field will induce a potential difference that opposes the change. In *each* of the secondary coils, therefore (since they surround the same magnetic field), an equal, but oppositely directed, potential difference will be induced. Since the coils are connected, these potential differences (voltages) add and the total voltage in the secondary is ten times the primary voltage. The step-up in voltage, therefore, depends on the ratio of the number of coils in the secondary to the number on the primary. It should be obvious that to step voltage down, one merely works the transformer backwards.

A final deduction: since energy is conserved (or nearly so since the heat losses

[2]In Edison's original generator, so the story goes, the design voltage was 100, but it actually produced 110 volts. Rather than rebuild the generator, he changed the filaments of the light bulbs and since then all small appliances have been designed for 110 volts.

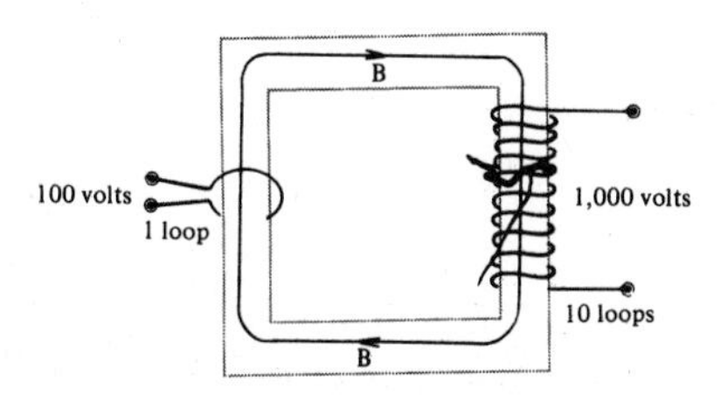

FIGURE A4-13 A transformer. The voltage in the single primary loop induces the same voltage in each of the ten secondary loops and they add up to produce ten times the primary voltage.

are small) the input and output power (energy per unit time) must be equal. If the voltage changes, something else must change. The current is the only candidate. Thus, on the left we have low voltage, high current, and on the right we have high voltage, low current.

With such a convenient device as a transformer to step voltages up and down, it was natural to go to ac power for transmission and distribution. Direct current voltage does not change. Thus, one has no varying B which can induce voltages. The transformer does not *convert* energy from one form to another. The energy remains electric kinetic energy. The only losses are i^2R losses in the windings and similar losses from currents induced in the iron. Properly designed transformers can be highly efficient, as high as 99 percent.

SUMMARY

We have just gone through the basic ideas of electricity, from charges, voltages, and currents to the ways of generating and transforming them. Because it can be carried by wires to the point of use, and converted efficiently and cleanly to most of the other useful forms of energy, electric energy is more and more the favored energy in our society. In Chap. 6 we described commercial electric energy: its generation, transmission, distribution and use. In further chapters we will look more deeply and see some of the hidden disadvantages of this popular form of energy.

APPENDIX 5

Energy from the Nucleus

In Chapter 10 we examine the potential and the danger inherent in our turning to the nucleus for energy. In this appendix we will ask why there is energy to be had from the nucleus and how we can get it out. We will also examine the meaning of radioactivity and what effects it has on living material.

THE NUCLEUS

At the center of the atom, the basic unit of a chemical element, is the nucleus. It is a small structure; where atomic sizes average around 10^{-8} cm, nuclei are about 10^{-12} cm in size. It is also incredibly solid and dense; 10^{12} times more dense than water. A cubic centimeter of nuclear matter would weigh 100 million tons.

The nucleus is composed of protons, electrically charged particles with the same amount of charge as the electron (although of opposite sign) but nearly 2,000 times more massive, and neutrons which are similar to protons but uncharged electrically.

The neutrons and protons are in motion inside the nuclear region, held there by the very strong nuclear force. In some ways the nucleus behaves like a constant density liquid, the neutrons and protons bouncing around inside and pulled back if they start to leave. This model, while very crude, will be helpful in understanding the mechanism of fission.

Since there is a nuclear force, there can be nuclear potential energy. Again in a crude analogy, a nucleus can be considered a "well" into which a nuclear particle "falls" in the same way a rock can fall into a real well. The dropping of a rock of mass m into a well of depth h converts mgh units of gravitational potential energy into kinetic energy which is then converted to heat energy as it strikes and sticks in the bottom. In a similar way a neutron or a proton, separated from a nucleus but within the range of the nuclear forces, has nuclear potential energy. If it is allowed to "fall" into the nucleus, this potential energy is converted to kinetic energy. If the particle is a proton, it must have enough energy to overcome the electrical repulsion of the other protons in the nucleus, to get over a "hill," so to speak, before it "falls" into the nucleus. The nucleus must then get rid of this extra energy.

From this analogy we see that energy is required to take a nucleus apart and that generally energy is released when we let neutrons and protons come together to form a nucleus. It is more complex than that, however, for the nuclei have a complex structure; some are more eager to accept neutrons or protons than others, and the energies with which these particles are bound to the various nuclei are not all the same.

Nuclear Terminology

In this appendix we will confine ourselves to the minimal nuclear physics which is needed to understand nuclear reactors and their byproducts. Nonetheless, we will have to introduce a little of the nuclear terminology to aid our discussion. We must first define isotopes.

An element is distinguished by the number of electrons in the atom or by the number of protons in the nucleus; these two numbers are the same. It is the attractive force of the positive charge of the nucleus that keeps the negative electrons in their atomic orbits, and, since atoms are electrically uncharged, there must be equal numbers of electrons and protons. The number of neutrons, however, does not effect the number of electrons (and, thus, does not effect the chemical properties). Nuclear species with the same proton content but different neutron content are called *isotopes.* Hydrogen, for instance, has a rare, heavier isotope, deuterium, with a neutron in the nucleus in addition to the proton. It also has a radioactive isotope, tritium, with two neutrons plus a proton in the nucleus. We will identify the isotopes by an initial which designates the element and give as a superscript, the number of particles–neutrons plus protons–in the nucleus. Thus, H^1 (hydrogen), H^2 (deuterium), and H^3 (tritium).

While we are introducing terminology, we must also define the energy unit commonly used in nuclear physics, million electron volts (Mev). The choice of this unit was made in the prenuclear physics days when the atom was the focus of physics. In studying atoms it was found that they could be taken apart by bombarding them with electrons accelerated across a potential difference (see Appendix 4) of a few volts. Charge multiplied by potential difference is energy. Thus, a natural unit of energy in atomic physics was the electron volt (eV). The nulcear energies are much larger than chemical energies because the nuclear force is so much stronger than the electrical force. The natural unit of nuclear energy is a million times larger, Mev instead of eV. This should suggest to you that the energies we can obtain from the nucleus are going to be much larger (about a million times larger) than the chemical energies we get from burning fuel.

Binding Energy and Nuclear Structure

A law of nature that we have referred to earlier is that systems seek to minimize their potential energy. Rocks fall, electrons arrange themselves in the atom, and nuclear particles arrange themselves in the nucleus so that their potential energies are as low as possible. Nuclei with this arrangement are *stable.* If they are sufficiently far from this optimum arrangement, they are unstable (radioactive) and they undergo a nuclear transformation, they change their internal structure in order to gain stability. We will discuss radioactivity later; let us now investigate stability further.

It is relatively easy to distinguish stable nuclei; they are the ones which exist in nature. Unstable (radioactive) ones transform eventually into stable ones. Thus, we can learn something about conditions of stability by looking at all the stable nuclei. Figure A5-1 shows all the naturally existing nuclei in a two-dimensional array with Z (the number of protons in the nucleus) displayed along the horizontal axis and N (the number of neutrons) along the vertical one. Each stable nucleus is represented by a dot on this chart. Dots which are lined up vertically (having the same proton number

Z) are isotopes of the same element. What we see from the arrangement in Fig. A5-1 is that the light nuclei seem to have N equal to Z, equal numbers of protons and neutrons, but as the number of particles grow, the region of stability (as one could call the broad line made by the dots) moves up on the chart to N greater than Z. There are several reasons for this, but the most important one is easy to understand. It is due to the electrical repulsion between protons. This force makes it easier to slip a neutron than a proton into a nucleus. At the high end of the chart, the number of protons Z gets so large that these elements are not really stable anymore and are weakly radioactive. It is in this region that the fissionable nuclei, U^{235}, U^{238}, T^{232}, and the like are found.

Another important structural fact about nuclei is that the energy required to take them apart, one nucleon (proton or neutron) at a time, is roughly constant; it does not change very much from nucleus to nucleus. The increase in potential energy obtained by separating all the nucleons in a nucleus (taking them all out of the "well" in our earlier analogy) is called the binding energy. What we have just said is that the binding energy *per* nucleon is almost constant; in fact, it is roughly 8 Mev per nucleon. In terms of our analogy it takes about 8 Mev of energy to lift one nucleon out of the nuclear well.

Fortunately for our energy needs, it is not *exactly* constant. We plot in Fig. A5-2 the binding energy per nucleon (BE/A) versus the total number of neutrons in the nucleus (A) (which is equivalent to N plus Z). What we see is that BE/A fluctuates for the very light nuclei but is always less than 8 Mev, then climbs to a broad maximum of a little more than 8 Mev from $A = 50$ to about $A = 100$, and then falls slowly as A gets larger. The reasons that BE/A is less than 8 Mev in the region of the light

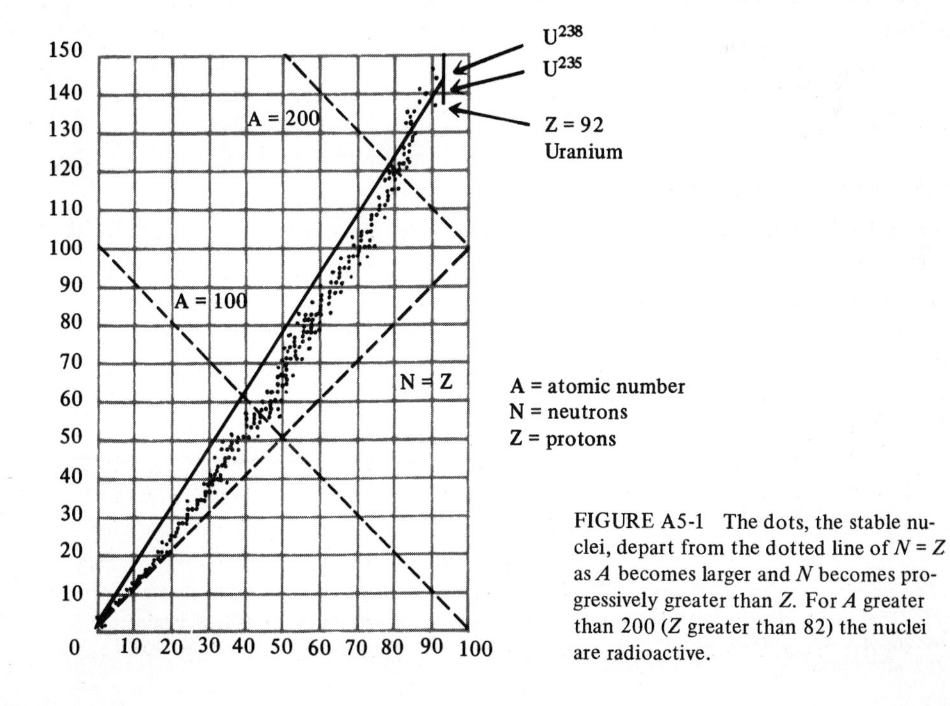

FIGURE A5-1 The dots, the stable nuclei, depart from the dotted line of $N = Z$ as A becomes larger and N becomes progressively greater than Z. For A greater than 200 (Z greater than 82) the nuclei are radioactive.

nuclei is too complex to discuss here but the main contribution to the falloff for large nuclei is the electrical repulsion of the large number of protons. The electrical force is in the direction opposite to the nuclear force, repulsive instead of attractive, and so it reduces the total potential or binding energy.

The way in which the nucleus can provide energy is summed up by the curve in Fig. A5-2. There are two ways to gain energy. One way is to take two light nuclei, say two H^2's and put them together to make He^4 which is at a larger A, and, therefore, has a larger binding energy per nucleon. Since, as we have said, the binding energy is the energy made available when a nucleus is put together, the difference between BE/A for two H^2's and BE/A for He^4 is excess energy released in the reaction. In a manner of speaking, the nucleons are bound more tightly in He^4 than in H^2, the well they fall into is a little deeper, and so energy is released. This is the fusion reaction whose potential for practical realization we discuss in Chap. 13.

Energy can also be gained by working from the other end. If we take a heavy nucleus, say, U^{235}, and split it into two lighter nuclei, strontium (Sr^{98}) and Xenon (Xe^{137}), for instance, we see that BE/A is again greater for either of the two than it was for the one and again the difference in BE/A is released. This amounts to a relatively large amount of energy. From Fig. A5-2 we see that the difference is about 1 Mev per nucleon. The *total* energy released by the 235 nucleons involved is therefore around 200 Mev.

There is another way to say the same thing in Einsteinian mass-energy terms. Einstein showed, as one of the consequences of his theory of relatively, that mass and energy were two forms of the same thing and related by the famous equation $E = Mc^2$ (where E is the energy, M the mass, and c the conversion constant, in this

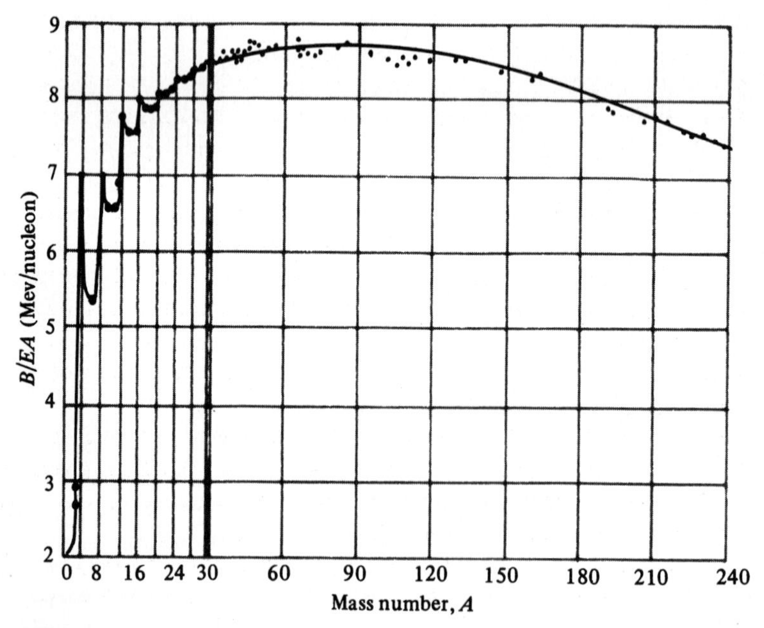

FIGURE A5-2 The BE/A fluctuates in a regular fashion for light nuclei (small A), peaks around $A = 60$ and then decreases. The difference between BE/A for $A \cong 235$ and $A \cong 100$ provides a measure of the energy available in fission.

case, the velocity of light).[1] What we have just said about the difference in binding energy could have been said in the following equivalent way. If (in the fusion example) we measure the mass of the two deuterium nuclei (H^2) and the mass of the helium nucleus (He^4) we find that the He^4 has a little bit less mass than the sum of the masses of the two H^2's. It is this missing mass that is converted into energy. Similarly, in the fission example the two nuclei, Sr^{98} and Xe^{137}, have less mass than did the original U^{235}; the difference has been converted to energy.

The 200 Mev per nucleus is not much energy in our terms. An Mev is equal to 1.6×10^{-13} joules. Thus, 200 Mev is equivalent to 3.2×10^{-11} joules. As a comparison, a grain of sand which has a mass of about 1/300 of a gram acquires 3.2×10^{-7} joules of energy in falling 1 cm—10,000 times larger than the energy released in fission. In a kilogram of uranium, however, there are an enormous number of nuclei, 2.6×10^{24} of them. Thus, the total energy available, in theory, from a kilogram of uranium is 2.6×10^{24} nuclei $\times$ 3.2×10^{-11} joules/nuclei, or 8.3×10^{13} joules, which is 7.9×10^{10} Btu or the energy equivalent of 3,000 tons of coal (at the rate of 26 M Btu per ton of coal). Let us look in more detail at this wonderous fission reaction.

DIVIDING FOR GAIN: THE FISSION REACTION

The elements to which we have turned for energy from fission are end products of the chain of events which began with the "Cosmic Egg" (see Chap. 2). They were produced, it is now believed, in thermonuclear reactions with heavy nuclei which occurred at the end of the sequence of such reactions in the giant stars. They were then picked up from the gas and dust in the next round of star formation. Our sun is such a new star, and, as a result, it has a full complement of these elements in its planets.

These heavy elements are rare. Only certain planets around a sun like ours will have traces of them. In addition, they are radioactive (unstable) and steadily but slowly disintegrate into lighter, more stable forms. But they are here, and we have learned to release some of their stored energy, first to destroy two Japanese cities, and now to produce electrical energy.

The trick is to make them split. We are helped in this by their shape. Returning to the model of a nucleus as a liquid which we mentioned earlier, we find by experiments that these heavy nuclei are not spherical, but rather more football-shaped with some of the protons concentrated at the ends. Because of the short range of the nuclear force it is difficult for them to overcome the repulsive force between the two ends. Anything that stretches the nucleus further can allow this repulsion to tear it apart. It is set up to fly apart; all it needs is a trigger.

The trigger is available in the form of a neutron. This uncharged particle is an ideal projectile; the strong electrical force of the highly charged nucleus can't keep it away. When it hits the nucleus it not only gives up its own kinetic energy, but when it is bound to the nucleus, that binding energy is given up.

The nucleus has several ways that it can get rid of the extra energy but we are interested only in the one which leads to fission. In this mode the nucleus begins to

[1] If M is expressed in kilograms and c is taken to be 3×10^8 meters/second, then the energy unit of E will be the joule.

vibrate like a lump of jelly. If the vibration goes through the sequence shown in Fig. A5-3, then fission takes place. As soon as it has "necked down" into the peanut-shell shape of Fig. A5-3*b*, the electric replusion between the two ends takes over and splits the nucleus into two fragments. Now the parts are separated, the nuclear force does not reach from one to the other, and the electrical repulsion sends them flying apart. At this stage the electric potential energy is about 200 Mev. This energy, the energy of fission, is quickly converted to kinetic energy.

Any heavy nucleus can be caused to fission if it is hit with a neutron of sufficient energy, say a few Mev. The nuclei which are used as the fuel in bombs and reactors, however, have a very special feature; because of their structure, they can bind a neutron tightly and thus release 6 Mev or so of binding energy. The neutron does not need to bring much kinetic energy of its own to such a nucleus. Even a very low energy neutron (a "slow neutron") wandering by can be captured, and, with the released binding energy start the fission-producing vibrations. Furthermore, a slow neutron stays in the vicinity of a nucleus longer and is, therefore, more easily captured than a fast (energetic) one.

We have now reached a level of detail that forces us to stop talking generally about uranium and look at the various isotopic forms in which it occurs. There are two, U^{238} and U^{235}, which occur naturally. U^{238} is far and away the most abundant form (99.3 percent of the total) while U^{235} makes up only 0.7 percent; in other terms, there is one U^{235} nucleus for every 140 U^{238} nuclei in an ordinary sample of uranium.

The picture should now be becoming clear; U^{238} must be the one that needs a fast neutron to make it fission; otherwise, any chunk of uranium ore would blow itself apart, triggered by the "slow" neutrons constantly bombarding earth from cosmic radiation. It is only the rare isotope, U^{235}, which can capture a slow neutron and fission.

Chain Reactions and Their Control

The best naturally occurring fuel for nuclear reactors is that rare nucleus U^{235}. We can also make two artificial nuclei, plutonium (Pu^{239}) and U^{233}, which are also fissionable with slow neutrons. (We will discuss the preparation of these fuels in our treatment of the breeder reactor.) To use U^{235} as a fuel, however, we must first gather enough of it together, then set it off, and most important of all, control the reaction.

A single U^{235} nucleus will capture a neutron, receive the 6 Mev binding energy, go into the vibrations we described and finally fission. The two pieces, the fission

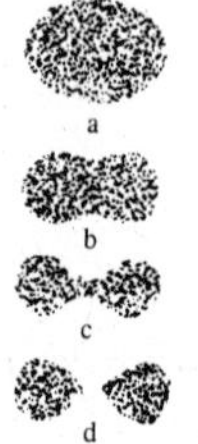

FIGURE A5-3 The steps in the fissioning of a nucleus.

products, carry off the 200 Mev of kinetic energy, but because they carry a large electric charge from their protons, they interact strongly with the matter around them and their kinetic energy is quickly converted to heat. The important conversion process in a fission reactor is, thus, nuclear potential energy to thermal energy.

The 200 Mev from one nucleus is, as we have said, not very much energy. To get usable amounts of energy we have to find some way to make essentially all the 10^{22} or so U^{235} nuclei in a chunk of uranium undergo fission. To see how this can occur we must take into account the other pieces, the neutrons, which also fly out when the nucleus fissions. These neutrons have a considerable amount of energy which they quickly lose by collisions with the matter around them. After they are slowed down they can strike other U^{235} nucleus and cause them to fission. They offer, therefore, the possibility of igniting a *chain reaction*: one nucleus releasing neutrons which cause other nuclei to fission, releasing more neutrons, and so on. It is of obvious importance to know how many neutrons are released per fission. If it is only one, on the average, then the reaction started by a blast of neutrons doesn't grow, but at least continues; with less than one it dies out. The explosive release of energy we are looking for needs more than one neutron released per fission. Fortunately for our need of electric power, the average number of neutrons released in the fission of U^{235} is 2.6. This is more than enough for a chain reaction.

The actual chain reaction is complicated by other things which can happen to the neutrons. In the first place they must be slowed down, for, as we have mentioned earlier, slow neutrons are much more readily captured by U^{235}. The most effective way to accomplish this is to have the neutron strike something its own size, a proton, for instance. In such a collision it loses half of its energy, while in striking a much heavier nucleus it will only lose a small fraction of its energy. Thus, water would seem to be a good substance to slow down neutrons, a good *moderator* in reactor terminology. Of course, the U^{238} which is always abundantly present in a nuclear fuel sample also acts as a moderator; however, it takes many more collisons with this heavy nucleus to slow the neutrons. Other commonly used moderators are heavy water (D_2O), in which the hydrogen is replaced by its chemically identical isotope deuterium (D is the H^2 we have earlier identified), and carbon (C^{12}), usually in the form of graphite. These two moderators have the advantage that they do not "capture" neutrons as hydrogen does and thereby take them out of the chain reaction.

The other parameter of importance to the success of a chain reaction is the size of the uranium fuel assembly. If it is too small, neutrons go out through the sides and are lost in that way. The size criterion is embodied in the *critical mass.* A certain amount of U^{235} of a given concentration is needed before the chain reaction can occur. When this critical mass is assembled, however, the chain reaction begins; a U^{235} nucleus fissions when struck by a slow neutron, the two or more neutrons released in the fission of this first nucleus fly off, are slowed down by collisions with the moderating substance, captured by other U^{235} nuclei, and cause them to fission and release more neutrons. The chain reaction grows and involves the entire fuel element in a few thousandths of a second. Uncontrolled, the critical mass explodes. This is roughly how an atomic bomb works. The fuel is kept in pieces smaller than the critical mass and then brought together when on target. What is needed for a nuclear reactor is some way to control, or slow down, the chain reaction.

Controlled Fission. To control a fission chain reaction, it is necessary to control the number of neutrons available for fission. What is needed is some way to take enough neutrons out of circulation, with each generation of fission events, so that the explosive buildup does not occur; in fact, we sometimes want to be able to take out so many neutrons that less than one per fission is available, thereby shutting off the chain reaction.

There are substances which very readily capture slow neutrons and bind them into a new (nonfissioning) nucleus. The reaction of neutrons with boron, one such element, is described symbolically by

$$B^{10} + n \longrightarrow Li^7 + \alpha + 2.8 \text{ Mev}$$

which says that the boron nucleus (B^{10}) captures a neutron n and the resulting nucleus (B^{11}) splits up into the lithium nucleus (Li^7) and an alpha particle (α). In the splitting up, energy (but no neutron) is emitted. Since the alpha particle cannot cause fission, this reaction effectively takes neutrons out of the chain reaction. The boron reaction thus allows not only the fine tuning of a reactor to the power level desired, but also provides a safety mechanism for quick shutoff, for "scramming" the reactor.

This technique of controlling the chain reaction by varying the rate of capture of the neutrons is simple in principle. In practice, the great speed with which the reaction builds up makes it much more ticklish. When a neutron is released it takes about 20 collisions with the protons of the moderator (and more for heavier moderators) for it to reach the low "thermal" energies needed for U^{235} capture. These collisions, however, happen in about 10^{-4} seconds. Thus, if two neutrons are allowed to be captured per fission, the chain reaction doubles its intensity in 10^{-4} seconds. Obviously, such a rate of increase cannot be allowed; a reactor operating near its maximum safe level could be suddenly at twice the level before a human or even a machine could react. At twice the safety level the great increase in temperature might destroy the safety equipment.

Even at lower neutron multiplication levels, the chain reaction increases with great speed. If, for instance, 1.001 neutrons per fission are allowed to interact, this means that the chain reaction grows by one part in a thousand for each burst of neutrons. With 10^{-4} seconds between bursts, however, the reaction still doubles in one tenth of a second. This quick buildup still requires extremely delicate settings and rapid response.

Fortunately, the real situation is better than this; some of the neutrons are delayed. They come from radioactive nuclei produced some time after the fission event by the series of decays that take place in fission products, and these neutrons, which, in U^{235} fission are about 1 percent of the total, do not appear until 10 seconds or more after the fission event (the average delay is about 14 seconds). This means that the reactor operator or the automatic control equipment has, in fact, 10 seconds or so to adjust to an increase. At the neutron ratio of 1.001, for instance, the equipment can be set to 0.99 on prompt neutrons—a ratio that would shut off the reaction—and then enough of the delayed neutrons can be allowed to contribute to bring the ratio up to the desired value. One of the difficulties with the plutonium-fueled reactors, which we shall discuss later, is that only 0.25 percent of the neutrons for that fission are delayed, which makes for a more delicate control situation.

We have made this excursion into a discussion of delayed neutrons in order to emphasize the precision of control needed to prevent major mishaps in a reactor. By providing the best in electronic equipment, building in overlapping and duplicate safety systems, and requiring strict licensing procedures for reactor operators, however, an excellent safety record has been maintained.

BREEDING NUCLEAR FUEL

The element uranium is relatively rare and U^{235} makes up only 0.7 percent of natural uranium. Even with the impressive amount of energy which fission releases from each ton of U^{235} there are not many thousands of tons of it around. With the onset of the nuclear age it became natural to look for other fissionable fuels, and, with the understanding of the nuclear physics of the fission reaction other fuels are being suggested.

What is required to make a nucleus fissionable by slow neutrons is a particular structure. It has to have a "hole," so to speak, for a neutron to fall into so that the probability of neutron capture will be high and a sufficient amount of energy will be released. U^{235} is the only naturally occurring nucleus which has that structure, but two fairly long-lived artificial ones, U^{233} and Pu^{239}, can be created.

These two artificial isotopes are made in neutron reactions. U^{233} is produced by the neutron bombardment of Th^{232} (thorium) and Pu^{239} in the same way from U^{238}. Since thorium is actually a little more plentiful than uranium and since U^{238} has been so far considered a waste material, the addition of these two elements promises to greatly extend the supply of fissionable fuels.

The promise comes from the possibility of breeding fissile materials, using neutrons that are surplus to the chain reaction to make new fuel which can replace that "burned up" in the core. The possibility of this occurring is obviously sensitive to the number of neutrons produced in fission. There must be more than two, for one is needed to continue the chain reaction and the other to breed a new fissionable nucleus: two is the "break-even" point. If there are more than two, then more fuel can be created than used. A nuclear reaction which accomplishes this is called a breeder reaction and the reactor using it a *breeder* reactor.

The reactions involving Th^{232} and U^{238} differ in an important way, the former, called a thermal breeder reaction, uses slow neutrons and the latter uses fast neutrons. These reactions are depicted in Figs. A5-4 and A5-5.

In Fig. A5-4 the reaction of Th^{232} with a thermal neutron is shown schematically. The capture of the slow neutron leads to the formation of Th^{233} which is unstable (radioactive). It has a surplus of neutrons and one of them changes into a proton by emitting a beta particle, an energetic electron. The change of a neutron to a proton transforms Th^{233} into Pa^{233} (protoactinium) which is also radioactive. The neutron to proton transformation is repeated and the longer-lived (but still mildly radioactive) nucleus U^{233} is formed. U^{233} has a structure similar to U^{235} and is an excellent fissile fuel.

A parameter of obvious importance in describing breeder reactors is the breeding ratio, the ratio of the fissile fuel produced to that consumed in the process. The

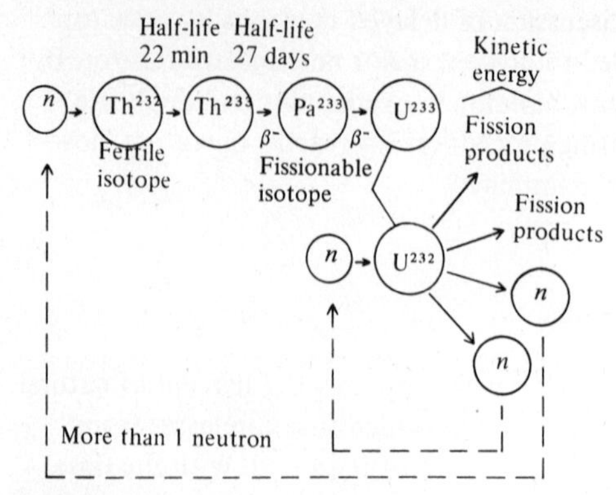

FIGURE A5-4 The thorium breeding cycle uses slow or "thermal" neutrons and produces U^{233} which is an excellent fission fuel.

breeding ratio must be greater than one if we are to gain fuel. Related to the breeding ratio is the doubling time, the time it will take a breeder reactor, under normal operating conditions, to produce enough fissionable material for a second reactor. In the thermal breeder reactors under development the breeding ratio is 1.05 and the doubling time 21 years (see Chap. 10).

Fast Neutron Breeders

At the present time the focus of the AEC reactor development effort is on the reaction by which U^{238} is converted to Pu^{239}. This reaction, which requires a fast (unmoderated) neutron, is shown schematically in Fig. A5-5. U^{238} captures this fast neutron and becomes U^{239} which is radioactive; it emits a beta particle (a neutron becomes a proton) and forms the nucleus Np^{239} (neptunium). This is also radioactive, beta emission occurs again, and Pu^{239} is formed. In this reaction the breeding ratio may be as high as 1.5 and the doubling time 8 to 10 years. This is of particular significance since it is about the doubling time of electrical energy consumption.

These breeding reactions have much to offer us in terms of resource energy; in fact, some of the advanced models being contemplated are expected to have doubl-

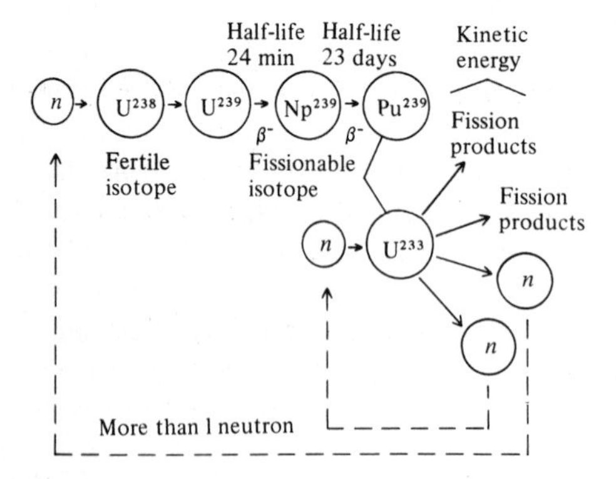

FIGURE A5-5 The U^{238} breeding cycle begins with the capture of an energetic neutron and then proceeds through a series of β-emission processed to form Pu^{239}

ing times of as short as 4 to 5 years. In full operation, they will not only fuel other breeders, but provide cheap fuel for light water "converter" reactors also. But they have serious disadvantages as well, which are given full consideration in Chap. 10, where we consider the reactors in which this breeding takes place.

RADIOACTIVITY

We have already mentioned the word "radioactive" several times. We will define it more specifically here, introduce some of the terminology associated with it, and then explain why so much radioactive material is produced in a fission reaction. In the final section we consider some of the biological effects of radiation.

A radioactive nucleus is unstable; it has either an excess of energy or is off the "region of stability" shown in Fig. A5-1. In either of these cases it will try to regain stability by ejecting radiation, the extra energy or excess particle.

There are three major radioactive emissions; gamma (γ) rays which are pure electromagnetic energy similar to X-rays but more penetrating, beta (β) particles, the fast electrons which we have already mentioned, and alpha (α) particles which are two neutrons and two protons clumped together, identical to the nucleus of a helium atom.

The γ-rays are usually emitted when only excess energy must be released. The β-rays are given off when it is necessary, in order to reach or approach stability, to change a neutron into a proton. The plus charge of the proton is balanced by the minus charge of the β so no new charge is created. There are similar processes which occur in some nuclei whereby a proton changes into a neutron and a positive electron, β^+, is emitted. Finally the α-particle is emitted, for instance, by the very heavy nuclei; all of those heavier than $A = 200$ (or $Z = 82$) on the chart in Fig. A5-1 are alpha emitters.

The most important fact to know about the dynamics of radioactive emission, or radioactive decay, as this transformation of a nucleus is called, is that the number of decays that take place in a sample of radioactive material is proportional to the number of radioactive nuclei in that sample. In symbols we write this

$$\frac{\Delta N}{\Delta T} = -\lambda N$$

which states that the number ΔN of nuclei which decay in a time interval ΔT is equal to a constant λ times the number N of radioactive nuclei present at the start of that time interval. (We recognize this as almost identical to the equation of the savings account in Appendix 1, with λ the interest. However, it has a minus sign which suggests that the exponential curve will go down instead of up. The suggestion is correct.)

The constant λ is a characteristic of each nuclear species. For an intensely radioactive material, λ will be large; for a weakly radioactive substance, λ will be small. (The minus sign indicates that the change in the total number ΔN is a *decrease* with time.)

This is a very interesting mathematical relationship. The plot on semilog paper of a curve of N versus T having this property looks like Fig. A5-6. It is exponential like

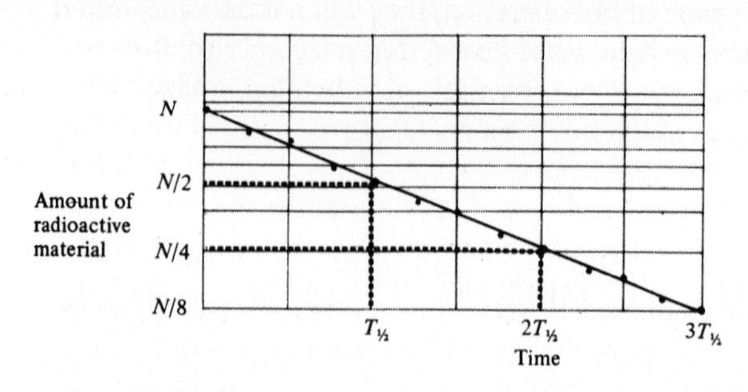

FIGURE A5-6 The decay of a radioactive nucleus is a decreasing exponential; half of the material transforms to something else (decays) in the time period of a half-life.

most of the curves we have looked at in this book, but the minus sign makes it a decreasing exponential. We admit to a perverse desire to show a decreasing exponential after all the ones which increase so inexorably in other parts of the text, but there is a more important reason. For increasing exponentials we found the doubling time an important descriptive quantity (see Appendix 1 and Chap. 5). For decreasing exponentials there is an analogous quantity called the half-life which is one of the most important parameters of radiation terminology. As Fig. A5-6 shows, the half-life ($T_{1/2}$) is the time it takes for a given number of radioactive nuclei to reduce down to $N/2$, for half of them to decay. Thus, after one half-life, half a given amount of radioactive material remains, after two half-lives, one quarter, after three, one eighth, and so on.

It is easy to show mathematically that $T_{1/2}$ is related to the constant λ we introduced earlier. This λ was a measure of the intensity of the radiation, of the decay rate, $\Delta N/\Delta T$. If a substance has a high decay rate then it will decay rapidly; it will have a small $T_{1/2}$. Thus, $T_{1/2}$ is inversely proportional to λ

$$T_{1/2} \propto 1/\lambda^{\dagger}$$

This will be of great importance as we look at the radioactivity of the fission products; those with large λ are radioactively "hot" and dangerous, but they die out rapidly. If they have half-lives of years, then they present a storage problem. Let us look at these fission products.

Fission Produced Radioactivity

Just as nuclear reactors bring us a new source of energy, they also bring a new form of pollution. To see what it is and where it comes from, we must go back to our description of fission and look at the nature of those "fission products" that fly apart when the U^{235} (or another) nucleus is split.

Typically, the $U^{235} + n$ splits into a heavier and a lighter nucleus; Sr^{98} (an isotope of strontium) and Xe^{138} (an isotope of Xenon) are examples. These isotopes are not the natural ones for these two elements: the heaviest stable forms are Sr^{88} and Xe^{136} The fission products have too many neutrons for stability and quickly begin to reduce this excess. The first step is to emit some neutrons. These are the "prompt" neutrons which contribute immediately to the chain reaction. In the case we have used for examples, $Sr^{98} \rightarrow Sr^{97} + n$ and $Xe^{138} \rightarrow Xe^{137} + n$, these are lighter isotopes

† In Appendix 1 we also pointed out that $T_{\text{double}} \propto 1/r$ [Eq. (A1-2)].

of the strontium and xenon. They are still very far from stability and must continue to change in order to achieve stability. In other words, they are highly radioactive.

The form of the radiation emitted varies from one fission product to another. Since their instability comes from an excess of neutrons the most prominent radiation is beta radiation. Gamma rays are often emitted as well to carry away extra energy. Occasionally, as we have said, a neutron will again be emitted at some step in the decay process, the delayed neutrons which are so useful in control procedures.

The most important parameter for the discussion of these radioactive leftovers is half-life, for this provides a measure of both the intensity of the radiation and of, how long they will stay around. Generally, the half-lives for the first steps are short —a few seconds—and get longer as they proceed down the path toward stability. The last step, the transformation (usually by beta emission) that changes the radioactive material into a stable one, is the longest one. Its half-life is usually measured in years. The half-lives of some of the fission products are given in Table A5-1 and the percentage distribution by the mass number A is shown in Fig. A5-7. On the basis of half-lives, we see that strontium (Sr^{90}) with a half-life of 28 years, and cesium (Cs^{137}) with a half-life of 30 years, will be the most troublesome. These two also have mass numbers, 90 and 137, near the peaks of the abundance curve in Fig. A5-7. The shorter-lived products, although intensely radioactive (large λ) will not be around for long, while technicium (Tc^{99}), for instance, with a half-life of 10^6 years, will be only weakly radioactive (λ small). While Table A5-1 does give us some idea of what

Table A5-1 Fission-Product Radioisotopes in Radioactive Wastes*

Element	*Atomic number*	*Half-life*	*Radiation emitted*
Krypton-85	36	4.4 hr (IT)† → 9.4 yr	$\beta,\gamma,e^- \rightarrow \beta,\gamma$
Strontium-89	38	54 days	β
Strontium-90	38	29 yr	β
Zirconium-95	40	65 days	β,γ
Niobium-95	41	90 hr (IT) → 35 days	$e^- \rightarrow \beta,\gamma$
Technetium-99	43	5.9 hr (IT) → 5×10^5 yr	$e^-,\gamma \rightarrow \beta$
Ruthenium-103	44	39.8 days	β,γ
Rhodium-103	45	57 min	e^-
Ruthenium-106	44	1 yr	β
Rhodium-106	45	30 sec	β,γ
Tellurium-129	52	34 days (IT) 72 min	β,γ
Iodine-129	53	1.7×10^7 yr	β,γ
Iodine-131	53	8 days	β,γ
Xenon-133	54	2.3 days (IT) → 5.3 days	$e^-,\beta \rightarrow \beta,\gamma$
Cesium-137	55	30 yr	β,γ
Barium-140	56	12.8 days	β,γ
Lanthanum-140	57	40 hr	β,γ
Cerium-141	58	32.5 days	β,γ
Cerium-144	58	590 days	β,γ
Praseodynium-143	59	13.8 days	β,γ
Praseodynium-144	59	17 min	β
Promethium-147	61	2.26 yr	β

*Source: Clark Goodman, "Science and Technology of the Environment," University of Houston, 1972.

β: beta particle, an electron.
e^-: internal electron conversion.
†IT: isomeric transition, internal.
γ: gamma ray, similar to x-rays.

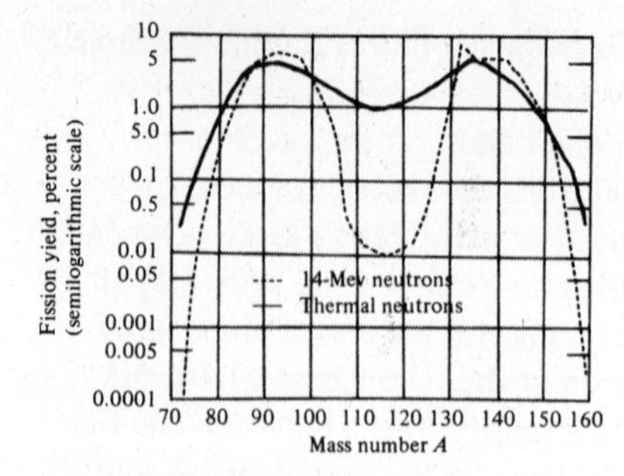

FIGURE A5-7 The douple-humped peak of fission particles. The most probable mass numbers, each accounting for almost 10 percent of the yield, are the nuclei in the regions $A = 90$-100 and $A = 134$-144. (The total percentage sums to 200 percent since there are two fragments per fission.)

materials are going to be around awhile, it says nothing about their biological effects. We turn to this in the final section.

BIOLOGICAL EFFECTS OF RADIATION

The particles (α, β, or γ) ejected by a radioactive nucleus have kinetic energy. As they traverse matter they lose that energy, either by a direct collision, by absorption, or, in the case of the electrically charged particles, by interacting with the electrons of the surrounding atoms. If the surrounding matter is living tissue, all these interactions are damaging to the cells; molecules are displaced, broken, or transformed.

It is difficult to make general statements about the degree of damage to living matter. Effects differ, on the one hand, according to the type of radiation (α, β, or γ, for instance) and, on the other hand, the effects depend on the nature of the struck cell—whether it is an ordinary body cell or a genetic cell.

An attempt has been made to take into account some of the first variation by defining a unit, the roentgen equivalent man (rem),[3] which is expressed in terms of energy absorbed per gram of tissue and has built into it the difference between various types of radiation in their effects on man.

We can also make some general statements about cell damage. We know what radiation does at high dosages. The absorbed energy disrupts and may kill the cell. If enough cells die the organ is killed; if crucial organs are killed the organism dies. Lethal radiation dosages for man are in the range of hundreds of rems, the LD 50 dose (lethal dose 50, lethal to 50 percent of the exposed sample) is 400 rem. The doses we are presently concerned with are much lower than this, in the region of a few rem. Here we are not dealing with instant death, but with delayed effects. These may be of two kinds; they may take the form of a disease such as some form of cancer or leukemia, or they may be genetic effects and only show up in future generations. These two effects must be dealt with separately.

Somatic Effects

If the damage is to the exposed individual, rather than his generations, the effect is called a somatic effect. Our chief concern here is with the carcinomas, cancers of various forms, and leukemia. There is no question that radiation can cause cancer, the difficulty is to assess the probability of this happening at very low levels of radiation. The assessment has to be a statistical one; there is no way to differentiate

[3] For the β- and γ- radiation we are dealing with, a dose of 1 rem corresponds to the absorption of 100 ergs of energy by a gram of tissue. An erg is 10^{-7} joules.

difficulty of obtaining meaningful information from this procedure can be seen in between a cancer caused by radiation and one caused by something else. What must be done is to compare populations exposed to known doses of radiation with unexposed populations, and look for the increase in the incidence of cancer. The search is complicated by the long delay, 15 to 25 years, between exposure and effect. The the case of leukemia. From data obtained at relatively high radiation doses, one is led to expect, from each rem of radiation, 1 to 3 additional cases of leukemia *per million persons exposed per year.* (The normal incidence is 60 cases per million). Medical statistics and the sizes of exposed populations are not equal to a clear analysis.

The major uncertainty in low-level radiation studies is over the effect of the dose. There are radiation biologists who believe that a *threshold* exists, that there is a level (see curve *A* of Fig. A5-8) below which there is no incidence of cancer; others believe that the incidence is proportional to the dose all the way down to zero (curve *B* of Fig. A5-8). In the absence of definitive data, the cautious decision is to assume the hypothesis of complete proportionality (curve B) and that position is taken by the regulatory agencies.

There are other complications. There may be a dependency on dose rate: the same dose spread out in time may be less damaging than if given all at once. There are certainly differences in effect which depend on diet, age, sex, and other factors. There are also concentrating and diluting mechanisms in nature which must be considered.

The task of setting limits for occupational radiation exposure and exposures of whole populations is assigned to an international group of radiological scientists, the International Commission on Radiological Protection (ICRP). Since they have accepted the linear dose dependence (curve *B*) as their starting point, they cannot set doses which will "cause no damage." By making some balances between risks and benefits and by comparing man-made exposures with natural exposure, they attempt to set maximum permissible doses (MPD) which will allow necessary activities to continue but keep exposures to a minimum. The present "whole body" MPD is 5 rems per year for those who are occupationally exposed. The ICRP recommends a "dose limit" for the general population of one tenth of the MPD, 0.5 rem per year.

In evaluating the risk of this MPD, the ICRP, in a separate publication,[4] estimates that exposure to 1 rem may cause from 20 to 120 additional cases of leukemia per million during a lifetime. Thus, if the entire United States population received the 0.5 rem dosage in one year, we might expect 2,000 to 12,000 new cases of these diseases to be added to the normal load (which is about 3,000 cancers per million or 600,000 total) during the entire lifetime of the present 200 million Americans. These numbers illustrate the character of radiation effect prediction. The risk in

[4]ICRP Publication Number 8, *The Evaluation of Risks from Radiation*, 1966.

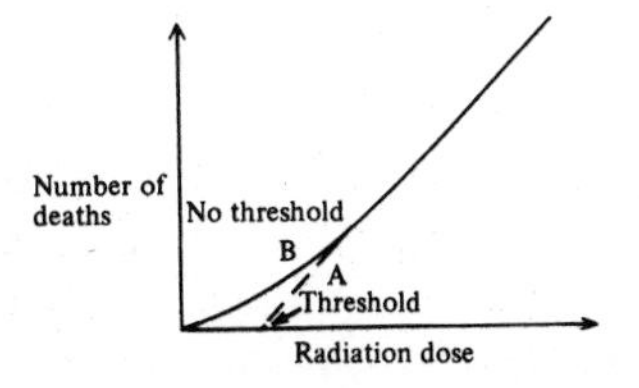

FIGURE A5-8 At very low radiation doses it is not known whether there is a threshold (curve A) below which there is no damage or whether damage persists down to zero dose (curve B).

terms of percentages is low: presently you have 3 chances in 1,000 of getting cancer; 0.5 rem of radiation exposure may increase that to as high as 3.1 chances in 1,000. The absolute numbers, thousands of deaths, are, however, large.

Genetic Effects

Radiation can also damage genetic material by causing gene mutations or damage to the chromosome. The effects may be serious enough to appear in the first generation (abortions, still births, congenital defects) or they may not appear until succeeding generations. Again, it is extremely difficult to establish specific cause and effect relations at low levels of radiation. Much of the information we use for man, for instance, has been extrapolated from data on mice. The ICRP recommendations are that exposure from all sources not exceed 5 rem during the reproductive lifetime of 30 years. They propose that this dose by apportioned as shown in Table A5-2. Chapter 10 offers a more detailed apportioning of this genetic dose.

It must be emphasized that there is an expectation of a certain number of genetic deaths, (the extinction of a gene line in the first or a succeeding generation) from even this low exposure. In addition, a genetic death may be preceded by several generations of weakened and unproductive individuals. Again drawing on ICRP data, the predicted number of genetic deaths from the 0.17 rems per year for 30 years is 1,055 per million births, or 4,850 for the United States total of 4.5 million births per year. It is true that this genetic dose will also cause an uncertain number of cancers; 2,500 a year if we again use the figure of 120 per million per rem and the United States population of 200 million.

If we ended our discussion of radiation here, we would do you a disservice. The numbers—thousands of deaths—are frightening, but thousands die every day. What we need to know to evaluate this cost of nuclear power is what radiation levels are to be expected and how they compare with the exposure we get from other sources of radiation.

Natural and Man-made Radiation

We live in a radioactive world; the soil is radioactive and we are constantly bombarded by cosmic rays from outside the solar system. To this we have added diagnostic and therapeutic X-rays, dental X-rays, and, more recently, fallout from the bomb tests of the 1950s and 1960s and radiation from reactors in the 1970s. All of these exposures vary considerably; people in Denver receive more radiation from cosmic rays than those who live in San Francisco; people in brick houses more than those in wooden houses.

Table A5-2 Illustrative Apportionment of 5 rem Lifetime Dose*

Type of exposure	*rads*
Occupational	1.0
Adult workers not directly engaged in radiation work	0.5
Population at large	2.0
Reserve	1.5

*Source: ICRP (International Committee for Radiological Protection), Publication 6, 1964.

Table A5-3 Average Radiation Exposure of U. S. Citizens*

Exposure	*Millirgms/Year*
Medical exposure–X-rays	
Gonad dose (diagnosis 1964)	55.0
Gonad dose (therapeutic 1964)	7.0
Bone marrow (diagnosis 1964)	125.0
Thyroid dose (diagnosis 1964) (Mainly dental x-rays)	1000.0
Weapons fallout dose (1968) (Ca-137: 0.54 mrem, Sr-90: 2.3 mrem)	3.0
Occupational exposure	
Nuclear energy gonad dose (1970)	0.8
All other occupations gonad dose (1966)	0.4
Other man-made sources Gonad dose (1966)	0.1
Natural background Whole body	100.0

*Source: Morgan, *Environment,* **13**:30, January/February 1971.

The best estimate of the average exposure from all sources is given in Table A5-3. In should be pointed out that the recommended dose limits, 170 mrem per year[5] are specifically limits of genetic exposure (gonadal) and that the 500 mrem per year to an individual or to sensitive population groups, is a "whole body" exposure. The allowable exposure to the thyroid, for instance, is (reduced to 1/10 the occupational exposure) 3,000 mrem. It must also be pointed out that the ICRP recommendations do not apply to medical X-rays; the decisions of what exposure a certain patient should receive are (properly) left to the individual doctor.

Table A5-3 is subject to different interpretations: one can focus on the population dose from nuclear reactors, calculate how many hundreds or thousands may die from cancer, and argue that this is reason enough to turn away from nuclear energy. One can, instead, compare the dose received from nuclear reactors to the unavoidable exposure to natural radiation and say that the effects of nuclear reactors can be ignored. One thing certainly should occur; in both the pro- and anti-nuclear camps a loud cry should go up calling for a reduction of exposure to medical X-rays. It has been estimated, for instance, that the average dose from diagnostic X-rays could be lowered by a factor of 3 if the exposure were only limited to the region of the film.

A neutral reading of Table A5-3 in the light of all that we now know seems to be the following. The present very inconclusive evidence is that low levels of radiation, comparable with that from natural sources, do cause somatic and genetic effects. Thus, each of those exposure numbers shown in Table A5-3 has some (unknown) number of deaths and genetic defects associated with it. The number associated with the nuclear reactor program is small but it can be expected to increase (in contrast to the death from other sources) as the number of reactors increase.

We face many risks in life: some, like cigarette smoking, are voluntary; some, like medical X-rays, are in exchange for definite benefits; some, like natural background radiation, are inescapable; and some, like the radiation from weapons tests

[5] mrem, or millirem represents thousandths of a rem.

and nuclear power plant operations, are involuntary and in exchange for controversial benefits. We properly focus attention on the involuntary ones, for they are, to some extent, controllable by the political process. These risks must be viewed against the background of others so that we can judge the worth of the investment we must make to decrease them. In the end, these deaths which will occur but go uncounted are added to the price of nuclear energy. Since we must know that price, the investigation of the biological effects of low radiation levels must be continued.

Bibliography

General References

Ayres, Eugene, and Charles A. Scarlott, *Energy Sources–The Wealth of the World*, (New York: McGraw-Hill, 1952).

"Energy," *Science*, **184**, April 19, 1974 (special issue).

"The Energy Crisis," *Bulletin of the Atomic Scientists*, **27**: 7-9, September, October, November 1971.

Fowler, John, "Energy and the Environment," *The Science Teacher*, **39**: 9: 10-22, December 1972.

"Energy and Power," *Scientific American*, **224**, September 1971 (single topic issue).

Landsberg, H.A., and S. H. Shurr, *Energy in the United States: Sources, Uses and Policy Issues*, (New York: Random House, 1967).

Wilson, Richard, and William J. Jones, *Energy, Ecology and the Environment*, (New York: Academic Press, 1974).

Chapter 3

Chalmers, Bruce, *Energy*, (New York: Academic Press, 1963).

Marion, Jerry B., *Energy in Perspective*, (New York: Academic Press, 1974).

Odum, Howard T., *Environment, Power and Society*, (New York: Wiley, 1971).

Oort, A. H., "The Energy Cycle of the Earth," *Scientific American*, **223**: 54, September 1970.

Reynolds, William C., *Energy: From Nature to Man*, (New York: McGraw-Hill, 1974).

Chapter 4

Angrist, Stanley W., *Direct Energy Conversion*, (Boston: Allyn and Bacon, 1965).

Davies, E. N., *Fuels and Power*, (Oxford, New York: Pergamon Press, 1969).

Dyson, Freeman, "Energy in the Universe," *Scientific American*, **224**: 50-59, September 1971.

Mott-Smith, Morton, *The Concept of Energy Simply Explained*, (New York: Dover Publications, 1964).

Sandfort, John F., *Heat Engines*, (New York: Doubleday, 1962).

Summers, Claude, "The Conversion of Energy," *Scientific American*, **224**: 3: 148-160, 1971.

Chapter 5

Darmstadter, Joel, *Energy in the World Economy: A Statistical Review of Trends in Output, Trade and Consumption Since 1925* (Baltimore: Johns Hopkins Press, 1971).

Darmstadter, Joel, "Energy Consumption: Trends and Patterns," in Sam H. Shurr, Ed., *Energy, Economic Growth and the Environment*, (Baltimore: Johns Hopkins Press, 1972).

"Energy Facts," Prepared for the Subcommittee on Energy of the Committee on Science and Astronautics, U. S. House of Representatives, 93rd Congress, First Session, November 1973, (Washington, USGPO, 1973).

Lapp, Ralph, *The Logarithmic Century*, (Englewood Cliffs, New Jersey: Prentice-Hall, 1973).

Morrison, Warren E., "An Energy Model for the United States," (Washington, D. C.: U. S. Bureau of Mines Circular 8384, USGPO, 1968).

Report to the Office of Science and Technology, "Patterns of Energy Consumption in the United States," (Washington, D.C.: Stanford Research Institute, USGPO, 1972).

Shurr, Sam H., and Bruce C. Netschert, *Energy in the American Economy, 1850-1825*, (Baltimore: Johns Hopkins Press, 1972).

Chapter 6

Chapman, D., and T. Mount, "Electricity Demand Growth and the Energy Crisis," *Science*, **178**: 703-708, November 17, 1972.
Energy and Human Welfare, A Critical Analysis, (New York: Macmillan, 1974).
Metcalf, Lee, and Vic Reinemer, *Overcharge*, (New York: McKay, 1967).
The 1970 National Power Survey, (Washington, D.C., Federal Power Commission, USGPO, 1971).
"Electric Power and the Environment," Report sponsored by the Energy Policy Staff, Office of Science and Technology, (Washington, D.C., USGPO, August 1970).
Vennard, Edwin, *The Electric Power Business*, (New York: McGraw-Hill, 1972).

Chapter 7

Brodine, Virginia, *Air Pollution*, (New York: Harcourt, Brace, Janovich, 1973).
An Economic Priorities Report, "The Price of Power," (New York: Council on Economic Priorities, 1972).
Hodges, Laurent, "Air Pollution," in *Environmental Pollution*, Chaps. 3-6 (New York: Holt, 1973).
Lave, L. B., and E. P. Seskin, "Air Pollution and Human Health." *Science* **169**: 723-733, 1970.
Williamson, Samuel J., *Fundamentals of Air Pollution*, (Reading, Mass.: Addison-Wesley, 1973).

Chapter 8

Brown, Theodore L., *Energy and the Environment*, (Columbus: Charles E. Merrill, 1971).
"Environmental Effects of Producing Electric Power," Hearings before the Joint Committee on Atomic Energy, 91st Congress of the U.S. 1st and 2nd Sessions, (Washington, D.C. USGPO, 1970).
Gustafson, Philip F., "Nuclear Power and Thermal Pollution," *Bulletin of the Atomic Scientists*, March 1970.
"Selected Materials on Environmental Effects of Producing Electric Power," U.S. Congress, Report Prepared by the Staff on the Joint Committee on Atomic Energy, (Washington, D.C.: USGPO, 1969).
"Thermal Pollution," Hearings before the Senate Public Works Committee, U. S. Congress, February 6-14, 1968, (Washington, D.C.: USGPO, 1968).
Woodson, R. D., "Cooling Towers," *Scientific American*, **224**: 70-78, 1971.

Chapter 9

Berkowitz, David A., *Power Generation and Environmental Charge*, (Cambridge: MIT Press, 1971).
Boesch, Donald F., Carl H. Hershuer, and Jerome H. Hilgram/Ford Energy Policy Project, *Oil Spills and the Marine Environment*, (Cambridge: Ballinger, 1974).
Caudill, Harry M., *Night Comes to the Cumberlands*, (Boston: Little, Brown, 1962).
Hodges, Laurent, *Environmental Pollution*, (New York: Holt, 1973).
Holdren, John, and Phillip Herrera, *Energy*, (San Francisco: Sierra Club, 1971).
Morris, Diane M., *Future Energy Demand and Its Effect on the Environment*, (Santa Monica: The Rand Corporation, 1972).
National Academy of Sciences/Ford Energy Policy Project, *Rehabilitation Potential of Western Coal Lands*, (Cambridge: Ballinger, 1974).
Report of the Study of Critical Environmental Problems, *Man's Impact on the Global Environment*, (Cambridge: MIT Press, 1970).

Chapter 10

Foreman, Harry, *Nuclear Power and the Public*, (Minneapolis: University of Minnesota Press, 1970).
Inglis, David Rittenhouse, *Nuclear Energy: Its Physics and Its Social Challenge*, (Reading, Mass.: Addison-Wesley, 1973).

Lapp, Ralph E., *A Citizen's Guide to Nuclear Power,* (Washington, D.C.: The New Republic, 1971).
Novick, Sheldon, *The Careless Atom,* (Boston: Houghton-Mifflin, 1969).
Seaborg, Glenn T., "Fast Breeder Reactors," *Scientific American,* **223**: 13-21, November 1970.
Seaborg, Glenn T., and William R. Corlis, *Man and Atom,* (New York: Dutton, 1971).
Willrich, Mason, and Theordore B. Taylor, *Nuclear Theft: Risks and Safeguards,* (Cambridge: Ballinger, 1974).

Chapter 11

Darmstadter, Joel, "Energy Consumption: Trends and Patterns," in Sam H. Shurr, *Energy, Economic Growth and the Environment,* (Baltimore: Johns Hopkins Press, 1971).
Lapp, Ralph E., *The Logarithmic Century,* (Englewood Cliffs, New Jersey: Prentice-Hall, 1973).
A Summary Report of the National Petroleum Council's Committee on U. S. Energy Outlook, *U. S. Energy Outlook,* (Washington, D. C.: National Petroleum Council, December 1972).
"A Survey of Energy Consumption Projections," Committee on Interior and Insular Affairs, U. S. Senate, (Washington, D. C.: USGPO, 1972).

Chapter 12

Halacy, D. S., Jr., *Fuel Cells: Power for Tomorrow,* (New York: World Publishing, 1966).
Hammond, Allen L., William D. Metz, and Thomas H. Maugh II, *Energy and the Future,* (Washington, D. C.: American Association for the Advancement of Science, 1973).
Hottel, H. C., and J. B. Howard, *New Energy Technology, Some Facts and Assessment,* (Cambridge: MIT Press, 1971).
Squires, Arthur M., "Clean Power from Dirty Fuels," *Scientific American,* **227**:26-35, October 1972.
Winsche, et al., "Hydrogen: Its Future Role in the Nation's Energy Economy," *Science,* **180**: 1325, June 19, 1973.

Chapter 13

Barnea, Joseph, "Geothermal Power," *Scientific American,* **226**: 70-77, January 1972.
Daniels, Farrington, *Direct Use of the Sun's Energy,* (New York: Ballantine, 1964).
Fenner, David, "Power from the Earth," *Environment,* **13**: 19-34, December 1971.
"Geothermal Energy," Hearings before the Subcommittee on Energy of the Committee on Science and Astronautics, U. S. House of Representatives, 93rd Congress, 1st Session, September 11, 13, and 18, 1973, (Washington, D. C.: USGPO, 1973).
Glaser, Peter E., "Power from the Sun: Its Future," *Science,* **162**: 557-861, November 22, 1968.
Rose, D. J., "Controlled Fusion, Status and Outlook," *Science,* **172**: 797-808, May 21, 1967.
"Solar Energy Research," Staff Report of the Committee on Science and Astronautics, U. S. House of Representatives, (Washington, D. C.: USGPO, December 1972).
Wood, Lowell, "Fusion Power," *Environment,* **14**: 29-33, May 1972.

Chapter 14

Barry, R. Stephen, "Recycling, Thermodynamics, and Environmental Thrift," *Bulletin of the Atomic Scientists,* May 1972.
Berg, Charles A., "Energy Conservation Through Effective Utilization," *Science,* **181**: 128-138, July 13, 1973.
Citizen Action Guide to Energy Conservation, Citizens Advisory Committee on Environmental Quality, (Washington, D. C.: CACEQ, September 1973).
"Conservation and Efficient Use of Energy," Joint hearings before certain subcommittees of the Committees on Government Operations and Science and Astronautics, House of Representatives, 93rd Congress, 1st Session, June 19, 1973, (Washington, D. C.: USGPO, 1973).
"Energy Conservation," U. S. Congress, Senate Committee on Interior and Insular Affairs, Hearings pursuant to S. Res. 45, the National Fuels and Energy Policy Study, 93rd Con-

gress, 1st Session, March 22 and 23, 1973, (Washington, D. C.: USGPO, 1973).
Hirst, Eric, and John C. Moyer, "Efficiency of Energy Use in the United States," *Science*, **179**: 1299-1304, March 30, 1973.
Large, David B., *Hidden Waste*, (Washington, D. C.: Conservation Foundation, 1973).

Chapter 15

Commoner, Barry, *The Closing Circle*, (New York: Knopf, 1973).
"Energy Policy Papers," Printed at the request of Henry M. Jackson, Chairman, Committee on Interior and Insular Affairs, U. S. Senate, pursuant to S. Res. 45, the National Fuels and Energy Policy Study, (Washington, D. C.: USGPO, 1974).
Ford Energy Policy Project, *A Time to Choose: America's Energy Future*, (Cambridge: Ballinger, 1974).
Freeman, David S., *Energy: The New Era*, (New York, Walker, 1974).
Macrakis, Michael S., Ed., *Energy: Demand, Conservation, and Institutional Problems*, (Cambridge: MIT Press, 1974).
Shurr, Sam H., *Energy, Economic Growth and Environment*, (Baltimore: Johns Hopkins Press, 1973).

Index